D0882996

MODERN SIZE-EXCLUSION LIQUID CHROMATOGRAPHY

Modern Size-Exclusion Liquid Chromatography

Practice of Gel Permeation and Gel Filtration Chromatography

W. W. Yau, J. J. Kirkland, and D. D. Bly

Central Research and Development Department
E. I. duPont de Nemours & Co.
Experimental Station
Wilmington
Delaware

A Wiley-Interscience Publication

JOHN WILEY & SONS
New York · Chichester · Brisbane · Toronto

Copyright © 1979 by John Wiley & Sons, Inc.

All rights reserved. Published simultaneously in Canada.

Reproduction or translation of any part of this work beyond that permitted by Sections 107 or 108 of the 1976 United States Copyright Act without the permission of the copyright owner is unlawful. Requests for permission or further information should be addressed to the Permissions Department, John Wiley & Sons, Inc.

***Library of Congress Cataloging in Publication Data*:**

YAU, W W 1937–
Modern size-exclusion liquid chromatography.

"A Wiley-Interscience publication."
Includes index.
1. Gel permeation chromatography. I. Kirkland, Joseph Jack, joint author. II. Bly, D. D., joint author. III. Title.

QD272.C444Y38 543′.08 79-12739
ISBN 0-471-03387-1

Printed in the United States of America

10 9 8 7 6 5 4 3 2 1

FOREWORD

When one agrees to take the unusual step of writing a foreword to a scientific book, he accepts the further responsibility of explaining to the reader what a foreward is supposed to do that the authors of a book could not do for themselves. Only one of the few books with forewords on my shelves provided a clue: an appreciative bystander, it said, might orient the reader as to the position of the book, not just within the narrow limits of its subject, but in the world as a whole.

The "world as a whole" seems a bit far-reaching, even for a book as good as this one, but the comment is appropriate. For I believe this book has unusual merit in bringing together and unifying, for the first time, three closely related subdisciplines, each of which has developed essentially independently to a position of considerable importance in its own field. They are:

Liquid chromatography of low-molecular-weight species, which has grown to be one of the most important techniques of analytical chemistry in a surprisingly short time.

Size-exclusion chromatography of (largely) synthetic macromolecules, which under the more familiar name of gel permeation chromatography provided the first practical technique for characterizing the distributions of molecular weights in important polymeric materials.

Gel filtration of (largely) biological macromolecules, which now takes its place as a quantitative, rather than merely a qualitative technique in the biological sciences.

In their unified treatment of these three aspects of separation science, the authors have produced a book that can be commended as transcending any single aspect of this broad and dynamic field. The reader will be correspondingly rewarded by a better understanding of how his particular interests apply, not just within the narrow limits of his subdiscipline, but in the world of chromatographic analysis as a whole.

FRED W. BILLMEYER, JR.

PREFACE

This book is concerned with high-performance size-exclusion column liquid chromatography (HPSEC), meaning high-performance gel permeation chromatography (HPGPC) and gel filtration chromatography (HPGFC).

Although size-exclusion chromatography (SEC) is a mature technique, significant new advances have recently been made in high-performance column packings and apparatus. These new systems have opened a significant performance gap between classical and modern SEC, so that the modern technique now requires only a few minutes instead of hours as do the traditional approaches.

The lack of an integrated treatment of gel permeation and gel filtration chromatography, and particularly of high-performance techniques, prompted us to prepare this book. Our approach has been to blend practical know-how and useful basic theory, and generally we have restricted our treatment to the modern, high-pressure aspects, a field in which we have been active. Therefore, this book is largely the result of personal experiences, and is in this sense our critical (and we hope, authoritative) account of modern SEC.

Both practice and basic principles are required for effective utilization of HPSEC. Frequently in books of this type there is a tendency to stress either the theoretical or the "cookbook" side of the subject. Books with primarily theoretical approaches can be so highly complex and tedious that workers shy away from attempting to establish a practical understanding of the subject. In this book we present the basic principles and practices so that the user can quickly determine the approaches needed to solve a problem. Our intent has been to write a book on HPSEC so that the average professional without previous experience in chromatography may use it to solve problems without the need of other literature.

Our treatment of band broadening and resolution in SEC (Chapts. 3 and 4) is more comprehensive than that of the other basic areas. This is deliberate, since we find that these important topics have not been discussed critically elsewhere. Rather than present only a specific SEC treatment, we describe SEC band broadening and resolution in relation to the general understanding of these concepts in all of liquid chromatography. Thus, comparisons between SEC and all of LC present a more complete and coordinated picture, hopefully one that eliminates a long-persisting confusion regarding band broadening and resolution in SEC.

Since much of the theory, equipment, and general laboratory technique for GPC and GFC are identical, we have integrated these two techniques, and have provided commonality of treatment in the areas where strong similarities exist. However, the applications of GPC and GFC generally are different, and we provide specific discussions of each in Chapters 12 and 13, respectively.

We would like to express our thanks to friends and associates who have influenced the formation of this book. We are grateful to Mr. C. R. Ginnard, Mr. H. J. Stoklosa, and Professor A. E. Hamielec for their helpful technical discussions. We especially appreciate the manuscript review by Professor Fred W. Billmeyer, Jr., who offered many helpful suggestions and comments. We are also grateful to several of our coworkers who reviewed some of the initial manuscript, and to Drs. R. E. Benson and E. C. Dunlop, who strongly supported this project from its inception. The loyalty and dedication of our secretary, Mrs. Patricia C. Lyons, in her skillful preparation of the manuscript are much appreciated. Finally, the generous support of the Central Research and Development Department of the E. I. du Pont de Nemours & Co. in this project is gratefully acknowledged.

W. W. Yau
J. J. Kirkland
D. D. Bly

May 1979
Wilmington, Delaware

CONTENTS

1. **BACKGROUND, 1**

1.1 Introduction, 1

1.2 History, 2

1.3 Utility of HPSEC, 3

Gel Filtration Chromatography, 3
Gel Permeation Chromatography, 4

1.4 Absolute Molecular Weight Methods: Types, Advantages, and Limitations, 12

1.5 The Literature, 14

2. **RETENTION, 19**

2.1 Introduction, 19

2.2 Solute Retention in SEC, 24

2.3 SEC Retention Mechanism, 27

2.4 Theoretical Models of SEC Separation, 31

Hard-Sphere Solute Model, 31
Rigid Molecules of Other Shapes, 37
Random-Coil Solute Model, 39

2.5 Other Considerations, 42

Factors Influencing SEC Retention, 42
Effect of Solute Conformation on SEC-MW Calibration, 43
Less Successful Attempts of SEC Retention Theory, 46
Size Separation in Small-Flow Channels, 50

3. **BAND BROADENING, 53**

3.1 Introduction, 53

Basic Column-Dispersion Processes, 55

Peak Variance, 57

3.2 LC Plate Theory, 59
Basic Plate Theory, 60
The van Deemter Equation, 63
Flow-Diffusion Coupling, 65
Reduced Plate Height, 69
3.3 Mechanism of SEC Band Broadening, 71
Experimental Verification, 72
Rate Theory, 78
Theoretical Inferences, 82
3.4 Influencing Factors, 85
Column Parameters, 85
Kinetic Factors, 88
Experimental Factors, 89
3.5 Experimental Methods, 91
Column Plate Count, 91
Column-Dispersion Calibration, 93

4. RESOLUTION, 97

4.1 Introduction, 97
Chromatographic Resolution, 97
Peak-Capacity Concept, 101
4.2 Resolution Concept in SEC of Polymers, 102
4.3 Molecular Weight Accuracy Criterion, 104
4.4 Applications of Column Performance Criteria, 108
4.5 Pore Geometry and Operational Effects, 114
Connecting Columns, 114
Separation Capacity of Single Pores, 114
Effect of Packing Pore-Size Distribution, 117
Effect of Operating Parameters, 119

5. EQUIPMENT AND DETECTORS, 123

A. Equipment, 123

5.1 Introduction, 123
5.2 Extracolumn Effects—General, 126

5.3 Mobile-Phase Reservoirs, 126
5.4 Solvent-Metering Systems (Pumps), 128
General Pump Specifications, 129
Reciprocating Pumps, 130
Positive-Displacement Pumps, 135
Constant-Pressure Pumps, 136
Comparison of Pumps, 138
5.5 Sample Injectors, 138
5.6 Miscellaneous Hardware, 142
5.7 Integrated Microprocessor-Controlled Instruments, 144
5.8 Laboratory Safety, 145

B. Detectors in HPSEC, 146

5.9 Introduction, 146
5.10 Differential Refractometer, 148
5.11 Ultraviolet Photometers and Spectrophotometers, 151
5.12 Other Detectors, 156

6. THE COLUMN, 165

6.1 Introduction, 165
6.2 Column Packings, 166
Semirigid Organic Gels, 167
Rigid Inorganic Packings, 173
Packings for Preparative SEC, 180
Modifying Siliceous Particles, 183
6.3 Column-Packing Methods, 186
Particle Technology, 186
Basis of Column-Packing Techniques, 189
Packing Rigid Solids, 191
Packing of Semirigid Gels, 199
6.4 Column Technology, 199
Column Materials, 199
Column Dimensions, Configuration, 200
Column Performance, 202
Techniques with Columns of Small Particles, 205

7. OPERATING VARIABLES, 209

7.1 Introduction, 209
7.2 Solvent Effects, 209
Sample Solubility, 209
Other Solvent Effects, 222
Flow Rate Effects, 224
Temperature Effects, 232
7.3 Substrate Effects, 234
7.4 Sample Effects, 239
Sample Volume, 239
Sample Weight or Concentration, 241

8. LABORATORY TECHNIQUES, 249

8.1 Introduction, 249
8.2 Solvent Selection and Preparation, 254
Convenience, 254
Sample Type, 254
Effect on Column Packing, 255
Operation, 255
Safety, 260
Solvent Purification and Modification, 261
8.3 Selection and Use of Standard Reference Materials, 262
8.4 Detector Selection, 264
8.5 Column Selection and Handling, 265
Optimum Single Pore-Size Separations, 265
Bimodal Pore-Size Separations: Optimum Linearity and Range, 267
Other Column Selection Guidelines, 268
Column Handling, 269
8.6 Chromatographic Design Considerations, 271
8.7 Making the Separation, 274
Dissolving the Sample and Standards, 274
Sample Solution Filtration, 275
Sample Injection, 276
Baseline Stability, 277
Obtaining and Using a Chromatogram Baseline, 277

8.8 Troubleshooting, 281
Too High Pressure, 281
Column Plugging, 281
Air Bubbles and Leaks, 282
Poor Resolution, 282
Low Solute Recovery, 282
Constancy of Separation, 282
Peak Shape, 283

9. CALIBRATION, 285

9.1 Introduction, 285
9.2 Calibration with Narrow-MWD Standards, 289
Peak Position Calibration, 289
Universal Calibration, 291
9.3 Calibration with Broad-MWD Standards, 294
Integral-MWD Method, 294
Linear Calibration Methods, 298
9.4 Accuracy of Calibration Methods, 302
9.5 Actual Molecular Weight Across the SEC Elution Curve, 307
9.6 Linear Calibration Ranges, 309

10. DATA HANDLING, 315

10.1 Introduction, 315
10.2 Simple Curve Inspection, 315
10.3 Curve Summation: Manual Method for Computing Molecular Weight Averages and MWD, 318
10.4 Manual Method for Correcting Molecular Weight Averages, 322
Application of the Correction Method for Molecular Weight Averages, 323
10.5 Automated Data Handling for Computing Molecular Weight Averages and MWD, 326
Real-Time Data Acquisition with Off-Line Processing, 327
Real-Time System, 331
10.6 Computer Method for Correcting Band-Broadened SEC Curves, 332

10.7 Special Methods and Information, 335
Use of Universal Calibration, 335
Calculation of Intrinsic Viscosity [η] from SEC Data, 338
Comparative Technique for MWD, 339
The Q-Factor Method, 340

11. SPECIAL TECHNIQUES, 343

11.1 HPSEC Separation of Small Molecules, 343
Experimental, 344
Molecular Weight Estimations, 346
Applications, 349
11.2 Preparative SEC, 357
Experimental, 358
Applications, 364
11.3 Recycle SEC, 367
Theory, 368
Equipment, 370
Uses of the Recycle Method, 372
11.4 Vacancy and Differential SEC, 376
11.5 High-Speed Process SEC, 378

12. GEL PERMEATION CHROMATOGRAPHY APPLICATIONS, 381

12.1 Introduction, 381
12.2 Value of the Strip-Chart Chromatogram, 382
12.3 Tetrahydrofuran-Soluble Polymers, 386
12.4 Polyolefins, 386
12.5 Polyamides and Polyesters, 390
12.6 Synthetic Water-Soluble Polymers, 394
12.7 Polyelectrolytes, 397
12.8 Branching and Chain Folding, 399
Branching, 399
Chain Folding, 401
12.9 Copolymers, 404
12.10 Miscellaneous, 413
Polymer and Additive Blending, 413
Quality Control, 414

Cleanup Procedures, 415
Particle Dispersions, 416

13. TECHNIQUES OF MODERN GEL FILTRATION CHROMATOGRAPHY, 419

13.1 Introduction, 419
13.2 Column Packings, 420
Preparation of Surface-Modified Substrates, 420
Properties of Surface-Modified Substrates, 423
13.3 Substrate/Mobile Phase Combinations, 423
Unmodified Porous Silicas, 424
Solute Complexation, 426
Surface-Modified Packings, 427
13.4 Operating Variables and Technique, 432
13.5 Utility, 434
Size Classification for Analysis, 434
Molecular Weight and Molecular Weight Distribution, 439
Isolation by GFC, 444
Prefractionation and Sample Cleanup, 445

SYMBOLS, 451

ABBREVIATIONS, 459

INDEX, 461

MODERN SIZE-EXCLUSION LIQUID CHROMATOGRAPHY

One

BACKGROUND

1.1 Introduction

This book is about modern, high-performance, size-exclusion chromatography (HPSEC). We differentiate it from traditional size-exclusion chromatography (SEC) by discussing only closed systems operated at high pressures. Additionally, we arbitrarily define HPSEC as a system in which the number of theoretical plates generated per second is about tenfold greater than that of traditional SEC systems. This means that separations commonly will take 10–20 min compared to the much longer (hours) separations of conventional size-exclusion chromatography. High pressures, rapid flow rates, and sophisticated equipment are characteristic of HPSEC. The short-time separations are made possible by small, porous-particle packings (e.g., 10–30 μm or less) that are packed into highly efficient columns to provide the improved resolution.

Size-exclusion chromatography is a liquid column chromatographic technique which sorts molecules according to their size. The sample solution is introduced onto the column, which is filled with a rigid-structure, porous-particle column packing, and is carried by solvent (mobile phase) through the column. The size sorting takes place by repeated exchange of the solute molecules between the bulk solvent of the mobile phase and the stagnant liquid phase within the pores of the packing. The pore size of the packing particles determines the molecular size range within which separation occurs.

1.2 History

Although high-performance size-exclusion chromatography (HPSEC) is a recent development, the technique is an offspring of traditional gel permeation chromatography (GPC) and gel filtration chromatography (GFC). These, in turn, have their roots in conventional liquid chromatography (LC). Ettre's interesting paper, "The Development of Chromatography" (1), describes how David Talbot Day demonstrated in 1897 that crude oil fractions could be separated through pulverized fuller's earth. Unfortunately, Day did not properly interpret the phenomenon that was occurring, and because of this the original founding of chromatography is often ascribed to Michael S. Tswett. In 1903–1906, Tswett clearly described the chromatographic separation of colored vegetable pigments in petroleum ether on calcium carbonate and recognized the method as a general process. From Tswett's early beginning, a large number of workers have continued to develop liquid chromatography into its present high-performance capabilities. Today high-performance liquid chromatography is used widely in various forms with many scientific disciplines (2).

The origin of gel filtration chromatography is generally attributed to J. Porath and P. Flodin (3). In 1959, these workers of the Biochemical Institute in Uppsala, Sweden, demonstrated that columns packed with cross-linked polydextran gels, swollen in aqueous media, could be used to size-separate various water-soluble macromolecules. The gels for this technique have been made commercially available and have been used extensively for biomolecule separations in low-pressure systems. The technique has been reviewed by Porath (4).

In 1964, J. C. Moore of the Dow Chemical Company disclosed the use of cross-linked polystyrene "gels" for separating synthetic polymers soluble in organic solvents (5) and, with this event, conventional gel permeation chromatography (GPC) was born. It was immediately recognized that, with proper calibration, gel permeation chromatography was capable of providing molecular weight (MW) and molecular weight distribution (MWD) information for synthetic polymers. Since these quantities were difficult to obtain by other methods (Sect. 1.4), gel permeation chromatography came rapidly into extensive use. The background and applications of conventional gel permeation chromatography have been reviewed by Bly (6).

The column packing materials used by Porath and Flodin for gel filtration and by Moore for gel permeation were particles of lightly cross-linked, porous, semirigid, organic-polymer networks. As such, they could be packed into columns and used with various mobile phases only at relatively low flow rates and pressures, <17 bar or 250 psi. At high pressures and flow rates, these packings collapse, and separations cannot be made. Because of these limitations, both conventional gel filtration chromatography and gel permeation chromatography are relatively slow techniques.

Modern high-performance size-exclusion chromatography (HPSEC) has resulted from the development of small, more rigid porous particles for column packings. The first small particles introduced commercially for HPSEC were μ-Styragel (a trade name for microparticle cross-linked polystyrene gel) by Waters Associates, Milford, Massachusetts. Packed into efficient columns, these semirigid 10 μm particles withstand relatively high pressure (e.g., 2000–3000 psi) and provide performance approximately 10 times better than that of the macroparticle cross-linked polystyrene (e.g., 70–150 μm Styragel) widely used previously. More recently, completely rigid inorganic-based particle packings have been introduced (Chapt. 6). Unger et al. (7, 8) and Kirkland (9, 10) have described porous silica particles, and Sato et al. (11) have discussed porous alumina for HPSEC.

1.3 Utility of HPSEC

The utility of HPSEC generally is the same as that of conventional size-exclusion chromatography except that the separation now may be made much faster or with much greater resolution.

Gel Filtration Chromatography

The separation of water-soluble macromolecules of biochemical origin by gel filtration chromatography normally is desired for one or more of the following reasons:

1. To prepare molecular fractions for characterization or further use.
2. To serve as a method for desalting or buffer exchange (i.e., to act as a substitute for dialysis).
3. To estimate molecular weight using calibration standards.
4. To estimate molecular association constants:
   (a) Complexes of small molecules with macromolecules.
   (b) Macromolecular aggregation.

Many examples of these uses are presented throughout this book, especially in Chapter 13.

The utility of high-performance gel filtration chromatography is illustrated in Figure 1.1, where the separation of a number of protein molecules is made in a matter of minutes. Traditionally, this analysis takes several

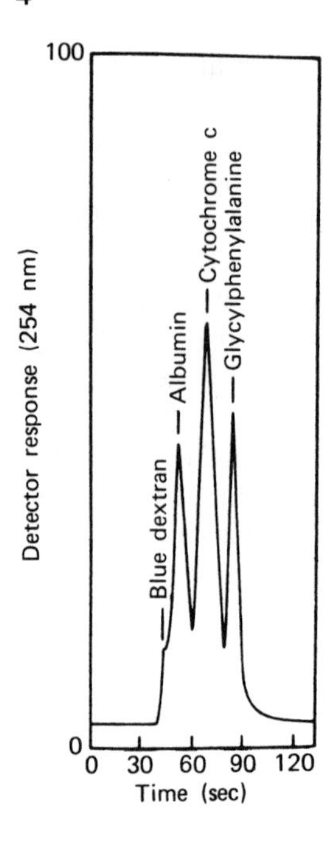

Figure 1.1 Chromatogram for size-exclusion chromatography of proteins.

Column, 30 × 0.41 cm stainless steel packed with 5–10 μm Glycophase G/CPG, 100 Å pore diameter; temperature, 25°C; velocity, 0.7 cm/s at 2700 psi; mobile phase, 0.1 M KH_2PO_4 (pH 6). (Reprinted with permission from Ref. 12.)

hours to perform. A calibration relating the molecular weight of carbohydrate-free globular proteins in water to their retention volume is shown in Figure 1.2. This calibration plot, which can be obtained in a few hours, normally would take at least a week to obtain by large-particle-based conventional gel filtration techniques. Ackers has provided a good review of the conventional gel filtration chromatographic separation of proteins (14).

Gel Permeation Chromatography

Gel permeation chromatography (GPC) normally is used as an analytical procedure for separating small molecules by their difference in size and to obtain molecular weight averages ($\overline{M}_n$, $\overline{M}_w$) or information on the molecular weight distribution (MWD) of polymers. At times, however, it is also used for preparing various molecular weight fractions for further use. The raw-data GPC curve is a molecular size distribution curve. If a concentration-sensitive differential refractometer is used as a detector, the GPC curve is really a size distribution curve in weight concentration. With calibration the raw data are converted to a molecular weight distribution curve and the respective molecular weight averages can be calculated. We present here a short

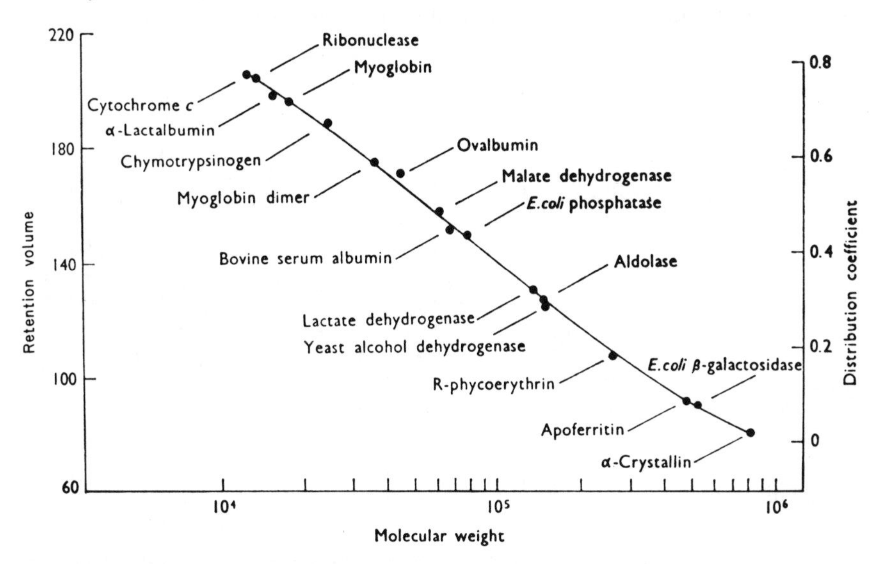

Figure 1.2 Relationship between molecular weight and retention volume for certain proteins in water.

(Reprinted with permission from Ref. 13.)

overview for polymers of the meaning of molecular weight distribution and molecular weight averages ($\overline{M}_n$ and $\overline{M}_w$) so that the reader can better understand this application of GPC.

Various reaction mechanisms are employed for the synthesis of high polymers. Examples are the addition reaction to form polyethylene from ethylene, and the condensation polymerization of hexanedioic acid and hexamethylenediamine to form the polyamide (nylon). During the course of a polymerization reaction, a large quantity of polymer chains are initiated, grow, and then are terminated (i.e., stop growing). The number and length (or weight) of the polymeric chains formed during the reaction vary with the reaction mechanism and the reaction conditions employed. At times the distribution of these chains is accurately predictable from statistical considerations; at other times (nonequilibrium processes) a priori predictions are not accurate. In either case GPC can be used to determine experimentally the distributions and the molecular weight averages of the polymer formed.

One convenient way of measuring the "average" chain length in a polymer sample provides a quantity known as $\overline{M}_n$, the number-average molecular weight. $\overline{M}_n$ is historically significant since for many years it has been a characterizing value obtained directly in the laboratory by colligative property methods (Sect. 1.4). $\overline{M}_n$ also has been correlated with a number of

Table 1.1 Examples of Effect of Molecular Weight or Molecular Weight Distribution on Various Polymer Properties

A. *General Correlations*

	Tensile Strength	Elongation	Yield Strength	Toughness	Brittleness	Hardness	Abrasion Resistance	Softening Temperature	Melt Viscosity	Adhesion	Chemical Resistance	Solubility
Increase molecular weight	+	+	+	+	+	+	+	+	+	−	+	−
Narrow the molecular weight distribution	+	−	−	+	−	−	+	+	+	−	+	0

Profile of performance property dependence on molecule-structure parameters for typical parameters. Key: +, property goes up; −, property goes down; 0, little change. Reprinted in part by permission of the publisher.[1]

B. *Specific Correlations*

Polymer	Property	Correlation
1. Poly(11-hydroxyundecanoic acid),[2] a polyester	Fiber and film strength, polymer solubility	Strength increases with increasing $\overline{M}_n$ while solubility decreases with increasing $\overline{M}_n$
2. Polyesters from ω-hydroxydecanoic acid[3]	Fiber strength	Increases with increase in $\overline{M}_n$
3. 66 Nylon[4]	Fiber tenacity	Increases with $\overline{M}_n$
4. Styrene butadiene rubber[5]	Die swell	Increases with increase in MWD
5. Poly(methyl methacrylate)[6]	Sensitivity as an electron resist	Increases with higher $\overline{M}_n$ and increases with narrower MWD
6. Polyalkylacrylates[7] 7. Polyolefins[7] 8. Polystyrenes[7]	Solution viscosity and shear stability index	Decrease with a decrease in $\overline{M}_w$ caused by shearing

Table 1.1 *(continued)*

Polymer	Property	Correlation
9. PE (polyethylene)[8]	Strength, toughness	Increase with increasing $\overline{M}_n$
	Melt fluidity, film friction	Decrease with increasing $\overline{M}_n$
10. PE[8]	Strength, toughness	Increase with decreasing $\overline{M}_w/\overline{M}_n$
	Fluidity (ease of processing)	Decrease with decreasing $\overline{M}_w/\overline{M}_n$
11. Epoxy resins[9]	"Acceptance quality" of circuit boards	Overall GPC curve (MWD) profile
12. Cellulose triacetate[10]	Density (d) and shrinkage (s) of films	d increases with MWD, s decreases with MWD

[1] E. A. Collins, J. Bareš, and F. W. Billmeyer, Jr., *Experiments in Polymer Science*, Wiley, New York, 1973, p. 312.
[2] V. V. Korshak and S. V. Vinogradovia, *Polyesters*, translated from the Russian by B. J. Hazzard, Pergamon Press, New York, 1965, p. 310
[3] W. H. Carothers and F. J. van Natta, *J. Am. Chem. Soc.*, **55**, 4715 (1933).
[4] J. Zimmerman, *Text. Manuf.*, **101**, 19 (1974).
[5] W. Mills and F. Giurco, *Rubber Chem. Technol.*, **49**, 291 (1976).
[6] J. H. Lai and L. Shepherd, *J. Appl. Polym. Sci.*, **20**, 2367 (1976).
[7] D. E. Hillman, H. M. Lindley, J. I. Paul, and D. Pickles *Br. Polym. J.*, **7**, 397 (1975).
[8] F. W. Billmeyer, Jr., *Textbook of Polymer Science*, Wiley, New York, 1972, p. 382.
[9] *Ind. Res.*, Jan. 1977, p. C1.
[10] N. P. Zakurdaeva and T. A. Ivanova, *Plast. Massy*, **9**, 68, 1976, *Chem. Abstr.*, **85**: 193430b.

polymer physical properties (Table 1.1), and is defined as the mass of the sample in grams $\sum W_i$, or $\sum N_i M_i$, divided by the total number of chains present N, which is $\sum N_i$. Here W_i and N_i are the weight and number of molecules of molecular weight M_i, respectively, and i is an incrementing index over all molecular weights present. Thus

$$\overline{M}_n = \frac{\sum N_i M_i}{\sum N_i} = \frac{\sum W_i}{\sum (W_i/M_i)}$$

and from GPC,

$$\overline{M}_n = \frac{\sum_{i=1}^{N} h_i}{\sum_{i=1}^{N} (h_i/M_i)} \tag{1.1}$$

where h_i is the GPC curve height at the ith volume increment and M_i the molecular weight of the species eluted at the ith retention volume. The equation assumes that h_i is proportional to solute concentration and M_i is sampled in equal volume increments.

Another molecular weight average that can be correlated with physical properties is the weight-average molecular weight, $\overline{M}_w$, which is determined in the laboratory from light scattering (Sect. 1.4) and ultracentrifugation measurements as well as from GPC. It is defined according to Equation 1.2:

$$\overline{M}_w = \frac{\sum N_i M_i^2}{\sum N_i M_i} = \frac{\sum W_i M_i}{\sum W_i}$$

and from GPC,

$$\overline{M}_w = \frac{\sum_{i=1}^{N} (h_i M_i)}{\sum_{i=1}^{N} h_i} \tag{1.2}$$

An example of the manual calculation of $\overline{M}_n$ and $\overline{M}_w$ from a GPC curve is shown in Chapter 10.

Some observations about the relative properties of $\overline{M}_n$ and $\overline{M}_w$ have been made (15). The value of $\overline{M}_w$ is always larger than $\overline{M}_n$ except that the values are identical for a monodisperse system. The ratio $\overline{M}_w/\overline{M}_n$ is a measure of the breadth of the polymer molecular weight distribution, being equal to unity for monodisperse systems, a value of 2 for a Flory most probable distribution, and exceedingly large for a cross-linked polymer. High-molecular-weight species particularly influence the value of $\overline{M}_w$, whereas the value obtained for $\overline{M}_n$ is influenced more by species at the lower end of the molecular weight distribution. If *equal weights* of molecules with $M = 10{,}000$ and $M = 1{,}000{,}000$ are mixed, $\overline{M}_w = 55{,}000$ and $\overline{M}_n = 18{,}200$; if *equal numbers* of each kind of molecule are mixed, $\overline{M}_w = 92{,}000$ and $\overline{M}_n = 55{,}000$ (15).

The molecular weight distribution (MWD) can be expressed graphically in the integral form as the cumulative weight fraction or cumulative number fraction versus molecular weight (MW) (or X, the number of repeat units in the chain). The MWD may also be in the differential form as the weight fraction or number fraction versus MW (or X). As used here, MW is a generic term for the molecular weight which is obtained by multiplying the repeat unit MW by the number of repeat units X. The true MWD can be deduced from the GPC curve only via calibration. Historically, before GPC became

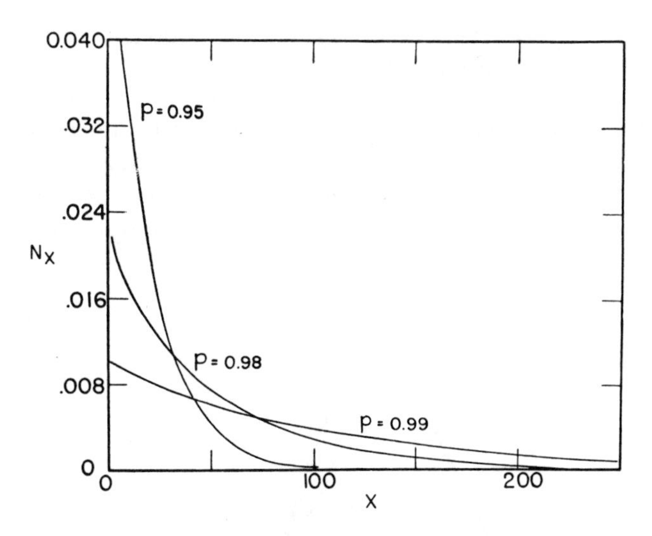

Figure 1.3 Mole fraction distribution of chain molecules in linear condensation polymers for several extents of reaction p.

(Reprinted with permission from Ref. 16.)

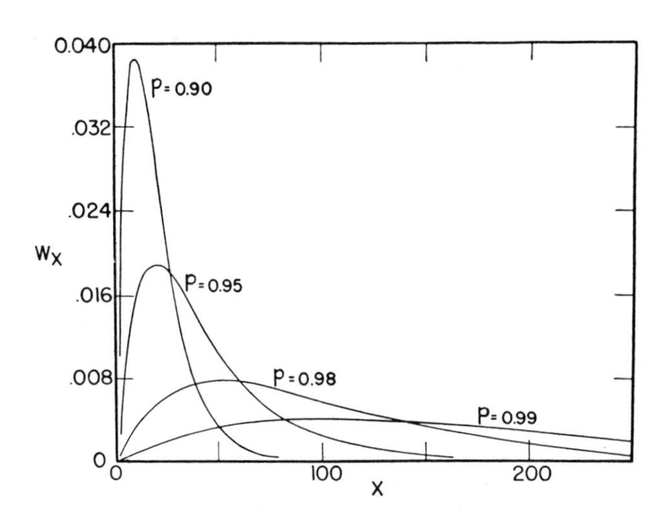

Figure 1.4 Weight fraction distributions of chain molecules in linear condensation polymers for several extents of Reaction p.

(Reprinted with permission from Ref. 16.)

available, these MWD curves were very difficult to obtain (Sect. 1.4). Examples of some of the various MW and MWD parameters are shown in Figures 1.3–1.5, which represent theoretical plots for condensation polymers (e.g., nylon) and other distribution functions. In the figures the extent of reaction p is defined as the mole fraction (of all functional groups available for polymerization both in monomer and in growing polymer chains) that has reacted at various times. The great utility of $\overline{M}_n$, $\overline{M}_w$, and MWD is

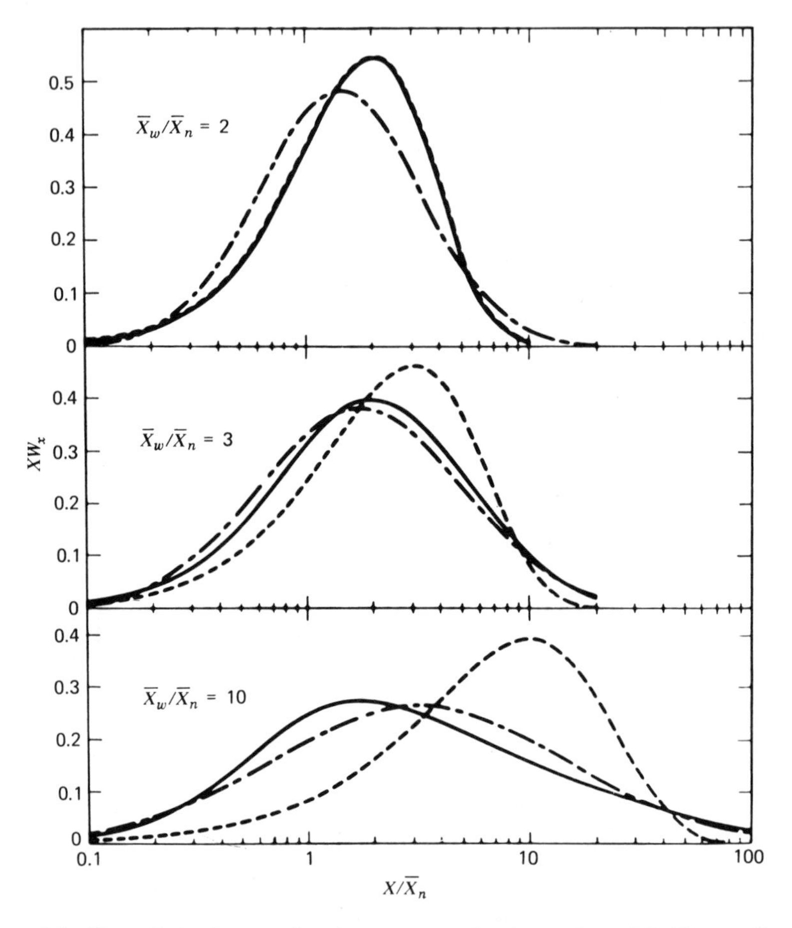

Figure 1.5 Theoretical gel permeation chromatograms for three values of X_w/X_n according to various distribution function formulations.

–·–·– is logarithmic normal, – – – is Schulz-Zimm, and ——— is modified Stockmayer; X_w is the weight-average chain length, and $\overline{X}_n$ is the number-average chain length. (Reprinted with permission from Ref. 17.)

shown in Table 1.1, where various correlations with physical properties for synthetic polymers are compiled. Calculations of $\overline{M}_n$, $\overline{M}_w$, and MWD are presented in detail in Chapter 10.

It is not always necessary to calculate the molecular weight averages or MWD to obtain useful information about a sample from the GPC curve. Simple inspection of chromatograms often reveals important information. For example, Figure 1.6 shows raw-data chromatograms of two batches of supposedly the same epoxy resin. Inspection immediately indicates, however, that batch 1443 is missing a significant amount of material on the low-molecular-weight side of the main peak. This absence of certain material could account for differences in sample properties. There also might be differences in $\overline{M}_n$ or $\overline{M}_w$ between these lots, but the values obtained would not indicate where differences occur in the overall MWD.

As mentioned above, values of $\overline{M}_w/\overline{M}_n$ have often been used traditionally to express the breadth of the molecular weight distribution. Figure 1.7 shows, however, that three different distribution curves can provide identical values of $\overline{M}_n$, $\overline{M}_w$, and $\overline{M}_z$. The parameter $\overline{M}_z$ is related to a higher moment of the distribution defined by

$$\overline{M}_z = \frac{\sum N_i M_i^3}{\sum N_i M_i^2}$$

At times $\overline{M}_z$ is correlated to polymer processing properties. If molecular weight values were obtained for these three distributions by light scattering, osmometry, or centrifugation, all the polymers would give identical $\overline{M}_n$ or

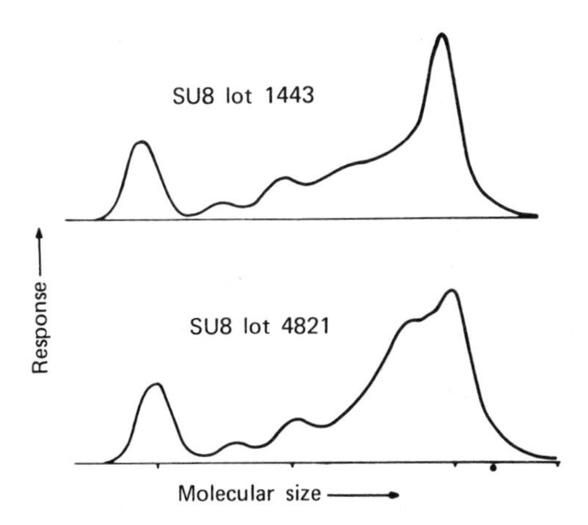

Figure 1.6 Comparison of two lots of SU-8 resin by GPC showing batch variations. (Reprinted with permission from Ref. 18.)

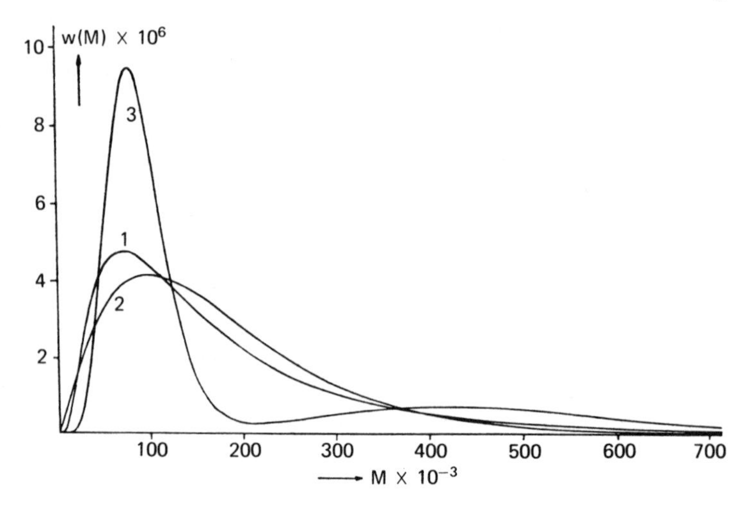

Figure 1.7 Three differential weight distribution curves corresponding to identical values of $\overline{M}_n$, $\overline{M}_w$, and $\overline{M}_z$.

Curve 1 is logarithmic normal function; curves 2 and 3 are sums of two exponential functions. (Reprinted with permission from Ref. 19.)

$\overline{M}_w$ or $\overline{M}_z$ values. Yet, clearly, the distributions are not alike, and physical properties of materials fabricated from these polymers could be different. This information illustrates the utility of the entire MWD profile as provided by GPC.

1.4 Absolute Molecular Weight Methods: Types, Advantages, and Limitations

As discussed in Section 1.3 and in detail in Chapters 9 and 10, both gel permeation and gel filtration chromatography are used to obtain information about the molecular weights and molecular weight distributions of synthetic or naturally occurring macromolecules. Calibration using standards of the same molecule type as the sample must be employed, however, to convert the chromatograms of the size-sorted molecules into molecular weight distribution curves and to calculate molecular weight averages. If the necessary standards for calibration are not available, it is necessary to resort to some other technique for preparing them or for directly determining the molecular weight distribution or molecular weight averages of the sample. We mention below several of these useful molecular weight methods to provide the reader with a starting point should such an independent measurement be required.

The number-average molecular weight ($\overline{M}_n$) is frequently determined by one of the colligative property methods, such as vapor-phase osmometry, high-speed membrane osmometry, boiling-point elevation, freezing-point depression, or by end-group determination from titration or infrared methods. The first four of these phenomena are interrelated, in part by Raoult's law, which relates the dependence of the vapor pressure of the solvent to the molar concentration (and thus the molecular weight) of the solute. Detailed discussions of the theory and measurements are found in various physical chemistry textbooks and other sources, including References 15, 16, and 20–22. The equipment used for the $\overline{M}_n$ determination is frequently homemade, but some commercial equipment is available (Utopia Instrument Co. and Wescan Instruments, Inc.).

The advantages of using the colligative property methods for determining $\overline{M}_n$ are that they are absolute and $\overline{M}_n$ can often be correlated with physical properties. The greatest disadvantage of using the colligative property techniques for $\overline{M}_n$ is that the respective apparatus responds poorly to high-molecular-weight materials, making it tedious to obtain accurate molecular weights for high polymers (e.g., $\overline{M}_n \geq 25{,}000$). Very careful experimental technique is required because the measurements are easily biased by the presence of oligomers and impurities such as water, catalyst residues, solvent residues, and antioxidants.

The weight-average molecular weight, $\overline{M}_w$, is normally determined from light-scattering (LS) measurements. When performed rigorously, the method is complex and the reader is referred to various texts for a description of the theory and process. (See, for example, Refs. 20 and 23–26). The LS method involves detection of light scattered by dissolved macromolecules in a highly clarified solution. The method favors large molecules and frequently cannot be used for samples below a molecular weight of 10^4. Monomers and other low-molecular-weight impurities do not influence the LS measurement of high polymers unless they are present at levels sufficient enough to act as a diluent or to change the refractive index of the medium. For large molecules (e.g., MW $> 10^5$), light-scattering measurements must be made at various angles and sample concentrations to eliminate multiple scattering effects. Commercial equipment is available (e.g., Chromatix, Inc.; C. N. Wood).

Prior to the advent of GPC, most methods for determining the molecular weight distribution of polymers required bulk fractionation of the whole polymer. Each polymer fraction had to be isolated, dried (sometimes purified), and characterized for $\overline{M}_n$, $\overline{M}_w$, or intrinsic viscosity $[\eta]$. Finally, a correlation of the molecular weight of each fraction had to be made with its weight fraction in the polymer sample. Although these fractionation techniques sometimes are simple to use and yield accurate data, they often take weeks to complete. The analysis times depend on the polymer type, solvent, temperature required for dissolution, accuracy needed, and degree of fraction characterization. Various fractionation approaches are discussed in References 27 and 28. These techniques are still being used, and their value depends on the problem and equipment available to the worker. These older methods have been supplanted largely by GPC and more recently by HPSEC.

1.5 The Literature

Special aspects of modern high-performance size-exclusion chromatography (HPSEC) are the subjects of several chapters of this book. However, HPSEC shares with conventional SEC some common theory and history. Consequently, the extensive literature on conventional GPC and GFC provides valuable background, theory, and insight for understanding the nature of the SEC process and the reasoning behind various HPSEC approaches. The literature for both conventional and high-performance studies is published in the same journals and abstracting media. Therefore, it is appropriate here to discuss the useful literature for all of SEC.

The SEC literature can be divided into primary sources, reviews, bibliographies, and current awareness sources and search methods. The primary sources are those journals which contain original research information published with GPC, SEC, LEC (liquid-exclusion chromatography), or equivalent titles. Journals that are important for GFC and GPC include:

Journal of Chromatography
Journal of Chromatographic Science
Chromatographia
Analytical Chemistry
Separation Science

Journals of specific interest to GPC include:

Journal of Applied Polymer Science
Journal of Polymer Science (Chemistry and Physics Sections)
Polymer Preprints (Division of Polymer Chemistry, American Chemical Society)
Macromolecules

Journals that are generally more specific for GFC are:

Analytical Biochemistry
Nature

Advances in Protein Chemistry

Biochemical Journal

Journal of Biological Chemistry

Biochemistry

Biochimica et Biophysica Acta

Books and reviews on SEC have been published periodically. However, most of these have dealt with conventional GFC and GPC. This book is the first to integrate the two areas conceptually and to focus on the high-performance aspects of the SEC method. References 6 and 29–32 include background discussions of SEC theory and development.

Committee D-20 of the American Society for Testing and Materials (ASTM) in 1974 generated a useful *Bibliography on Liquid Exclusion Chromatography*, subtitled "Gel Permeation Chromatography" (33). This bibliography includes most of the literature on GPC from its inception until 1972. A supplement, published in early 1977, covers the literature from 1972 through 1975 and contains all references from *Chemical Abstracts*, Vols. 77–84 (34). Both the original bibliography and the supplement are divided into eight subject categories and contain a permuted title index and an author index. A portion of a sample page of the permuted title index is shown as Figure 1.8.

STUDYING THE MUTUAL ACTION OF
. DETERMINATION OF MW AND MWD OF LOW-CIS
PC V. DETERMINATION OF MW AND MWD OF POLYSTYRENE. .
.. DETERMINATION OF MW AND MWD OF LOW-CIS LINEAR POLYBUTA
V. DETERMINATION OF MW AND MWD OF POLYSTYRENE. /ION OF MOLE
OLY-2-VINYLPYRIDINE SAMPLES IN N,N-DIMETHYLFORMAMIDE
OLY-2-VINYLPYRIDINE SAMPLES IN N,N-DIMETHYLFORMAMIDE
GPC OF BENZENE, NAPHTHALENE, AND ANTHRACENE
AND BRANCHED MOLECULAR STRUCTURES IN NATURAL AND BUTADIENE-STYRENE RUBBERS
A BINARY SOLUTIONS ON SILICA GELS AND NATURAL CLAY MINERALS IN THE ABSENCE OF G.
GPC OF NATURAL RUBBER
A LIPOPHILIC DEXTRAN GEL NATURALLY OCCURING DIOL LIPIDS. PART 6. FRA
HEM CHANGE IN THE HYDROPHILIC NATURE OF SYNTHETIC GELS DURING ADSORPTION
GPC. PART 2. NATURE OF THE SEPARATION.
ANALYSIS OF THE NEGATIVE PEAKS IN GPC
TION OF RIGID MOLECULES IN INERT POROUS NETWORKS STATISTICAL THEORY
INTERACTIVE GEL NETWORKS. I. TREATMENT OF SIMPLE COMPLEXATIO
GPC OF NEUTRAL HYDROXY LIPIDS ON SEPHADEX LH-20
DIOL LIPIDS. PART 6. FRACTIONATION OF NEUTRAL LIPIDS ON A LIPOPHILIC DEXTRAN GEL
IN COMPLEXES BY GEL CHROMATOGRAPHY AND NEUTRON ACTIVATION ANALYSIS
ON OF GEL CHROMATOGRAPHY TO CELLULOSE NITRATE
ND CELLULOSE. 1. EFFECT OF DEGREE OF NITRATION OF CELLULOSE ON MOLECULAR WEIG
DATION OF POLYETHYLENE WITH FUMING NITRIC ACID. I. SINGLE CRYSTALS.
ATION OF POLYETHYLENE WITH FUMING NITRIC ACID. II. BULK POLYETHYLENE.
TION OF POLYETHYLENE WITH FUMING NITRIC ACID. III. FIBROUS STIRRING-
POLYMER CRYSTALS USING FUMING NITRIC ACID. PART-1. MOLECULAR WE
THE POLYDISPERSION OF SOME NITROAROMATIC DERIVATIVES OF C
NITROCELLULOSE CHROMATOGRA
POLYETHYLENE IN A NITROGEN ATMOSPHERE OF
RATION AT LOW NOISE LEVELS WITH

Figure 1.8 Sample page from bibliography on liquid (size) exclusion chromatography, ASTM-AMD-40. (Reprinted with permission from Ref. 33.)

Although the available ASTM bibliography and supplements are very valuable to workers in GPC, unfortunately no such bibliography exists for GFC. Some information on GFC is contained in the GPC work; however, it is not complete.

Bibliographic compilations can also be obtained from abstracting services that are very useful for maintaining current awareness in SEC. Specific subscriptions are available (35–39). In addition, subscriptions can be obtained from current awareness service sources that use *Chemical Abstracts Condensates* computer tapes. We have had good experience with two of these: The Aerospace Research Application Center, Polars Research and Conference Center, Indiana University Foundation, Bloomington, Indiana 47401, and the Georgia Information Dissemination Center Library, Georgia Institute of Technology, Atlanta, Georgia 30332. Each center is provided with a machine-readable tape corresponding to an up-to-date weekly issue of *Chemical Abstracts.* For input, the scientist prepares a profile or "search strategy" using key words or phrases. The computer system then uses these inputs to search the tape for the desired material. Search strategies may also be developed to permit bunching of terms, weighting of terms, specifications on the weight or author ordering to be used in citing the articles found, and so on. The Indiana awareness system is being used to generate the ASTM's AMD-40 Supplements. (33, 34)

The advantage of current awareness systems is primarily one of convenience. They replace searching the primary literature and yet keep the worker relatively current. The disadvantage of current awareness systems is that no matter how carefully the strategy is prepared, articles not wanted are invariably listed and occasionally the method misses a wanted article because of poor indexing or titling.

REFERENCES

1. L. S. Ettre, *Anal. Chem.*, **43**, 20A (1971).
2. L. R. Snyder and J. J. Kirkland, *Introduction to Modern Liquid Chromatography*, Wiley, New York, 1974.
3. J. Porath and P. Flodin, *Nature*, **183**, 1657 (1959).
4. J. Porath, *Lab. Pract.*, **16**, 838 (1967).
5. J. C. Moore, *J. Polym. Sci., Part A*, **2**, 835 (1964).
6. D. D. Bly, "Gel Permeation Chromatography in Polymer Chemistry," in *Physical Methods in Macromolecular Chemistry*, Vol. 2, B. Carroll ed., Dekker, New York, 1972, Chapt. 1.
7. J. Probst, K. Unger, and H. J. Cantow, *Agnew. Makromol. Chem.*, **35**, 177 (1974).
8. K. Unger, R. Kern, M. C. Ninou, and K. F. Krebs, *J. Chromatogr.*, **99**, 435 (1974).
9. J. J. Kirkland, *J. Chromatogr. Sci.*, **10**, 593 (1972).
10. J. J. Kirkland, *J. Chromatogr.*, **125**, 231 (1976).

11. S. Sato, Y. Otaka, N. Baba, and H. I. Iwasaki, *Bunseki Kagaku*, **22**, 673 (1973).
12. S. H. Chang, K. M. Gooding, and F. E. Regnier, *J. Chromatogr.*, **125**, 103 (1976).
13. P. Andrews, *Br. Med. Bull.*, **22**, 109 (1966).
14. G. K. Ackers, *Adv. Protein Chem.*, **24**, 343 (1970).
15. F. W. Billmeyer, Jr., *Textbook of Polymer Science*, 2nd ed., Wiley, New York, 1971.
16. P. J. Flory, *Chem. Revs.*, **39**, 137 (1946).
17. H. L. Berger and A. R. Shultz, *J. Polym. Sci., Part A*, **3**, 3643 (1965).
18. T. D. Zucconi and J. S. Humphrey, *Polym. Eng. Sci.*, **16**, 11, 1976.
19. R. Koningsveld, *Adv. Polym. Sci.*, **7**, 1 (1970).
20. H. Morawetz, *Macromolecules in Solution*, Vol. XXI, *High Polymers*, 2nd ed., Wiley-Interscience, New York, 1975.
21. H. Coll, *Makromol. Chem.*, **109**, 38 (1967).
22. W. R. Krigbaum and R.-J. Roe, "Measurement of Osmotic Pressure," in *Treatise on Analytical Chemistry*, Part 1, *Vol. 7*, I. M. Kolthoff and P. J. Elving, eds., Wiley-Interscience, New York, 1967, p. 4461.
23. J. Brandrup and E. H. Immergut, eds., *Polymer Handbook*, 2nd ed., Wiley, New York, 1975.
24. E. A. Collins, J. Bares, and F. W. Billmeyer, Jr., *Experiments in Polymer Science*, Wiley-Interscience, 1973.
25. M. B. Huglin, ed., *Light Scattering from Polymer Solutions*, Academic Press, New York, 1972.
26. C. Tanford, *Physical Chemistry of Macromolecules*, Wiley, New York, 1961.
27. M. J. R. Cantow, ed., *Polymer Fractionation*, Academic Press. New York, 1967.
28. D. E. Blair, *J. Appl. Polym. Sci.*, **14**, 2469 (1970).
29. H. Determann, *Gel Chromatography*, Springer-Verlag, New York, 1968.
30. K. H. Altgelt and L. Segal, eds., *Gel Permeation Chromatography*, Dekker, New York, 1971.
31. J. Cazes, *Gel Permeation Chromatography*, American Chemical Society (Short Course), Washington, D.C., 1972.
32. A. Lambert, "Review of Gel Permeation Chromatography," *Br. Polym. J.*, **3**, 13 (1971).
33. *Bibliography on Liquid Exclusion Chromotography*, Atomic and Molecular Data Series AMD-40, ASTM, 1916 Race St., Philadelphia, Pa., 19103.
34. *Bibliography on Liquid Exclusion Chromatography*, Atomic and Molecular Data Series AMD-40-S1, ASTM, 1916 Race St., Philadelphia, Pa., 19103.
35. *Chemical Abstracts*, Physical and Analytical Chemistry Sections, 79–80, American Chemical Society, Columbus, Ohio.
36. *Gas and Liquid Chromatography Abstracts*, Applied Science Publishers, Ripple Road, Barking, Essex, England [annual since 1958, LC begins in Vol. 13 (1971].
37. *Liquid Chromatography Abstracts*, Science and Technology Agency, 3 Harrington Road, South Kensington, London SW7, 3ES, 787 High Road, North Finchley N12 8JT, England (quarterly since 1973).
38. *Liquid Chromatography Literature Abstracts and Index*, Preston Technical Abstracts, 6366 Gross Pt. Road, Box 312, Niles, Ill. 60648 (bimonthly since 1972).
39. Institute for Scientific Information, 325 Chestnut St., Philadelphia, Pa., 19106 (also uses a profile approach called ASCA and *Current Contents*, published weekly, uses key word/cross reference approach).

Two

RETENTION

2.1 Introduction

In column chromatography, sample components migrate through the column at different velocities and elute separately from the column at different times. As a solute moves along with the carrier fluid (mobile phase), it is at times momentarily held back either by the surface or a contained stagnant phase of the column packing (stationary phase). Since solutes move only when they are in the mobile phase, the distribution of solute molecules between the mobile and the stationary phases determines the average solute migration velocity. Molecules that favor the stationary phase migrate more slowly and elute from the column later.

All forms of chromatography are, therefore, simply differential migration separation processes where sample components are selectively retained to different degrees by a stationary phase. The mobile phase in the chromatographic process normally is a gas (gas chromatography) or a liquid (liquid chromatography, LC). The LC stationary phase can be a solid surface as for liquid-solid chromatography or a stagnant liquid for liquid-liquid chromatography (LLC). According to the mechanism of solute retention, LC methods can be classified into four categories: ion-exchange, adsorption, liquid-partition, and size-exclusion chromatography (SEC). The last category includes the techniques of gel filtration chromatography (GFC) for biopolymer size separations and gel permeation chromatography (GPC) for analyses of synthetic polymer molecular weight distribution (MWD). As we will see, the retention mechanism in SEC is unique in that solute distribution between phases is established by entropy instead of enthalpy differences.

Since both the basic separation mechanism and the information obtainable from SEC are quite different from those of other LC methods, technologies have been developed specially for SEC in analyzing polymer molecular weight (MW). Many early advances in SEC were made by biochemists and polymer chemists. Thus, GFC and GPC publications have been more prominent in biochemistry and polymer journals than in analytical and chromatographic publications. As a result, nomenclature and conventions derived for GFC and GPC are often not consistent with those for other LC methods. Since the general LC instrumentation and column techniques are becoming an integral part of high-performance SEC (HPSEC), it is useful that practitioners be acquainted with the equivalences and the differences in SEC and general LC terminology. The general nomenclature and conventions for LC peak retention are reviewed in the following, while special SEC peak retention terminology is discussed in the next section.

There are four ways of reporting conventional LC peak retention: [1] retention time, t_R; [2] retention volume, V_R; [3] peak capacity factor, k'; and [4] solute distribution coefficient, K_{LC}. The term t_R can be measured most directly by experiment, but it is the least definitive parameter for identifying sample components. On the other hand, the term K_{LC} is the most difficult parameter to measure, but it is the most fundamental quantity for describing peak retention.

The simple experimental value of t_R, measured by the time required for a peak to elute from the column following sample injection (see the bottom of Fig. 2.1), is useful only for comparing peaks that have appeared in the same chromatogram. The value of t_R is sensitive to changes in experimental conditions such as flow rate and the specific columns used; therefore, it is not very specific for defining sample components. The retention volume V_R is a more fundamental quantity in that it accounts for flow rate differences. To calculate V_R, the mobile-phase volume flow rate, F, must be known as well as the t_R values, since $V_R = Ft_R$. While peak retention reported as V_R is not subject to flow rate change, it can still vary with differences in column size and instrumental dead volume. Such variations are inherently compensated for with the more basic retention parameter, k'. Physically, k' represents the ratio of the *weight of solute* in the stationary phase to that in the mobile phase. Thus the weight fraction of solute remaining in the mobile phase is $1/(1 + k')$. For an unretained peak, $t_R = t_0$, $k' = 0$, and the value for the solute weight fraction in the mobile phase equals unity, meaning that the solute resides only in the mobile phase. Since solutes migrate only when in the mobile phase, the retention time should be inversely proportional to this weight fraction:

$$1:\frac{1}{1 + k'} = t_R:t_0$$

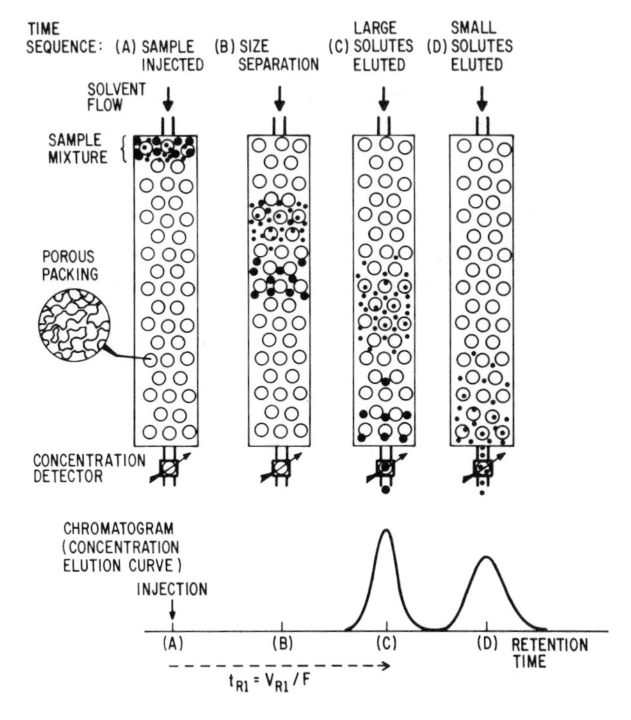

Figure 2.1 Development and detection of size separation by SEC.

or

$$k' = \frac{t_R - t_0}{t_0} \tag{2.1}$$

and

$$k' = \frac{V_R - V_M}{V_M} \tag{2.2}$$

assuming constant flow rate, where $V_M = Ft_0$ for the retention volume of the unretained solute. [There are several ways to determine t_0 and V_M. One simple method is to inject a monomer having similar structure and size as the mobile-phase solvent (e.g., pentane when hexane is the mobile phase) and to detect the unretained peak with a differential refractometer.] Although widely used for comparing conventional LC data, values of k' still do not compensate for differences in the stationary-phase concentration caused by the difference in the surface area and porosity of the column packing. Peak retention, or value of k', increases with increasing stationary-phase loading.

To account for differences in stationary-phase loading, the parameter K_{LC} should be used to define retention. In fact, K_{LC} is the only parameter that can uniquely define the retention characteristics of different organic compounds in conventional LC experiments with a specified column packing, mobile phase, and temperature. Physically, K_{LC} is the ratio of *solute concentration* in the stationary phase to that in the mobile phase. For a given mobile phase and column packing, values of K_{LC} uniquely reflect the basic thermodynamic balance of solute between the phases. Assuming that the equivalent liquid volume for a stationary phase is V_s (the actual liquid volume for LLC, or the volume equivalent to the surface effect retention in adsorption, or the weight of absorbent in ion exchange), K_{LC} is related to k' by

$$k' = \frac{K_{LC} V_s}{V_M} \tag{2.3}$$

Inserting this relationship into Equation 2.2, one can show that

$$V_R = V_M + K_{LC} V_s \tag{2.4}$$

Equation 2.4 represents the equilibrium theory of conventional LC peak retention. It explains why the experimental value of V_R is determined solely by the thermodynamic balance of solute distribution between phases. The validity of the equilibrium LC retention theory is supported by experimental observations. However, since V_s of conventional LC is difficult to determine accurately by experiment, values of K_{LC} are difficult to measure and are not commonly used in practice. The volume elements in Equation 2.4 are illustrated in Figure 2.2*a* for partition LLC as an example where the mobile-phase volume V_M is subdivided into two parts: the moving mobile-phase volume, V_o, and the stagnant mobile phase, V_i.

Direct use of terminology traditional to LC (henceforth called LC terminology) in SEC applications can sometimes be awkward. In SEC (Fig. 2.2b), the size separation occurs only within the mobile-phase volume, V_M, where different-size solutes distribute differently between V_o and V_i, that is, between the solvent moving outside the packing and the stagnant solvent inside the pores of the packing. The distribution favors V_o more for larger solutes. According to LC terminology and Equation 2.2, the SEC chromatogram would have to be interpreted with awkward negative values of k', since $V_R \leq V_M$ in SEC as solute elutes before the solvent peak (i.e., $t_R < t_0$ and $k' < 0$, according to Equation 2.1). This is why the distribution coefficient K_{SEC} is used in SEC as the peak retention index instead of k' as in conventional LC (Sect. 2.2). (For the same reason, the separation factor α, defined in conventional LC as the ratio of k' for two solutes, is not used in SEC.)

(a)

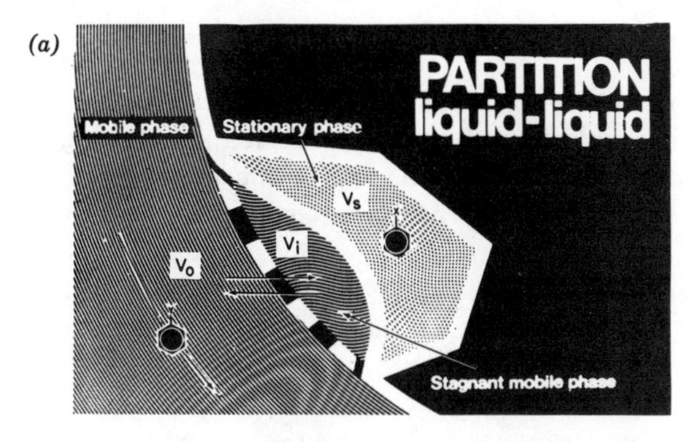

(b)

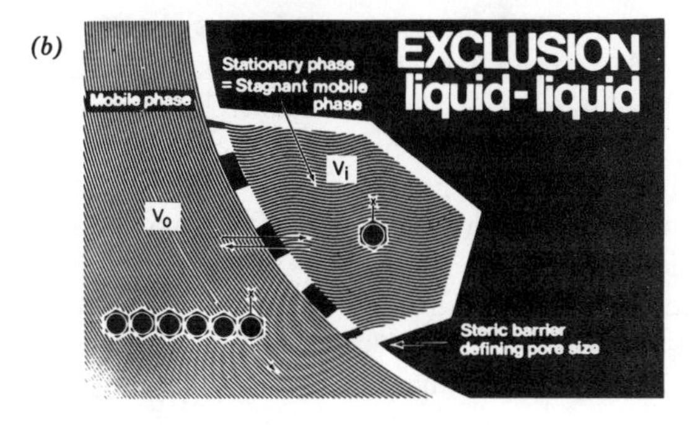

Figure 2.2 Liquid chromatographic retention mechanism.
(a) partition; (b) exclusion. (Reprinted with permission from Ref. 2.)

According to Equation 2.2, failure to distinguish the moving and the stagnant parts of the mobile phase does not affect the estimation of k'. However, it does cause an error in the calculation of solvent velocity, v:

$$v = \frac{L}{t_0} = L\left(\frac{F}{V_M}\right) = L\frac{F}{V_o + V_i} \tag{2.5}$$

where L is the column length. For porous packings that contain stagnant mobile phase, calculation of mobile-phase velocity according to Equation 2.5 will underestimate the true solvent velocity in the column. The calculated value in this case is actually the volume-averaged velocity of the moving and the stagnant mobile phases (Fig. 2.2). (The true solvent linear velocity is given by Equation 2.9, discussed in the next section.)

2.2 Solute Retention in SEC

In LC methods other than SEC, sample components are retained by the column packing and elute after the unretained solvent peak. However, in SEC, solutes are partially excluded from the column packing and elute ahead of the solvent peak. As a solute band moves along with the solvent down the column and around the packing particles, the solute molecules repeatedly permeate or diffuse in and out of the pores of the packing. The driving force for this process is the concentration gradient between the phases. The development and detection of a size separation in SEC are illustrated in Figure 2.1. Here we show that larger solute molecules elute faster than the smaller molecules because they have less penetration into the pores of the packing. Solutes of two distinct sizes can be resolved into two peaks as shown in the chromatogram. All SEC peaks detected at the end of the column are of finite width, as illustrated in the figure. Even for solutes of only one size the elution peak is still necessarily broader than that expected from the finite injected sample volume because of the mixing effects in the column, detector, and connecting tubing. Peak broadening processes affect the performance of SEC analyses, and Chapters 3 and 4 are devoted to these particular subjects.

While in the GFC separation of naturally occurring macromolecules, or in the SEC separation of small molecules, several distinct peaks may be obtained, in synthetic polymer analyses the SEC chromatogram or elution curve usually is just a broad, continuous elution pattern. To extract polymer MWD information from an SEC chromatogram, the exact MW versus V_R calibration relationship of the SEC column is required. Pertinent calibration methods and data-handling techniques for SEC-MWD calculations are discussed in Chapters 9 and 10.

In discussing LC retention, the volumes of the mobile phase inside and outside the pores of column packing are grouped into one volume term, V_M, the retention volume of the solvent peak (Eq. 2.4). Since all peaks in the other LC methods elute after V_M, it is not so important to distinguish between the stagnant versus the moving parts of the mobile phase volume, V_M. Subdivision of V_M is necessary for explaining SEC (where the term "mobile phase" simply means, in SEC, the carrier solvent), because the stagnant part of the "mobile" phase residing in the pores is, in effect, the "stationary" phase for SEC separation (Fig. 2.2). To avoid confusion with the stationary-phase volume V_s in the other LC methods, the stagnant solvent in the porous packing structure in SEC is designated as V_i, the internal pore volume. The remaining liquid volume in an SEC system is designated as the void volume V_o, which is mainly the intersititial liquid volume between the packing particles. By definition, then,

$$V_M = V_o + V_i \tag{2.6}$$

Size separation in SEC is the result of differential solute distribution between the solvent spaces outside and inside the pores of the column packing. This solute distribution can be described by the SEC distribution coefficient K_{SEC} (or more specifically, K_{GPC} or K_{GFC}), which represents the ratio of the average solute concentration in the pores to that outside the pores. Because of the size-exclusion effect, not all the pore volume V_i is accessible to large solutes. Solute concentration inside the pore decreases with increasing solute size. In effect, then, the total accessible liquid volume for different-size solutes is not $(V_o + V_i)$ but $(V_o + K_{SEC} V_i)$. Substitution of this accessible liquid volume for V_M in Equation 2.4 leads to the general retention equation

$$V_R = V_o + K_{SEC} V_i + K_{LC} V_S \tag{2.7}$$

In SEC practice it is important that the last term in Equation 2.7 be minimized by using inert column packings to avoid interference of surface effects on SEC solute retention. With negligible surface effects, SEC retention can be approximated as

$$V_R = V_o + K_{SEC} V_i \tag{2.8}$$

The functional dependence of Equation 2.8 on solute MW constitutes the SEC calibration relationship, as illustrated by Figure 2.3 and discussed below. To cover wide molecular weight ranges in SEC separations the SEC calibration curve is plotted conventionally with molecular weight in the logarithmic scale of base 10. Peak retention in SEC should be recorded in V_R (not t_R) units to minimize the distortion of the elution curve shape due to possible flow rate variations. (For high-speed HPSEC analyses, using real-time data acquisition, it is important to compensate adequately for flow rate variation to assure the accuracy in molecular weight of the SEC results; see Sect. 7.2.) The detailed features of the SEC elution curve are important because they are used in direct interpretation of polymer sample MWD.

In Figure 2.3, a high-molecular-weight solute, designated as solute A, elutes at the void or exclusion volume, V_o, of the SEC column. This solute migrates down the column only through the interstitial spaces between the packing particles. The velocity of this solute, which can be calculated as column length divided by (V_o/F), provides a true measure of the solvent linear velocity:

$$v(\text{true}) = L\left(\frac{F}{V_o}\right) \tag{2.9}$$

Since $V_o < V_M$, the true solvent velocity calculated by Equation 2.9 is necessarily larger than the average solute velocity according to Equation 2.5.

As the molecular weights of the polymer solutes decrease (peaks B and C in Fig. 2.3), the fraction of the pore volume accessible to the solutes increases,

NMU LIBRARY

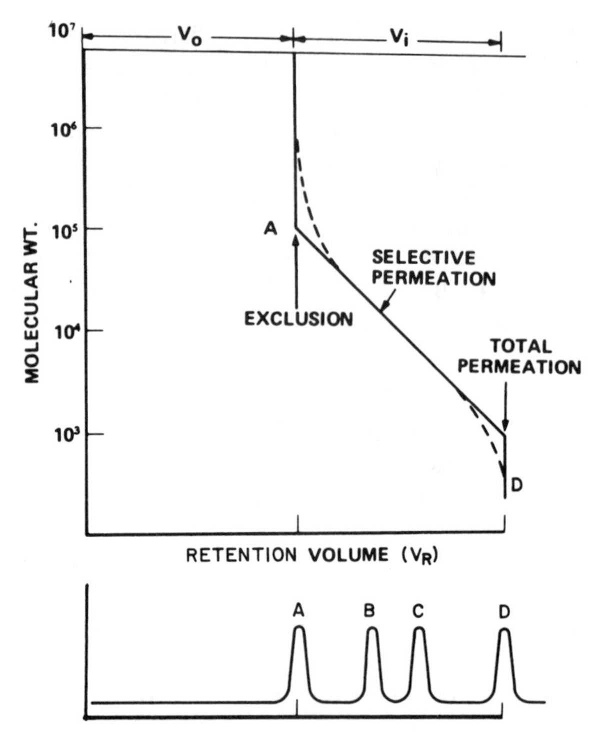

Figure 2.3 **SEC calibration and separation range.**
(Reprinted with permission from Ref. 1.)

causing peaks to elute later. For solute D, which is small enough to access all the pore volumes, elution occurs at the total permeation limit. The fact that the retention volume for peak D is interpreted as "total permeation" in SEC but as "unretained peak" in conventional LC reflects an interesting contrast of conventions and viewpoints. When comparing data from different SEC experiments or in discussing SEC theory, values of K_{SEC} calculated according to Equation 2.8 are preferable to values of V_R for describing SEC peak retention. The values of K_{SEC} compensate for column-size variations. For any SEC column regardless of size, $K_{SEC} = 0$ at exclusion and $K_{SEC} = 1$ at total permeation. The dashed line in Figure 2.3 illustrates the gradual approach of the usual experimental SEC calibration curve to the column exclusion and permeation limits. The solid straight line, the linear approximation to the calibration curve, is commonly used in SEC to facilitate MWD calculations (Chapt. 9).

The information in Figure 2.3 also suggests that SEC intrinsically is a low-resolution technique. Unlike other LC methods, which can be developed to resolve up to hundreds of component peaks representing many column volumes and extended retention times, SEC separations are constrained to occur within the limits of the packing pore volume. Thus only a few peaks can be fully resolved in SEC. Limited SEC peak capacity is a practical constraint to the SEC analyses of small molecules (Sect. 11.1). However, the relatively low resolution of SEC does not prevent using the technique to obtain important polymer molecular weight information. The individual molecular weight components of a sample need not be well resolved for determining the MWD features of the whole polymer. The concept of SEC resolution and molecular weight accuracy is a subject of discussion in Chapter 4.

The large difference in peak capacity between SEC and other forms of LC can also be explained in terms of basic retention parameters. The value of K_{SEC} is constrained to be between 0 and 1, which means that solute distribution favors the unrestricted space outside the pore. On the other hand, values of K_{LC} are unlimited, which means that solute distribution favors the stationary phase, as is the case for most LC peaks. This difference in peak capacity between SEC and the other LC methods is indicative that different thermodynamic balances are involved in controlling solute distribution. As will be described next, SEC is uniquely different from the other LC methods in that it is a chromatographic process controlled by entropy, not enthalpy.

2.3 SEC Retention Mechanism

As solute molecules migrate through the chromatographic column, they transfer back and forth between the moving and stationary phases, constantly redistributing themselves between the phases to satisfy the thermodynamic equilibrium. Under normal chromatographic conditions, solute distribution approximating thermodynamic equilibrium is achieved. (This is true even for the large, slowly diffusing solutes in SEC, as proved by the flow rate study and the static mixing experiment described later in this section.) Thermodynamic equilibrium of solute distribution is defined as the condition in which the chemical potential of each solute component is the same in the two phases (3). For dilute solutions at equilibrium, solute distribution can be related to the standard free energy difference (ΔG°) between the phases at constant temperature and pressure:

$$\Delta G^\circ = -RT \ln K \tag{2.10}$$

with

$$\Delta G^\circ = \Delta H^\circ - T\,\Delta S^\circ \tag{2.11}$$

where K is the solute distribution coefficient, R the gas constant, T the absolute temperature, ln the natural logarithm (base e), and ΔH° and ΔS° standard enthalpy and entropy differences between the phases, respectively.

Solute partitioning in the other forms of LC occurs largely because of solute/stationary phase interactions. Whether absorption or adsorption is involved, the transfer of solutes between the phases is associated with intermolecular forces and substantial enthalpy changes. The entropy change in the other LC methods is generally small and usually can be ignored. Therefore, by combining Equations 2.10 and 2.11 and neglecting the ΔS° term, one can derive K_{LC} as

$$K_{LC} \simeq e^{-\Delta H^\circ/RT} \tag{2.12}$$

The value for ΔH° is usually negative (corresponding to an exothermic sorption for an attractive solute/stationary phase interaction), resulting in K_{LC} values being larger than unity, according to Equation 2.12, and LC peaks eluting later than the solvent peak. On the other hand, solute distribution in SEC is governed mainly by the entropy change between phases (4). Again by combining Equations 2.10 and 2.11 but with $\Delta H^\circ \simeq 0$, K_{SEC} is derived as

$$K_{SEC} \simeq e^{\Delta S^\circ/R} \tag{2.13}$$

Since solute mobility becomes more limited inside the pores of the column packing, solute permeation in SEC is associated with a decrease in entropy, or a negative value of ΔS° (see more discussion in Sect. 2.4). This effect causes K_{SEC} values to be less than unity, according to Equation 2.13, and solutes to elute before the solvent peak.

While Equation 2.12 indicates that a direct temperature dependence exists for peak retention with the other LC methods, the temperature independence of SEC peak retention is predicted by Equation 2.13. This theory is well substantiated by experimental observations (e.g., see Fig. 7.9. Here we see that a large temperature change from 25 to 150°C had only a small effect on the SEC retention characteristics for polystyrenes and polyisobutenes of different molecular weights). Temperature changes have only a small indirect effect on SEC retention as they affect the size of the polymer solute molecules, which in turn affects the ΔS° value. In good solvents, the size of the polymer molecules changes very little with temperature. This is in agreement with the observed small shifts of SEC retention at different temperatures. The effect of temperature on the calibration curves is small only in the context of verifying that the SEC separation is an entropy-controlled process. Even these small curve shifts will significantly affect the accuracy of the molecular weight results in polymer MWD analyses. Besides, temperature does have a significant influence on SEC peak broadening as with all LC peaks (Sect. 3.4). Therefore, large temperature fluctuations in SEC experiments should be avoided.

The validity of explaining SEC retention only in terms of thermodynamic considerations requires that the solute distribution in the SEC experiment be close to thermodynamic equilibrium. The fact that this occurs in SEC columns is supported by two studies (5), one showing that SEC retention is independent of flow rate and the other providing measurement of equilibrium solute distribution through a series of simple static mixing experiments. Flow rate has little effect on SEC retention even for large 37–42 μm SEC packings. For skewed SEC peaks at very high flow rates, the values of V_R at the center of mass, not the maximum, of each peak should be used in studying the effects of flow rate on SEC retention. Experimental results indicate that the kinetic or mass transfer process does not influence the retention mechanism in the SEC separation (6, 7) (see Sect. 7.2 and Figs. 7.7 and 7.9, which verify the preceding observations).

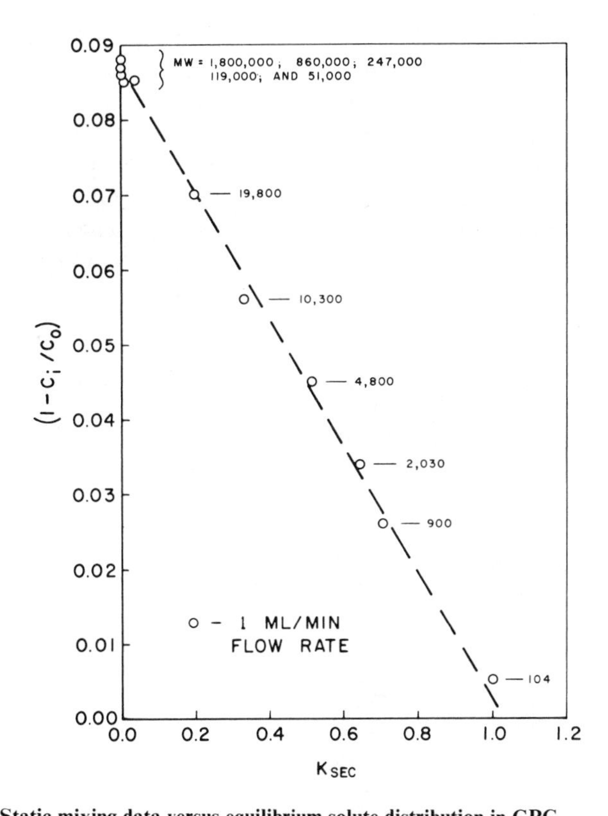

Figure 2.4 Static mixing data versus equilibrium solute distribution in GPC.
Bio-Rad porous glass, 200 Å designation; GPC experiment, 0.5 % polystyrene in $CHCl_3$; mixing experiment, 0.5 % polystyrene in $CHCl_3$, 20 ml solution with 10 g of porous glass. (Reprinted with permission from Ref. 8.)

Thus SEC separation is controlled by the differential *extent* of permeation rather than by the differential *rate* of permeation. Further proof for this contention is provided by static polymer/porous packing mixing experiments (8, 9). Here a polymer solution of a known volume and initial concentration, C_i, is mixed with a known amount of dry, porous packing material. The mixture is allowed enough time for complete solute permeation. The concentration, C_o, of the final solution is then measured and compared with C_i. The change in solution concentration is a direct measure of the equilibrium solute distribution. If solute distribution in SEC separations reaches thermodynamic equilibrium, the experimental values of K_{GPC} for solutes of different molecular weights should vary linearly with the corresponding values of $(1 - C_i/C_o)$ obtained in the independent mixing experiment. The data shown in Figure 2.4 fully support this proposition and the equilibrium theory.

The results of the temperature, flow rate, and static mixing experiments clearly show that *SEC retention is an equilibrium, entropy-controlled, size-exclusion process.* This mechanistic model indicates that solute diffusion in and out of the pores is fast enough with respect to flow rate to maintain equilibrium solute distribution. Thermodynamic size exclusion is the fundamental basis common to all the SEC theories discussed in Section 2.4, where models for different-shaped solutes are considered in the quantitative prediction of the SEC calibration curve. The basic features of the thermodynamic theory of SEC retention are summarized in Table 2.1, which also shows the fundamental differences between SEC and other LC methods.

Table 2.1 Thermodynamics of LC Retention[a]

Size Exclusion (GPC, GFC)

$K_{SEC} = e^{-\Delta G/RT} \simeq e^{\Delta S/R}$
Entropy (S)-controlled process
ΔS = negative for all solutes; S(stationary) $<$ S(mobile)
$K_{SEC} < 1$; solute elutes before solvent; k' = negative
Temperature independent

Other LC Methods (Partition, Adsorption, Ion Exchange)

$K_{LC} = e^{-\Delta G/RT} \simeq e^{-\Delta H/RT}$
Enthalpy (H)-controlled process
ΔH = negative for most solutes; H(stationary) $<$ H(mobile)
$K_{LC} > 0$; solute elutes after solvent; k' = positive
Temperature dependent

[a] $\Delta G = \Delta H - T\Delta S$.

2.4 Theoretical Models of SEC Separation

The theoretical models described in this section are attempts to explain quantitatively K_{SEC} and SEC calibration as a function of the size and shape of the solute and the pore. The models are based on the equilibrium steric SEC theory described above. They are sometimes referred to as the equilibrium theories of SEC separation. Variations among the equilibrium theoretical models are related to the forms and the structures of the solute molecules. For solutes of different conformation, K_{SEC} can have different physical significance such that different approaches to the theoretical interpretation are needed. The conformations of the pore structures are also important factors that affect only the value of K_{SEC}, not the nature of the size-exclusion effect. The hollow cylindrical pore shown in Figure 2.5 illustrates the exclusion effect of three types of solute molecules: the hard-sphere, the rigid-rod, and the random-flight coiled-chain models. The utility of the solute model varies, depending on the true shape of particular macromolecule of interest. The random-coil model is usually appropriate for synthetic polymers, while rigid rod and sphere models find applications mostly in biopolymer studies. SEC theories for the three solute models are discussed separately in this section.

Hard-Sphere Solute Model

The exclusion effect of hard spheres is illustrated in Figure 2.5*a*, which shows a spherical solute of radius r inside a cylindrical cavity of radius a_c. Here the exclusion process can be accounted for by straightforward geometrical

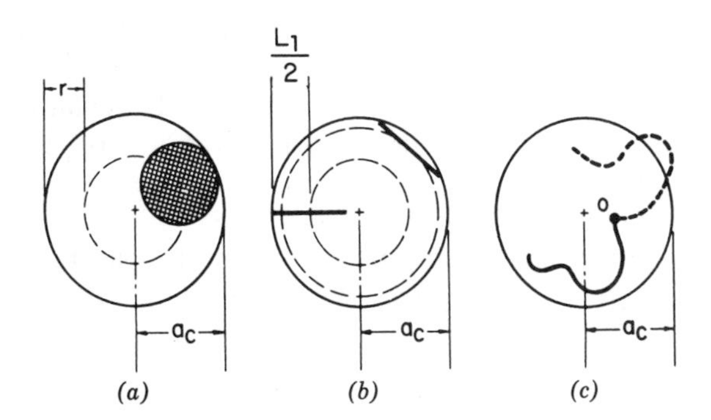

Figure 2.5 Exclusion effect in cylindrical void of radius a_c:
(*a*) hard sphere of radius r; (*b*) thin rod of length L_1 in two orientations in the plane of the cross section; (*c*) random-flight chain with one end at point 0, showing allowed conformation (solid curve) and forbidden conformation (dashed curve). (Reprinted with permission from Ref. 10.)

considerations of the solute exclusion from the walls of the cavity. The center of the sphere (solute) cannot approach the cavity wall closer than a distance r away. Effectively, all the sphere sees is a smaller cylindrical volume of radius $(a_c - r)$ described by the dashed circle rather than the entire cavity volume of radius a_c. (It is assumed that the cylindrical cavity is infinitely long, that is, has negligible end effects.) The center of the hard sphere has free access to the space inside the dashed circle but cannot enter the space outside the dashed line. During solute distribution, the solute molecules can permeate only into the inner space of the cavity. At equilibrium solute distribution there will be a step change of solute concentration across the dashed circle: the solute concentration inside this circle will be constant and equal to that in the open space outside the cavity, and the concentration outside the dashed circle to the cavity wall will be zero. Therefore, the average solute concentration of the entire cavity will be less than that outside the cavity. The ratio of the concentrations inside and outside the cavity is equal to the fraction of the cavity area inside the dashed circle. The situation is equivalent to one with a distribution coefficient:

$$K_e = \left(\frac{a_c - r}{a_c}\right)^2$$

or

$$K_e = \left(1 - \frac{r}{a_c}\right)^2 \qquad \text{(cylinder-shaped pores)} \qquad (2.14)$$

In this case the distribution coefficient is physically equivalent to the fraction of the pore volume accessible to the spherical solute molecules. The equilibrium solute distribution is represented here by K_e with subscript e to distinguish it from K_{SEC}, which is defined as the solute distribution coefficient in the SEC experiment. Thermodynamically, this exclusion process can be considered as the restriction of the solute spatial freedom inside the cavity due to the infinite energy barrier at the dashed circular line. Since the solute is geometrically symmetrical, considerations of configurational changes (rotational freedom) and conformational changes (intramolecular structural changes) are not necessary in this case.

Similarly, equations for K_e for spherical solutes with other simple pore shapes can also be derived using accessible pore volume considerations (11):

$$K_e = 1 - \frac{2r}{\bar{a}} \qquad \text{(slab-shaped pores)} \qquad (2.15)$$

$$K_e = \left(1 - \frac{2r}{3\bar{a}}\right)^3 \qquad \text{(spherical pores)} \qquad (2.16)$$

$$K_e = \left[1 - \left(\frac{2r}{\bar{a}}\frac{1}{1+P}\right)\right]\left[1 - \left(\frac{2r}{\bar{a}}\frac{P}{1+P}\right)\right] \qquad \text{(rectangular pores)} \qquad (2.17)$$

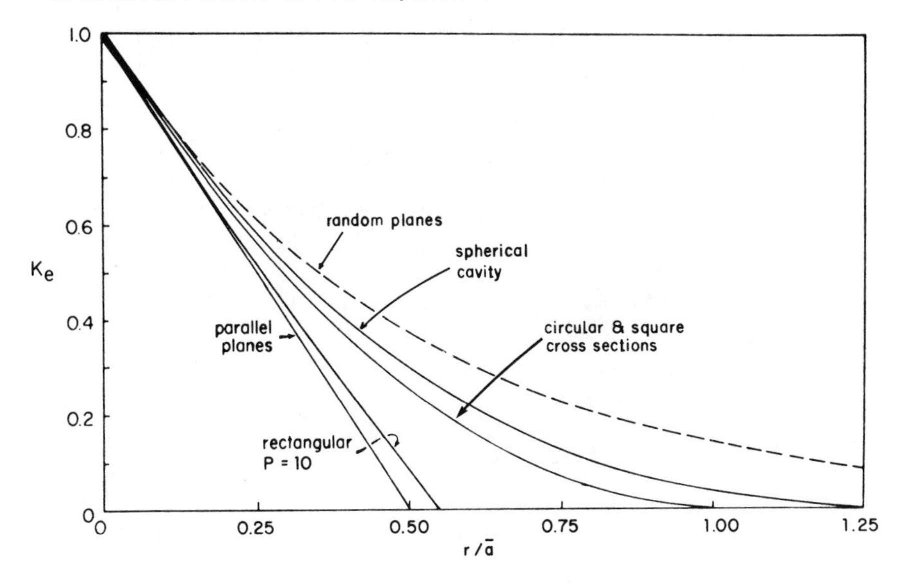

Figure 2.6 Distribution coefficient K_e for spherical molecules of radius r.
K_e for various types of pores versus the ratio of r over the effective radius $\bar{a}$ of the pores. (Reprinted with permission from Ref. 11.)

In the equations, P is the ratio of the long to the short side of the rectangular cavity and $\bar{a}$ is defined as the effective radius by the equation

$$\bar{a} \equiv 2 \times \frac{\text{pore volume}}{\text{pore surface area}} \tag{2.18}$$

The use of $\bar{a}$ to define the pore size of different pore shapes aids in the meaningful comparison of K_{SEC} for different pore geometries. This parameter, $\bar{a}$, is also a good fundamental quantity for describing chromatographic pore size. The following equivalences exist: $\bar{a}$ = radius of a cylinder; $\bar{a} = P/(1 + P)$ times the short side of a rectangle. [One finds $\bar{a} = 2/s$, where s is defined as the surface area/unit pore volume, or the reciprocal of the hydraulic radius defined as the volume/surface area ratio (11). Both $\bar{a}$ and s are experimentally measurable parameters regardless of pore shapes.]

The plot of Equations 2.14 to 2.17 is shown in Figure 2.6. The dashed curve in the figure represents the separation of spherical solutes by a random-planes model suggested by Giddings et al. to describe the porous network structure of SEC packings (11). The curve was calculated by the equation

$$K_e = \exp\left(-\frac{2r}{\bar{a}}\right) \quad \text{(random-planes pore model—spherical solute)} \tag{2.19}$$

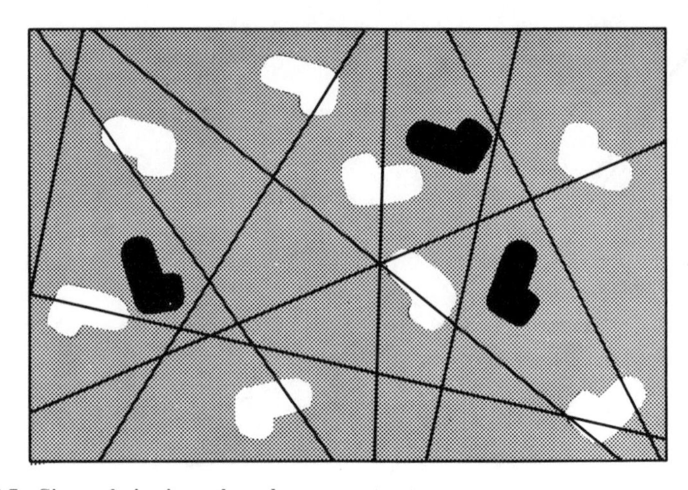

Figure 2.7 Size exclusion in random-planes pore structure.
Unshaded bodies, excluded solute molecules; shaded bodies, permeating solute molecules. (Reprinted with permission from Ref. 11.)

A sketch of the random-planes pore model is shown in Figure 2.7. Pores in this model are formed randomly by intersecting planes.

The curves for different-shaped pores in Figure 2.6 differ considerably, except that as the solute radius decreases, all K_e curves approach unity along a common line. Further examination shows that all curves converge to $K_e = 1 - 2r/\bar{a}$ as the $r/\bar{a}$ value approaches zero. For sufficiently small spherical solutes, all elements of the pore inner surface appear as plane areas, and wall curvature and the corners of pores of different shapes become unimportant. The fact that the curves have a common convergence at small values of $r/\bar{a}$ strongly indicates that $\bar{a}$ [or s in the Giddings expressions (11)] is a fundamental chromatographic pore-size parameter. The curve for random planes has a more gradual change in slope and spreads over a wide range of $r/\bar{a}$ than the other curves. This is expected, since the pores of the assumed random pore structure are not uniform in size, and the presence of a size distribution of pores tends to extend the $r/\bar{a}$ range. There have been other theories for hard-sphere solutes using different random porous network models, including the random-rods pore model (12, 13) and the random-spheres pore model (14). The random-rod model is used in the historical Laurent–Killander–Ogston theory of GFC retention. The random-sphere pore model approximates the pore shapes inside a SEC packing particles by the voids between randomly arranged microspheres. The model is most suited for describing the porous silica microsphere (PSM) packings because of the expected similarity in pore structure (Chapt. 6). A random-pore geometry is more realistic than uniform-pore-shape models. Pore shapes in actual SEC packings are not uniform. Variations in pore shape and cross section are, in effect, a form of pore size distribution (PSD). Random-pore models account for these pore geometry variations.

The advantage of hard-sphere SEC theory is its simplicity. There have been many attempts to explain the SEC separation of random-coil polymers with the "equivalent-sphere" approach. Reference 14 gives an example of this, in which the combined interpretation of equivalent-sphere solute and random-spheres pore models has provided a quantitative fit to an experimental polystyrene SEC calibration curve. The study provided a reasonable estimate that the radius of the equivalent hard sphere, $\bar{r}$, equals 0.89 times the value of the radius of gyration of the polymer R_g.

The conversion factor between R_g and r of a true hard sphere is 0.78 (Eq. 2.30). The value of R_g can be estimated from polymer molecular weight using Equations 2.20 to 2.23:

$$R_g = kM^{\alpha} \tag{2.20}$$

where M is polymer molecular weight and k and α are constants having specific values for different polymer/solvent systems (the usual value for α is between 0.5 and 0.6). Values for k and α are obtained from known Mark-Houwink constants **K** and **a** (Tables 8.2 and 10.2 and Ref. 15) and the intrinsic-viscosity equation derived from a statistical theory of polymer solutions (16):

$$[\eta] = \mathbf{K}M^{\mathbf{a}} \tag{2.21}$$

and

$$[\eta] = \frac{\Phi_0(1 - 2.63\varepsilon + 2.86\varepsilon^2)(\sqrt{6}R_g)^3}{M} \tag{2.22}$$

where Φ_0 is a universal constant, equal to 2.86×10^{23}, and $\varepsilon = (2\mathbf{a} - 1)/3$. With the published **K** and **a** values in Reference 17, the conversion between R_g and MW for polystyrene in tetrahydrofuran takes the form

$$R_g = 0.137M^{0.589}\ \text{Å} \tag{2.23}$$

Table 2.2 lists the corresponding values of $\bar{r}$ and R_g for the commonly used narrow-MWD polystyrene standards. Table 2.3 shows an example of the temperature effect on R_g. From Equations 2.20 to 2.22, one can show that $\alpha \simeq \frac{1}{3}(1 + \mathbf{a})$.

Table 2.2 Estimated Radius of Equivalent Hard Sphere ($\bar{r}$) and Radius of Gyration (R_g) for Polystyrene Standards in Tetrahydrofuran

Molecular Weight	R_g (Å)	r (Å)
1,800,000	660	585
860,000	428	379
411,000	277	246
160,000	159	141
97,200	119	105
51,000	81	72
20,400	47	42
10,300	32	28
4,000	18	16
2,250	13	12
1,250	9	9

Taken from Reference 14.

Table 2.3 Effect of Temperature on the Radius of Gyration of Polystyrene[a] in Cyclohexane

Temperature (°K)	R_g (Å)
305.7	494
307.2	518
311.2	576
318.2	625
328.2	665
333.2	690

[a] The molecular weight of this sample was 3.2×10^6.
Taken from Reference 18.

Rigid Molecules of Other Shapes

Exclusion effects of rigid molecules are illustrated in Figure 2.5*b*, which shows a thin rod of length L_1 inside a cylindrical cavity of radius a_c. Quantifying the exclusion process here is much more complicated than for the hard-sphere model. For the rigid rod, the walls of the cavity restrain both the spatial and the rotational freedom of the rod. When the center of the rod is within the small dashed circle in the sketch (i.e., the rod is more than a distance $L_1/2$ away from the wall), the rod will have full freedom to rotate without touching the wall. As the center of the rod is moved closer to the wall, certain rotational angles in the plane of the cross section are no longer allowed because the ends of the rod hit the wall. Finally, as the rod reaches to the position illustrated at the upper right corner of the sketch, it has *no* rotational or angular (configurational) freedom in the plane of cross section. The final theoretical account of this exclusion effect is complicated further by the necessity of considering rods situated not only in the plane of the cross section but also tilted at all allowed angles out of the plane.

This statistical problem for SEC solute distribution has been studied in detail by Giddings et al. (11). The study suggested the following general expression for the statistical theory of equilibrium solute distribution:

$$K_e = \frac{\int e^{-u(q)/kT}\, dq}{\int dq} \tag{2.24}$$

where q represents the generalized coordinates indicating the solute position, orientation, and internal structural geometries that are needed for describing the changes of spatial, configurational, and conformational freedoms of the solute molecules. The energy $u(q)$ is infinitely large when a geometric configuration (q) intersects with the wall of the cavity; $u(q)$ is equal to zero otherwise. For rigid molecules, the conformational considerations are ignored, since there can be only one fixed solute conformation. (It may be recalled that in the case of hard-sphere model, both the configurational and the conformational considerations were ignored.) The thesis of the statistical theory basically is a surface-overlapping phenomenon that forbids configurations that cause any part of the solute to intersect with the wall of the cavity.

Exact expressions for K_e for simple rods in cavities of even very simple shapes are quite complex. The equations of Reference 11 are not reproduced here, but the resulting curves are shown in Figure 2.8. The general shapes of the curves are similar to those in Figure 2.6. The main difference is that thin-rod curves have less well defined exclusion limits than those for hard spheres. Except for the random-planes model, the curves in Figure 2.6 all intersect

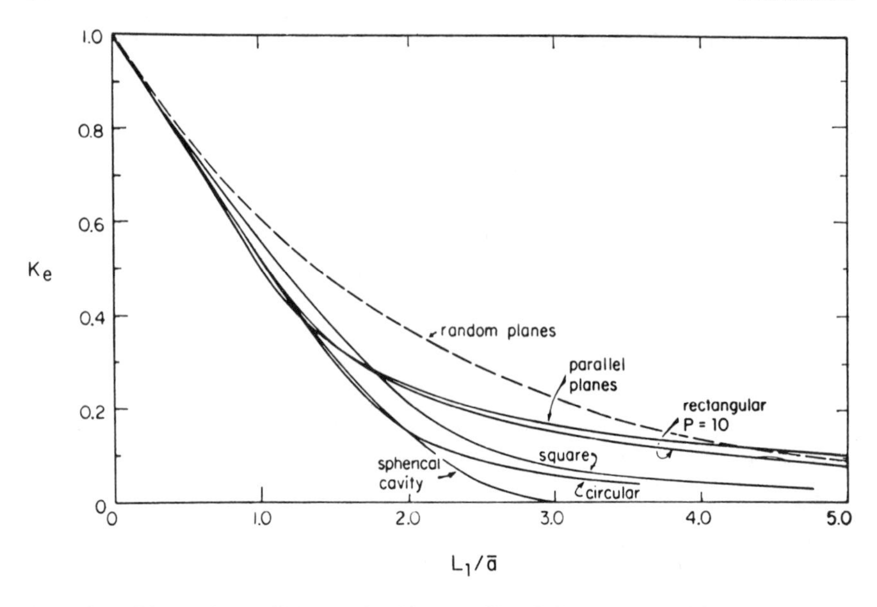

Figure 2.8 Distribution coefficient K_e for thin rods of length L_1.
K_e for various pores versus the dimensionless parameter $L_1/\bar{a}$. (Reprinted with permission from Ref. 11.)

with the $r/\bar{a}$ axis at $K_e = 0$. On the other hand, the plots in Figure 2.8 take a gradual asymptotic approach to the exclusion limit $K_e = 0$, with the exception of the spherical-pore-shape curve. The small, yet finite K_e value at large $L_1/\bar{a}$ in Figure 2.8 can be attributed to the permeation of the finite number of rods being oriented in the direction of the long axis of the assumed infinitely long pores. The curve for the random-planes model in Figure 2.8 was calculated from

$$K_e = \exp\left(-\frac{\bar{L}}{\bar{a}}\right) \qquad \text{(random-planes pore model— rigid solutes in general)} \qquad (2.25)$$

where $\bar{L}$ is the mean external length of the solute, defined as the average length of the projection of the solute molecule along the axes of random orientations. For thin rod-shaped solutes, $\bar{L}$ in Equation 2.25 was replaced by $L_1/2$ in calculating the values of K_e.

The random-planes pore model illustrated in Figure 2.7 can be pictured as an initially free volume partitioned into pores by solid planes inserted at random location and orientation. If a molecule of given configuration in the free space is intersected by one or more of the inserted planes, that orientation of the solute molecule represents a forbidden state which is automatically excluded from the porous network [i.e., infinite $u(q)$ for that state]. In Figure 2.7 the randomly positioned bodies represent molecules initially in equilibrium in bulk fluid. Those molecules (unshaded) cut by the superimposed random surfaces are excluded from the hypothetical pore network created by the randomly oriented surfaces. The partition coefficient K_e is the ratio of the number of uncut (shaded) molecules to the total. This kind of statistical consideration leads to Equation 2.25. Mathematically, Equation 2.25 results from the more general distribution expression, Equation 2.24, after proper integration over the spatial coordinates. Equation 2.25 is generally applicable to rigid molecules of any shape, including ellipsoids, capsules, and doughnut-shaped solutes as well.

The SEC theory for rigid molecules is ideal for interpreting GFC separation of biological polymers or SEC of small molecules, since these molecules lack the internal conformational degrees of freedom. An interesting case of a "once broken" (bent) rod has been examined (19), the result showing that K_e is rather insensitive to the presence of one universal joint at the center of an otherwise rigid rod. The random-coil model is more realistic in representing flexible synthetic polymers.

Random-Coil Solute Model

In the present discussion of the equilibrium SEC theory of random-coil polymers, we follow the explanations provided by Casassa (10, 20–24). Figure 2.5*c* illustrates two conformations of a flexible polymer chain with one end fixed inside the cylindrical cavity. Even with one end fixed, the chain can still assume many conformations. The presence of the wall makes some conformations no longer possible, however, as, for example, the dashed conformation shown in the sketch. This restraint of conformational freedom causes a decrease in both entropy and solute concentration inside the cavity. Calculation of K_e directly from Equation 2.24 is very difficult in this case. However, the problem can be solved with a second-order partial differential equation for a particle undergoing Brownian motion, subject to the boundary condition that at no time is the particle allowed to step out of the confines of the cavity wall (20). The result of this approach for a cylindrical-pore model gives

$$K_e = 4 \sum_{m=1}^{\infty} \beta_m^{-2} \exp\left[-\left(\frac{\beta_m R_g}{\bar{a}}\right)^2\right] \tag{2.26}$$

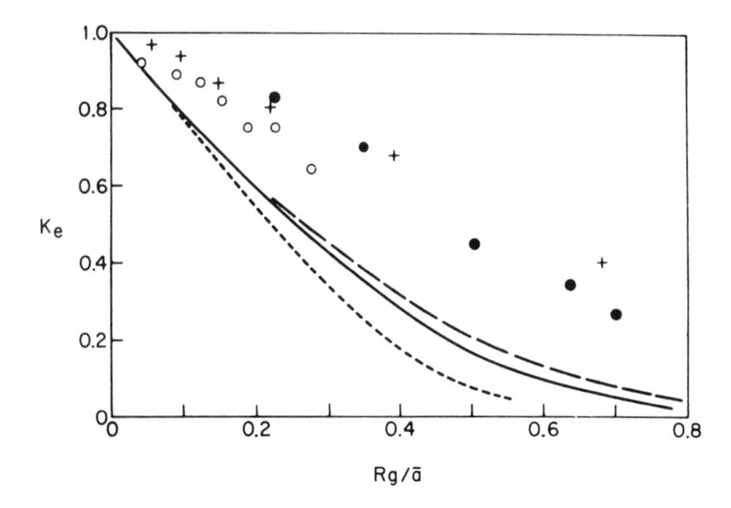

Figure 2.9 Dependence of the distribution coefficient on the molecular size/pore size ratio $R_g/\bar{a}$ for linear flexible-chain polymers.

The lower curves are theoretical results for random-coiled solute in slab-shaped (short dashes), cylindrical (thick solid curve), and spherical (long dashes) cavities. The experimental points are the polystyrene data from Reference 25 (open and filled circles) and from Reference 23 (crosses). (Reprinted with permission from Ref. 24.)

where β_m is a numerical constant that equals the *m*th root of the Bessel function of the first kind of order zero. The values of K_e from Equation 2.26 are shown in Figure 2.9 along with the plots for the slab and the spherical pore models plotted as a function of $R_g/\bar{a}$ in the linear scale. The experimental points in Figure 2.9 are calculated using the apparent pore radius from mercury intrusion data.

The large quantitative discrepancy between the theoretical curves and experimental data shown in Figure 2.9 can be explained by the fact that mercury intrusion has underestimated the packing pore radius. A persistent hysteresis loop is usually observed in the mercury intrusion-depressurization cycles used in studying pore size (23). The observed hysteresis shown in Figure 2.10 suggests the presence of "ink-bottle" structures in the porous packing (26). In Figure 2.10 it is possible that the apparent pore radius of 210 Å for the mercury intrusion branch in both cycles may correspond to the narrow entrance of ink-bottle pores, and the method probably seriously underestimates the actual pore size of the packing. Calculation of the effective radius $\bar{a}$ from Equation 2.18 using the measured pore volume (e.g., mercury penetration) and the measured surface area (e.g., BET) gives $\bar{a}$ = 412 Å. This

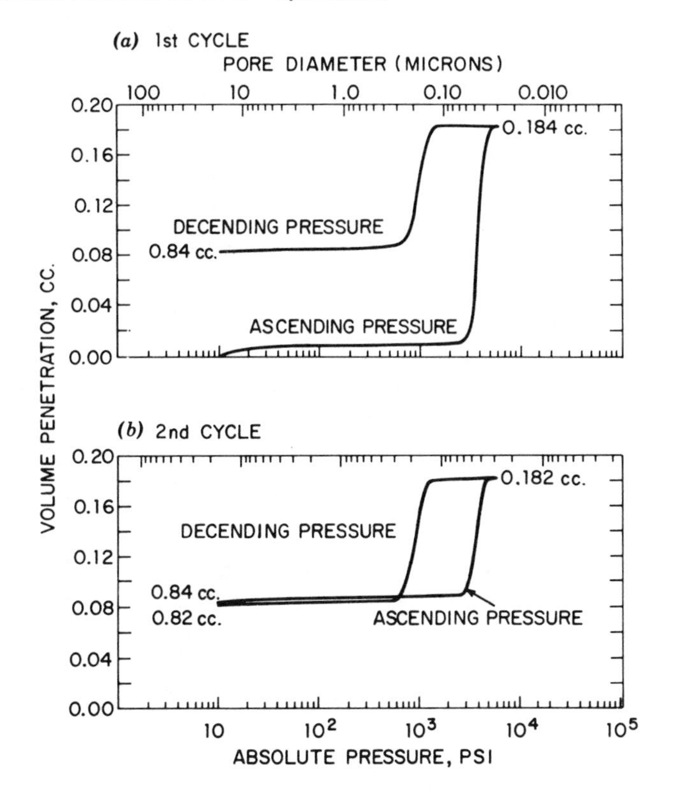

Figure 2.10 Mercury porosimetry curves.
Bio-Glass 500 porous glass SEC packing: (*a*) first mercury intrusion-depressurizing cycle; (*b*) second consecutive mercury intrusion-depressurizing cycle. (Data from Ref. 47.)

new value of $\bar{a}$ brings about a much closer fit between Casassa's random-coil SEC theory and experiment. The good fit is shown in Figure 2.11, where the curves of K_e are plotted versus $R_g/\bar{a}$, now on a logarithmic scale. The success in explaining SEC separation using the value of $\bar{a}$ again verifies $\bar{a}$ as a basic SEC pore-size parameter. It should be noted that the same GPC data in Figure 2.9 from Reference 23 are used in Figure 2.11 to illustrate the concept.

Comparison of the random-coil solute model in Figure 2.9 with the hard-sphere solute model in Figure 2.6 shows that, for the cylindrical pore shape, there is not much difference between the two corresponding curves. This indicates that the equivalent-sphere approximation of flexible polymers holds up quite well in interpreting SEC retention.

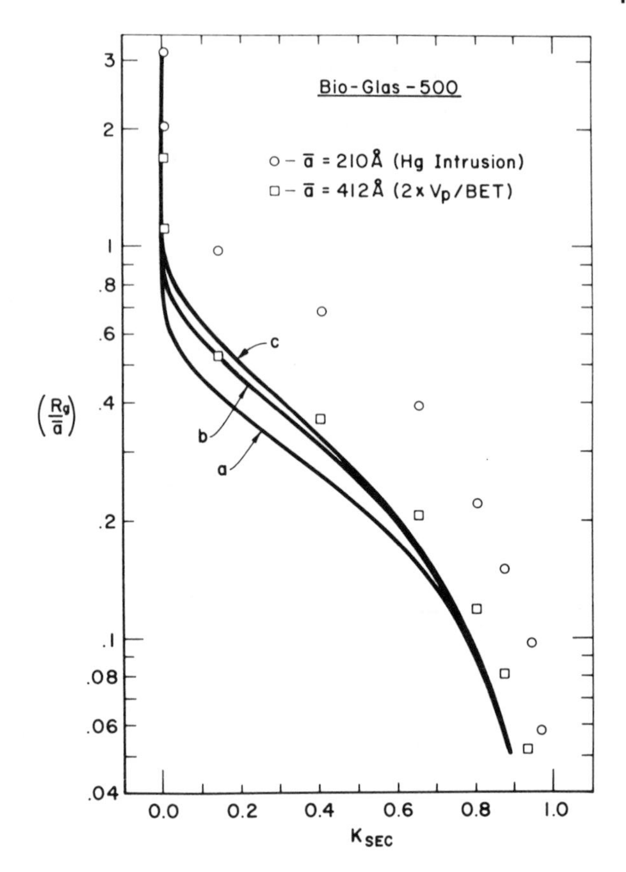

Figure 2.11 Single-pore-size SEC calibration curves.
Theoretical curves, a = slab, b = cylindrical, c = spherical pore models; ○ and □, experimental data. (Reprinted with permission from 23.)

2.5 Other Considerations

Factors Influencing SEC Retention

All the theoretical SEC models discussed above express K_e or K_{SEC} as a function of the size ratio of the solute and pore (e.g., $R_g/\bar{a}$ for the case of coiled molecules). One can expect, therefore, that factors which affect either R_g or $\bar{a}$ will influence K_{SEC}.

The chromatographic factors that affect K_{SEC} through their influence on $\bar{a}$ are pore size, pore shape, and pore-size distribution. (The pore volume of the packing affects SEC retention but not K_{SEC}.) The effect of pore-size distribution on SEC separation can be studied using Equation 2.28 discussed below. Pore-size parameters can be utilized to combine columns effectively for optimizing SEC performance and accuracy of sample molecular weight results (Sects. 4.4, 8.5, and 9.3). Factors that affect K_{SEC} by their influence on R_g are solvent power, polymer branching, and copolymer composition. [The effects of flow rate and temperature on R_g are usually small and are not important considerations in optimizing SEC separation. However, these kinetic factors do affect SEC peak broadening and resolution, as discussed in Chaps. 3 and 4. Only in a poor solvent is the polymer R_g appreciably affected by temperature (Table 2.3). In rare cases when SEC analyses in poor solvents are necessary, care needs be taken to control column temperature.]

The fact that Casassa's theory (10) suggests R_g as the fundamental SEC solute-size parameter is in agreement with the universal calibration concept (Sect. 9.2). From Equation 2.22 one can see that the product $[\eta] \cdot M$ is directly proportional to R_g to the third power (i.e., the hydrodynamic volume of the polymer molecule). Interpretation of SEC results based on hydrodynamic volume should unify SEC information obtained on different polymer/solvent systems. This is, of course, the basis of the universal SEC calibration method of studying polymer branching and copolymer composition (Sect. 12.8).

There are also factors that interfere with SEC retention processes and perturb analytical information. For example, a successful SEC experiment should be free from surface interaction, polymer aggregation (26a), and *in situ* shear degradation of the polymer in the columns. Small gel particles should be filtered out of the polymer sample solutions. Concentration overloading should be kept to a minimum (27). The effects of many of these complicating factors are not well understood and until they are, the best practice is to avoid them. These problems are also discussed in Chapters 7 and 8.

Effect of Solute Conformation on SEC-MW Calibration

The dependence of the size of macromolecules on molecular weight varies as a function of the conformation of the macromolecule. From simple geometry considerations, one can see that $L_1 \propto M$ for rod-shaped macromolecules and $r \propto M^{1/3}$ for solid-spherical macromolecules. For random-coil polymers, $R_g \propto M^{\alpha}$, where the exponent $\alpha \simeq \frac{1}{2}$ (Eq. 2.20). The value of the molecular weight exponent determines the rate at which the size of macromolecule change with molecular weight. For example, a tenfold increase in molecular weight roughly corresponds to a $10 \times L_1$, $2 \times r$, or $3 \times R_g$, change in molecular sizes of different solute conformations. Solute conformation therefore affects the slope and the molecular weight range of the SEC-MW calibration curve (Figure 2.3). A SEC column containing a single-pore-size packing has about two decades' molecular weight separation range for random-coil polymers. This same column will separate solid-spherical solutes with three decades of molecular weight range, but rodlike molecules with only one decade of molecular weight range. Actually, the slopes of single-pore-size SEC-MW calibration curves provide useful clues regarding the conformation of the solute macromolecules.

The SEC retention characteristics for solutes of very different conformations can be compared on a common ground when the size of the solute is expressed in R_g. For flexible polymers, R_g is actually the root-mean-square radius of gyration of all chain configurations. For rigid structures of homogeneous density, R_g is simply the radius of gyration of the structure and is defined by Equation 2.27 (18):

$$R_g \equiv \sqrt{\sum_i^N \frac{X_i^2}{N}} \tag{2.27}$$

where N is the number of mass elements in the structure and X_i is the distance of the ith element to the center of mass of the solid body. For solid spheres of radius r, Equation 2.27 is reduced to

$$R_g^2(\text{sphere}) = \frac{\int_{X=0}^{r} 4\pi X^4 \, dX}{\int_{X=0}^{r} 4\pi X^2 \, dX} = \frac{3}{5}(r)^2 \tag{2.28}$$

For rigid rods of length L_1, one obtains

$$R_g^2(\text{rod}) = \frac{\int_{X=0}^{L_1/2} X^2 \, dX}{\int_{X=0}^{L_1/2} dX} = \frac{L_1^2}{12} \tag{2.29}$$

Since the molecular weight of a solid-sphere molecule is proportional to the volume of the sphere (i.e., to the third power of r),

$$R_g(\text{solid sphere}) = 0.78r \propto M^{1/3} \tag{2.30}$$

For rigid-rod and random-coiled molecules,

$$R_g(\text{rigid rod}) = 0.29L_1 \propto M \tag{2.31}$$

and

$$R_g(\text{flexible coil}) \propto M^{\alpha} \qquad \alpha \simeq \tfrac{1}{2} \tag{2.32}$$

As mentioned above in the context of discussing the SEC-MW calibration curve shape, the molecular weight dependence of R_g can provide a powerful tool to assist in the determination of molecular structure.

Table 2.4 Radii of Gyration of Macromolecules of Different Morphology

Morphology	Molecular Weight	R_g (Å)
Serum albumin (solid sphere)	66,000	29.8[a]
Catalase (solid sphere)	225,000	39.8[a]
Myosin (more like rod than coil)	493,000	468
Polystyrene fraction (flexible coil)	3.2×10^6	494
DNA (rigid rod)	4×10^6	1170
Bushy stunt virus (solid sphere)	10.6×10^6	120[a]
Tobacco mosaic virus (rigid rod)	39×10^6	924

Taken from Reference 18.

[a] Obtained by x-ray scattering; all other R_g values are from light scattering.

Care must be taken in interpreting GFC retention of solutes of different conformation. Retention characteristics in GFC are affected by the shape of the macromolecule as much as by its molecular weight. As illustrated in Table 2.4, the sizes of natural macromolecules are greatly influenced by their shapes. Although the table lists the macromolecules in the order of increasing molecular weight, the R_g values fluctuate greatly, reflecting the expected large effect of solute conformation of R_g, according to Equations 2.30 to 2.32.

Most synthetic polymers in dilute polymer solutions are close to being random-coil, rather than rod, conformation. Proper GPC-MW calibration is needed to compensate for the variation of the value of α for different polymer/solvent systems at different temperatures. A small change in value of α, between 0.5 and 0.6 for flexible polymers, can cause large molecular weight errors in GPC. It is important that the GPC-MW calibration is obtained under the same GPC conditions and using molecular weight standards of the same polymer structure as the samples (Chapt. 9). For a specific polymer/solvent system, the value of α is expected to change very little with molecular weight. This is the foundation of the fact that GPC elution curves can directly provide sample MWD information.

The fact that the same functional dependence of R_g on molecular weight applies to different molecular weight fractions of a particular polymer can be used to elucidate structure. This is shown, for example, by the data of Table 2.5. For poly-γ-benzyl-L-glutamate in chloroform/formamide solvent, R_g is seen to be proportional to the first power of molecular weight. This is

Table 2.5 Dependence of Radius of Gyration on Molecular Weight

Molecular Weight	R_g (Å)	$\frac{R_g}{M} \times 10^3$	$\frac{R_g}{M^{1/2}}$	$\frac{R_g}{M^{0.55}}$
Poly-γ-benzyl-L-glutamate in Chloroform/formamide, 25°C				
262,000	528	2.02	1.03	—
208,000	408	1.96	0.89	—
130,000	263	2.02	0.73	—
Polystyrene in Butanone, 22°C				
1,770,000	437	0.25	0.33	0.160
1,630,000	414	0.25	0.32	0.158
1,320,000	367	0.28	0.32	0.158
940,000	306	0.33	0.32	0.159
524,000	222	0.42	0.31	0.159
230,000	163	0.71	0.34	0.183

Taken from Reference 18.

clear evidence that the molecule is a rigid rod. On the other hand, the data confirm the fact that polystyrene is of the random-coil type, since R_g is roughly proportional to the square root of molecular weight.

Less Successful Attempts of SEC Retention Theory

A brief review of several less successful SEC theories is presented here so that readers will not be misled by wrong concepts for explaining SEC retention. Correct concepts of SEC retention are strengthened by explaining some misconceptions about SEC retention.

Some early unsuccessful attempts to explain SEC separation used a differential diffusion model (28, 29), which assumed a nonequilibrium separation process resulting from the different diffusion rates of different molecular weight solutes. The differenttial diffusion assumption later was demonstrated as inappropriate by flow rate data and simple static mixing experiments (Figs. 7.7 and 2.4). After the inception of the now-accepted steric size-exclusion model, it was suggested that the differential diffusion rate effect be included as a secondary mechanism in an attempt to explain the speculated flow rate dependence of SEC retention (30, 31). However, differential diffusion is definitely not an influencing factor in SEC retention.

There is no flow rate dependence of SEC retention. Peak skew at high flow rates caused the false impression of SEC peak retention being flow rate dependent.

A mechanism called "separation by flow" or SBF (32–35) has also drawn considerable attention in challenging the equilibrium size-exclusion model. The original version of SBF (32, 33) assumes that size separation is the result of the larger solutes being more excluded from the surface of the packing, and therefore forced into the faster-moving mobile-phase stream near the center of the solvent flow channels in the interstitial space. This mechanism suggests that the main elements of SEC separation are the solvent velocity profile and the solute-to-wall exclusion in the flow channels (but *not* the pores in SEC packings). Arguments against this model were later provided by an experiment using a column packed with nonporous glass beads of sizes comparable to usual SEC porous packings (5, 36). The fact that no size separation was observed in this experiment indicated that the pores in the packings are essential to solute size separation. Polymer molecules are too small compared to the space between SEC packings to cause enough of a wall-exclusion effect for significantly influencing the average solute velocity.

The deficiencies of the original SBF model were recognized and remedied in later versions (34, 35). Additional small-flow channels were introduced (by concept) inside the packing particles with sizes comparable to the polymer solute molecules. This later model also allowed solute diffusion in and out of the small channels to avoid the prediction of an unreasonably large peak broadening effect. With all these modifications, the new version of SBF is essentially that of the equilibrium size-exclusion model, except for the insignificant assumption of solvent flow through the packing particles. (Flow largely goes around the packings anyway because of the large disparity of the large and small channel sizes.) Therefore, the modified SBF model adds nothing significant to the basic equilibrium size-exclusion interpretation of SEC separation.

A stochastic model has also been used to explain SEC, using an analogy of general adsorption chromatography (37, 38). Here the SEC peak retention and peak broadening are described in terms of the rate constants for assumed stationary-phase entrapping and releasing processes. The difficulty with this model is that the proposed analogy between the parameter of adsorption rate in adsorption chromatography to the parameter of polymer solute size in SEC is arbitrary and confusing and fails to demonstrate significance toward the understanding of SEC.

A somewhat oversimplified steric, size-exclusion model assumes that the size distribution of the pores alone is directly responsible for the size-separating ability of the SEC packings (39–42). This model suggests that individual pores of the packing have a one-to-one size correspondence with the individual polymer solute molecules being fractionated by the packing. In other words, this approach considers that pore-size distribution (e.g., from mercury porosimetry) directly predicts the accessibility of pore volume for the different-sized permeating polymer molecules. The contention is that SEC retention and the SEC-MW calibration curve are directly predictable from the pore-size distribution information. Basically, this model assumes (42) that

$$V_R = V_o + \int_{\bar{r}}^{\infty} \Phi(r)\, dr \tag{2.33}$$

where V_R and V_o are the SEC retention and void volumes, $\bar{r}$ in the integral limit is the equivalent radius of the polymer solute, and $\Phi(r)\,dr$ is the total volume of pores with radius between r and $r + dr$. This basic prediction is not supported by experimental data (Figs. 2.12 and 2.13). The dashed line in Figure 2.12 is the predicted SEC calibration curve calculated from Equation

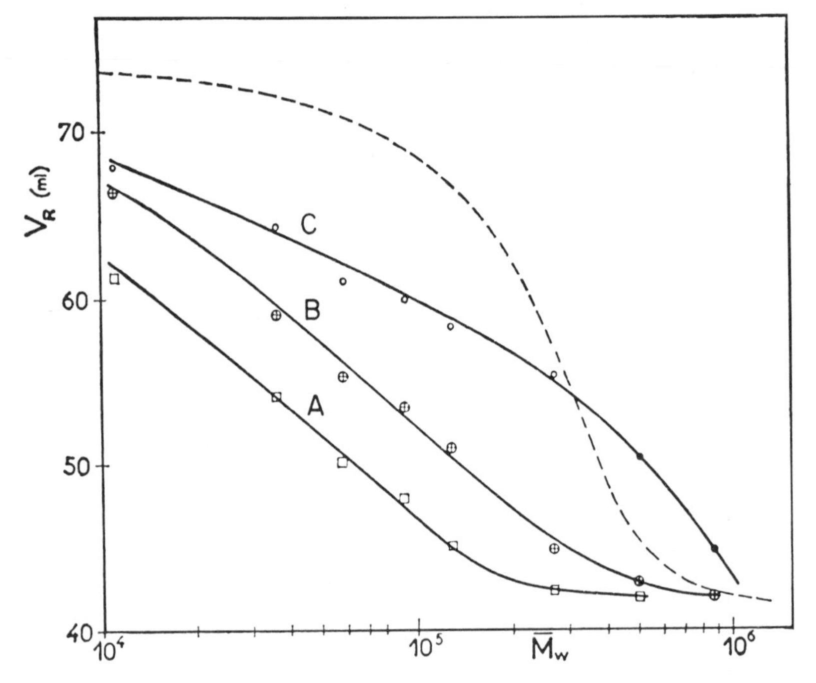

Figure 2.12 Retention volume versus logarithm of weight-average molecular weight of polystyrene calibration samples.
Column length, 2 m; packing, silica beads of different pore-size distribution. Dashed line is calculated curve for silica according to Equation 2.33. A–C, experimental data. (Reprinted with permission from Ref. 42.)

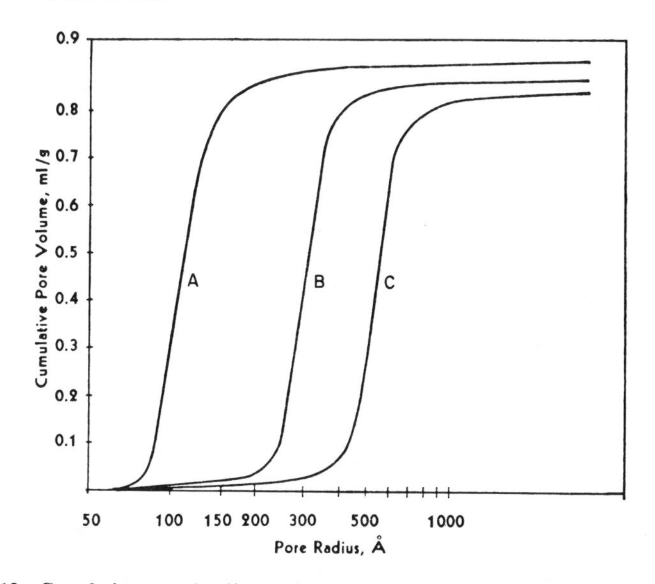

Figure 2.13 Cumulative pore-size distributions.
Determined by mercury porosimetry for three types of silica beads, A–C. (Reprinted with permission from Ref. 42.)

2.33 using the pore-size distribution data of packing *B* in Figure 2.13. The predicted curve obviously is far different from the experimental SEC calibration curve of packing *B*, which is plotted in the solid line in Figure 2.12. It appears that all the experimental SEC calibration curves have a much more gradual change in slope than the corresponding pore-size distribution curves.

The shortcoming of this model is in its basic assumption, which mistakenly supposes that a broad spectrum of pore sizes is the essential element for achieving fractionation of polymer solute molecules of widely different sizes. The model has mistakenly neglected the exclusion effect of different-size solute molecules inside a single pore—the single-pore exclusion effects illustrated in Figure 2.5. The fraction of the pore volume available to a solute is equivalent to K_{SEC} $(\bar{r}, r)$, a function of solute and pore radii, as described in Section 2.4. Thus the correct expression for retention should be (43).

$$V_R = V_o + \int_{\bar{r}}^{\infty} K_{SEC}(\bar{r}, r)\Phi(r)\, dr \qquad (2.34)$$

The fact that *the accessible volume of a pore varies for different-size solutes* is a simple, very important, but often neglected, concept. It affects the understanding and interpretation of SEC resolution and the proper selection of

SEC column sets (Sects. 4.4, 8.5, and 9.3). [Equation 2.34 is the basic convolution equation for studying the effect of pore-size distribution on SEC calibration and resolution (Chapt. 4).]

Size Separation in Small-Flow Channels

Discussions above on the separation by flow (SBF) object to the use of the SBF concept to explain SEC retention. However, these arguments are not advanced to contest the feasibility of SBF separation itself. In fact, separation by SBF has been detected using ultrafine packings (44), although resolution is much poorer than in SEC. It is unlikely that the SBF approach will ever challenge SEC for polymer and small-molecule analyses. However, the use of SBF for small, submicron particle analyses is a more practical but still limited possibility. This form of separation is sometimes called "hydrodynamic chromatography," and it is considered as a rather specific analytical technique useful for limited particle-size ranges (45).

An alternative way of utilizing the differential velocity streams in a flow channel to separate molecular sizes is the new technique called "thermal field flow fractionation" TFFF (46). The technique works with a differential thermal field to force macromolecules of different sizes to segregate in different velocity streams near the wall of the flow channel. In TFFF, larger solutes elute after the smaller ones, in contrast to SEC separation. In an another form of FFF, sedimentation FFF for particle-size analyses, particulates are forced into different velocity streams by the gravitational field resulting from an applied centrifugal force (46).

REFERENCES

1. L. R. Snyder and J. J. Kirkland, *Introduction to Modern Liquid Chromatography*, Wiley, New York, 1974, p. 339.
2. Spectra-Physics, Technical Manual, *Basics of Liquid Chromatography*, 3rd ed., Spectra-Physics, Santa Clara, Calif., 1977.
3. F. T. Gucker and R. L. E. Seifert, *Physical Chemistry*, Norton, New York, 1966.
4. J. V. Dawkins, *J. Polym. Sci., Part A-2*, **14**, 569 (1976).
5. W. W. Yau, C. P. Malone, and H. L. Suchan, *Sep. Sci.*, **5**, 259 (1970).
6. W. W. Yau, H. L. Suchan, and C. P. Malone, *J. Polym. Sci., Part A-2*, **6**, 1349 (1968).
7. J. J. Hermans, *J. Polym. Sci., Part A-2*, **6**, 1217 (1968).
8. W. W. Yau, C. P. Malone, and S. W. Fleming, *J. Polym. Sci., Part B*, **6**, 803 (1968).
9. T. L. Chang, *Anal. Chim. Acta*, **42**, 51 (1968).
10. E. F. Casassa, *J. Phys. Chem.*, **75**, 3929 (1971).
11. J. C. Giddings, E. Kucera, C. P. Russell, and M. N. Myers, *J. Phys. Chem.*, **72**, 4397 (1968).

12. T. C. Laurent and J. Killander, *J. Chromatogr.*, **14**, 317 (1964).
13. A. G. Ogston, *Trans. Faraday Soc.*, **54**, 1754 (1958).
14. M. E. Van Kreveld and N. Van Den Hoed, *J. Chromatogr.*, **83**, 111 (1973).
15. H. Coll and D. K. Gilding, *J. Polym. Sci., Part A-2*, **8**, 89 (1970).
16. O. B. Ptitsyn and Y. E. Eizner, *Zh. Fiz. Khim.*, **32**, 2464 (1958).
17. J. M. Evans and L. J. Maisey, in *Industrial Polymers: Characterization by MW*, J. H. S. Green and R. Dietz, eds. Transcripta Books, London, 1973, p. 89.
18. C. Tanford, *Physical Chemistry of Macromolecules*, Wiley, New York, 1961, Chapts. 3 and 5.
19. E. F. Casassa, *J. Polym. Sci., Part A-2*, **10**, 381 (1972).
20. E. F. Casassa, *J. Polym. Sci., Part B*, **5**, 773 (1967).
21. E. F. Casassa and Y. Tagami, *Macromolecules*, **2**, 14 (1969).
22. E. F. Casassa, *Sep. Sci.*, **6**, 305 (1971).
23. W. W. Yau and C. P. Malone, *Polym. Prepr.*, **12**, 797 (1971).
24. E. F. Casassa, *Macromolecules*, **9**, 182 (1976).
25. J. C. Moore and M. C. Arrington, International Symposium on Macromolecular Chemistry, Tokyo and Kyoto, 1966, paper VI-107.
26. J. W. McBain, *J. Am. Chem. Soc.*, **57**, 699 (1935).
26a. A. H. Abdel-Almin and A. E. Hamielec, *J. Appl. Polym. Sci.*, **16**, 1093 (1972).
27. D. Berek, D. Bakos, L. Soltes, and T. Bleha, *J. Polym. Sci., Part B*, **12**, 277 (1974).
28. W. W. Yau and C. P. Malone, *J. Polym. Sci., Part B*, **5**, 663 (1967).
29. G. K. Ackers, *Biochemistry*, **3**, 723 (1964).
30. W. W. Yau, *J. Polym. Sci., Part A-2*, **7**, 483 (1969).
31. M. Kubin, *J. Chromatogr.*, **108**, 1 (1975).
32. E. A. DiMarzio and C. M. Guttman, *J. Polym. Sci., Part B*, **7**, 267 (1969).
33. E. A. DiMarzio and C. M. Guttman, *Macromolecules*, **3**, 131 (1970).
34. C. M. Guttman and E. A. DiMarzio, *Macromolecules*, **3**, 681 (1970).
35. H. F. Verhoff and N. D. Sylvester, *Macromol. Sci.—Chem.*, **A4**, 979 (1970).
36. R. N. Kelley and F. W. Billmeyer, Jr., *Anal. Chem.*, **41**, 874 (1969).
37. J. B. Carmichael, *J. Polym. Sci., Part A-2*, **6**, 517 (1968).
38. J. B. Carmichael, *Polym. Prepr.*, **9**, 572 (1968).
39. M. J. R. Cantow and J. F. Johnson, *J. Polym. Sci., Part A-1*, **5**, 2835 (1967).
40. M. J. R. Cantow, R. S. Porter, and J. F. Johnson, *J. Polym. Sci., Part A-1*, **5**, 987 (1967).
41. M. J. R. Cantow and J. F. Johnson, *Polymer*, **8**, 487 (1967).
42. A. J. de Vries, M. LePage, R. Beau, and C. L. Guillemin, *Anal. Chem.*, **39**, 935 (1967).
43. M. LePage, R. Beau, and A. J. de Vries, *J. Polym. Sci., Part C*, **21**, 119 (1968).
44. S. Mori, R. S. Porter, and J. F. Johnson, *Anal. Chem.*, **46**, 1599 (1974).
45. H. Small, *Chem. Tech.*, March, p. 196 (1977).
46. J. C. Giddings, *J. Chromatogr.*, **125**, 3 (1976).
47. W. W. Yau, E. I. DuPont de Nemours & Co., unpublished results, 1974.

Three

BAND BROADENING

3.1 Introduction

In column chromatography, a small volume of the sample solution is injected to form a band at the top of the column. As this band migrates downstream, its width increases. The sample solution in the band becomes increasingly more dilute as the band becomes more spread out in the direction of the flow, parallel to the axial (or longitudinal) direction of the column. Band broadening of a pure component can be used to measure the efficiency of the chromatographic system. (The term "band broadening" is commonly used with the implication that the solute band consists of a pure component.) Column band broadening is measured experimentally by the width of single chromatographic peaks such as those illustrated at the bottom of Figure 2.1. Gross overestimation of column band broadening can occur if the probe chromatographic species is not pure but contains partially separated components, or is a polymer species having an appreciable MWD.

The nomenclature used in reporting chromatographic band broadening in the literature is quite varied and sometimes confusing. Readers should be aware of the many near synonyms for band broadening that appear in the literature, such as peak broadening, zone spreading, and instrumental, axial, longitudinal, or column dispersion.

All forms of band broadening are detrimental to chromatographic resolution. Basically, chromatographic separation is a demixing phenomenon; for example, in the LC analysis of small molecules, different molecular species in the original sample solution are demixed as they elute from the column in

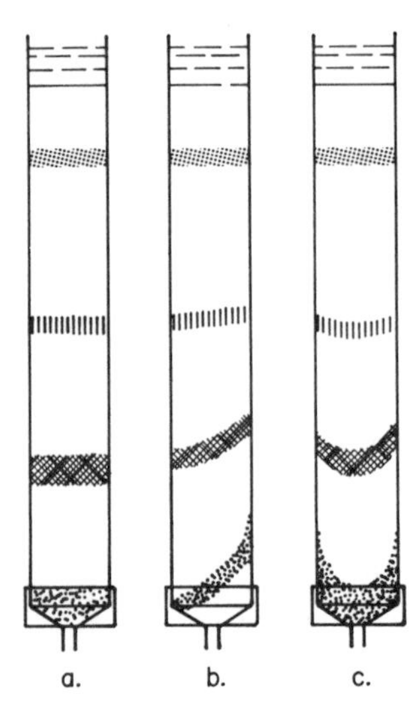

Figure 3.1 Effect of packing inhomogeneity on band distortion.

(*a*) uniform migration of a band in a well-packed column; (*b*) band distortion due to uneven packing density across the column; (*c*) band distortion due to radial packing inhomogeneity.

separated peaks. On the other hand, band broadening is a back-mixing or remixing phenomenon which causes the LC peaks to spread out and overlap. The effect is to make peak identification and peak-size analysis more difficult. Band broadening is also detrimental to SEC. In the analysis of broad MWD polymers by SEC, the effect of band broadening is to interfere with the integrity of the MWD information as displayed by the elution curve profile. A small distortion of the SEC curve shape by band broadening can cause large MW errors in the SEC analysis. The problems of MW accuracy and SEC resolution are discussed in detail in Chapter 4. Proper methods of correcting for the band-broadening effect in SEC calibration and MW computation procedures are discussed in Chapters 9 and 10.

Excessive peak broadening can result from poorly packed columns. A uniformly packed column is illustrated in Figure 3.1*a*, in which the entire band across the column is shown to migrate evenly through the column. In columns with large packing inhomogeneity as illustrated in Figure 3.1*b* and *c*, the solute band can become grossly distorted as the band migrates through the column. This band distortion is observed as excessive band broadening. Macroscopic channeling is another packing defect that can cause large peak broadening because of the fingering effect of the solute band in the packed bed. A detailed consideration of packing homogeneity and packing techniques is discussed in Chapter 6.

Gross band broadening can also result from excessive extracolumn volume that is present in the chromatographic instrument. Large-volume elements such as a flow filter or pulse dampener must be installed before the sample injector to avoid excessive band broadening. Details for minimizing extra-column band-broadening effects are discussed in Chapter 5.

In this and the next section, general LC band-broadening effects are discussed in detail. This basic information is not only useful to the understanding of band broadening of polymers in SEC, but is directly applicable to the practice of SEC separations of small molecules and oligomers.

Basic Column-Dispersion Processes

In a well-planned SEC experiment where well-packed columns and an efficient instrument are used, ultimate SEC column efficiency will depend on the inherent band-broadening processes occurring in the column. The uniform band broadening illustrated in Figure 3.1*a* is caused by molecular mass transfer processes and microscopic flow irregularities inherent in the column packing structure. These band dispersion effects constitute a large part of the overall band broadening observed in usual HPSEC experiments. Accordingly, a large part of this chapter is devoted to discussion of this important subject. An understanding of the basic column dispersion processes and their dependence on SEC operating variables is needed for making efficient SEC separations with the best compromises among time, accuracy, effort, and convenience.

The basic elements of SEC column dispersion are illustrated in Figure 3.2. Figure 3.2*a* represents a cross section of the solute band profile at the column inlet just after sample injection. Figure 3.2*b* shows one of the three fundamental processes leading to band broadening in SEC: *eddy diffusion*. This process arises because sample molecules take separate routes through the packed bed, as illustrated by the various arrows. Since the solute moves at different speeds in wide and narrow flow paths, some solute molecules move faster downstream than the others within a given time span. As a result of this eddy-diffusion phenomenon, a spreading of the solute molecules occurs from the initial narrow band in *a* to a broader band in *b*.

A second contribution to band broadening occurs as a result of the resistance of solute to *mobile-phase mass transfer* (Fig. 3.2*c*). This broadening process is caused by the velocity gradient profile which exists in a single flow stream. Since liquid near the surface of the column packing particle moves relatively more slowly than the liquid at the center of the flow stream, solute molecules at the center migrate farther downstream than the others. Band broadening due to this dispersion process decreases with increasing lateral diffusion rate of the solute molecules between the fast- and the slow-moving liquid regions. At times this dispersion process is called "mobile-phase lateral diffusion," or "extraparticle mass transfer."

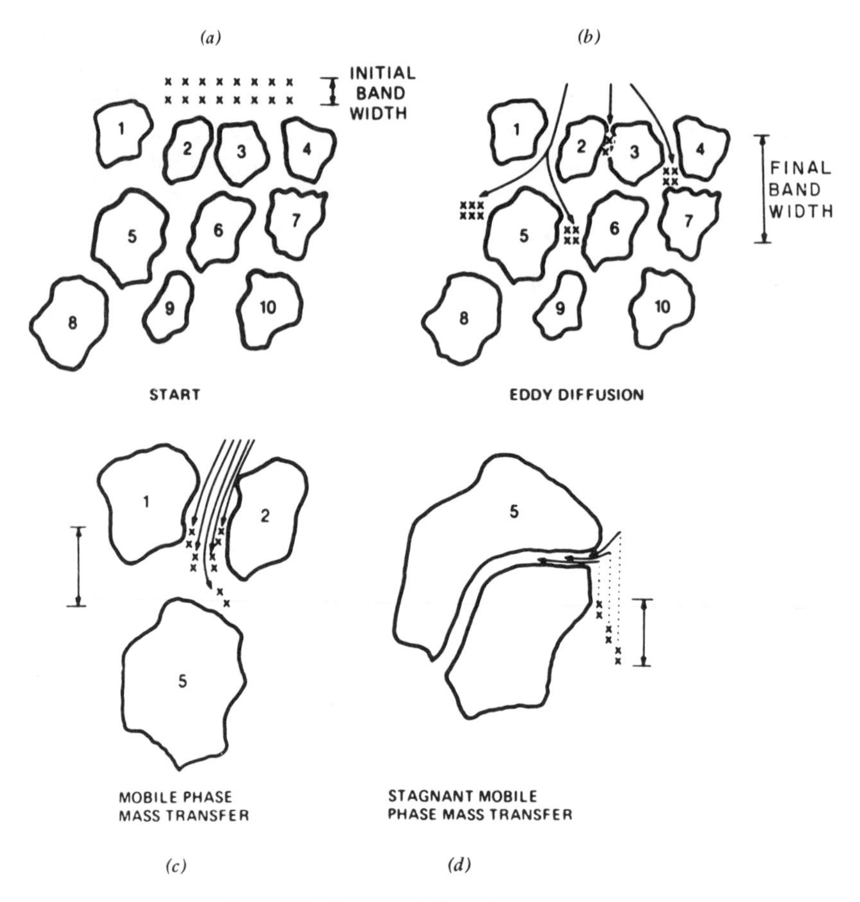

Figure 3.2 Basic peak dispersion processes.

Band broadening because of the resistance of the solute to *stationary-phase mass transfer* is illustrated in Figure 3.2*d* (for simplicity, a packing particle with a single pore instead of a complex pore structure is shown). This process of band broadening arises from the slow solute diffusion in and out of the pores of the packing particles. While some molecules are diffusing into the pores, others move with the solvent farther downstream. For large solute molecules with low diffusion coefficients this type of solute downstream migration will cause extensive band broadening. Therefore, this dispersion process, which has also been called "stationary-phase lateral diffusion," "intraparticle mass transfer," or "stationary-phase nonequilibrium mass transfer," is the major contributor to band broadening in the SEC analysis of macromolecules. [For general LC methods other than SEC, this dispersion process is better known in terms of "stagnant mobile-phase mass transfer," the term "stationary-phase mass transfer" being reserved to describe the dispersion effect due to the LC stationary phase (not shown in Fig. 3.2).]

Longitudinal diffusion is another basic band-broadening process (not shown in Fig. 3.2) in which the band is broadened along the column's axis parallel to the flow direction by molecular diffusion of the solute. This form of band dispersion is important in GC but is insignificant in large-molecule SEC because of slow diffusion.

Peak Variance

The phenomenon of chromatographic band broadening is a random statistical process of solute mixing and is therefore subject to statistical analyses. In statistics, the fundamental parameter for describing the width of a statistical distribution is the variance of the distribution. This basic concept is adopted in chromatography, where the variance of single chromatographic peaks is the fundamental parameter for evaluating column band-broadening effects. Mathematically, peak variance in its most general form (Var) is defined as the second central moment of the peak (described in terms of a continuous distribution of normalized chromatogram height h):

$$\mathrm{Var} \equiv \sigma_x^2 = \int_{-\infty}^{\infty} h(V - V_R)^2 \, dV \tag{3.1}$$

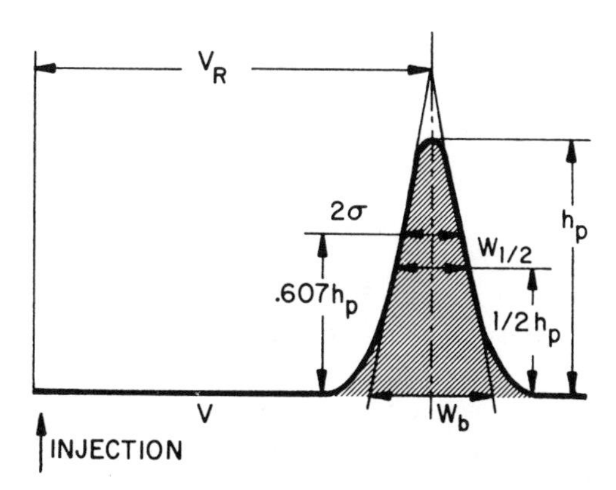

Figure 3.3 Band-broadening parameters; Gaussian peak model.

or, defined as the mean-square deviation of V from V_R for the peak (described by discrete chromatogram heights h_i):

$$\mathrm{Var} \equiv \overline{(V - V_R)^2} = \frac{\sum_i h_i(V_i - V_R)^2}{\sum_i h_i} \tag{3.2}$$

where σ_x is the standard deviation of a general statistical distribution, V and V_i the retention volume variable, V_R the retention volume of the chromatographic peak (see Fig. 3.3), and the subscript i the sequence index for the discrete equally spaced data points used in the variance calculation.

A very important property of the statistical variance is the *additivity rule of the variances* (1), which states that the overall peak variance is the sum of the individual variances resulting from each of the independent band-broadening effects occurring in the chromatographic process. Thus for *mutually independent dispersion processes*

$$\mathrm{Var} = \sum_i \mathrm{Var}_i \tag{3.3}$$

where Var_i represents variance contributions from the various dispersion processes. Equation 3.3 is very useful in chromatographic band-broadening studies because it allows different dispersion effects and individual columns and volume elements to be evaluated separately or as an integral part of the chromatographic system. Specific forms of Equation 3.3 are used in later chapters for describing specific SEC dispersion effects.

The variance formulations of Equations 3.1 to 3.3 are universal expressions for chromatographic peaks in general, regardless of peak shape. These equations are the most accurate expressions for evaluating chromatographic band broadening. However, without a peak model for reference, it is difficult to visualize the physical significance of the peak variance concept and to establish a tie between the mathematic symbols in these equations and the observable parameters in the chromatographic experiment.

Conventional band-broadening parameters are developed from the Gaussian-peak-shape model, as shown in Figure 3.3. The contour of a Gaussian elution peak is described by the equation

$$h = \frac{A}{\sigma\sqrt{2\pi}} e^{-(V - V_R)^2/2\sigma^2} \tag{3.4}$$

where A is the area of the peak, σ the standard deviation of the Gaussian peak in retention volume units, and h, V, and V_R are as defined in Equations 3.1 and 3.2. (Peak standard deviation is sometimes reported in retention time units; however, this practice is not recommended. Unless flow rate is stated, the standard deviation in the time units gives only incomplete information about band broadening.) It can be shown that $\sigma = 0.43W_{1/2} = W_b/4$, where

$W_{1/2}$ and W_b, also in volume units, are the peak width at half-height and at the base, respectively.

Substitution of h from Equation 3.4 into Equation 3.1 leads to the result that the variance of a Gaussian peak is equal to the square of the peak standard deviation:

$$\text{Var} = \sigma^2 = \left(\frac{W_b}{4}\right)^2 \quad \text{(Gaussian peak)} \tag{3.5}$$

With Equation 3.5, the tie is established between variance and the experimental quantities shown in Figure 3.3. According to Equation 3.5, peak variance increases linearly with the square of the peak width.

With the Gaussian peak model, the variance additivity rule in Equation 3.3 becomes

$$\sigma^2 = \sum_i \sigma_i^2 \tag{3.6}$$

Because band-broadening effects are summed according to the square of the σ values, the effect of one relatively large dispersion element is greatly magnified in the overall chromatographic band broadening. A single element with large dispersion in a system can dominate the total band broadening and damage the efficiency of the entire system. For example, the high-performance features of HPSEC columns cannot be realized (and thus the potential of the method will be wasted) if such columns are used with conventional SEC instruments which generally exhibit large extracolumn band-broadening effects. To achieve high-quality system performance, care must be excercised to avoid the use of any element in the chromatographic system which causes excessive band dispersion.

Equations 3.5 and 3.6 derived from the Gaussian peak model provide good predictions of column performance with usual HPSEC experiments. The use of the Gaussian peak model to study band broadening is supported by both the plate and the rate theories of band broadening (see the discussions of Eqs. 3.8, 3.10, and 3.30). However, for studying dispersion processes that cause large peak skewing, there will be errors in calculating variance by using Equations 3.5 and 3.6 instead of Equations 3.1 to 3.3. The development of the band-broadening parameters with a skewed peak model is discussed in Section 3.5.

3.2 LC Plate Theory

There are two ways of approaching the theoretical interpretation of chromatographic band broadening. In the kinetic or the *rate theory* considered in the next section, band broadening is explained in terms of realistic models involving molecular diffusion and flow mixing. The other is the *plate theory*, which is a simplified, phenomenological approach. It explains band broadening by random fluctuations around the mean retention volume by a simulated partitioning model in a chromatographic column. The plate theory was first applied to LC studies by Martin and Synge (2), and many early advances in gas chromatography also owe a great deal of credit to the development of this insight. Because of its simplicity, the plate theory will continue to be a useful, general model for studying chromatographic band broadening. The basic derivation of the general plate theory can be found in many GC and LC books (3, 4), and only a brief explanation is given below.

Basic Plate Theory

In the plate model the chromatographic column is pictured as being divided into N number of adjoining separation zones, with each zone having such a length that there can be complete equilibrium of the solute between the mobile and the stationary phases within the zone. Each zone is called a "theoretical plate," and its length in the column is called the "height equivalent to a theoretical plate," HETP, or simply the "plate height," H.

To illustrate the plate concept, a rudimentary five-plate column ($N = 5$) is shown in Figure 3.4, where the sequence of the plates is indexed by the serial number r. The feature of equilibrium partition in each plate is indicated in the figure by the balance between q and p, which are the fractions of the total solute in the mobile and the stationary phases, respectively, with $q + p = 1$. In this picture the flow of the carrier liquid is simulated by the sequential displacement of the entire top mobile-phase section to the right, one plate at a time. The number of times that this volume displacement has taken place following the introduction of a sharp band into the first plate is designated by the index number n. With each volume displacement, only a fraction q of the solute in each plate is carried to the next plate, leaving a fraction p behind. The solute in each plate reequilibrates in the new situation, and the displacement process repeats. This repetitive partition process leads to a solute distribution among many neighboring plates that follows the binomial

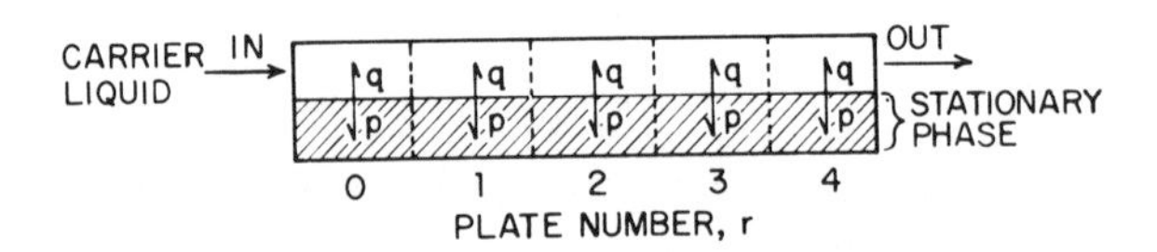

Figure 3.4 Hypothetical column of five theoretical plates.

distribution function. According to binomial statistics, the fraction of the original solute being in the rth plate following n displacements is

$$W(n, r) = \frac{n!}{r!(n-r)!} q^r p^{n-r} \tag{3.7}$$

In chromatography the solute concentration detector monitors the fraction q of the Nth (last) plate as a function of n. The elution curve is therefore described by $q \times W(n, N)$, where n is proportional to retention volume. For the usual large number of plates in chromatographic columns ($N > 50$), the binomial solute distribution becomes indistinguishable from the Gaussian distribution function (5). With algebraic transformation, the Gaussian peak elution profile as predicted by the plate model can be expressed in terms of the experimental quantities of concentration (c), retention volume (V), peak retention volume (V_R), sample weight W, and p, the fraction of solute in the stationary phase:

$$C = \frac{W}{\sqrt{2\pi p V_R^2/N}} e^{-N(V-V_R)^2/2pV_R^2} \tag{3.8}$$

By comparing Equation 3.8 with the general Gaussian function, (Eq. 3.4), one finds that

$$\sigma = V_R \sqrt{\frac{p}{N}} \quad \text{or} \quad N = \frac{pV_R^2}{\sigma^2} \tag{3.9}$$

As p approaches unity,

$$\sigma = \frac{V_R}{\sqrt{N}} \quad \text{or} \quad N = \frac{V_R^2}{\sigma^2} \tag{3.10}$$

Other results of the general plate theory are:

$$H = \frac{L}{N} = L\left(\frac{\sigma^2}{V_R^2}\right) \tag{3.11}$$

and

$$H = \sum_i H_i = \left(\frac{L}{V_R^2}\right) \sum_i \sigma_i^2 \tag{3.12}$$

where L is the column length and H_i the individual plate height contribution of independent column dispersion effects. Equation 3.12 is derived directly from Equations 3.6 and 3.11.

In summary, the predictions resulting from the general plate theory are:

1. Gaussian peak shape (Eq. 3.8).
2. Peak width increases linearly with retention volume (Eq. 3.10).
3. Each peak in a chromatogram has approximately the same values of N and H (Eq. 3.10).
4. N increases linearly with column length (Eq. 3.11).

(Items 2 and 3 of these predictions are not observed in SEC. The SEC column dispersion has many unique features as discussed in the next section.) The predicted dependence of band broadening on peak retention according to the general theory for GC and LC of small molecules is illustrated in Figure 3.5. Early peaks are tall and spikelike; later peaks are short and broad. The peaks in the figure were calculated from Equation 3.8 for a hypothetical column of 400 plates. Equal peak areas and $p = 1$ were assumed in the calculation.

The success of the plate theory can be attributed to the fact that experimental observations in GC and LC (other than SEC) are in good agreement with the theoretical predictions. Approximate constancy of N and H for various probe peaks in a chromatogram is usually found experimentally. For peaks that are only very slightly retained ($p < 1$, Eq. 3.9), the value of N can vary with V_R of the probe peak. In some cases experimental values of N calculated according to Equation 3.10, which assumes that $p = 1$, are often somewhat larger for peaks of low retention (6). For low retention peaks, Equation 3.10 used in the experimental N-value calculation overestimates the true column plate count, because the actual value of p is less than the implied p value of 1 in the equation.

The number of theoretical plates N is a dimensionless quantity. The value of N is a fundamental measure of the system efficiency, independent of

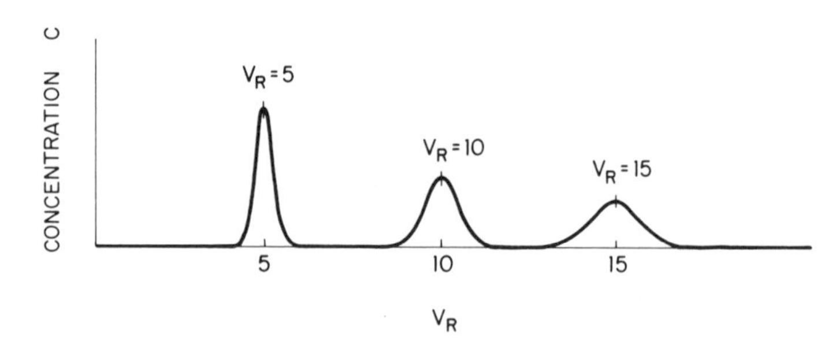

Figure 3.5 Theoretical peak shapes for a hypothetical column of 400 plates.

whether the chromatographic results are reported in retention volume or retention time units. The same is true for plate height H, which is in the units of column length; the same value of N is obtained whether it is calculated as $(V_R/\sigma)^2$ or $(t_R/\sigma_t)^2$, where σ_t is the peak standard deviation in time. While N measures system efficiency, H measures the specific column efficiency. For systems with low extracolumn dispersion, H is a measure of the intrinsic efficiency of the column packing. In chromatography a plate is only a fictitious model which does not actually exist for chromatographic columns. However, in practice, the terms plate count N and plate height H are used as if they are real physical quantities.

Plate height equations derived from the basic plate height expression in Equation 3.12 permit critical evaluation of various dispersion processes in terms of their relative importance to system efficiency for different forms of chromatography. Each independent process is associated with a variance σ_i^2 and the corresponding plate height contribution H_i. The problem is to identify important dispersion processes and express the corresponding H_i contributions in terms of physical and experimental parameters. One approach to the problem is to use the random-walk model suggested by Giddings (7). This model considers each dispersion process as being a random displacement of the solute molecules back and forth among flow streams of different velocities. In the random-walk model, the variance of each dispersion process can be expressed as

$$\sigma_x^2 = nl^2 \tag{3.13}$$

where n and l are the number and the mean characteristic length of the random steps, respectively. This semiempirical approach to derive the plate height equation can usually provide the correct functional dependence of H_i on important physical parameters. Since this model does not necessarily give a realistic description of the actual dispersion process, semiempirical adjustable constants are commonly included in the derived plate height expressions for explaining the experimental band-broadening data. The random-walk model is most useful for analyzing complex dispersion processes from complicated multichannel flow irregularities and mass transfer considerations. Many of these complex dispersion effects have been discussed in Reference 7.

The van Deemter Equation

For dispersion effects that involve simple flow and diffusion processes, exact expressions for H_i can be derived from rigorous mass transfer differential equations from the rate theory approach. A classical example is the theoretical work of van Deemter et al. (8), which led to the successful prediction of the

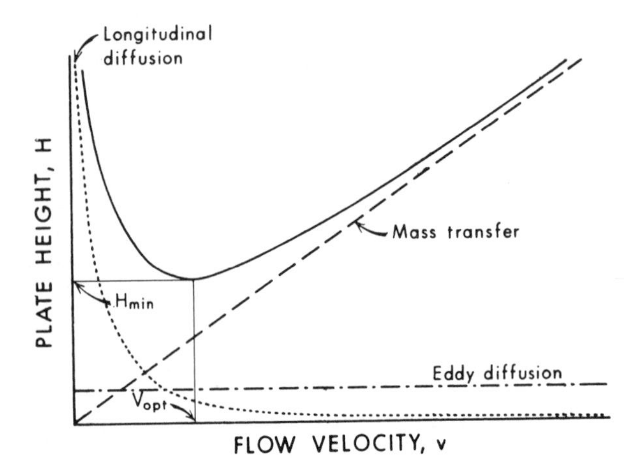

Figure 3.6 Theoretical van Deemter plot.
(Reprinted with permission from Ref. 9.)

dependence of GC column efficiency on carrier gas velocity. The now well-known van Deemter equation is highly instructive in illustrating basic peak dispersion processes.

For a general discussion of the effect of flow rate on plate height, the van Deemter equation can be simply represented by

$$H = A + \frac{B}{v} + Cv \tag{3.14}$$

where v is the flow velocity and the constants A, B, and C are associated with the plate height terms due to eddy diffusion, logitudinal diffusion, and mass transfer, respectively. A graphical representation of the parameters in Equation 3.14 is shown in Figure 3.6. Since Equation 3.14 is well known among chromatographers, reference to the three dispersion processes simply as the A, the B, and the C term is a commonly accepted practice. The dispersion process due to eddy diffusion (A term, Fig. 3.2*b*) is a simple flow splitting phenomenon which is not expected to vary with flow velocity. Band broadening due to simple molecular diffusion in the long axis of the column is the B term. This term decreases (Fig. 3.6) with increasing flow rate since a shorter time is available for longitudinal diffusion in a faster chromatographic separation. For the mass transfer or lateral diffusion processes (C term, Fig. 3.2*c* and *d*), an increase in flow rate emphasizes the velocity differences between flow streams, which results in an increase in plate height. The solid line in Figure 3.6, which is the sum of all three dispersion processes, shows a minimum in plate height (H_{min}) which corresponds to the "optimum" velocity v_{opt}; at this velocity the column has maximum efficiency. In practice, flow rates somewhat higher than v_{opt} are often used for reasonably fast chromatographic separations. Band broadening in most LC and SEC separations is controlled by the mass transfer terms since the longitudinal effect (B term) is insignificant, and except for small molecules, H_{min} is not observed in SEC.

The C term in Equation 3.14 is the sum of plate height contributions from three possible processes: [1] the C_M term from the extraparticle effects, as illustrated in Figure 3.2*c* (this term is present even for nonporous solid packings), [2] the C_{SM} term from stagnant mobile-phase effects, as illustrated in Figure 3.2*d* (this is an important SEC term often called the "stationary" mass transfer term in SEC), and [3] the C_S term from conventional LC stationary phase mass transfer effects involving the basic sorption processes.

Historically, all LC dispersion processes were considered as being independent of each other. This concept constitutes the classical interpretation of LC band broadening, as expressed in the expanded van Deemter equation:

$$H = A + \frac{B}{v} + C_M v + C_{SM} v + C_S v \tag{3.15}$$

This equation predicts a linear increase of plate height with increasing flow velocity at high-flow-rate regions, where the overall plate height is dominated by the C term. (This is expected to occur in LC at moderately high flow rates due to the relatively small A and B terms.) However, in practice, increase of plate height is found to taper off at the high flow rates. A plausible explanation for this is provided by the Giddings coupling theory (7), which is discussed next.

Flow-Diffusion Coupling

The coupling concept is in contrast to the assumed independence of the eddy- and lateral-diffusion terms in the classical plate height theory (Eq. 3.15). The coupling theory (7) maintains that both the eddy flow or stream-splitting effect and lateral diffusion can effectively move solute molecules from one flow stream to another. Thus the combined effect of eddy and lateral diffusion provides more chances for each solute molecule to experience the different velocities in the various flow channels. The more frequently the individual molecules can sample the various flow velocities while traveling downstream in the column, the more likely it is that they can attain the same statistical mean velocity and all can elute from the column more closely together.

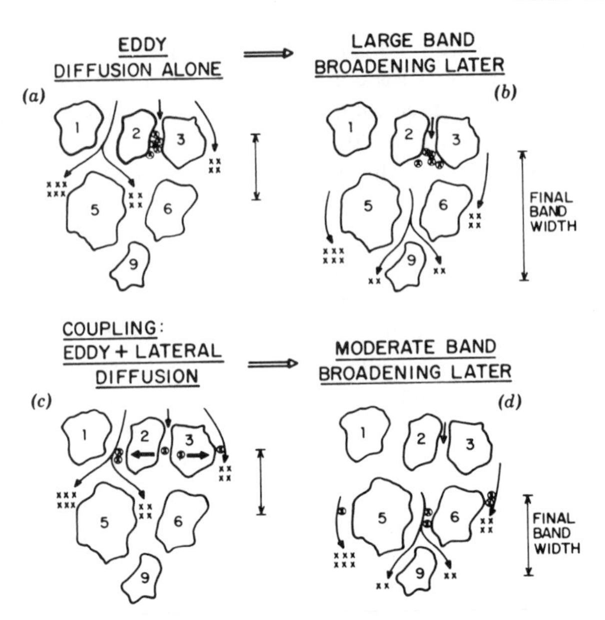

Figure 3.7 Reduced band broadening through coupling of eddy and lateral diffusion.

The end result of coupling is reduced band broadening compared to that of eddy diffusion alone. A simplified explanation of these concepts is illustrated in Figure 3.7, where band broadening due to eddy diffusion alone is compared to that of coupled eddy-lateral diffusion. In this figure the locations of solute molecules are pictured in two time frames. The frames at the right (*b* and *d*), taken a short time after the ones on the left (*a* and *c*), show that the molecules have moved farther downstream with respect to the packing particles and formed a broader band. All the solute molecules in the figure are considered structurally identical, but the slower-moving ones have been encircled for identification. The motions of these molecules are the focus of this discussion. When eddy diffusion works alone (*a* and *b*), these slow molecules lag far behind the others and contribute greatly to the overall band broadening. In coupling (*c* and *d*), these molecules have a chance to escape from the slow flow stream via lateral diffusion around the packing particles (indicated by the two oppositely pointed arrows in *c*). These diffusion-coupled molecules can thus follow faster-moving streams and elute closer to the other molecules, resulting in a reduced band width, as illustrated in *d*. While Figure 3.7 shows how the diffusion-coupling effect of the microscopic flow irregularities can reduce band broadening, an analogous coupling situation can exist for band broadening due to nonuniform velocity profile over

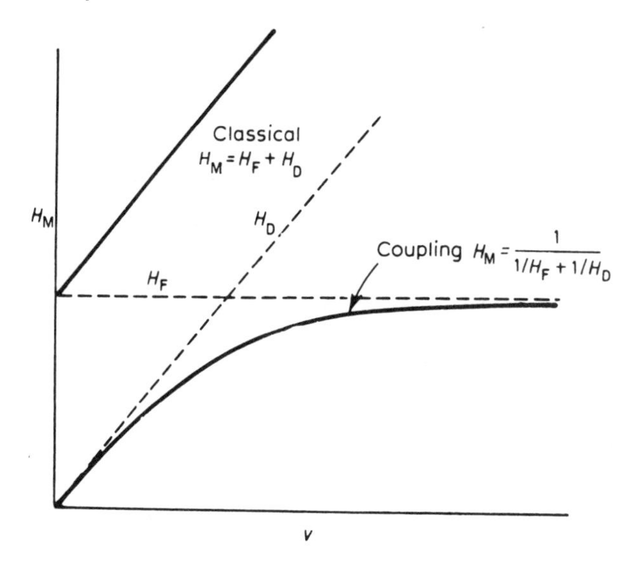

Figure 3.8 Extraparticle mobile-phase plate height contributions; classical versus coupling theory.

(Reprinted with permission from Ref. 7.)

the column cross section (3, 10). Of course, the velocity profile contribution to plate height is less in well-packed columns.

Based on the random-walk model of the coupling concept, the combined mobile-phase plate height H_M can be expressed as (7)

$$H_M = \frac{1}{1/A + 1/C_M v} \tag{3.16}$$

As shown in Figure 3.8, the plate height contribution of the coupled term calculated from Equation 3.16 is smaller than that of its individual component terms. At high flow rates, H_M approaches the eddy-diffusion term (the A or H_F term). The quantity H_D in Figure 3.8 reflects the C_M band-broadening term.

The plate height equation that incorporates the coupling concept can be expressed as

$$H = \frac{B}{v} + C_{SM} v + C_S v + \frac{1}{1/A + 1/C_M v} \tag{3.17}$$

The general curve shape of the H versus v plot predicted by this equation has been confirmed by many experimental studies and is also supported by data obtained on nonporous packings, as discussed in the next section.

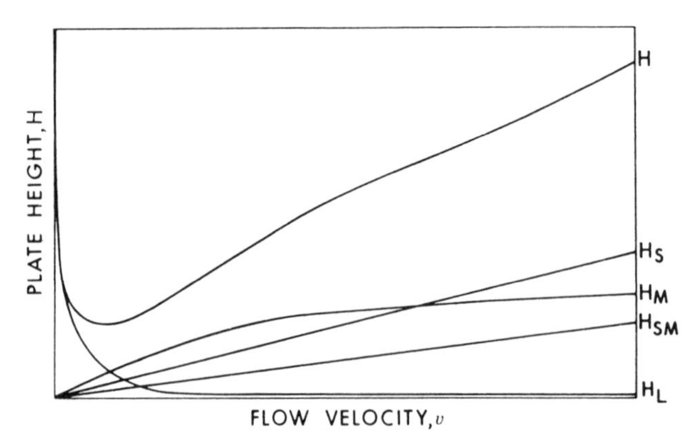

Figure 3.9 Dependence of plate height on mobile-phase velocity.
(Reprinted with permission from Ref. 9.)

The coupling theory is generally considered theoretically more sound than the classical van Deemter expressions. However, it should be noted that the focus of the LC coupling theory is on extraparticle dispersion effects. This consideration is important to SEC only for small molecules. For the SEC of macromolecules, band broadening is dominated by the C_{SM} term, which is not subject to coupling.

Although the magnitude and the relative importance of each plate height contribution from various dispersion mechanisms vary from one form of LC to another, the general functional dependence of each contribution to plate height on flow rate can be depicted by the plot shown in Figure 3.9. This figure shows the H versus v characteristics of each plate height component, and also the overall plate height:

$$H = H_L + H_{SM} + H_S + H_M \tag{3.18}$$

where H_L, H_{SM}, H_S, and H_M are the plate height contributions due to longitudinal-diffusion, stationary-mobile-phase, stationary-phase, and interparticle-mobile-phase mass transfer processes, respectively. They describe the corresponding terms in Equation 3.17.

The plate height factors given in Equation 3.18 represent a general rather than a comprehensive account of the column dispersion processes. The extracolumn dispersion effect, which is not included, is expected to behave much like the independent mass transfer terms, with its plate height contribution increasing linearly with flow velocity. The shape of the overall H versus v plot can vary greatly depending on the particular chromatographic technique used. When there is a single dispersive effect dominating in a particular chromatographic system, the shape of the overall H versus v plot will bear a resemblance to this component dispersion effect. In practice, it is desirable to have an experimental H versus v plot of the working chromatographic system. Such data can provide valuable insights into the relative importance of different plate height components and permit compromises in the experimental conditions to be made to obtain high resolution or separation speed.

For more elaborate chromatographic design considerations, a more detailed plate height equation with explicitly expressed dependence on packing particle size and solute-diffusion coefficients is more appropriate (7, 11):

$$H = b\frac{D_M}{v} + c_{SM}\frac{vd_p^2}{D_{SM}} + c_S\frac{vd_f^2}{D_S} + \frac{1}{1/ad_p + D_M/c_M vd_p^2} \tag{3.19}$$

where d_p = particle diameter of the packing
d_f = film thickness of the LC stationary phase
D_M, D_{SM}, D_S = solute-diffusion coefficients corresponding to extra-particle, stagnant mobile phase, and the stationary phase, respectively

with a, b, c_M, c_{SM}, and c_S being the coefficients of the respective dispersion terms in the plate height equation. The magnitudes of these coefficients are generally a function of the nature and the loading of the stationary phase, as well as the geometry of the packing and its pore structure. (The explicit expression for c_{SM} in the context of SEC band broadening is described in the next section.) Implicitly, the plate height is a function of many other operating variables such as temperature, solvent viscosity, and so on, as discussed in Section 3.4 for the case of SEC.

Reduced Plate Height

The plate height equation can also be expressed in terms of dimensionless quantities, reduced plate height **h**, and reduced velocity $\mathbf{v}$ (7):

$$\mathbf{h} = \frac{H}{d_p} \tag{3.20}$$

$$\mathbf{v} = \frac{vd_p}{D_M} \tag{3.21}$$

The value of $\mathbf{v}$ is often several times larger in LC than in GC because of lower solute-diffusion rates in liquids (D_M values in liquids of about 10^{-5} cm^2/sec). Even larger values of $\mathbf{v}$ are typical in SEC for macromolecules which have

Table 3.1 Equations Describing the Plate Height Contribution of the Extraparticle Mobile-Phase Effects

Giddings (7)	$\mathbf{h} = \frac{b}{\mathbf{v}} + a \Big/ \left(1 + \frac{a}{c_M} \mathbf{v}^{-1}\right)$	(1)
Huber (13)	$\mathbf{h} = \frac{b}{\mathbf{v}} + a \Big/ \left(1 + \frac{a}{c_M} \mathbf{v}^{-1/2}\right)$	(2)
Horvath and Lin (12)	$\mathbf{h} = \frac{b}{\mathbf{v}} + a \Big/ \left(1 + \frac{a}{c_M} \mathbf{v}^{-1/3}\right)$	(3)
Done and Knox (14)	$\mathbf{h} = \frac{b}{\mathbf{v}} + c_M \mathbf{v}^{1/3}$	(4)
Snyder (15)	$\mathbf{h} = c_M \mathbf{v}^n, 0.3 \leq n \leq 0.7$	(5)

very small D_M values ($\sim 10^{-7}$ cm^2/sec). A typical value of **h** for a monomer with an efficient column is approximately 2–3; a poorly packed column usually has an **h** value > 10 (Sect. 6.4).

To study flow rate effects, the use of the reduced **h** and **v** values permits column efficiency data collected from different chromatographic studies to be compared effectively. An example of this is found in a band-broadening study of the extraparticle mobile-phase effects (12), where the Giddings coupling expression for the extraparticle effects ($H_L + H_M$ in Eq. 3.18) is tested against several empirical equations to explain experimental data. Table 3.1 lists these equations with the original references. Each equation given in the table has a characteristic slope in the linear region of the log **h** versus log **v** plot. The predicted slope is 1 for equation (1), $\frac{1}{2}$ for equation (2), and $\frac{1}{3}$ for equations (3) and (4) in Table 3.1. The slope of equation (5) is equal to the variable exponent n. The difference in the predicted slope is clearly seen among the theoretical curves shown in Figure 3.10, where equation (3) is chosen to illustrate the case for the slope of $\frac{1}{3}$. The curves in the figure were calculated from the selected values of a, b, and c_M to best fit experimental data obtained from a column packed with solid glass beads. Clearly, the experimental data are best fitted by a slope of $\frac{1}{3}$ from equations (3) or (4). The rather poor agreement between equation (1) (Giddings' coupling theory) and the experimental data suggests a need for further theoretical development on the subject. Fortunately, the lack of a quantitative theory poses much less of a problem to SEC than to other LC methods, because the extraparticle effect contributes little to the SEC plate height, especially for macromolecules.

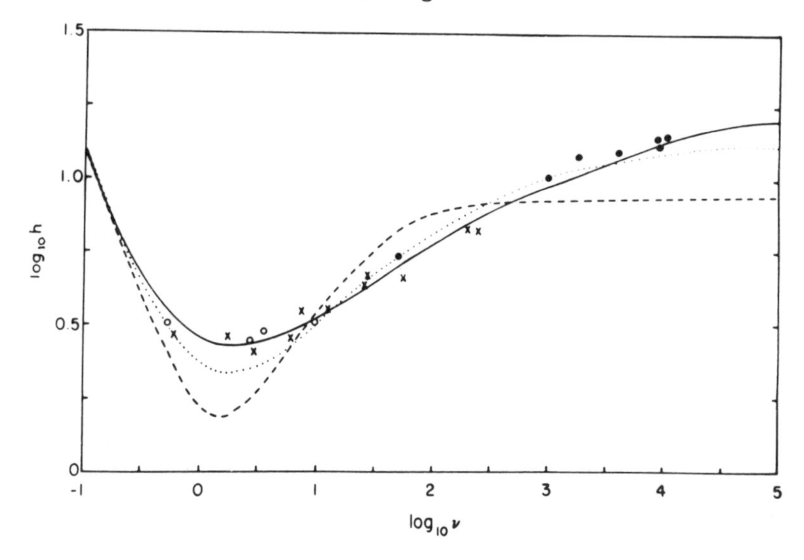

Figure 3.10 Plots of extraparticle mobile-phase effects.
The data points were obtained with a single glass bead column by using acetone in *n*-hexane (×), benzene in *n*-hexane (○), and benzoic acid in ethylene glycol (●). The curves represent the equations given in Table 3.1 with the parameters that gave the best fit to the experimental data: dashed curve, Equation (1); dotted curve, Equation (2); solid curve, Equation (3). (Reprinted with permission from Ref. 12.)

In the next section we consider the characteristic features of band broadening in SEC, which are mainly [1] the porous, nonsorptive nature of SEC packings, and [2] the slow, restricted, and molecular-weight-dependent diffusion coefficient of macromolecules.

3.3 Mechanism of SEC Band Broadening

While the volume of the solvent inside the porous packing does not affect solute selectivity in other LC methods, it in fact serves as the stationary phase in SEC, in the sense that it causes the differential elution of solutes. Accordingly, while this liquid volume is described as the stagnant mobile phase in general LC discussions, it is called the "stationary phase" in SEC. This subtle difference in basic concept has caused much confusion and many inconsistencies between SEC and general LC terminology. Thus a clarification of band-broadening terminology is presented here prior to the discussion of the SEC band-broadening mechanism.

The meaning of the phrase "stationary phase mass transfer" is different when used in SEC versus general LC discussions. The phrase means the H_{SM} term in SEC, but the H_S term in other LC methods (Eq. 3.18). In a classical sense the LC stationary term H_S defines the dispersion effect of a distinct, separate LC stationary phase, but this does not at all apply to SEC separations involving nonsorptive packings. In SEC the primary concern is the H_{SM} term, which is called the stagnant-mobile-phase dispersion in LC discussions. Since the phrase "stagnant mobile phase" is somewhat confusing in SEC discussions, we have adopted the convention of calling H_{SM} the "stationary-phase" effect in the following discussions of SEC band broadening. Where conflict exists, the H_S term will be called the "LC stationary-phase" effect for distinction.

With regard to band broadening in SEC, the plate height contribution due to longitudinal diffusion H_L is insignificantly small because the large solute molecules commonly encountered in SEC have very small diffusion coefficients. With H_L and H_S dropped from Equations 3.18 and 3.19, we have, for SEC.

$$H = H_{SM} + H_M \tag{3.22}$$

or

$$H = c_{SM}\frac{vd_p^2}{D_{SM}} + \frac{1}{1/ad_p + D_M/c_M vd_p^2} \tag{3.23}$$

Since the diffusion coefficients D_M and D_{SM} in Equation 3.23 are dependent on solute molecular weight, band broadening is a function of sample molecular weight. This poses a practical problem for the accurate interpretation of SEC data for broad-MWD samples.

Experimental Verification

The validity of Equation 3.23 is well substantiated by the data shown in Figures 3.11 to 3.15 (10, 16, 17). In the studies cited, the H_M mobile-phase coupling term and the H_{SM} permeation term were successfully isolated for separate evaluation by using both porous and nonporous (nonpermeating) column packings in the experiments. The plate height data in Figure 3.11, obtained with nonporous packings, clearly show the coupling characteristics of the mobile-phase dispersion effects, the second term in Equation 3.23. Here plate height data are plotted against the Reynolds number, vd_p/η, where η is the kinematic viscosity. For small v or large D_M, the coupling term behaves much like C_M, the mobile-phase mass transfer term alone, and is expected to increase steadily with increasing flow rate. This effect is observed in Figure 3.11 for the monomer solutes: hexane, cyclohexane, and n-$C_{36}H_{74}$.

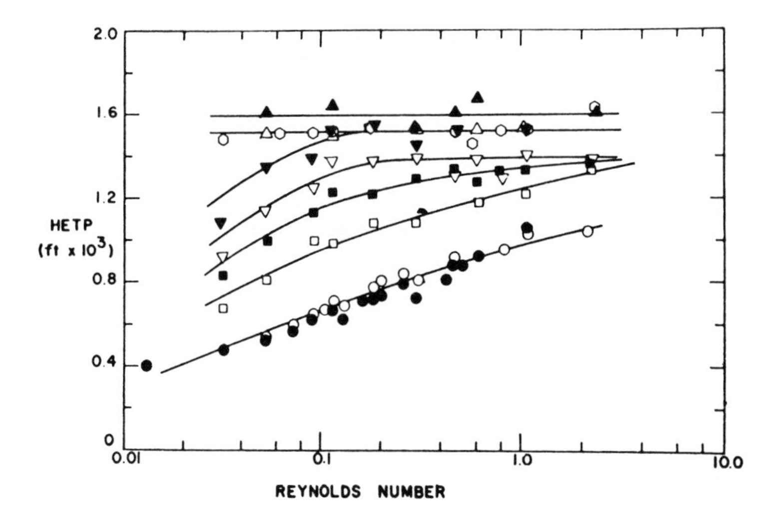

Figure 3.11 Plate height versus Reynolds number (vd_p/η) for 105–125 μm nonporous glass bead column.

●, Hexane; ○, cyclohexane; □, n-$C_{36}H_{74}$; ■, 2000 PS (polystyrene); ▽, 3600 PS; ▼, 10,300 PS; ⬡, 19,800 PS; △, 97,200 PS; ▲, 160,000 PS. (Reprinted with permission from Ref. 10.)

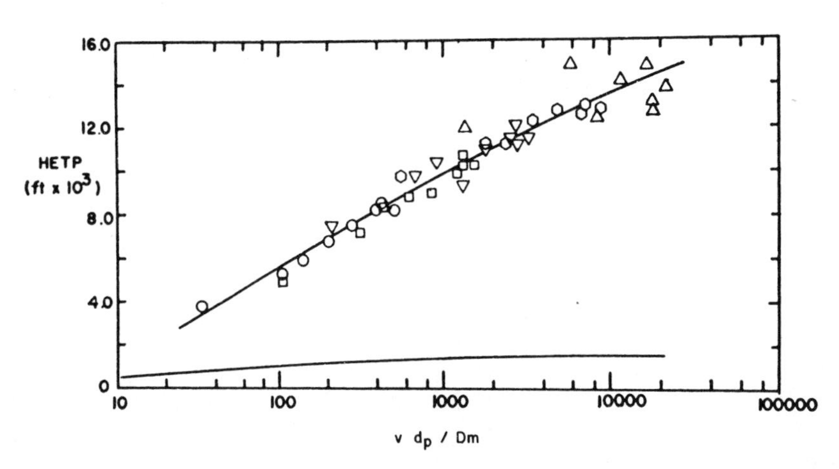

Figure 3.12 Plate height versus reduced velocity (vd_p/D_M) for 350–420 μm nonporous glass bead column.

Symbols as in Figure 3.11. Data with 105–125 μm particles from Figure 3.11 are represented as a line near the bottom of the figure. (Reprinted with permission from Ref. 10.)

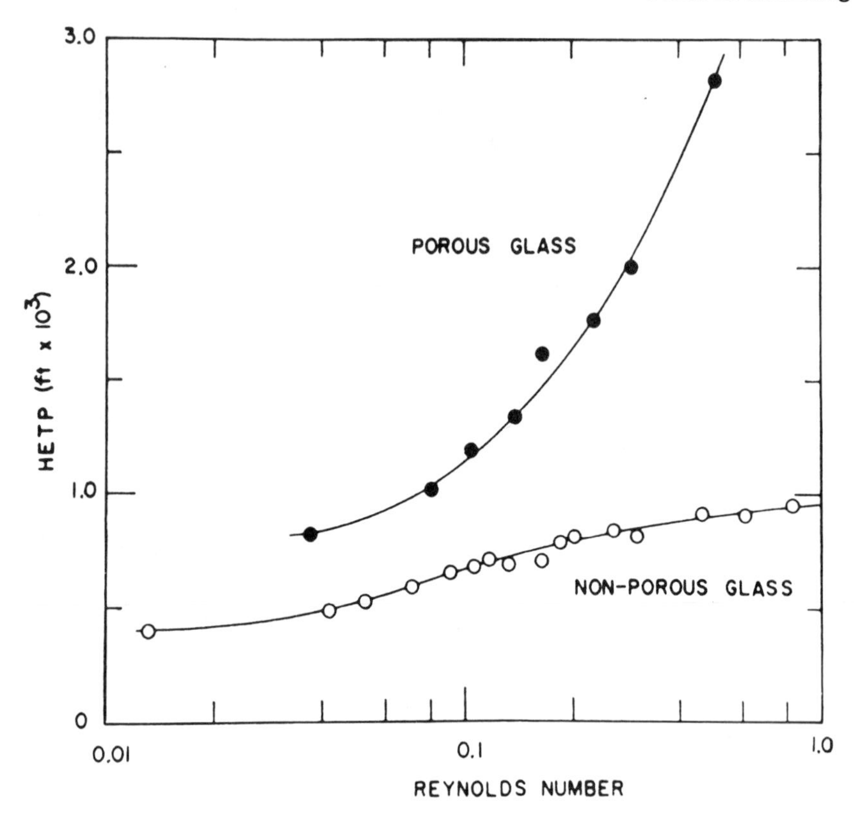

Figure 3.13 Comparison of band dispersion for porous and nonporous column packings. Particles, 105–125 μm; solute, cyclohexane. (Reprinted with permission from Ref. 16.)

With decreasing diffusion rate D_M and increasing flow velocity v, the chance for lateral solute exchange by diffusion is reduced, which brings out more of the eddy-diffusion characteristics (see the pictures of eddy and coupling effects in Fig. 3.2). As shown in Figure 3.11 for polymer solutes, at the high flow rates the observed plate height approaches a constant value which is the limiting eddy-diffusion plate height. For higher-molecular-weight solutes, this limiting condition is reached at a lower flow velocity (lower Reynolds number), as expected. This definitive illustration of the extraparticle coupling effect is made possible through the use of polymer samples with large variations in diffusion coefficients. Actually, the polymer data in Figure 3.11 are more illustrative for demonstrating the LC "mobile-phase" coupling effects illustrated in Figure 3.10.

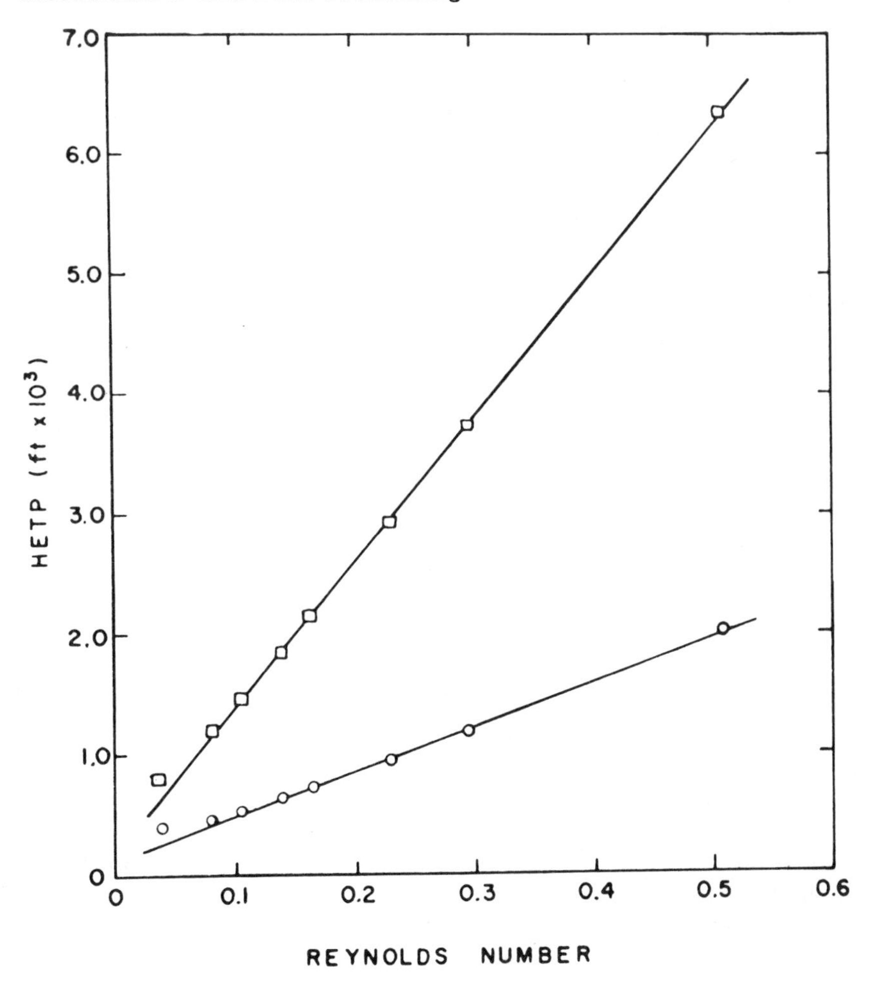

Figure 3.14 Effect of permeation on plate height as a function of Reynolds number (vd_p/η). Data with 105–125 μm Porasil A. ○, Cyclohexane; □, hexatriacontane. (Reprinted with permission from Ref. 16.)

To illustrate the effect of particle size on the H_M term, the HETP data in Figure 3.11 were replotted in Figure 3.12 against the reduced velocity $\boldsymbol{v}$ ($\boldsymbol{v} = vd_p/D_M$) to be compared with data obtained from a nonporous packing of much larger particle size. The figure shows the expected large increase in the plate height and slope of the plate height curve for the larger particle column packing.

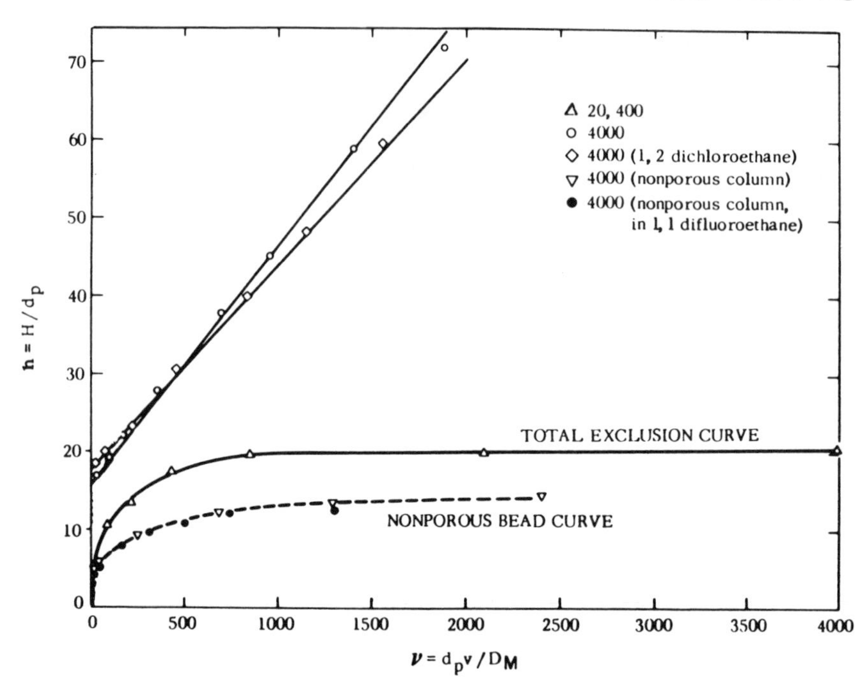

Figure 3.15 Effect of permeation on reduced plate height versus reduced velocity. Solvent, 1,1-dichloroethane, except as noted. (Reprinted with permission from Ref. 17.)

Under identical operating conditions, band broadening with a porous column packing is much larger than that with a nonporous packing. The additional band broadening is due to the "SEC stationary mass transfer" or the "permeation" plate height contribution. This permeation contribution, which is the excess plate height of porous glass over nonporous glass of the same particle size, is shown in Figure 3.13 to increase steadily with Reynolds number, or flow rate. According to theory (the first term in Eq. 3.23), this excess plate height due to permeation should vary linearly with flow rate, with the rate of increase being inversely proportional to the solute-diffusion coefficient. This is indeed observed experimentally as illustrated in Figure 3.14. Note that the HETP curve for the larger solute (hexatriacontane, smaller D_{SM}) increases faster with flow rate than that of the smaller solute (cyclohexane, larger D_{SM}).

The drawback of the preceding method of extracting the permeation contribution from the SEC plate height is the assumption of equally well packed columns. This assumption may not be realistic, since different columns never pack identically, especially those filled with porous versus nonporous packing

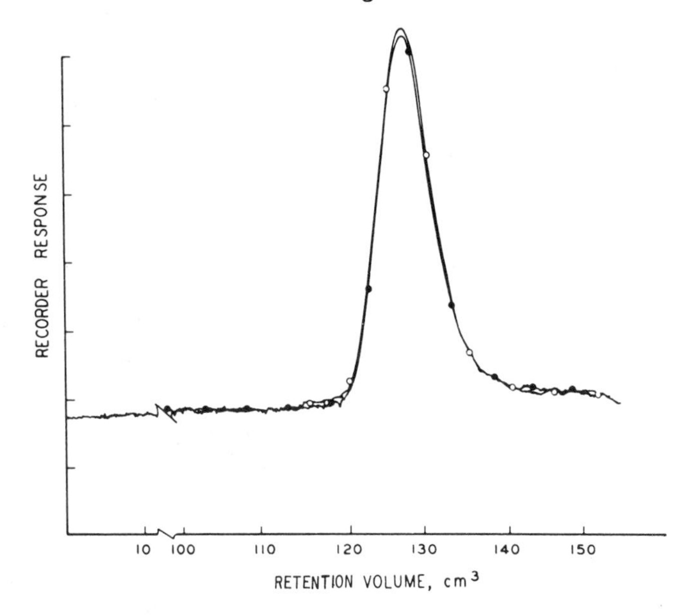

Figure 3.16 Effect of interrupted flow on SEC chromatograms.
Polystyrene sample, 2030 MW. ●, Normal sample injection and elution; ○, elution delay, 17 days. (Reprinted with permission from Ref. 18.)

materials. This potential problem can be obviated by using a nonpermeating species in the same column of porous packing to obtain the H_M term (i.e., to use a solute larger than the pores of the packing). The only dispersion experienced by bands of totally excluded solutes is due to the extraparticle mobile-phase effect. The successful use of nonpermeating solute to study the H_M term is illustrated in Figure 3.15, where, as expected, the total exclusion curve behaves much like the H_M curve of nonporous packings. Besides illustrating the large plate height and the flow rate dependence of the permeation contribution, Figure 3.15 also shows that small chemical differences in the mobile phase have only secondary effects on the characteristics of SEC dispersion, provided that the different solvents are of comparable viscosity.

In using Equation 3.23 for SEC band broadening, the longitudinal diffusion contribution to dispersion is ignored. This contention is supported by the data of an interrupted flow SEC experiment shown in Figure 3.16. The two nearly superimposable elution curves shown in the figure were obtained with one curve eluted immediately after sample injection and the other curve obtained after holding the sample on the column for 17 days with the solvent flow stopped. This is a good illustration of the very slow longitudinal diffusion rate of polymer molecules in packed columns. This means that, in the usual SEC experiments, the increase in plate height at low flow rates, like that shown by the theoretical curves in Figure 3.9, is of no practical concern.

The data presented in Figures 3.11 to 3.16 provide general support for the validity of Equation 3.23 and also provide the following insights. In the useful SEC separation range where partial solute permeation is expected, the SEC peak broadening is more dominated by the permeation process itself than by mobile-phase effects. When studying SEC band broadening the emphasis must therefore lie heavily on understanding the permeation term in Equation 3.23. To explain the total exclusion curve in Figure 3.15, one has to assume that c_{SM} goes to zero (as $K_{SEC} = 0$) at total exclusion; that is, the coefficient c_{SM} in the permeation term must be a function of the extent of permeation (K_{SEC}, Chapt. 2). In addition, the quantity D_{SM} in the permeation term which represents the solute-diffusion coefficient in the pores of SEC packings can vary greatly with solute molecular weight. Because solutes in SEC analyses are often of sizes comparable to those of the pores, the "restricted diffusion" effect must be considered in interpreting D_{SM} (see the discussion in Sect. 3.4). The rate theories of SEC dispersion described below provide more nearly quantitative understanding for some of these effects.

Rate Theory

In developing a rate theory, differential equations are derived to describe solute mass balance in a differential column section such as that illustrated in Figure 3.17 (8, 19, 20). The four classical, independent mass transfer processes are shown in their partial differential form: longitudinal diffusion (D_M

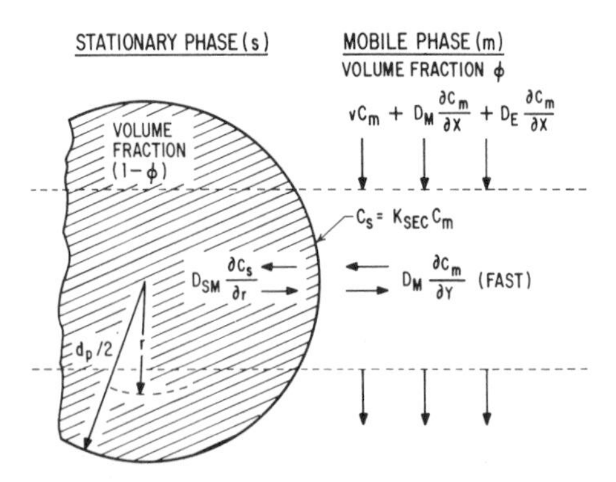

Figure 3.17 Mass transfer of solute in a thin section of SEC column.

$\partial C_m/\partial x$), eddy diffusion ($D_E\, \partial C_m/\partial x$), stationary mass transfer ($D_{SM}\, \partial C_s/\partial r$), and mobile-phase lateral diffusion ($D_M\, \partial C_m/\partial y$), where x is the column-length variable along the column axis, y the distance variable in the lateral direction, and C_m and C_s the solute concentrations in the mobile and the stationary phase, respectively. The quantity D_E is the eddy-diffusion coefficient, which is expected to be proportional to d_p and v (21).

A rate theory for SEC dispersion employs the following partial differential equations (24–26) (Fig. 3.17). For stationary mass transfer (permeation),

$$\frac{\partial C_s}{\partial t} = D_{SM}\left(\frac{\partial^2 C_s}{\partial r^2} + \frac{2}{r}\frac{\partial C_s}{\partial r}\right) \tag{3.24}$$

where r is the radial distance from the center of the spherical porous particle. For mobile-phase mass transfer,

$$\frac{\partial C_m}{\partial t} + v\frac{\partial C_m}{\partial x} - D_M\frac{\partial^2 C_m}{\partial x^2} = -\sigma D_{SM}\left(\frac{\partial C_s}{\partial r}\right)_{r=d_p/2} \tag{3.25}$$

where $\sigma = 6(1 - \Phi)/d_p\Phi$, with Φ being the volume fraction of the extra-particle solvent volume. For the boundary condition, at $r = d_p/2$,

$$C_s = K_{SEC}\, C_m \tag{3.26}$$

In these expressions, coupling of eddy diffusion to mobile-phase transfer and longitudinal diffusion effects are neglected. These assumptions are appropriate for SEC, since overall mobile-phase dispersion is usually small compared to permeation.

Although the complete solution of these differential equations (Eqs. 3.24 and 3.25) under the specific boundary conditions is not known, an approximate solution can be obtained for a limiting case where near-equilibrium solute distribution exists between the phases. This limiting condition closely approximates most SEC experiments and predicts a near-Gaussian elution peak shape.

With the help of a digital computer, a numerical solution to the differential equations for SEC dispersion can be obtained (25). The computed elution curves are shown in Figures 3.18 and 3.19 to demonstrate the predicted effect of packing-particle diameter (d_p) and flow rate on SEC peak shapes. In Figure 3.18*a*, the simulated monomer SEC peak becomes increasingly broader without changes in peak symmetry as the packing particle size increases. Figure 3.18*b* shows the particle-size effect on an earlier eluted SEC peak of a high-molecular-weight polymer. Owing to the lower diffusion coefficient, the polymer peak is more sensitive to the increase in d_p. With increasing d_p, the polymer peak becomes broader, and more skewed, with its peak maximum leaning more toward the direction of low retention volume.

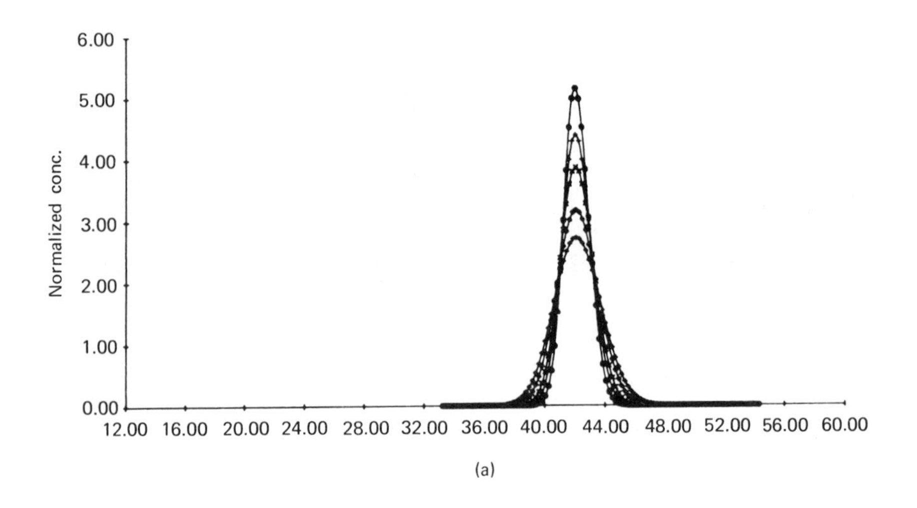

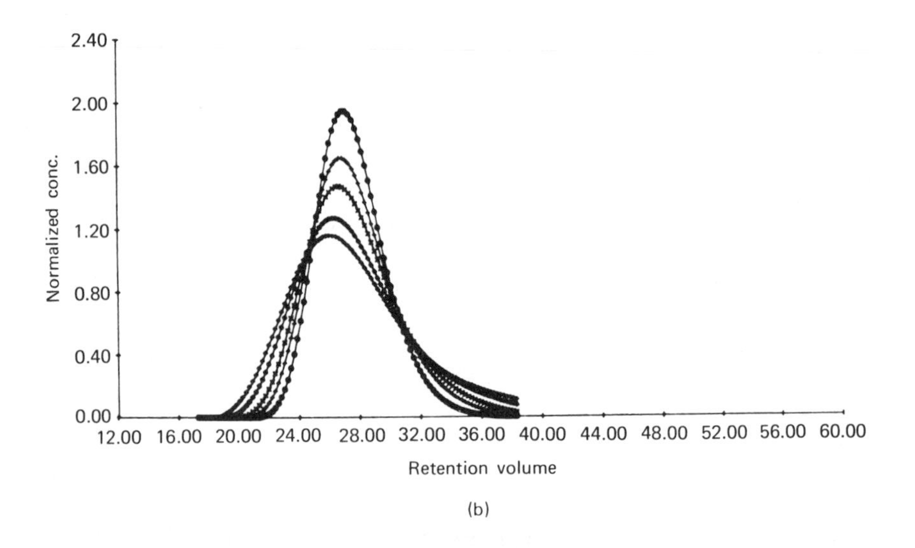

Figure 3.18 Predicted effect of particle diameter on SEC peak shapes.
(*a*) ethylbenzene peak; (*b*) peak for polystyrene, 160,000 MW. The d_p values (in units of μm): ○, 30; +, 40; ★, 50; ∅, 70; ▲, 90. (Reprinted with permission from Ref. 25.)

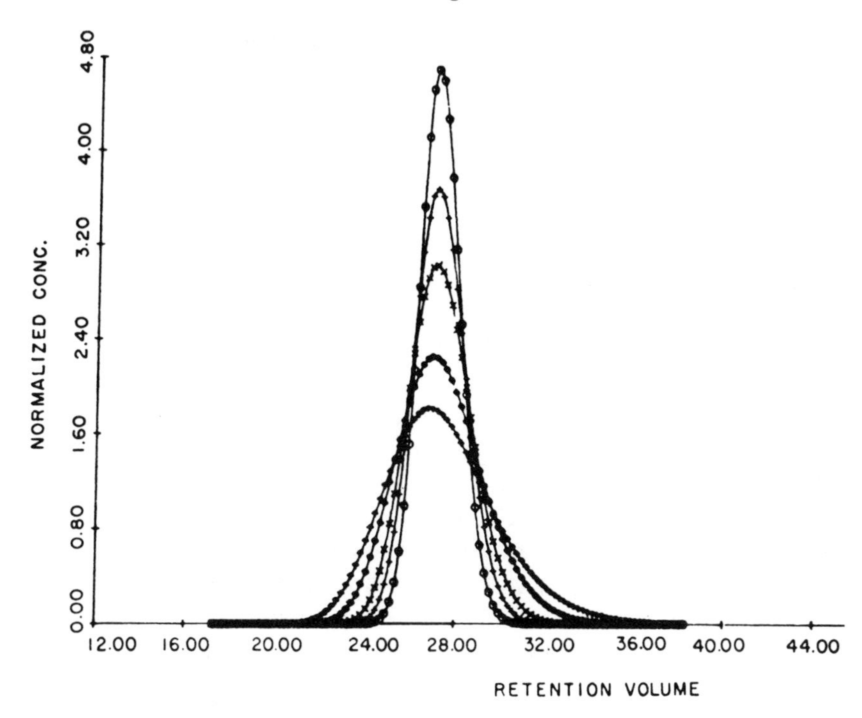

Figure 3.19 Predicted effect of flow rate on SEC peak shape.
Particle diameter d_p, 60 μm; solute, polystyrene, 160,000 MW. Flow rates: ○, 2 ml/min; +, 3 ml/min; ×, 4 ml/min; ●, 6 ml/min; ▲, 8 ml/min. (Reprinted with permission from Ref. 25.)

Figure 3.19 shows the effect of flow rate on the theoretical chromatogram of a polymer sample. As flow rate increases, the width as well as the skewness of the peak also increases, with the peak maximum again shifting to lower retention volume. All the theoretical predictions described above are quite commonly observed in SEC experiments, again demonstrating that SEC dispersion is a permeation-rate-limited process. [The curves in Figs. 3.18 and 19 were computed assuming that $D_M \propto (\text{MW})^{-\alpha_m}$ and $D_{SM} \propto (\text{MW})^{-\alpha_s}$, with $\alpha_m = 0.6$ and $\alpha_s = 1.0$. The MW dependence of diffusion coefficients is considered in Sect. 3.4.]

Explicit expressions of peak shape in terms of statistical moments (1) can be derived from the original differential equations by using Laplace transformations without having to actually solve the equations or making any assumptions about the limiting cases (24). The results obtained for the permeation dispersion process are described below. As an approximation, for the first three moments, μ_1 represents mean retention, μ_2 represents peak variance, and μ_3 represents peak skewness. These terms are quantitatively expressed as

$$\mu_1 \equiv V_R = \left(1 + \frac{1-\phi}{\phi} K_{\text{SEC}}\right) V_o = V_o + K_{\text{SEC}} V_i \tag{3.27}$$

$$\mu_2 \equiv \overline{(V - V_R)^2} = \frac{V_o V_i}{30L} K_{\text{SEC}} \frac{vd_p^2}{D_{SM}}$$

$$= \frac{1}{30} K_{\text{SEC}} V_i \frac{Fd_p^2}{D_{SM}} \tag{3.28}$$

$$\mu_3 \equiv \overline{(V - V_R)^3} = \frac{1}{420} K_{\text{SEC}} V_i \left(\frac{Fd_p^2}{D_{SM}}\right)^2 \tag{3.29}$$

where ϕ, V_o, V_i, L, and F are the void-volume fraction, void volume, internal pore volume, column length, and volume flow rate, respectively.

Since $H = L/N = L\mu_2/\mu_1^2$ (Eq. 3.11), the exact expression for the H_{SM} term (SEC stationary-phase dispersion) in Equations 3.22 and 3.23 can be obtained by combining Equations 3.27 and 3.28 to give the expression

$$H_{SM} = c_{SM} \frac{vd_p^2}{D_{SM}} = \frac{K_{\text{SEC}} V_i/V_o}{30(1 + K_{\text{SEC}} V_i/V_o)^2} \frac{vd_p^2}{D_{SM}} \tag{3.30}$$

Theoretical Inferences

Many important features of SEC analyses can be explained by the SEC rate theory (Eqs. 3.27 to 3.30). For example, Equation 3.27 indicates that the peak retention volume V_R, when defined as the first moment or the center of gravity of the peak, is not a function of flow rate. This prediction, demonstrated by the theoretical elution curves in Figure 3.19, is also verified experimentally by the data in Figure 3.20. Although the retention volume at the peak apex of a polymer sample decreases with increasing flow rate, the average retention volume of the peak is unchanged. At low flow rates, the retention volume at peak maximum approaches that of the first moment (or center of gravity) of the peak when the peak becomes more symmetrical. An example for a limiting case of Equation 3.30 is found in the earlier discussion of H_M coupling (see Figs. 3.13 and 3.15), where the H_{SM} term is forced to zero by using either a nonporous packing ($V_i = 0$) or a nonpermeating solute ($K_{\text{SEC}} = 0$).

Equation 3.28 predicts SEC peak dispersion to increase linearly with flow rate, and the slope of an H versus v plot to increase with solute MW (except

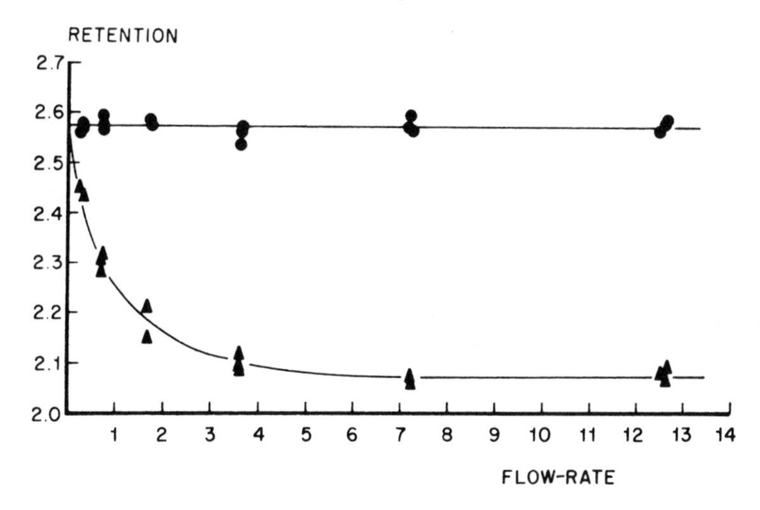

Figure 3.20 Effect of flow rate on peak apex and first moment position.
Sample, polystyrene, 51,000 MW. ●, First moment (V_R); ▲, peak maximum. Retention and flow rate are reported in terms of the weight of solvent eluted. (Reprinted with permission from Ref. 26.)

for nonpermeating species). This is generally observed in SEC, except in those experiments which involve very high flow rates and high-molecular-weight solutes of low diffusion coefficient D_{SM}. Deviations from the predicted linearity follow a unique pattern, as shown in Figure 3.21. The characteristic shape of the H versus v curve for high-molecular-weight solutes shown in *a* and *b* of the figure suggests the possibility of flow-diffusion interaction within the pore structure of the SEC packing. In fact, the data can be quantitatively fitted by an intraparticle interaction expression including diffusion and an empirical intraparticle flow, or convection velocity (26). This empirical interaction term provides the solid curves plotted in Figure 3.21. No rigorous interpretation of this intraparticle convection is yet available, although it may be attributed either to eddy currents or to flow through all or part of the pore structure (27). This flow-diffusion interaction term required to explain the data in Figure 3.21 is not to be confused with the Giddings coupling term (Eq. 3.16), which only plays a role in the extraparticle mobile phase involving plate height values orders of magnitude smaller than those of concern in the present discussion.

The predicted effect of MW and flow rate on the skew of SEC peaks as shown in Figures 3.18 and 3.19 can be studied quantitatively by using peak

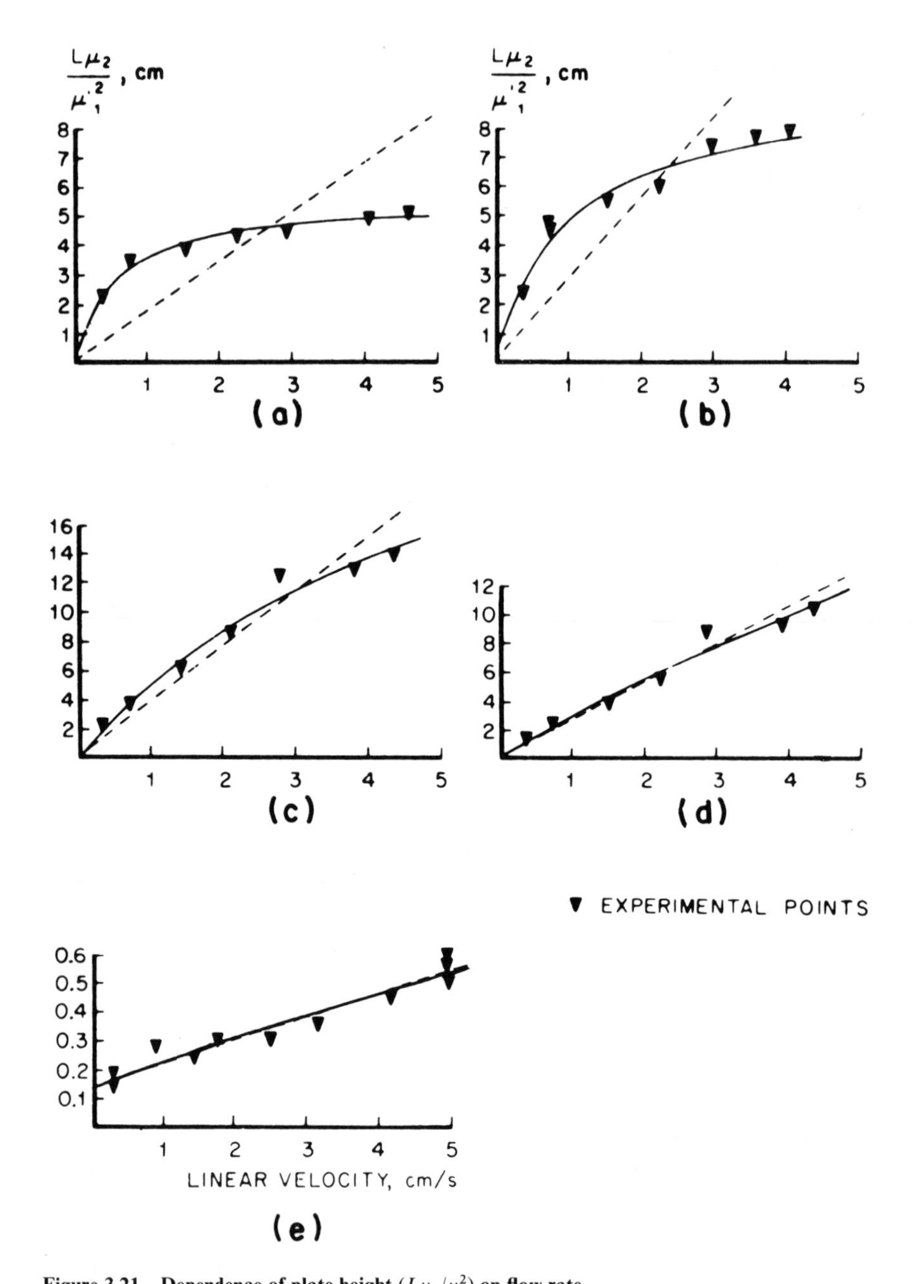

Figure 3.21 Dependence of plate height ($L\mu_2/\mu_1^2$) on flow rate.

Samples, polystyrene standards of following MW: (*a*) 160,000; (*b*) 97,200; (*c*) 51,000; (*d*) 20,400. For (*e*), the solute is toluene. Abscissa: velocity, cm/sec. (Reprinted with permission from Ref. 26.)

moments from Equations 3.28 and 3.29. Peak skew γ_{SM}, which is due to SEC stationary-phase lateral diffusion, can be expressed as

$$\gamma_{SM} = \frac{\mu_3}{\mu_2^{3/2}} = \left(\frac{15}{98K_{\text{SEC}}V_i}\frac{Fd_p^2}{D_{SM}}\right)^{1/2} \tag{3.31}$$

Both Equation 3.31 and the theoretical curves in Figure 3.19 predict that peak skewing increases with increasing flow rate, particularly for large solutes.

Since the stationary-phase mass transfer term H_{SM} is the most important dispersion factor in SEC, further implications of Equations 3.30 and 3.31 are considered in the next section. Many of the fundamental concepts discussed in this chapter serve as the foundation for optimizing columns (Chapt. 6) and operating variables (Chapt. 7).

3.4 Influencing Factors

This section directs attention to experimental parameters that can affect peak dispersion in SEC stationary phases. Other dispersion processes are not considered in this section since they are relatively unimportant in SEC, even for small-molecule SEC analyses. Adequate general information about these other dispersion processes has been given in earlier sections. Specifically, in this section, Equations 3.30 and 3.31 are considered more carefully in terms of the effects of [1] column parameters: V_i/V_o, K_{SEC}, and d_p; [2] kinetic factors: v, F, and D_{SM}; and [3] other experimental parameters: temperature, solvent viscosity, sample concentration, and so on.

Column Parameters

The column parameter V_i/V_o, the ratio of pore volume to void volume, is basically a SEC retention parameter which is directly proportional to the porosity of SEC packings. Although the plate height contribution H_{SM} increases with V_i/V_o (Eq. 3.30), SEC column packings with large porosity are still generally preferred because of their better separating ability. Large values of V_i/V_o mean more useful pore volume available for the MW separation and better overall SEC resolution. Since V_i/V_o is independent of column dimensions, so is H_{SM}, according to Equation 3.30. Thus neither column length nor column diameter variations should have much effect on the SEC plate height contribution H_{SM} (except as they lead to packing differences).

The quantity K_{SEC}, the SEC distribution coefficient, is also a retention parameter that is dependent on the size of the solute molecules relative to the packing pore size. Like V_i/V_o, K_{SEC} is usually optimized for MW selectivity and SEC resolution rather than for peak broadening. The retention volume

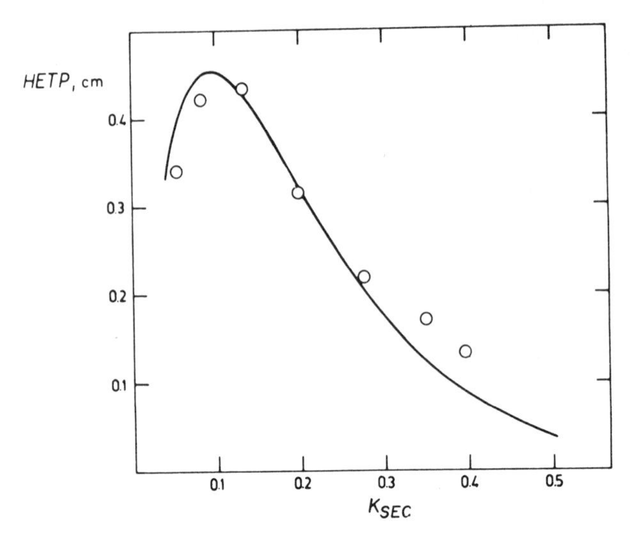

Figure 3.22 Variation of SEC plate height as a function of K_{SEC}.
Points, data from Tung and Runyon (30); curve, semiempirical prediction by Kubin (31). (Reprinted with permission from Ref. 31.)

dependence of SEC plate height approximated by H_{SM} is the result of the combined effects of the extent and rate of permeation determined by K_{SEC} and D_{SM}. A small K_{SEC} value means low solute retention V_R, large solute MW, and small solute-diffusion coefficient D_{SM}. Since D_{SM} changes more than K_{SEC} as a function of V_R, the diffusion coefficient (D_{SM}) usually dominates SEC band broadening (H_{SM}) in the most part of the SEC separation range. Only when K_{SEC} approaches zero (the total exclusion limit) will H_{SM} reverse the trend and start to decrease. This reversal predicts a maximum in the H versus K_{SEC} plot, as experimentally verified by the data in Figure 3.22 (28–30). The solid line in the figure describes the theoretically predicted curve (31).

Packing particle diameter d_p is the most influential of all experimentally adjustable parameters affecting chromatographic band broadening. The strong d_p dependence of H_{SM} is predicted by the squared term d_p^2 in Equation 3.30. This square dependence on particle size also predicts a line with a slope of 2 for the H versus d_p data in a log-log plot, as verified in Figure 3.23 (32). The data in the figure suggest a slight decrease of the slope at small d_p values which may indicate poorer packed columns at small d_p, or alternatively, sufficiently small H_{SM} terms that become less dominating over the overall SEC plate height. The impact of new HPSEC packings of $d_p < 10$ μm on modern separations can be appreciated by the extrapolation of the data in

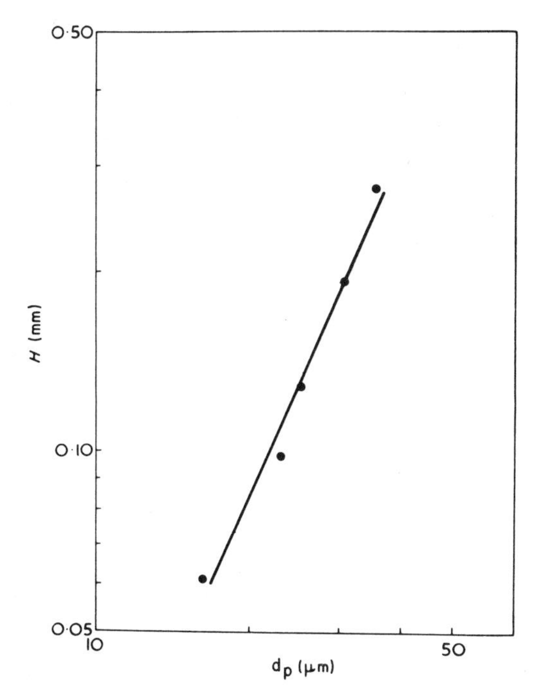

Figure 3.23 Dependence of SEC plate height on particle diameter.
Sized fractions of the same SEC gel packing particles are used. (Reprinted with permission from Ref. 32.)

Figure 3.23. Plate height values are at least an order of magnitude smaller in HPSEC than those in conventional SEC using large porous packings (e.g., $d_p > 35$ μm). Practical implications of the effect of d_p on SEC efficiency are discussed in Sections 6.4 and 7.3.

In HPSEC, peak skewing due to H_{SM} peak dispersion is much reduced with the use of small d_p packings. Equation 3.31 indicates that peak skew should increase with decreasing K_{SEC} and V_i, but decrease with decreasing d_p. For a 10 μm particle and usual HPSEC conditions (assume that $F = 1$ ml/min, $K_{SEC} V_i = 10$ ml), the estimated γ_{SM} values for peak skew according to Equation 3.31 are very small: $\gamma_{SM} = 0.05$ for a polymer solute of a diffusion coefficient of $D_{SM} = 10^{-7}$ cm^2/s, and $\gamma_{SM} = 0.005$ for a monomer solute of $D_{SM} = 10^{-5}$ cm^2/s. This calculation indicates that the peak skew observed in HPSEC experiments must be due to restricted diffusion in the pore structure so that D_{SM} actually becomes much smaller than 10^{-7} cm^2/s.

The discussion above has considered the effect of totally porous particles on peak dispersion. However, pellicular or superficially porous particles (solid core, porous shell) can also be used for HPSEC, to reduce the influence of H_{SM} peak broadening. For pellicular packings, d_p is approximated by the thickness of the porous surface layer. For larger polymer solutes that are prone to shear degradation, the use of large pellicular packings of 10–20 μm with a 2–3 μm porous layer is preferred to achieve the column efficiency which would otherwise require impractically small porous particles ($d_p < 5$ μm) and high shear forces in the flow streams.

Kinetic Factors

Among kinetic parameters, the effect of solute diffusion coefficient (D_{SM}) on band broadening is not as well studied as that of solvent velocity (v) or flow rate (F). Classically, flow rate studies are used in LC and SEC as a practical tool for studying column dispersion and for evaluating the performance of chromatographic systems. Since flow rate effects were discussed in detail in the last section, the focus of the following discussion is on D_{SM}.

The slow permeation in SEC is sometimes called a "nonequilibrium" process. This is only to indicate a microscopic nonequilibrium state of solute distribution between phases in the column. Since most of the local solute nonequilibrium is averaged out during the course of solute migration through the entire column, this effect results only in band broadening without affecting solute retention.

The molecular weight dependence of D_{SM} is the main cause for the MW dependence of the SEC plate height, as pointed out in the discussion of Figure 3.22. The reason that D_{SM} varies with solute MW is twofold: [1] solute size increases with MW to cause D_M (diffusion coefficient in open space) to decrease, and [2] restriction of diffusion inside the SEC pore structure increases with increasing solute size [i.e., D_{SM} is a function of MW; $D_{SM} = D_M f(\text{MW})$]. A rigorous expression is available to explain the influence of solute size on D_M (33):

$$D_M = \frac{RT}{6\pi\eta_0 N_0}\left(\frac{10\pi N_0}{3\mathbf{K}}\right)^{1/3}(\overline{M}_v)^{-(1+\mathbf{a})/3} \tag{3.32}$$

where R is the gas constant; T the absolute temperature; η_0 the solvent viscosity; N_0 Avogadro's number, $\mathbf{K}$ and $\mathbf{a}$ constants of the Mark–Houwink viscosity equation $[\eta] = \mathbf{K}\overline{M}_v^{\mathbf{a}}$, with $[\eta]$ the solute intrinsic viscosity; and $\overline{M}_v$ the viscosity-average molecular weight of the solute. The usual value of $\mathbf{a}$ for random-coil polymers varies from 0.5 in a poor solvent to 0.8 in a good solvent. Therefore, D_M is expected to vary with $(\text{MW})^{-\alpha_m}$, with α_m falling between 0.5 and 0.6.

No rigorous expression for the restricted diffusion of molecules in real pores is available. This is understandable in view of the difficulty of defining a realistic pore model to account for the restriction of solute diffusion in the irregular pores of SEC packings. However, some attempts have been made to describe restricted diffusion using simple models. For instance, the following expression was derived from considerations of wall friction on the motion of a solid sphere in a cylindrical pore (34):

$$D_{SM} = \frac{D_M}{\tau}(1 - 2.104\lambda + 2.09\lambda^3 - 0.95\lambda^5) \qquad (3.33)$$

with λ the solute diameter/pore diameter and τ the tortuosity factor ($\sim$2.1–2.4). For polymers, the size parameter λ is a function of solute MW [i.e., $\lambda \propto (\mathrm{MW})^{0.5 \sim 0.6}$]. Diffusion data for simple organic solutes in fine-pore structures suggest the following simple empirical equation (34):

$$\log_{10}\left(\frac{D_{SM}\tau}{D_M}\right) = -2\lambda \qquad (3.34)$$

Experimental Factors

Both Equation 3.33 and 3.34 indicate that D_{SM} is directly proportional to D_M. Therefore, the restricted diffusion coefficient D_{SM} should increase as D_M increases with increasing temperature T according to Equation 3.32, and result in an improved SEC resolution. This prediction is verified by the experimental data given in Figure 7.8 (35, 36). The same equations also predict that D_{SM} should increase in solvents of lower viscosity η_0 or small **a** (poor solvent for the polymer). However, solvent choices for many important polymers are often very limited. Therefore, optimization of SEC efficiency via solvent selection is not a common practice. The use of poor solvent to gain SEC efficiency is generally not recommended; very poor solvents may increase the chance of solute adsorption on packing surface. Many of these considerations are discussed in Chapters 7 and 8.

Solute concentration overloading can add to band broadening (see Fig. 7.16). The data show that polymer solutes at higher concentrations elute later, suggesting a more compact polymer conformation at the overloading concentrations. This overloading effect poses an SEC precision problem and an upper limit to the sample concentration that can be used to improve the signal/noise ratio of SEC analyses (Sect. 7.4).

Extracolumn dispersion effects are of special importance to the overall performance of HPSEC as compared to conventional SEC because of the high-efficiency (low-dispersion) feature of HPSEC columns. Since the dispersion effect of each element in an HPSEC instrument is different (37), the subject is discussed in more detail in Chapter 5, with special attention

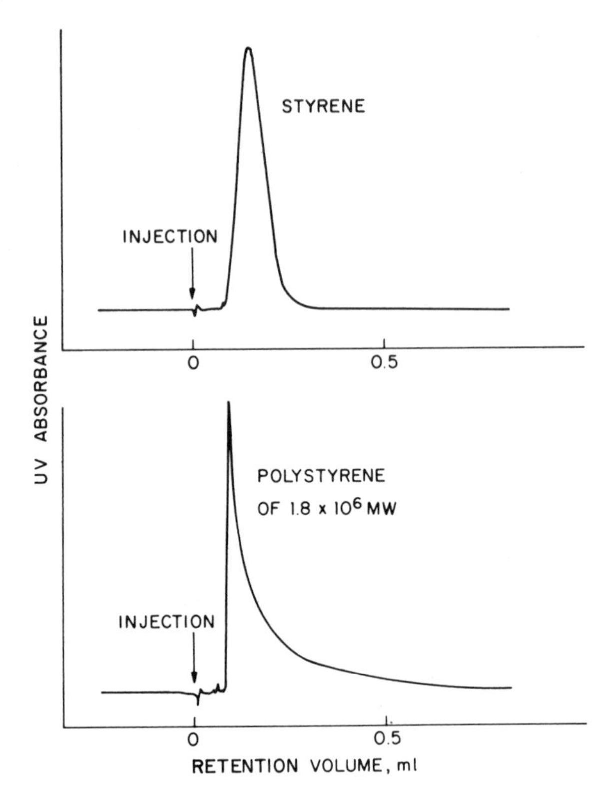

Figure 3.24 Extracolumn peak dispersion in SEC.
Straight tubing, 0.018 in. i.d., 1 m long; solvent, chloroform; flow rate, 0.5 ml/min. (Reprinted with permission from Ref. 20.)

given to individual effects. In general, these dispersion effects have characteristics similar to those of highly "nonequilibrium" lateral mass transfer processes, as illustrated by the highly skewed elution peak profile in Figure 3.24 for a polymer sample eluted from long tubing.

While the retention parameters void volume V_o, pore volume V_i, and distribution coefficient K_{SEC} determine the extent of solute permeation and therefore control solute retention and separation, they are rather unimportant factors in SEC band broadening. The opposite is true for the quantities d_p, v, F, and D_{SM} and such D_{SM}-related factors as T, η_0, and **a**. While these latter parameters (d_p, v, etc.) can affect the rate of solute permeation into the pores and control the column dispersion, they have no or very little effect on SEC solute retention and separation. This is evident in the expression for the first moment of the peak (Eq. 3.27), which contains none of these latter parameters.

3.5 Experimental Methods

Column Plate Count

There are several methods available for measuring column plate count N, defined as V_R^2/σ^2. The absolute method of using Equation 3.2 to calculate σ^2 is too tedious for hand calculation and usually is best done by computer. However, if the peaks are symmetrical and close to Gaussian shape (Eq. 3.4), Equation 3.2 can be expressed in variables that are more easily measured experimentally. The most often used approximations include (38)

$$N = 16\left(\frac{V_R}{W_b}\right)^2 \tag{3.35}$$

where W_b is the baseline width formed by the tangents of the peak intersecting the baseline (approximately 4σ; see Fig. 3.3),

$$N = 5.54\left(\frac{V_R}{W_{1/2}}\right)^2 \tag{3.36}$$

where $W_{1/2}$ is the peak width at one-half the peak height, and

$$N = 2\pi\left(\frac{h_p V_R}{A}\right)^2 \tag{3.37}$$

where h_p is the peak height and A the peak area. Although Equations 3.35 to 3.37 are widely used as a measure of chromatographic performance, it is not well recognized that serious errors in the calculation can result if the peaks of interest are not close to Gaussian shape.

The error in the calculation of plate count by the simplified methods of Equations 3.35–3.37 was determined quantitatively in a computer simulation study using the skewed peak model of the exponentially modified Gaussian function (39). The peak contour of this model is described by the convolute integral between a Gaussian constituent having the standard deviation σ and an exponential modifier having the time constant τ:

$$h = \frac{A}{\tau\sigma\sqrt{2\pi}} \int_0^\infty \exp\left[-\left(\frac{V - V_R - V'}{\sqrt{2}\sigma}\right)^2 - \frac{V'}{\tau}\right] dV' \tag{3.38}$$

The variance of the peak (σ_x^2) equals ($\sigma^2 + \tau^2$) and peak skew or tailing increases with the ratio τ/σ:

$$\text{peak skew} = \gamma \equiv \frac{\mu_3}{\mu_2^{3/2}} \simeq \gamma' \equiv \frac{2(\tau/\sigma)^3}{[1 + (\tau/\sigma)^2]^{3/2}} \tag{3.39}$$

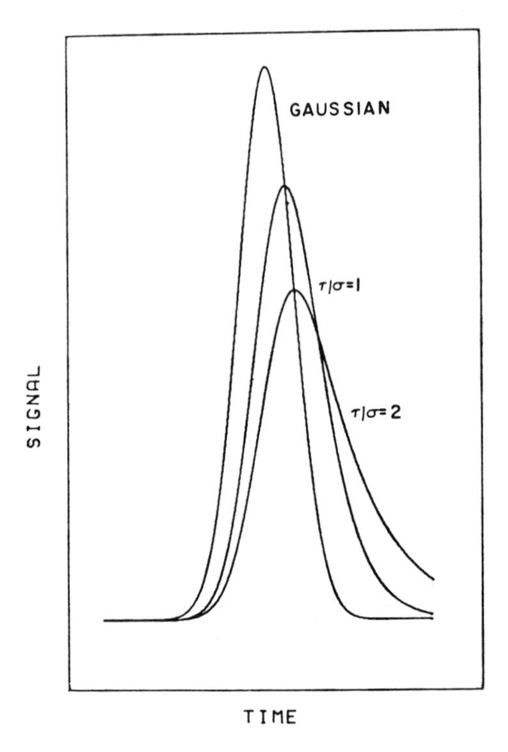

Figure 3.25 Effect of τ/σ on peak shape.
(Reprinted with permission from Ref. 40.)

Figure 3.25 shows several synthesized chromatographic peaks with varying τ/σ ratios in which the value of σ for the peaks was held constant and the value of τ was increased. Table 3.2 shows the results of the plate count calculations obtained on the series of peaks with constant σ and increasing τ. With values $\tau/\sigma > 1$ (peak skew > 0.7; peak asymmetry > 1.2; Eq. 3.39), methods other than the moment method give significant positive errors in plate count. The exponentially modified Gaussian peak model has been used in a more precise method (compared to the statistical moments calculations, Eq. 3.2) to isolate σ and τ constituents for characterizing the variance and the skewing of experimental chromatographic peaks (41). It is also the basic model of the GPCV3 calibration method described in Section 9.3. Peak skew values of the σ-τ model are related to the more practical peak asymmetry factors described in Section 6.5 (see Fig. 6.24).

The observed variance of a column series is often larger than predicted from the sum of the variances of the individual columns. This error arises

Table 3.2 Plate Count Calculation (Constant σ^2, Increasing τ^2)

			Plate Count			
τ/σ	Peak Skew, γ	True Value	Moment Method (Eq. 3.2)	Area Method (Eq. 3.37)	Tangent Method (Eq. 3.35)	Peak Half-Width Method (Eq. 3.36)
0.1	0.002	7670	7690	7680	7330	7630
0.5	0.18	6200	6210	6440	6230	6410
1.0	0.71	3870	3880	4740	4470	5060
1.5	1.15	2380	2410	3580	3850	4010
2.0	1.43	1550	1620	2760	3060	3200

[a] From peaks in Figure 3.25.
Taken from Ref. 40a.

because Equations 3.35–3.37, which do not account for peak skewing, are often used with tailing peaks to estimate the variance of columns. The additivity of plate count for a column set is described more fully in Section 8.6.

Column-Dispersion Calibration

Column dispersion is a major factor that causes inaccuracy in quantitative SEC interpretations, because it distorts the elution curve and affects the calibration and the molecular weight calculations. Compensation for the dispersion effect in SEC calibration and molecular weight calculation is considered in Section 9.3. To account for column dispersion by using the calibration methods developed in Section 9.3, one needs first to know how much peak broadening has been imposed on the experimental SEC elution curve. Unfortunately, it is difficult to determine true SEC column dispersion, because all polymer SEC peaks, except a totally excluded polymer peak, are somewhat broadened by MW separation as well as by column dispersion. (An example of the broadening due to molecular weight separation even for narrow-MWD polystyrene standards can be seen in Figure 7.4. The relatively flat H versus v plot of the PS 3600 MW standard suggests that most of the band width of this standard must be due to molecular weight separation, since the plate height contribution due to column dispersion is expected to change with changing flow rate while that due to molecular weight separation is not.) Peak broadening caused by molecular weight separation can vary from one polystyrene standard to another, so it is not possible to distinguish peak broadening due to molecular weight separation from broadening caused by column dispersion. This situation causes problems in the accurate characterization of SEC column dispersion over the entire separation volume range. Although there are two techniques that can be used to solve this problem, the reverse-flow experiment (28, 30) and the recycle technique (42, 43), these are rather complicated methods, used only to obtain very accurate calibration for SEC column dispersion or values of polydispersity ($\overline{M}_w/\overline{M}_n$) for narrow-MWD polymer standards.

In the reverse-flow technique, the polymer sample is injected in the normal way, but when the sample peak is halfway through the column, the flow is reversed. The MW separation processes are now reversed, but band broadening due to dispersion effects continues. When the peak reaches the detector, now located at the top of the column, it reflects only the band broadening due to dispersion processes. Molecular weight separation has been completely canceled by the flow reversal, assuming equal elution time each way. The results of such an experiment are shown in Figure 3.22; such data can be used to obtain the σ-calibration curve for a column. Once determined, this curve is expected to be independent of the nature of the polymer sample and can be used directly in the SEC calibration methods described in Section 9.3 to compensate for the column-dispersion effect. The recycle technique of characterizing SEC column dispersion is described in Section 11.3.

REFERENCES

1. W. Feller, *An Introduction to Probability Theory and its Applications*, 2nd ed., Wiley, New York, 1957, p. 216.
2. A. J. P. Martin and R. L. M. Synge, *Biochem. J.*, **35**, 1358 (1941).
3. A. B. Littlewood, *Gas Chromatography*, 2nd ed., Academic Press, New York, 1970, Chapts. 5 and 6.
4. S. Dal Nogare and R. S. Juvet, Jr., *Gas-Liquid Chromatography*, Wiley, New York, 1962, Chapt. 3.
5. C. S. G. Phillips, *Gas Chromatography*, Academic Press, New York, 1956, p. 95.
6. L. R. Snyder and J. J. Kirkland, *Introduction to Modern Liquid Chromatography*, Wiley, New York, 1974, p. 29.
7. J. C. Giddings, *Dynamics of Chromatography*, Dekker, New York, 1965.
8. J. J. van Deemter, F. J. Zuiderweg, and A. Klinkenberg, *Chem. Eng. Sci.*, **5**, 271 (1956).
9. B. L. Karger, L. R. Snyder, and C. Horvath, *An Introduction to Separation Science*, Wiley, New York, 1973, Chapt. 5.
10. R. N. Kelley and F. W. Billmeyer, Jr., *Anal. Chem.*, **41**, 874 (1969).
11. R. J. Hamilton and P. A. Sewell, *Introduction to High Performance Liquid Chromatography*, Wiley, New York, 1977, Chapt. 2.
12. C. Horvath and H. J. Lin, *J. Chromatogr.*, **126**, 401 (1976).

13. J. F. K. Huber, *J. Chromatogr. Sci.*, **7**, 85 (1969).
14. J. N. Done and J. H. Knox, *J. Chromatogr. Sci.*, **10**, 606 (1972).
15. L. R. Snyder, *J. Chromatogr. Sci.*, **7**, 352 (1969).
16. R. N. Kelley and F. W. Billmeyer, Jr., *Anal. Chem.*, **42**, 399 (1970).
17. J. C. Giddings, L. M. Bowman, Jr., and M. N. Meyers, *Macromolecules*, **10**, 443 (1977).
18. A. R. Cooper, A. R. Bruzzone, and J. F. Johnson, *J. Appl. Polym. Sci.*, **13**, 2029 (1969).
19. L. Lapidus and N. R. Amundson, *J. Phys. Chem.*, **56**, 984 (1952).
20. P. R. Kasten, L. Lapidus, and N. R. Amundson, *J. Phys. Chem.*, **56**, 683 (1952).
21. A. Klinkenberg and F. Sjenitzer, *Chem. Eng. Sci.*, **5**, 258 (1956).
22. M. Kubin, *Collect. Czech. Chem. Commun.*, **30**, 1104 (1965).
23. M. Kubin, *Collect. Czech. Chem. Commun.*, **30**, 2900 (1965).
24. J. J. Hermans, *J. Polym. Sci., Part A-2*, **6**, 1217 (1968).
25. A. C. Ouano and J. A. Barker, *Sep. Sci.*, **8**, 673 (1973).
26. M. E. van Kreveld and N. van den Hoed, *J. Chromatogr.*, **149**, 71 (1978).
27. C. M. Guttman and E. A. DiMarzio, *Macromolecules*, **3**, 681 (1970).
28. L. H. Tung and J. C. Moore, "Gel Permeation Chromatography," in *Fractionation of Synthetic Polymers*, L. H. Tung, ed., Dekker, New York, 1977, Chapt. 6.
29. W. W. Yau, C. P. Malone, and H. L. Suchan, *Sep. Sci.*, **5**, 259 (1970).
30. L. H. Tung and J. R. Runyon, *J. Appl. Polym. Sci.*, **13**, 2397 (1969).
31. M. Kubin, *J. Chromatogr.*, **108**, 1 (1975).
32. J. V. Dawkins, T. Stone, and G. Yeadon, *Polymer*, **18**, 1179 (1977).
33. A. Rudin and H. K. Johnston, *J. Polym. Sci., Part B*, **9**, 55 (1971).
34. C. N. Satterfield, C. K. Colton, and W. H. Pitcher, Jr., *Am. Inst. Chem. Eng. J.*, **19**, 628 (1973).
35. G. Trenel, M. John, and H. Delleweg, *FEBS Lett.*, **2**, 74 (1968).
36. J. Y. Chuang, A. R. Cooper, and J. F. Johnson, *J. Polym. Sci., Part C*, **43**, 291 (1973).
37. J. C. Sternberg, *Advances in Chromatography*, Vol. 2, J. C. Giddings and R. A. Keller, eds., Dekker, New York, 1966, p. 205.
38. A. T. James and A. J. P. Martin, *Analyst*, **77**, 915 (1952).
39. E. Grushka, *Anal. Chem.*, **44**, 1733 (1972).
40. R. E. Pauls and L. B. Rogers, *Sep. Sci.*, **12**, 395 (1977).
40a. J. J. Kirkland, W. W. Yau, H. J. Stoklosa and C. H. Dilks, Jr., *J. Chromatogr. Sci.*, **15**, 303 (1977).
41. W. W. Yau, *Anal. Chem.*, **49**, 395 (1977).
42. J. L. Waters, *J. Polym. Sci., Part A-2*, **8**, 411 (1970).
43. Z. Grubisic-Gallot, L. Marais, and H. Benoit, *J. Polym. Sci., Part A-2*, **14**, 959 (1976).

Four

RESOLUTION

4.1 Introduction

Chromatographic Resolution

Traditionally, chromatographic column performance has been expressed in terms of the number of theoretical plates N (Eqs. 3.11, 3.35 to 3.37), the plate height H, or the *column resolution* R_s:

$$R_s = \frac{2(V_{R2} - V_{R1})}{W_{b1} + W_{b2}} \tag{4.1}$$

where V_R is the peak retention volume; W_b is the chromatogram peak width formed by intersection of the tangents to the curve inflection points with the baseline in retention volume units, $W_b = 4\sigma$ (Fig. 3.3); and σ is the peak standard deviation (proportional to peak width) caused by column dispersion and expressed in volume units (ml, or number of syphon dumps). The subscripts 1 and 2 serve to identify two closely eluting solutes. The plate height H (or HETP, height equivalent to a theoretical plate) is equal to L/N, where L is the column length (Sect. 3.2).

Equation 4.1 may also be written as

$$R_s = \frac{V_{R2} - V_{R1}}{2(\sigma_1 + \sigma_2)} \approx \frac{\Delta V_R}{4\sigma} \tag{4.2}$$

The values of σ are determined experimentally from the chromatograms of single molecular species (see Sect. 3.5), and to a first approximation,

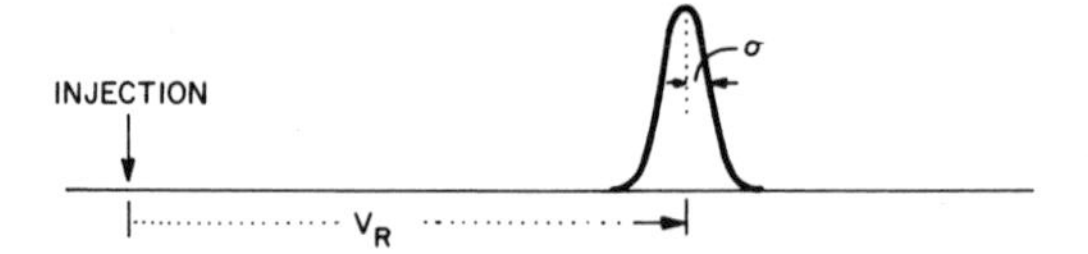

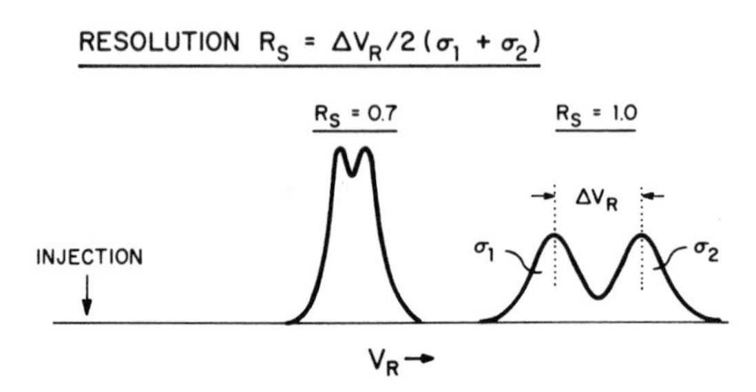

Figure 4.1 Traditional column performance parameters.

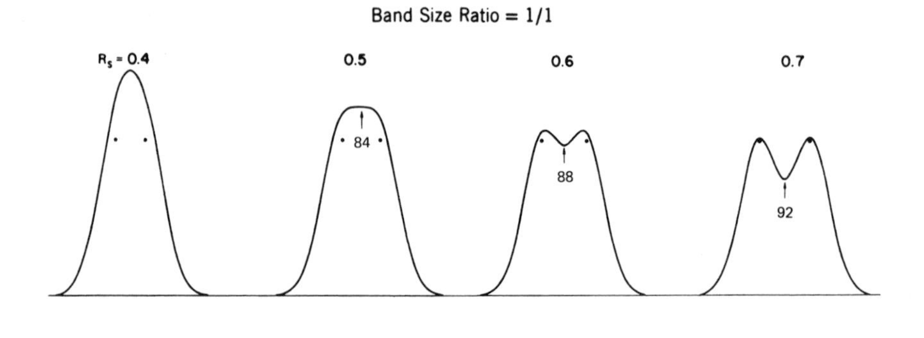

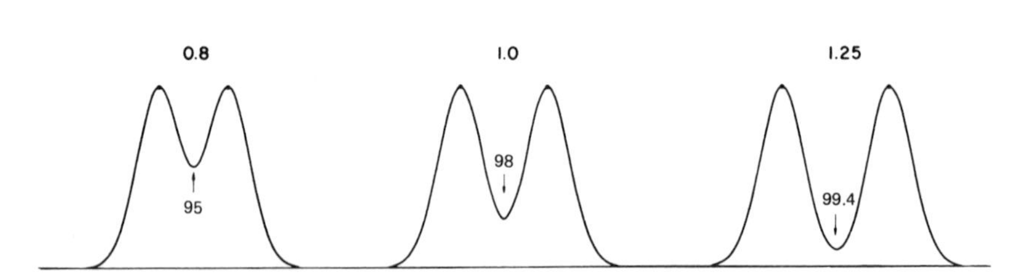

Figure 4.2 Standard resolution curves for a band-size ratio of 1 : 1. Values of R_s, 0.4–1.25. (Reprinted with permission from Ref. 1.)

$\sigma_1 = \sigma_2 = \sigma$. The resolution factor R_s is a more meaningful column performance parameter than plate count N. R_s accounts for peak broadening (σ, N) as well as the selectivity of the column (ΔV_R). In LC separations other than SEC, the value of R_s calculated by Equation 4.2 depicts how well peaks are resolved. An illustration of this is shown in Figure 4.1 for the cases of $R_s = 0.7$ and 1.0. However, the R_s factor is still not a totally adequate general column performance parameter, because its value varies with the particular choice of peaks. Since R_s is a dimensionless quantity, the resolution of a particular pair of solute peaks has the same value of R_s whether the peak elution is recorded in retention volume or in retention time units. As for the calculation of plate count N (Sects. 3.1 and 3.2), the resolution relationships (Eqs. 4.1 and 4.2) implicitly assume the Gaussian (symmetrical) peak shape.

To provide a visualization of resolution units, standard resolution curves calculated for theoretical Gaussian peaks are shown in Figures 4.2 and 4.3. In Figure 4.2 all the component peaks are of equal height, simulating equal

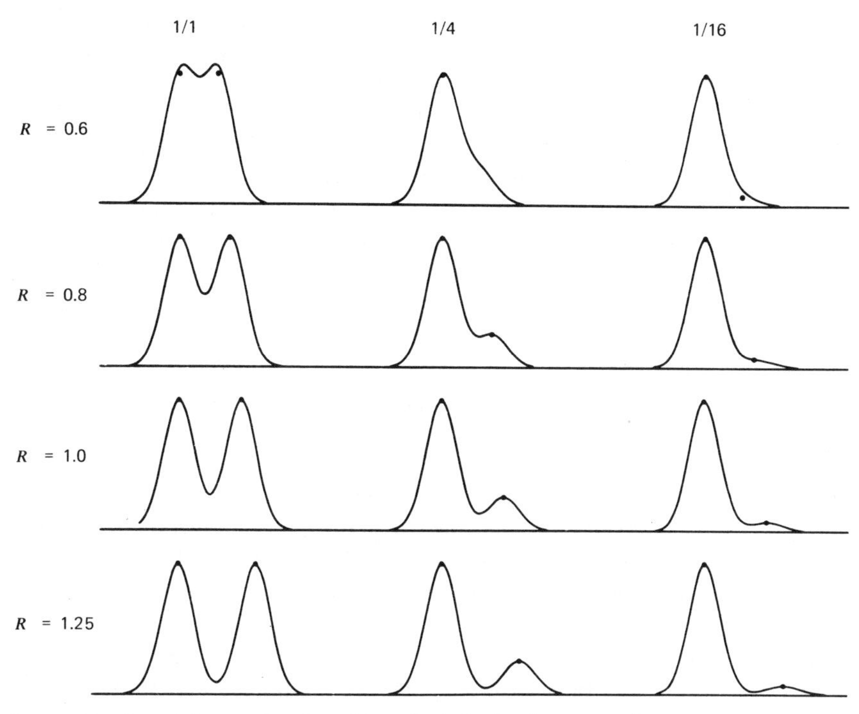

Figure 4.3 Separation as a function of R_s and relative band concentration.
(Reprinted with permission from Ref. 1.)

concentrations of component solute species. The evidence of a double peak begins at $R_s = 0.5$, which is sometimes called 2σ resolution because it corresponds to $\Delta V_R = 2\sigma$ in Equation 4.2. At $R_s = 1.0$ (4σ separation), the peaks are reasonably well resolved. Complete peak separation to baseline resolution occurs at $R_s = 1.5$ (6σ separation).

Actual solute overlapping or mixing between the elution peaks of equal size is not as extensive as it might appear from casual observation. At $R_s = 0.5$, there is only a 16% actual solute overlap. The overlap of one solute on the other is only 2% at $R_s = 1.0$. In other words, even at the low-resolution case of $R_s = 0.5$, fractions of each component species of 84% purity can be recovered at the equal purity cut point indicated by the arrows in Figure 4.2. At $R_s = 1.0$, the two recovered fractions are 98% pure for equal-height peaks. As a result of curve overlapping, the peak apexes of the composite chromatogram at low resolution (e.g., $R_s = 0.6$) are taller and closer to each other than those of the component peaks. The apexes of the original component peaks are indicated by the dots shown in Figures 4.2 and 4.3.

Similar sets of the theoretical standard resolution curves are available for other band concentration ratios (1, 2). A selection of such curves is shown in Figure 4.3. Such standard reference curves are very useful for estimating the values of R_s of experimental peaks. With a recollection of the various R_s curve shapes, a quick estimate of the value of R_s can be made on the spot by glancing at the features of the experimental chromatogram. Since the expression for R_s in Equation 4.2 is independent of the individual peak heights, the same value of R_s can correspond to resolution curves very different in shape, depending on band ratios. As shown in Figure 4.3, as the band ratio increases, the features of the smaller peak are less distinguishable because of the increased interference of the larger peak. This effect makes the quantitative detection of smaller peaks on the tailing edge of larger peaks more difficult. Under these circumstances, the standard R_s curves can be very helpful for comparison with experimental chromatograms to detect the presence and estimate the areas of the smaller peaks, or to locate the proper cut point if a fraction of the smaller peak of a certain purity is desired. For a pair of peaks with a large band concentration difference, the equal-purity cut point shifts toward the smaller peak, since the solute molecules of the larger peak spread more into the smaller peak. For a further discussion of the use of standard resolution curves, see References 1 and 2.

Since column dispersion and thus N for polymer solutes in SEC varies as a function of retention volume (Sects. 3.3 and 3.4), the traditional LC resolution and peak-capacity expressions described below are of little use in SEC of polymers. However, they are generally applicable to SEC separations of small molecules. Special considerations are required for studying SEC resolution in polymer MWD analyses, as described in Section 4.2.

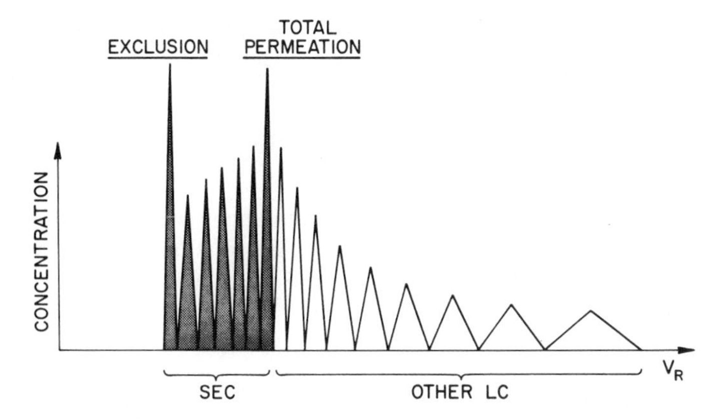

Figure 4.4 Characteristics of column dispersion and peak capacity in SEC and the other LC methods.

For LC methods other than SEC, experimental values of plate count are nearly independent of the retention volumes of the individual solutes; that is, according to Equation 3.11, peak width increases linearly with increasing retention volume. This is illustrated in Figure 4.4 by the unshaded peaks in the region marked "other LC." Constant values of N mean $\sigma \propto (1 + k')$, where $k' = (V_R - V_t)/V_t$, with k' being the usual LC peak-capacity factor described in Section 2.1 and V_t being the retention volume of the total permeation peak (often called the "unretained peak" in discussing conventional LC separations). In terms of basic LC retention parameters, the resolution in Equation 4.2 can be expressed as

$$R_s = \frac{\Delta k' \sqrt{N}}{4(1 + k')} = \frac{(\alpha - 1)\sqrt{N}}{4}\left(\frac{k'}{1 + k'}\right) \tag{4.3}$$

where α is the separation factor, which equals the k' ratio of the two adjacent peaks (i.e., $\alpha = k'_2/k'_1$), and the plate count N is assumed constant. Equation 4.3 is very useful for the design and optimization of LC methods other than SEC, since the resolution of LC peaks can be controlled by independently changing separation selectivity α, efficiency N, or capacity k' (2).

Peak-Capacity Concept

For small molecules, the quality of separation can also be effectively described in terms of *peak capacity n*. This term is defined as the maximum number of peaks that can be resolved within a specified range of retention volume. For cases in which solute peaks having the same plate count are to be separated with 4σ resolution, it has been shown that (3)

$$n = 1 + \frac{\sqrt{N}}{4} \Delta \ln V_R \tag{4.4}$$

where $\Delta \ln V_R$, the difference between the logarithms of the retention volumes, specifies the retention range of interest. The relationship described by Equation 4.4 for constant N is illustrated in Figure 4.4 by the closely spaced peaks beyond the total permeation volume.

As with Equation 4.3, a constant plate number N must also apply in Equation 4.4 to all the peaks of interest in a chromatogram. This is usually the case in LC methods other than SEC. As a result of the significant decrease in solute diffusion with increasing solute size, the earlier peaks in SEC actually suffer more band broadening due to column dispersion. This is a trend directly opposite to that of the other LC methods (see Fig. 4.4 for illustration and Sects. 3.3 and 3.4 for further discussion).

The term "peak resolution" is not commonly used in SEC because it does not fit properly in the context of describing SEC column performance. A major use of SEC is not to resolve and identify species but to retrieve MWD information from the chromatogram. Special concepts of SEC resolution and MW accuracy are required to define SEC column performance in polymer analyses. However, these concepts are derived from the general LC resolution considerations discussed above.

4.2 Resolution Concept in SEC of Polymers

A quantitative expression of SEC resolution for polymer MWD analyses is needed to determine where the performance of HPSEC stands relative to conventional SEC and with respect to theoretical and instrumental performance limits, and to determine practical goals for HPSEC performance relative to cost and time.

Since the dependence of SEC peak separation ΔV_R on solute molecular weight is known via the SEC calibration curve (e.g., Figure 4.5), a unique opportunity exists in SEC for eliminating the dependence of the resolution factor R_s on the probing solutes (4–6). The useful portion of the SEC calibration curve can be approximated by a straight line of slope D_2 and intercept D_1:

$$M = D_1 e^{-D_2 V_R} \tag{4.5}$$

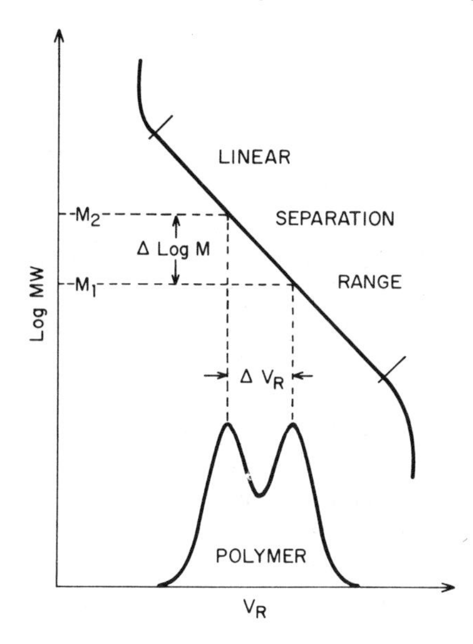

Figure 4.5 Dependence of SEC peak separation on solute MW.

By taking the natural logarithm and rearranging, Equation 4.5 becomes

$$V_R = \frac{\ln D_1 - \ln M}{D_2} \tag{4.6}$$

or

$$\Delta V_R = \frac{\Delta \ln M}{D_2} = \frac{\ln (M_2/M_1)}{D_2} \tag{4.7}$$

Substitution of Equation 4.7 into basic resolution expression, Equation 4.2, gives

$$R_s = \frac{\ln (M_2/M_1)}{2D_2(\sigma_1 + \sigma_2)} \simeq \frac{\Delta \ln M}{4\sigma D_2} \tag{4.8}$$

This equation describes how well the SEC column can distinguish between two molecules of the same polymer type but differing by a molecular weight factor M_2/M_1. This description of SEC resolution is useful but too specific to allow the data of different columns or different laboratories to be compared. However, in SEC we are interested in the resolution pertained in the elution curve as a whole, not so much that between specific pairs of eluted fractions. To provide a general measure of SEC resolution, the concept of *specific resolution* R_{sp} has been developed (4). Dividing Equation 4.8 by $\Delta \log M$ leads directly to the expression for SEC specific resolution:

$$R_{sp} = \frac{R_s}{\Delta \log M} = \frac{0.58}{\sigma D_2} \tag{4.9}$$

[the conversion between the natural and the base 10 logarithm expressions (i.e., $\ln M = 2.303 \log M$) is accounted for in Eq. 4.9]. Note that in Equation 4.9 the explicit dependence of SEC resolution on sample molecular weight is now eliminated in the expression for R_{sp}. Specifically, Equation 4.9 states that the resolution factor R_{sp} in the linear calibration region is equal to the usual chromatographic resolution R_s (Eq. 4.1) for a pair of peaks having a decade of molecular weight difference (Sect. 8.6). Experimental values of R_{sp}, which are not expected to vary much with the selection of probe samples, can be used as a SEC column performance parameter for evaluating and comparing SEC columns or column sets.

To provide a performance factor for comparison of different column packings, the expression for R_{sp} must be compensated for column length. Since D_2 is proportional to the reciprocal of the column length L, and σ is proportional to the square root of L, Equation 4.9 can be normalized for column length to give the packing resolution factor, which is equivalent to R_{sp} for a 1 cm column:

$$R_{sp}^* = \frac{0.58}{\sigma D_2 \sqrt{L}} \tag{4.10}$$

The advantage of using the resolution factors R_{sp} and R_{sp}^* instead of plate count N in evaluating SEC columns and column packings is further illustrated by the experimental results in Tables 4.2 and 4.3.

4.3 Molecular Weight Accuracy Criterion

The quality of the SEC results in polymer analyses should be assessed in terms of the accuracy of the final calculated values of MW. It is important that the molecular weight accuracy of the SEC polymer analyses can be predicted from measurable column parameters. The resolution concept in SEC still does not provide the same utility as in GC and the other LC methods, where a resolution value can unequivocally define the system efficiency as well as the quality of the final results. A simple resolution value simply does not provide the desired molecular weight accuracy information about a system for polymer analyses. Fortunately, a relationship exists between the SEC resolution and the MW accuracy, as described below.

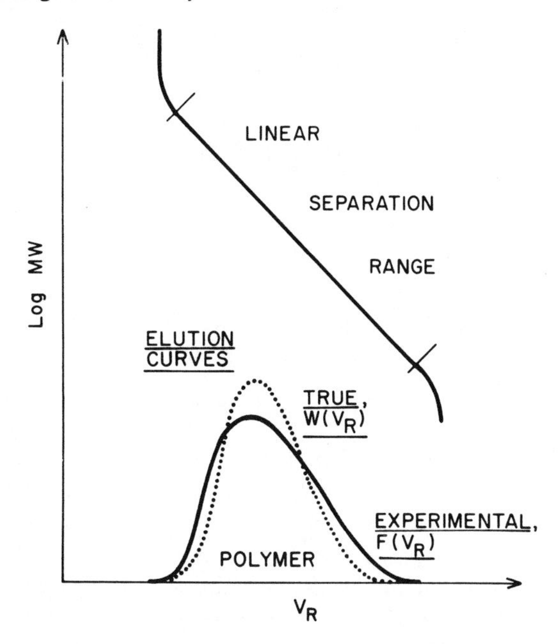

Figure 4.6 Effect of instrumental band broadening on SEC elution-curve shape.

In SEC the elution curve is broadened by column dispersion as illustrated in Figure 4.6. The SEC-MW accuracy problem resulting from column dispersion is directly related to the differences between the experimental, $F(V_R)$, and the true, $W(V_R)$, elution curves. The elution curves F and W are related by a convolution integral (7):

$$F(V_R) = \int_{-\infty}^{\infty} W(y)G(V_R - y)\,dy \tag{4.11}$$

where $G(V_R - y)$ is an instrument-column-dispersion function which describes the weight fraction of a solute that should have been eluted at the retention volume y but is actually dispersed and detected at the retention volume V_R. The true values of $\overline{M}_w$ and $\overline{M}_n$ of a polymer sample for linear calibration (Eq. 4.5) can be written as (8–11)

$$(\overline{M}_n)_{\text{true}} = \frac{1}{\sum_{V_R}[W(V_R)/D_1 e^{-D_2 V_R}]} \tag{4.12}$$

and

$$(\overline{M}_w)_{\text{true}} = \sum_{V_R} W(V_R) D_1 e^{-D_2 V_R} \tag{4.13}$$

On the other hand, observed molecular weight averages are calculated from the experimental elution curves $F(V_R)$ instead of $W(V_R)$:

$$(\overline{M}_n)_{\text{exp}} = \frac{1}{\sum_{V_R}[F(V_R)/D_1 e^{-D_2 V_R}]} \tag{4.14}$$

and

$$(\overline{M}_w)_{\text{exp}} = \sum_{V_R} F(V_R) D_1 e^{-D_2 V_R} \tag{4.15}$$

The true and experimental molecular weight averages can be directly related to each other by a single correction factor X (8–11):

$$\overline{\text{MW}}_{\text{true}} = (X)\overline{\text{MW}}_{\text{exp}} \tag{4.16}$$

Equation 4.16 represents an important theoretical advance in modern SEC data reduction. The values of X for various molecular weight averages derived from two instrument dispersion functions, one for symmetrical and one for skewed peak shapes, are summarized in Table 4.1, where $\overline{M}_v$ and **a** are the viscosity-average molecular weight and the exponent constant, respectively, for the Mark-Houwink viscosity/molecular weight relationship (Sect. 2.4; see Table 8.2 for published values of **a**). These correction factors, which are the same as those used in the linear calibration methods (i.e., Hamielec, GPCV2, and GPCV3), are noted in the column headings of the table. The delta function in the second column simply indicates that zero column dispersion is assumed. The Gaussian function in the third column

Table 4.1 SEC Dispersion Correction Factors: $\overline{\text{MW}}_{\text{true}} = (X)\overline{\text{MW}}_{\text{exp}}$

	Values of X for Various Instrument Dispersion Functions			
	Delta Function (Hamielec)	Gaussian Function (GPCV2)	Exponentially Modified Gaussian Function (GPCV3)	
$\overline{M}_n$	1	$e^{(D_2\sigma)^2/2}$	$e^{(D_2\sigma)^2/2}\left[\left(\frac{1}{1-D_2\tau}\right)e^{-D_2\tau}\right]$	for $D_2\tau < 1$
$\overline{M}_w$	1	$e^{-(D_2\sigma)^2/2}$	$e^{-(D_2\sigma)^2/2}(1 + D_2\tau)e^{-D_2\tau}$	for $D_2\tau > -1$
$\overline{M}_z$	1	$e^{-3(D_2\sigma)^2/2}$	$e^{-3(D_2\sigma)^2/2}\left[\left(\frac{1+2D_2\tau}{1+D_2\tau}\right)e^{-D_2\tau}\right]$	for $D_2\tau > -\frac{1}{2}$
$\overline{M}_v$	1	$e^{-\mathbf{a}^2(D_2\sigma)^2/2}$	$e^{-\mathbf{a}^2(D_2\sigma)^2/2}(1 + \mathbf{a}D_2\tau)e^{-\mathbf{a}D_2\tau}$	for $D_2\tau > -\mathbf{a}$

simulates symmetrical peak dispersion. The exponentially modified Gaussian function used to develop the fourth column is the same generalized skewed peak model, as previously described in Section 3.5 (12, 13).

Based on Equation 4.16, a molecular weight error expression can now be derived in terms of column parameters only. The general expression of this molecular weight error, normalized by the value of the molecular weight average, is

$$\frac{\overline{\mathrm{MW}}_{\mathrm{exp}} - \overline{\mathrm{MW}}_{\mathrm{true}}}{\overline{\mathrm{MW}}_{\mathrm{true}}} = \frac{1}{X} - 1 \tag{4.17}$$

Using Equation 4.17 and Table 4.1, fourth column, the errors in $\overline{M}_w$ and $\overline{M}_n$ can be predicted for various band-broadening situations (different values of σ and τ). The calculated molecular weight errors are plotted in Figure 4.7. It is shown that molecular weight errors increase with increasing column dispersion as measured by the term $(\sigma^2 + \tau^2)$ (Sect. 3.5). The effect of increased peak skewing as measured by τ/σ is to cause more peak tailing into the longer retention volume region and larger error in experimental values of $\overline{M}_n$.

The level of molecular weight error or inaccuracy directly reflects the performance of SEC columns. This fact underlies the basic concept of the

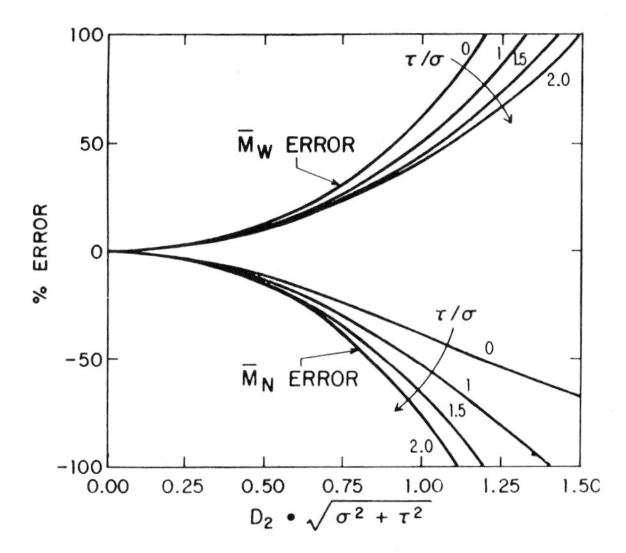

Figure 4.7 Predicted SEC-MW error due to column dispersion.
Curves for τ/σ of 0, 1, 1.5, and 2.0 are calculated according to Equation 4.17 using the correction factor X in Table 4.1, fourth column. (Reprinted with permission from Ref. 9.)

molecular weight accuracy criterion for SEC column performance. Specifically, the molecular weight accuracy criterion is defined as M^*, the molecular weight error averaged for $\overline{M}_w^*$ and $\overline{M}_n^*$, which are derived from the Gaussian dispersion function:

$$\overline{M}_n^* = e^{-(1/2)(\sigma D_2)^2} - 1 \tag{4.18}$$

and

$$\overline{M}_w^* = e^{(1/2)(\sigma D_2)^2} - 1 \tag{4.19}$$

Note that the value of $\overline{M}_n^*$ is always negative and the value of $\overline{M}_w^*$ is always positive according to Equations 4.18 and 4.19. The absolute value of $\overline{M}_n^*$ is to be used for the M^* calculation. These equations have practical utility, since they serve to predict molecular weight accuracy directly from experimental column parameters σ and D_2. Also, these equations can be used to specify the values of column σ and D_2 required to achieve a desired SEC-MW accuracy. Both σ and D_2 are positive quantities in SEC analysis. Familiarity with the basic properties of D_2 and σ in these equations (see Table 4.4) is needed to make the best practical usage of the molecular weight accuracy criterion.

It is important to note that both the SEC specific resolution (R_{sp}) and molecular weight accuracy (M^*) factors are uniquely defined by the value of σD_2 for the chromatographic system. Therefore, the product of σ and D_2 is by itself a fundamental SEC column performance parameter. Inherently, columns of different individual values of σ and D_2 can perform equally well as long as they have the same combined value of σD_2. In practice, SEC systems with *small* values of σD_2 are sought to achieve high resolution and MW accuracy. Note also that the values of R_{sp}, M^*, and σD_2 are all dimensionless and are therefore valid for studying SEC systems in general, regardless of whether retention volume, syphon counts, or retention time is used in defining the SEC calibration and elution curves. These performance parameters provide the interesting feature that they are independent of sample MWD. Sample MWD is not used in the derivation and does not appear in the final expression for these parameters. Therefore, values of R_{sp}, M^*, and σD_2 reflect properties of the column alone and should be nearly the same for a particular column set, regardless of differences in probe sample MWD (whether it is a single, bimodal, broad, or narrow distribution).

4.4 Applications of Column Performance Criteria

The validity of the SEC performance concept above is in practice dependent on the basic premise that σD_2 is reasonably constant and independent of solute molecular weight and retention volume. The experimental value of

D_2 can be calculated from Equation 4.7 if narrow-MWD polymer standards are available, or by the broad-MWD standard calibration methods described in Section 9.3. By connecting columns of different pore-size packings, the value of D_2 (or the slope of the SEC calibration curve) of the assembled column set can be made essentially invariant over a wide molecular weight separation range (see discussion in Sect. 4.5 below and in Sects. 8.5 and 9.6 for a bimodal-pore-size column set). Near the exclusion and the total permeation volumes, D_2 approaches infinity, which forces the SEC resolution to zero. Therefore, in evaluating SEC performance, polymer standards that elute too close to the exclusion or the total permeation volume should be avoided.

The value of σ or column dispersion is determined, to a first approximation, as the experimental value of σ for a very narrow MWD polymer standard. The value of σ for a monomer peak should not be used, since it usually grossly underestimates the true column dispersion. For obtaining more accurate values of column σ, special SEC experiments such as recycle (Sect. 11.3) and reverse-flow techniques are required (Sect. 3.5).

Usually, a constant value of σ is not observed experimentally for use in the σD_2 resolution concept. The value of column σ in SEC is dependent on the retention volume (Fig. 3.22). However, in practice, the value of σ used is the average of the smaller σ values as determined for narrow-MWD polystyrene standards. It would be most accurate to account for this σ variation. However, this is difficult to accomplish and the dependency of σ on V_R is small relative to the total magnitude of σ.

One way to test the molecular weight accuracy criterion (Eqs. 4.18 and 4.19) is to vary the value of σ of a column set by changing solvent flow rate and then to compare the observed and predicted values of M^*. The results of such an experiment are shown in Figure 4.8, where the experimental values of M^* (open circles) are shown to correspond closely to the theoretical values (dashed curves) that are calculated from various values of σ and the measured value of column D_2. Experimental values of $\overline{M}_w$ and $\overline{M}_n$ for the test polystyrene standard were calculated directly by the usual point-by-point summation of the elution curves observed at different flow rates (Eqs. 4.14 and 4.15). These values are compared as in Equation 4.17 to the "true" value of MW of the standard supplied by the vendor to calculate the experimental $\overline{M}_w^*$ and $\overline{M}_n^*$ errors. The particular column set used in this experiment was arbitrarily chosen for illustration. However, similar results were obtained for column sets of other packing materials, listed in Table 4.2. They also support the general utility of the SEC-MW accuracy criterion.

Table 4.2 also verifies that plate count N, measured by the value of σ for a monomer peak (Sect. 3.2), is a poor indicator of SEC column performance in terms of resolution or polymer molecular weight accuracy. For example, for the 2700-plate Porasil column set, the molecular weight

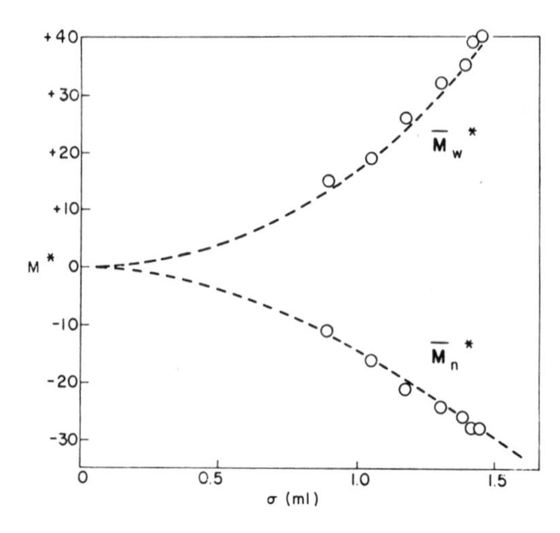

Figure 4.8 Effect of column dispersion on MW accuracy.
Columns, Vit-X column set (see Table 4.2); sample, polystyrene standard, 97,200 MW. O, experimental; ---, theoretical values of $\overline{M}_w^*$ and $\overline{M}_n^*$. (Reprinted with permission from Ref. 4.)

error (M^*) caused by column dispersion is 7%, compared with 11% for a Styragel column set of $N = 7500$. The better molecular weight accuracy of the Porasil column set in this case is partly due to its smaller value of D_2 as compared to the Styragel column set. These data support the contention that column plate count measured from a monomer peak does not accurately reflect the capability of the SEC system for polymer molecular weight analyses and that the R_{sp} and M^* accuracy values are more useful quantitative criteria for SEC column performance.

The performances of the various column sets in Table 4.2 are compared directly in Figure 4.9, which represents a master plot of M^* versus σD_2. This plot, the basic SEC-MW accuracy criterion, can be used universally for comparing the performances of all SEC column sets. The data in Table 4.2 and Figure 4.9 show, for example, that the PSM column set studied exhibits an excellent level of molecular weight accuracy of 2% for a 15-min analysis. Column-packing particle size is the most significant factor which differentiates the 15-min "high-performance" SEC from the 3-hr conventional SEC analysis.

The utility of the R_{sp}^* factor (Eq. 4.10) for comparing SEC packings is demonstrated in Table 4.3. Note that R_{sp}^* is independent of column dimensions and that LiChrospher and PSM particles have comparable values of R_{sp}^* for polymer analyses. The data also verify that values of R_{sp}^* for solutes at total permeation (toluene in this case) are not very useful for

Table 4.2 Performance Comparison of Several Column Sets Using Various GPC Column-Packing Materials

Column Packing	Number of Columns	Total Length (cm)	Flow Rate (ml/min)	Sample Analysis Time (min)	Particle Size (μm)	Linear Calibration Range (MW)	Plate Count, N, Toluene	σD_2^a	R_{sp}	M^{*b} (%)
(1) Styragel	4	488	1	180	50	10^3–10^6	7,500	0.45	1.27	11
(2) Porasil	4	488	1	180	75–150	2×10^4–10^6	2,700	0.37	1.56	7
(3) Vit-X	4	200	2	15	30	5×10^3–10^6	3,500	0.59	0.97	18
(4) μ-Styragel	4	120	2	15	10	2×10^3–10^6	13,000	0.50	1.14	13
(5) LiChrospher	4	100	1.5	15	10	5×10^3–10^6	5,800	0.23	2.50	3
(6) PSM	5	60	1.25	15	7	10^3–2×10^6	24,500	0.21	2.72	2

Taken from Reference 4.

[a] σ measured with 97,200 MW polystyrene.

[b] M^* is obtained from Equations 4.18 and 4.19 using the measured values of σ and D_2.

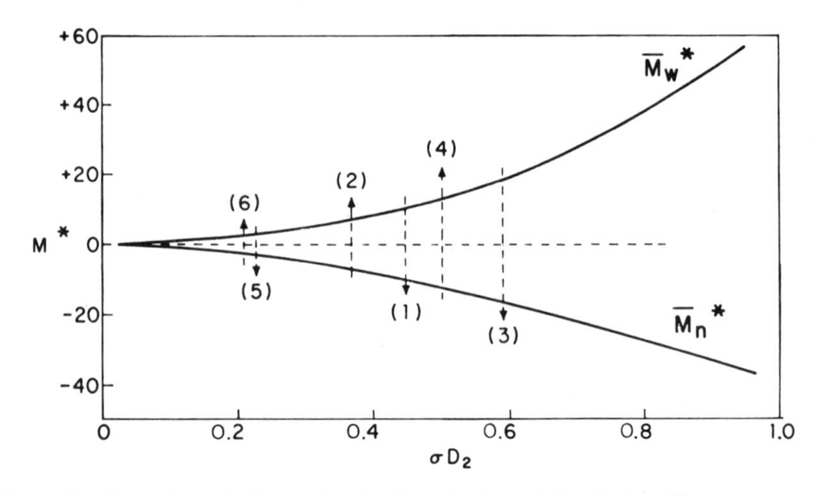

Figure 4.9 Comparison of column sets using the molecular weight criterion M^*.
1, Styragel; 2, Porasil; 3, Vit-X; 4, μ-Styragel; 5, LiChrospher; 6, PSM. (Reprinted with permission from Ref. 4.)

defining the performance of SEC packings for polymer analyses. While LiChrospher packings show greater selectivity (smaller D_2), presumably because of a generally larger porosity, PSM packings have the advantage of higher efficiency (smaller H and σ) because of smaller particles and narrower particle and pore-size ranges.

It should be noted at this point that Equations 4.18 and 4.19 (the M^* criterion) predict the accuracy of the values of $\overline{M}_w$ and $\overline{M}_n$, not the accuracy of the entire MWD curve. The requirement of *MWD accuracy* on column performance is more stringent than that of the *molecular weight accuracy*. Predictions of satisfactory accuracy of the average molecular weights does not necessarily mean acceptable accuracy for the entire observed MWD curve. The value of M^* is the error caused by column dispersion alone and does not include errors in values of MW assigned to polymer standards, errors due to flow rate variation, operator errors, and so on. In actual polymer sample analyses, molecular weight errors due to column dispersion can be corrected by using the appropriate SEC calibration and molecular weight calculation method (Sect. 9.3). Methods for correcting column dispersion in the MWD curve are discussed in Section 10.6. It should be emphasized that the validity of the SEC resolution calculations depends on the accuracy of the following approximations: the linearity of the calibration curve, the appropriateness of using a Gaussian instrument function, and a constant value of column σ.

Table 4.3 Comparison of Unmodified LiChrospher and PSM Packings for HPSEC

Packing	Pore Size (Å)	Volume Porosity (%)	Particle Diameter (μm)	Linear MW Fractionation Range	D_2	σ (ml)			R^*_{sp}	
						Toluene	PS[a]	MW	Toluene	PS[a]
LiChrospher 100	100	72.5	10	3×10^3–5×10^4	1.28	0.087	0.229	5,000	1.05	0.39
PSM-500	125	45.8	9	5×10^3–4×10^4	2.17	0.067	0.147		0.80	0.36
LiChrospher 500	500	63.7	10	1.5×10^4–1.5×10^5	1.32	0.107	0.275	51,000	1.01	0.32
PSM-800	300	48.3	6	6×10^3–2×10^5	2.25	0.054	0.118		0.95	0.43
LiChrospher 1000	1000	63.7	10	3×10^4–2×10^6	1.98	0.096	0.142	97,000	0.61	0.41
PSM-1500	750	50.2	9	4×10^4–2×10^6	4.56	0.030	0.068		1.30	0.60
LiChrospher 4000	4000	63.7	10	10^5–$>7 \times 10^6$	3.84	0.092	0.144	390,000	0.33	0.21
PSM-4000	3800	53.0	6	7×10^4–$>7 \times 10^6$	5.76	0.052	0.136		0.50	0.19

Taken from Reference 14.
[a] PS, polystyrene.

4.5 Pore Geometry and Operational Effects

Properties of SEC separating systems that are of great practical importance can be deduced from theoretical insights into σ and D_2. For clarity, we discuss only the conclusions and practical guidelines in this section. Detailed discussions of properties of D_2 and σ can be found in Chapters 2 and 3, respectively.

Connecting Columns

The total value of σ for a column series can be calculated from those for the individual columns according to the additive property of peak variance (Sect. 3.1):

$$\sigma^2 = \sum_i \sigma_i^2 \tag{4.20}$$

On the other hand, it can be shown from the additivity property of peak retention that (15)

$$\frac{1}{D_2} = \sum_i \left(\frac{1}{D_2}\right)_i \tag{4.21}$$

or

$$C_2 = \sum_i (C_2)_i \tag{4.22}$$

where C_2 is the linear calibration constant when the SEC calibration is expressed as $V_R = C_1 - C_2 \log \mathrm{MW}$, where $C_2 = \ln 10/D_2$ (Sect. 9.3). Note that for solute molecular weight values outside the linear separation region, the value of D_2 for the individual columns approaches to infinity (value of C_2 approaches zero). As predicted by Equation 4.20, one poor column with an exceptionally large value of σ_i can dominate the value of σ of a column set and degrade column performance as a whole (Sect. 8.6).

Separation Capacity of Single Pores

By the nature of the size-exclusion mechanism, there is a finite minimum slope to the calibration curve, that is, a lower limit to the value of D_2 (smaller D_2 means better resolution) even when there is no pore-size distribution (PSD) in the SEC column packing (15). It often is mistakenly assumed that a broad spectrum of pore sizes is required for the SEC packing to effectively fractionate broad-MWD polymers. In fact, however, pores of a single pore size are capable of fractionating polymer molecules over a substantial

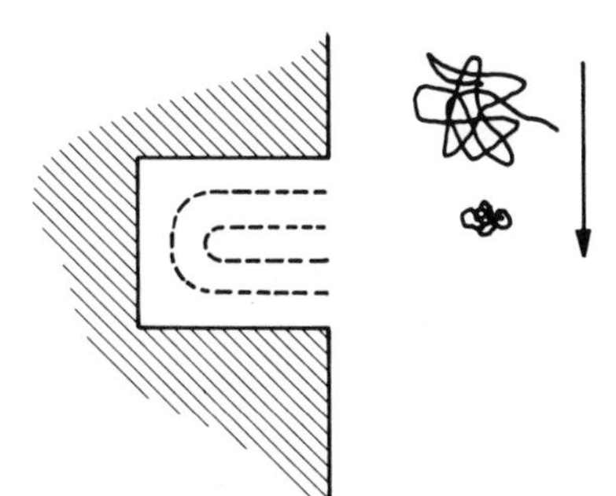

Figure 4.10 Size-exclusion effect in a single pore. Larger solute sees smaller effective pore volume.

molecular weight range (1.5–2 decades for random-coil polymer solutes). Figure 4.10 illustrates simply how a single pore can separate solute molecules of differing sizes by means of a solute-to-wall exclusion effect inside the pore. Because of steric interference, the centers of large incoming solute molecules are kept away from the interior walls of the pore, as illustrated by the inner broken line. However, smaller molecules can approach closer to the wall, as represented by the outer broken line in the figure. Thus a larger fraction of the pore volume is accessible to smaller molecules than to larger molecules. The progression from total permeation to total exclusion does not occur abruptly (if it did, it would produce a horizontal SEC calibration line with $D_2 = 0$) but takes place gradually (with a finite value of D_2 over a substantial size range for solute molecules.

The limiting values of D_2 for single pores can be predicted from the basic retention theory presented in Section 2.4. For random-coil-type solutes (Eq. 2.26):

$$\text{limiting } D_2 \simeq \frac{1}{3 \times \text{pore volume}} \tag{4.23}$$

Thus the limiting value of D_2 is inversely proportional to the column pore volume and, therefore, to the internal porosity of the SEC packing particles.

The effect of pore shape on the limiting D_2 value is small. The large shape differences between the cylindrical and slab pore models cause only a 20% difference in D_2 (Fig. 2.11). Particles with equal pore volume but different size have identical values of D_2 according to the SEC retention theory (Sect. 2.4). A change in pore size only shifts the SEC calibration curve up or down along the molecular weight scale, without affecting its slope or value of D_2.

On the other hand, the limiting value of D_2 of a single pore is strongly dependent on the shape of the solute molecule. For a certain chemical

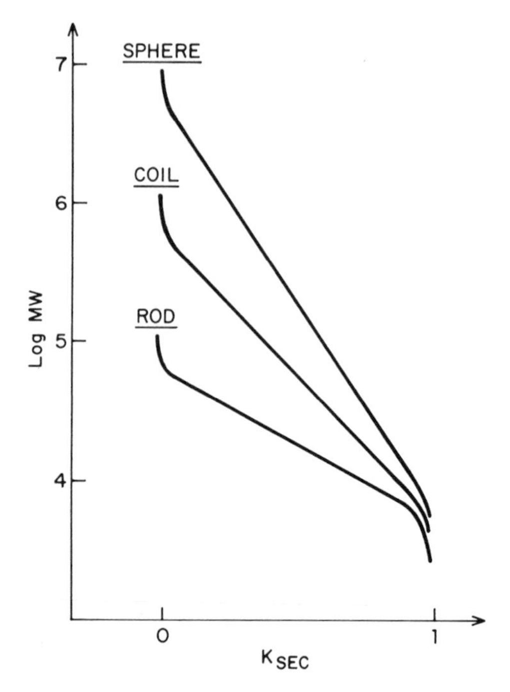

Figure 4.11 Effect of solute geometry on SEC-MW calibration curve slope. Rigid rod, $R_g \propto M$; flexible coil, $R_g \propto M^{\alpha}$, $\alpha \simeq \frac{1}{2}$; solid sphere, $R_g \propto M^{1/3}$.

structure and molecular weight, the more extended is the conformation of a macromolecule, the more it will be excluded from the pores of SEC packings. Therefore, SEC calibration curves for different solute conformations behave like those illustrated in Figure 4.11. The more open structure of the rigid-rod shape elutes before the random-coil structure of the same molecular weight, with the calibration curve of the rod molecule falling below that of the coiled molecule. On the other hand, the more compact structure of the hard-sphere type will elute after the coiled molecules of the same molecular weight, producing a calibration curve that lies above that for the coiled molecule.

The dependence of solute size on molecular weight varies with solute conformation. According to Equations 2.30, 2.31, and 2.32, solute size varies in proportion to the molecular weight raised to a power of about 1, $\frac{1}{2}$, and $\frac{1}{3}$ for the rod-like, the coil-like, and the sphere-like solutes, respectively. It is to be expected, therefore, that the calibration curve will be the steepest for the sphere-like solutes, with its value of D_2 being $\frac{3}{2}$ that of the coiled solute. The curve for the rod-like solutes has the lowest slope, with its value of D_2 being only $\frac{1}{2}$ of that of the coiled solute. The expected molecular weight separation range of a single pore is about one decade for rod-like molecules and three decades for spheres, as compared to the usual approximately two-decade MW separation range for random-coil solutes.

Effect of Packing Pore-Size Distribution

Since the pores in actual SEC packings have irregular cross sections and finite pore-size distributions (PSD), the observed value of D_2 and molecular weight separation range are always larger than predicted by the theoretical limits even for the "single pore-size" columns. However, the effect of pore geometry on D_2 and molecular weight range has often been overestimated. It has sometimes been mistakenly assumed that the shape of the SEC calibration curve is dictated entirely by the PSD curve of the packing, leading to the misconception that the SEC separation capacity (value of D_2) can be greatly improved by the use of packings with a very narrow PSD. This fallacy is caused by a failure to recognize the theoretical limit on the value of D_2 as described by Equation 4.23. The value of D_2 and molecular weight range of HPSEC columns of a single pore size are usually only 30–60% higher than the theoretical limits. However, the theoretical limits are based on a simplified model, and in practice irregularities in pore cross sections are unavoidable, so it is not possible to recover this 30–60% loss in SEC separation capacity by minimizing the PSD of the SEC packing. A quantitative theory for the effect of PSD on SEC calibration curve is discussed in Section 2.5 (Eq. 2.34).

Since SEC separation capacity is limited by the available pore volume of the column packing (Fig. 4.4) the design of an SEC experiment involves a trade-off between resolution and versatility. With SEC columns of only one pore size, all the SEC separation capacity is concentrated in a narrow molecular weight range to give a minimum D_2 (or maximum resolution). However, the linear molecular weight range of a single-pore-size column is too narrow to provide accurate analyses for broad-MWD polymers in general. For example, the MWD of a typical condensation polymer as described by the Flory MWD curve is quite broad, extending over two decades in molecular weight. The narrow molecular weight range of single-pore-size columns can force the wings of a Flory MWD curve into the nonlinear calibration region, causing distortion of the polymer elution curve and over 20% error in calculated molecular weight values (15). A preferable approach to the SEC analysis of polymers is the use of an SEC column set of different pore sizes to provide a wide molecular weight separation range (Sect. 8.5). The increased convenience and versatility can usually justify the use of a wide-molecular-weight-range column set for general-purpose SEC. A column set with a wide linear molecular weight separation range when used in conjunction with the broad standard linear calibration methods (Sect. 9.3) gives good molecular weight accuracy in SEC analyses.

Proper SEC column selection is a compromise between two goals: *wide-molecular-weight calibration range* for convenience and versatility, and a calibration curve with *good linearity* for maximum accuracy in MW determination. The best compromise is obtained by using columns with packings of only two pore sizes (i.e., the *bimodal PSD approach*) (15).

By simulating the conventional method of connecting columns of many similar pore sizes, Figure 4.12 shows how the SEC-MW calibration curve (right) broadens in range as the PSD of the packing increases from zero (single pore size) to 0.15 and 0.65. Here PSD is expressed as standard deviation of the log-normal PSD curve. The dashed lines in the figure are the linear approximation of the calibration curves. The plot at the left of Figure 4.12 shows that the separation range (I_R) increases with increasing PSD; however, the linear fit (I_L) rapidly becomes poorer beyond a PSD of 0.15. The separation range, I_R, is given in decades of the solute radius of gyration, R_g. The value of I_R is calculated as the difference in the logarithm of the limiting values of R_g near exclusion and total permeation. The "goodness" of the linear fit between the dashed line and the calibration curve at the right of

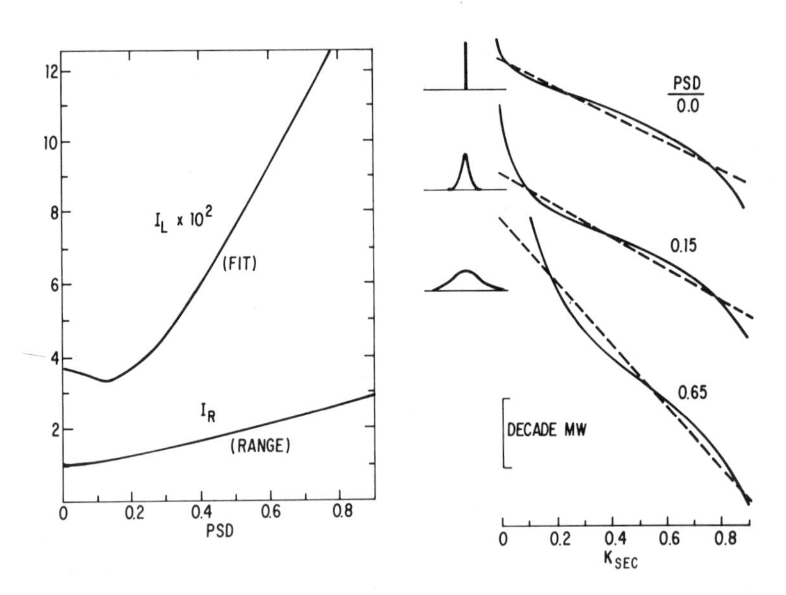

Figure 4.12 Effect of pore-size distribution on calibration linearity and molecular weight range for SEC: Monomodal.

I_R and I_L in units of decades of radius of gyration. (Reprinted with permission from Ref. 15.)

Figure 4.12 is measured by the root-mean-square derivation of the fit I_L, in the same units as I_R. The situation is much improved in the case of the bimodal PSD approach. Figure 4.13 shows that as the difference in pore size increases (increasing Δ log PS), I_R increases steadily. However, I_L goes through a minimum at Δ log PS = 1 (with two pores of about one decade difference in size), representing the best linear calibration fit. The calibration curve at the right of Figure 4.13 has a four-decade of molecular weight range with an excellent linear fit. (See Sect. 8.5 for selecting bimodal PSD columns.)

Effect of Operating Parameters

The dependence of SEC column dispersion on retention is quite complex, as discussed in Section 3.3. Since SEC peak dispersion is a mass-transfer-limited process, it is very sensitive to most experimental parameters, including packing particle size, flow rates, solvent viscosity, sample concentration, extracolumn effects, and packing inhomogeneity (Sect. 3.4). Because of the dependence on all these factors, values of column σ should be determined each time a change in experimental conditions is made (Sect. 3.5). Illustrations of the loss of resolution by increasing flow rate and sample concentration are shown in Figures 4.14 and 4.15, respectively. Since SEC resolution

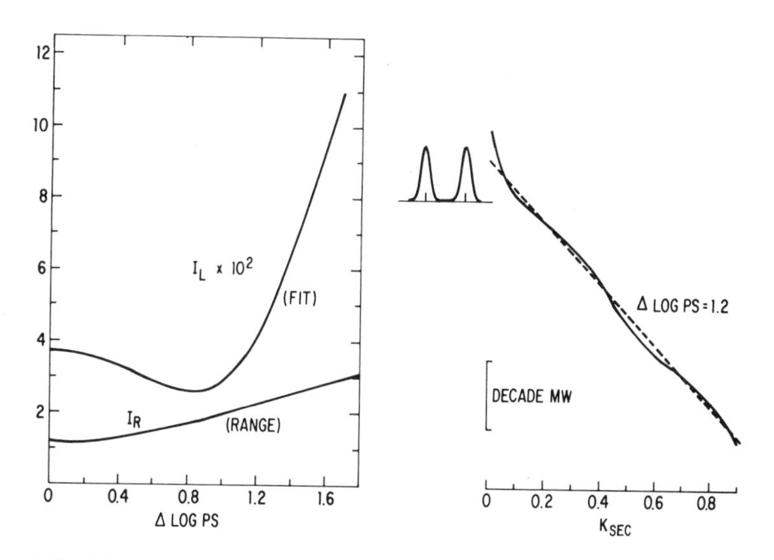

Figure 4.13 Effect of pore-size distribution on calibration linearity and molecular weight range: Bimodal.

Pore-size distribution, 0.15; pore volume ratio, 1.0. Units as for Figure 4.12. (Reprinted with permission from Ref. 15.)

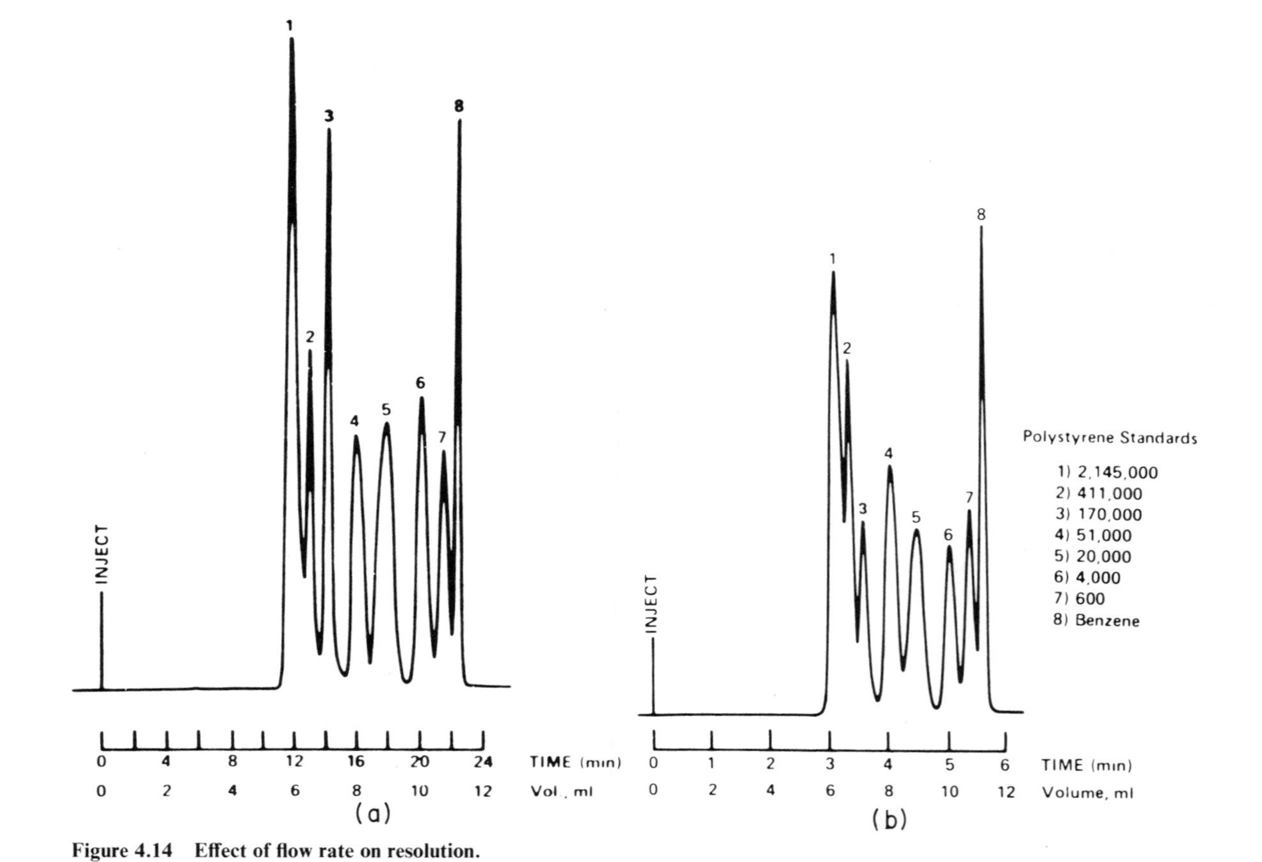

Figure 4.14 Effect of flow rate on resolution.

Separation of polystyrene standards on μ-Bondagel columns. Columns, 125, 300, 500, and 1000 Å; mobile phase, methylene chloride. Flow rate: (*a*) 0.5 ml/min; (*b*) 2.0 ml/min. (Reprinted with permission from Ref. 16.)

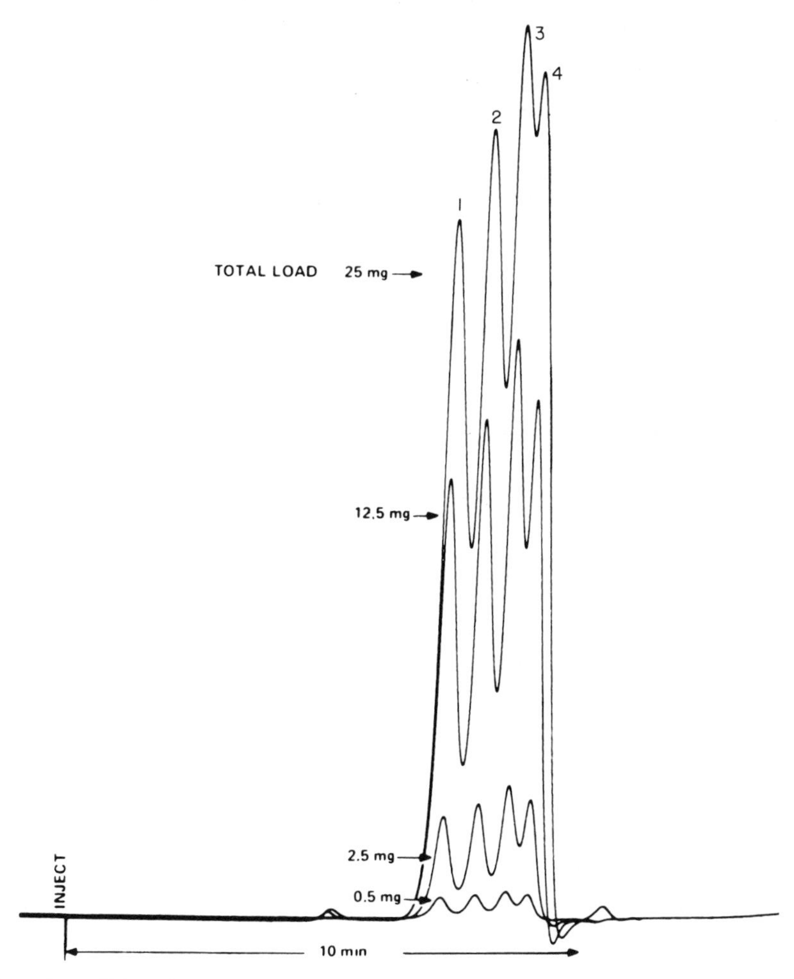

Figure 4.15 Effect of sample load on resolution.
Column, two μ-Styragel 100 Å; mobile phase THF; flow rate 2.0 ml/min. Solutes: 1, dioctylphthalate; 2, dibutylphthalate; 3, diethylphthalate; 4, dimethylphthalate. (Reprinted with permission from Ref. 16.)

depends on the product of σ and D_2, any operating parameter that affects either σ or D_2 will affect the resolution. The expected effect of some common SEC experimental parameters are summarized in Table 4.4.

In summary, both experience and theory have shown that σ, D_2, R_{sp}, and M^* are the most accurate and effective terms for expressing SEC column performance.

Table 4.4 Influence of Operating Parameters on SEC Performance

Parameters	D_2	σ	R_{sp}, M^*
Column volume	++	++	++
Particle size	–	++	++
Particle porosity	+	–	+
Particle shape	–	–	–
PSD	+	–	+
Pore size	–	–	–
Pore shape	–	–	–
Solute conformation	++	+	++
Flow rate	–	++	++
Solvent viscosity	–	+	+
Temperature	–	+	+

– Negligible (or unsubstantiated) effect.
\+ Moderate effect.
++ Large effect.

REFERENCES

1. L. R. Snyder and J. J. Kirkland, *Introduction to Modern Liquid Chromatography*, Wiley, New York, 1974, Chapt. 2 and 3.
2. L. R. Snyder, *J. Chromatogr. Sci.*, **10**, 200 (1972).
3. J. C. Giddings, *Anal. Chem.*, **39**, 1027 (1967).
4. W. W. Yau, J. J. Kirkland, D. D. Bly, and H. J. Stoklosa, *J. Chromatogr.*, **125**, 219 (1976).
5. D. D. Bly, *J. Polym. Sci., Part C*, **21**, 13 (1968).
6. A. E. Hamielec, *J. Appl. Polym. Sci.*, **14**, 1519 (1970).
7. L. H. Tung, *J. Appl. Polym. Sci.*, **13**, 775 (1969).
8. W. W. Yau, H. J. Stoklosa, and D. D. Bly, *J. Appl. Polym. Sci.*, **21**, 1911 (1977).
9. W. W. Yau, H. J. Stoklosa, C. R. Ginnard, and D. D. Bly, 12th Middle Atlantic Regional Meeting, American Chemical Society, April 5–7, 1978, paper PO 13.
10. A. E. Hamielec and W. H. Ray, *J. Appl. Polym. Sci.*, **13**, 1319 (1969).
11. T. Provder and E. M. Rosen, *Sep. Sci.*, **5**, 437 (1970).
12. E. Grushka, *Anal. Chem.*, **44**, 1733 (1972).
13. W. W. Yau, *Anal. Chem.*, **49**, 395 (1977).
14. J. J. Kirkland, *J. Chromatogr.*, **125**, 231 (1976).
15. W. W. Yau, C. R. Ginnard, and J. J. Kirkland, *J. Chromatogr.*, **149**, 465 (1978).
16. R. V. Vivilecchia, B. G. Lightbody, N. Z. Thimot, and H. M. Quinn, *J. Chromatogr. Sci.*, **15**, 424 (1977).

Five

EQUIPMENT AND DETECTORS

A. EQUIPMENT

5.1 Introduction

HPSEC requires somewhat different equipment from that used in traditional GPC and radically different apparatus from that used in conventional GFC. Although elegant separations have been carried out in the past with relatively modest apparatus, more sophistication is featured in HPSEC, whether in determining accurate molecular weights of synthetic polymers, or analyzing samples with biologically active macromolecules. Extensive developments in high-performance liquid chromatography have made available a variety of equipment which also can be utilized satisfactorily for HPSEC (see especially Refs. 1 and 1a).

To provide high-quality results, HPSEC equipment must be designed according to many of the criteria listed in Table 5.1. Although to meet various goals (e.g., high analysis speed) particular equipment designs are required, analytical accuracy necessitates the greatest range and control of operating parameters. Thus an apparatus that provides good analytical accuracy often will meet the design requirements of any HPSEC method. An apparatus constructed to meet all the criteria listed in Table 5.1 should be useful for any HPSEC separation. A general schematic for HPSEC equipment is shown in Figure 5.1. Additional components may be needed for specialized analyses.

Table 5.1 Criteria for HPSEC Equipment

	Goal			
Equipment Design Feature	Analytical Accuracy	Retention Reproducibility	Analysis Speed	Separation Versatility
Precise flow rate	×	×		
Temperature control	×	×		×
Precise sampling	×	×		
Stable detection	×			
High-signal/noise detection	×		×	×
Fast detection	×		×	
High-pressure pumping	×		×	×
Efficient columns	×		×	
Automatic data handling	×		×	
Low-dead-volume system	×		×	
Flow rate sensing	×			
Range of column packings				×
Chemically resistant				×
Variety of detectors				×

Differences between the basic equipment for HPGPC and HPGFC are relatively minor. Both of these techniques require high-pressure liquid pumps, efficient chromatographic columns, sensitive and versatile detectors, and precise sample injection. The HPGPC technique does especially require an appropriate data-handling system for calculating molecular weights of synthetic polymers. All-glass (or glass-Teflon) equipment may be required in specialized areas of HPGFC which involve corrosive aqueous mobile phases, and components of this construction with some performance

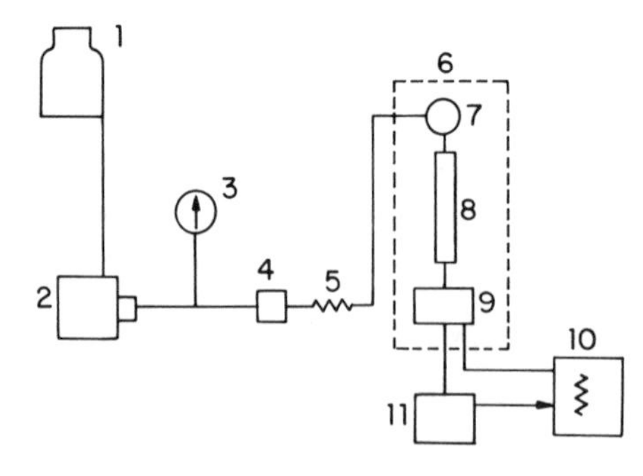

Figure 5.1 Schematic diagram of typical HPSEC apparatus.
1, reservoir; 2, pump; 3, pressure gauge; 4, line filter; 5, pulse dampener; 6, thermostatted oven; 7, sample injector; 8, chromatographic column; 9, detector; 10, recorder; 11, data-handling equipment.

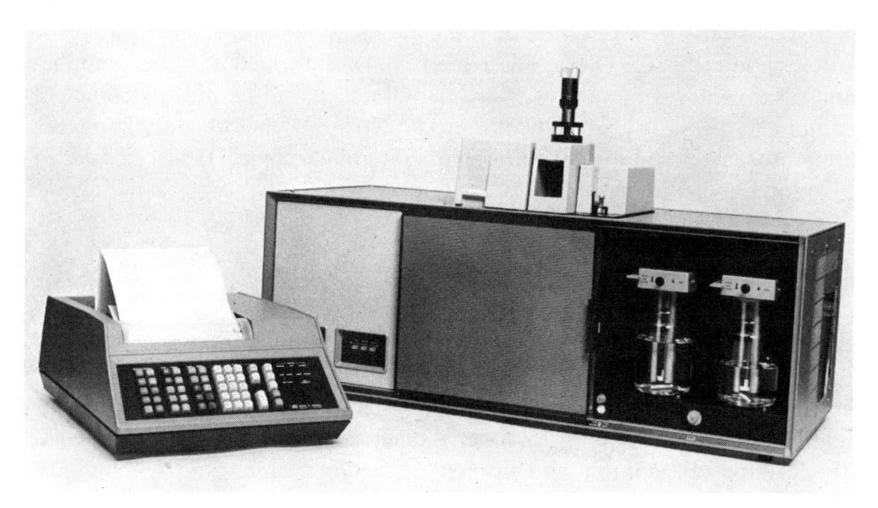

Figure 5.2 Modern digitally controlled liquid chromatograph.
(Reprinted with permission of Hewlett-Packard Co.)

limitations, are commercially available. Except for special areas (e.g., high-temperature GPC, as discussed in Sect. 12.4), modern LC equipment frequently can be used for both high-performance gel permeation and gel filtration chromatography without modification. Names and addresses for suppliers of modern LC equipment are found in recent annual buyer's guide editions of several scientific journals (e.g., *Analytical Chemistry*, *American Laboratory*).

Whether to choose modular HPSEC equipment (i.e., equipment assembled from components) or a completely integrated apparatus depends on the anticipated application. If great versatility or range of applicability (e.g., operation at higher temperatures) is not required, or if there are budget limitations, simple modular equipment may well be adequate. On the other hand, integrated commercial instruments generally provide for better performance and convenience, and are particularly attractive when methods are to be exchanged between laboratories.

Sophisticated, integrated LC instruments which are applicable to HPSEC are now widely available. Some of these are even completely digitally controlled, as is the equipment shown in Figure 5.2. In this instrument all the desired operating parameters (e.g., temperature, flow rate) are entered on the controlling console by the operator. A microprocessor then controls all these operating parameters; it also accepts the detector output and functions as an on-line data processor.

In this chapter we describe in some detail the components required for an effective HPSEC system. The reader is also informed of the advantages and disadvantages of various instrumental designs to permit the choice of equipment to satisfy a particular need. However, before describing the apparatus, a general discussion of extracolumn effects in HPSEC is needed to aid in understanding this important parameter.

5.2 Extracolumn Effects—General

In addition to the inherent band broadening which occurs within the chromatographic column (Chapt. 3), additional broadening also occurs outside the column. This extracolumn band broadening results from the sample injection, and also from other elements of the SEC apparatus, such as the detector cell, column end fittings, connectors, and so forth. Thus, the observed total band width W_t is a function of the sample band volume (width at the baseline, $\sim 4\sigma$) due to column dispersion, W_c, and the sample injection volume W_i, plus the extracolumn band broadening that occurs within the SEC apparatus. This peak broadening relationship may be expressed in terms of peak volume:

$$W_t^2 = W_c^2 + W_i^2 + W_d^2 + W_j^2 + W_x^2 \tag{5.1}$$

The quantities W_d, W_j, and W_x represent the increased peak widths (volumes) associated with the extracolumn effects in the detector, endfittings, and connecting tubing, respectively. Extracolumn band broadening should be kept to a minimum so that the observed peak volume W_t closely approximates the actual peak volume W_c. As a rule of thumb this means that the total of the injected volume and the other extracolumn peak volumes should be less than one-third of W_c for a monomer peak in the chromatogram. This then limits the increase in W_t to about 10%.

Because the volume W_c of an HPSEC band can be quite small (e.g., $\sim 40\ \mu l$ in extreme cases), it is particularly important that extracolumn effects be minimized. Particular attention must be placed on the design of all equipment components to ensure that these cause insignificant broadening of the true band width. As discussed in Section 4.3, significant band broadening can cause large errors in molecular weight results. The origin of extracolumn effects for the various equipment components are discussed in the individual sections.

5.3 Mobile-Phase Reservoirs

Since flow rates in HPSEC typically are 1–3 ml/min, and separations usually are completed in a half-hour or less, the volume of mobile phase used for a single analysis is relatively small. As a result, the total volume used in a workday is moderate, and reservoirs for HPSEC typically hold about 1 liter. For preparative applications involving large-diameter columns, larger volume reservoirs are needed (e.g., several liters).

Reservoirs usually are made of stainless steel or glass, but should be inert to the mobile phase and not easily broken. For convenience, some reservoirs are designed so that the mobile phase may be degassed *in situ* to prevent bubbles from forming in the detector during the separation. Elimination of

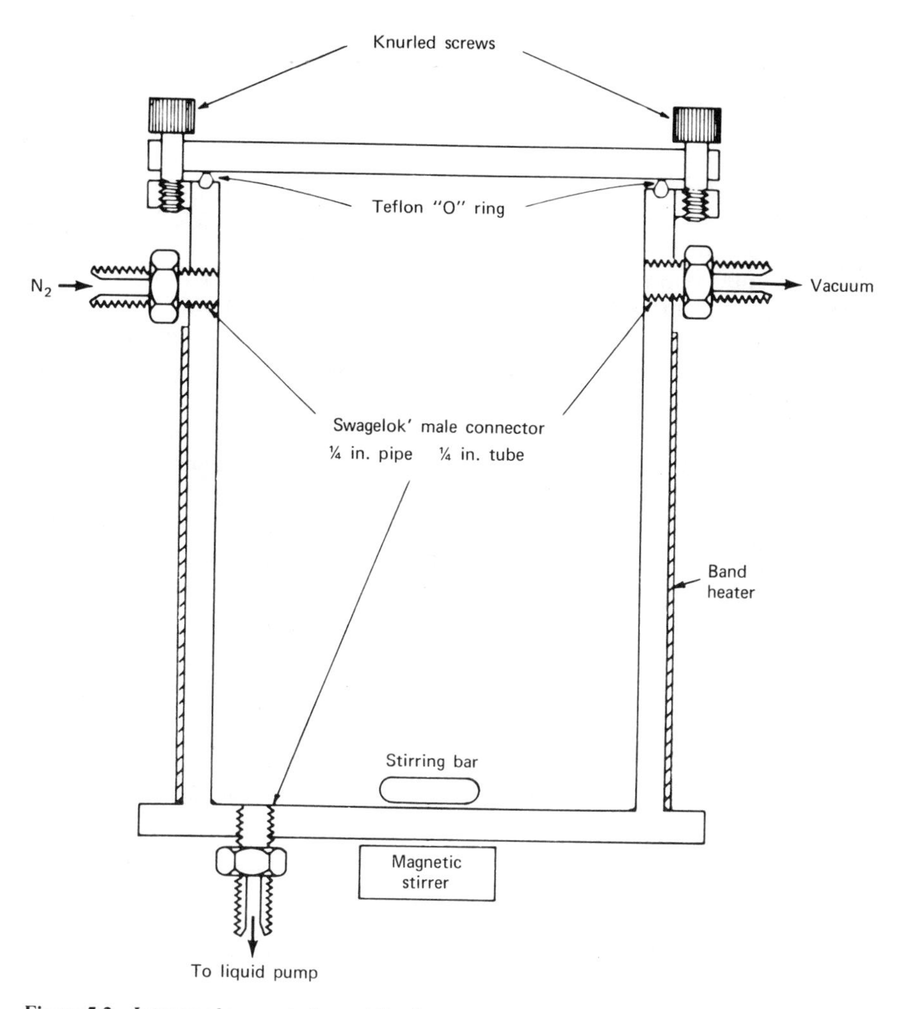

Figure 5.3 Integrated reservoir for mobile phase.
(Reprinted with permission of DuPont Instrument Products Division.)

oxygen also at times is required to prevent its reaction with certain mobile or stationary phases. To facilitate *in situ* degassing, reservoirs may be equipped with a heater, a stirring mechanism (e.g., magnetic stirring bar), and separate inlets for vacuum or nitrogen purge. The stainless steel reservoir system illustrated in Figure 5.3 permits degassing by applying a vacuum to the filled reservoir for a few minutes with stirring; warming the liquid aids this operation. Degassing can also be effectively achieved by thoroughly purging the mobile phase with helium, which has a very low solubility in virtually all liquids. After initial sparging by a fast flow of helium for a few minutes, a slow purge of helium is then used to maintain the mobile phase. A helium or nitrogen purge also prevents oxygen from redissolving in a sensitive mobile phase after degassing, and in some instances improves safety by preventing accidental ignition of flammable vapors.

5.4 Solvent-Metering Systems (Pumps)

Providing a constant, reproducible supply of mobile phase to the column is the most important function of the HPSEC solvent-metering system. In all HPSEC instruments, relatively high pump pressures are needed to overcome resistance to flow offered by the small particles used in the columns (Sect. 7.3). The general operational requirements for a solvent-metering system in HPSEC are listed in Table 5.2.

Most modern solvent-metering systems for chromatography are constructed of stainless steel and Teflon for a maximum resistance to chemical attack. Pump seals made from virgin or filled Teflon resist all solvents that have been used for GPC, including concentrated sulfuric acid. Sapphire pistons are often used where applicable. Pumps constructed entirely of Teflon and glass are available commercially and may be substituted in rare instances where stainless steel cannot be tolerated; however, the pressure output of these devices generally is limited to 500 psi.

Table 5.2 General Requirements for HPSEC Pumps

Deliver mobile-phase volume flow rate with an overall precision of better than 1%
Have pressure output of at least 4000 psi
Provide pulse-free or pulse-damped output
Give flow rates of at least 3 ml/min, preferably up to 10 ml/min
Be chemically resistant to a wide range of mobile phases
Have small hold-up volume for rapid solvent changes and recycle operation (desirable but not essential)

General Pump Specifications

The solvent-metering or pumping system can often be the limiting factor for accurately determining the performance of the chromatographic separation, particularly when molecular weight information is desired. Constancy of flow rate is especially important, as elaborated in Section 7.2. Certain specifications become dominant when considering solvent-metering systems for HPSEC: [1] pump resettability, [2] short-term precision, [3] pump pulsation or "noise," [4] drift, and [5] flow rate accuracy. By resettability (or repeatability), we mean the ability to reset the pump to the same flow rate repeatedly. Short-term precision is a measure of the reproducibility of the volume output by the pump over a few minutes. Pump "noise" or pulsation arises from flow changes as result of operational functions such as piston movement and check valve operation. Drift is a measure of a generally continuous increase or decrease in the pump output over relatively long periods (e.g., hours). Pumping accuracy relates to the ability of pumps to deliver exactly the flow rate indicated by a particular setting. While all the foregoing considerations regarding pumps are important in HPSEC, pump resettability and drift are usually the most critical (Sect. 7.2). In addition to these performance features, operational convenience, durability and serviceability also should be considered when selecting a pump for HPSEC.

Two general types of solvent-metering systems are available: positive-feed and flow-feedback systems, both with and without compressibility and pulsation compensation. With flow feedback, some function of the flow rate is monitored continuously, and adjustments are made to maintain a constant flow rate in a closed-loop operation. A block diagram of flow feedback for a pneumatic amplifier pump is shown in Figure 5.4. Flow-feedback solvent metering is more costly but can provide the more precise flow rates needed for HPSEC.

Commercially available pumps can be classified into three groups: reciprocating, positive displacement, and constant pressure. Before describing the various types for HPSEC, we should note that there are various levels of sophistication available in pumps, each having certain advantages (and disadvantages) for particular applications. The most precise pumps generally are not required when determining the molecular weight distribution of polymers, *provided* that a computer and an on-line mobile-phase flow-rate-measuring device (e.g., syphon counter) are used. In this case, automatic corrections for pumping variations are made by the data-handling system during the run. On the other hand, when no computer-compensating system is available, precise pumping systems are sought for all types of HPSEC applications.

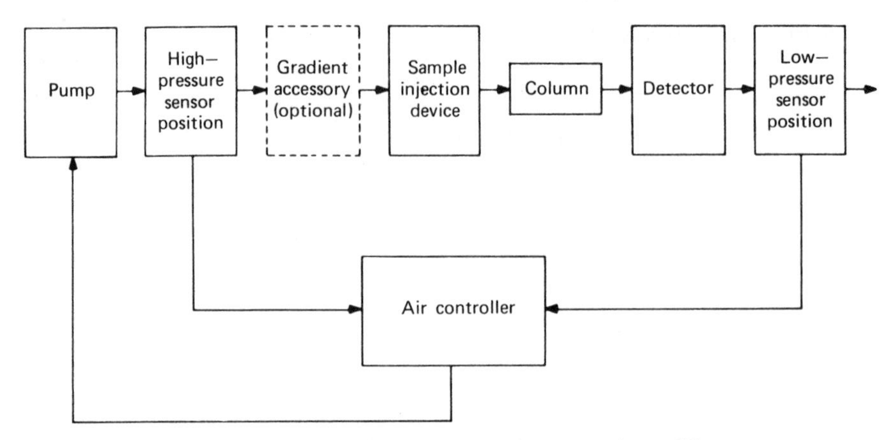

Figure 5.4 Block diagram of a flow-controlled system for pneumatic-amplifier pump.
(Reprinted with permission of DuPont Instrument Products Division.)

Reciprocating Pumps

Reciprocating pumps are the most widely used pumps in HPSEC because of their generally satisfactory performance. Models with output pressures up to about 10,000 psi (680 bar), and maximum volumetric outputs of 10–20 ml/min are typical of pumps used for analysis. Various commercial models are distinguished by the techniques used to minimize pulsating flow output and by the mode of compensation used for solvent compressibility and flow rate changes.

Simple, single-head, reciprocating pumps, such as those shown schematically in Figure 5.5 (e.g., Milton Roy Model 196-31), are relatively inexpensive.

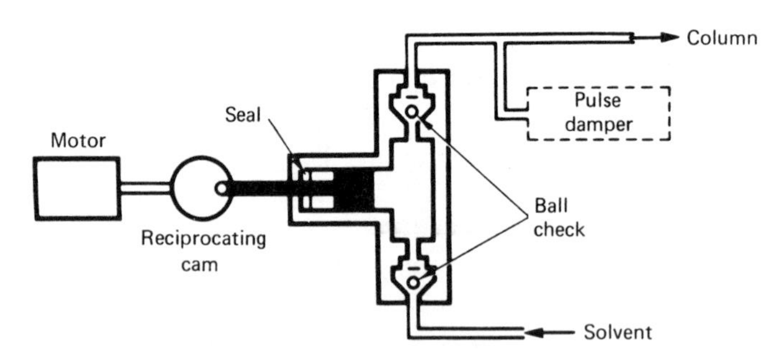

Figure 5.5 Schematic diagram of simple reciprocating pump.
(Reprinted with permission from Ref. 2.)

However, with these simple types, changes in solvent or column back pressure cause minor flow rate changes because of solvent compressibility. Pulsations are also greatest with simple reciprocating pumps, resulting in considerable detector noise, which increases with increasing flow rate delivery. A pulse damper generally is used to minimize this noise, the extent of dampening being a function of the detector. Ultraviolet photometers are more tolerant to pulses than a refractometer. Pulse-damping devices represent a compromise with convenience, however, since they increase the volume of the system between the pump and the chromatographic column and require additional purging when changing the mobile phase.

More sophisticated single-head pumps (e.g., Altex Model 110, Varian Model 5000) utilize a sinusoidal cam to drive the pump piston in the pumping and refill cycles, so that pulsations are minimized. One approach (e.g., Altex Model 110) uses a circuit design which recognizes the approaching end of the pumping stroke so that the motor driving the piston suddenly speeds up to deliver extra liquid in anticipation of the upcoming fast (200 ms) refill stroke when no liquid flows. The motor torque, which is proportional to the volume output of the pump, is monitored so that the motor speed is returned to the level operating before the back-fill stroke. As a result of such special devices, sophisticated single-head pumps generally exhibit lower pumping noise and improved pumping accuracy relative to simple types.

Dual-head pumps with pistons controlled by circular cams operated at 180° out of phase produce reduced flow pulsations (Fig. 5.6*b*) relative to single-head circular cam pumps (Fig. 5.6*a*) but are more expensive. With these pumps, one chamber fills while the other provides flow to the column. Additional reduction in pump pulsation is obtained with dual-head pumps

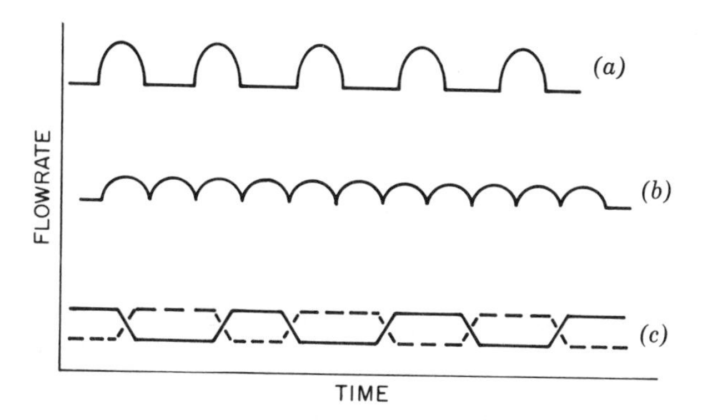

Figure 5.6 Reciprocating pump output patterns.

(*a*), Simple single-head reciprocating pump; (*b*), dual-head, circular-cam pump; (*c*), dual-head, sinusoidal-cam output (with changeover ramp).

which are driven by sinusoidal cams designed to produce a linear displacement of the piston. If the cam-activated strokes are perfectly matched, an essentially pulse-free output flow should result (Fig. 5.6*c*). However, in practice, some mismatch occurs, which produces a slight pulse at the end of the changeover points in the pumping cycle. Pulses are minimized by arranging a piston-driving cycle of slightly more than 180°, to include gradual take-over periods of one pump head relative to the other. This type of pump is often quite satisfactory for HPSEC and represents one of the most widely used types at present. Several commercial versions are available (e.g., Waters Model 6000B, DuPont Model 860).

One of the most sophisticated of the piston-type pumps (DuPont Model 850, Jasco) uses a three-head rotary cam plate, actuating three pistons that are arranged so that they are phased 120° apart. The swept volume of each pumping head is the same for each stroke, since the three pistons are driven by a single-face cam which actuates each piston in turn. The high precision of flow output is a function of each pump head operating at full stroke, with the total pump output being digitally controlled by the speed of the pump motor. The 120° phasing allows essentially a continuous flow in both solvent output and input streams, as illustrated in Figure 5.7. The flow output of

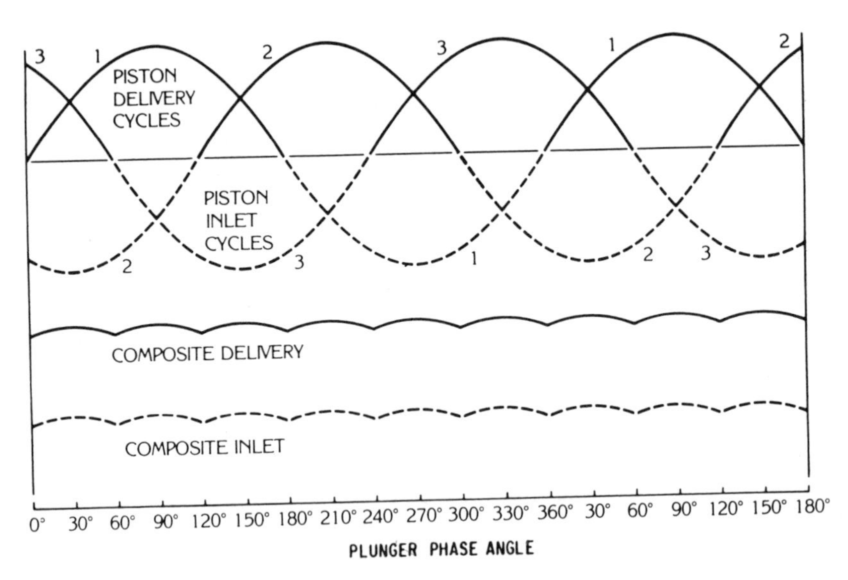

Figure 5.7 Output of three-headed reciprocating pump.
Numbered lines represent flow profile from single pump heads 1, 2, and 3; solid line is composite flow output profile; dashed line is composite pump input (suction). (Reprinted with permission of Du Pont Instrument Products Division.)

this type of pump is a succession of closely overlapping solvent pulses, the resulting total output exhibiting such small pulsations that pulse dampening is not required. The flow output of this design is one of the most precise of pumps available for HPSEC. In addition, the very small volume of each pumping chamber (35 μl) represents a distinct advantage in solvent change-over or in recycle applications.

Several manufacturers provide flow-feedback systems to adjust and correct the imperfect flow of a reciprocating pump. The general approach is the continuous measurement of flow rate by an appropriate transducer, which produces a signal when the flow rate varies from the preset value. The signal is then used to electronically adjust the pump to deliver more or less solvent to maintain an essentially constant flow of solvent relatively pulse free. For example, in one method for flow rate control, a differential pressure transducer measures the pressure drop across a restricter (Altex Model 100A). The pressure (and therefore the flow) is maintained constant by electronically controlling the rate of the pump motor which controls solvent output. For highest accuracy, this particular method requires individual calibration for each solvent.

The use of special systems to compensate for flow rate variations resulting from mobile-phase compressibility and pulsations is schematically illustrated in Figure 5.8. Output *a* represents the uncorrected flow output from a dual-head reciprocating pump as a result of the cam rotation driving the piston

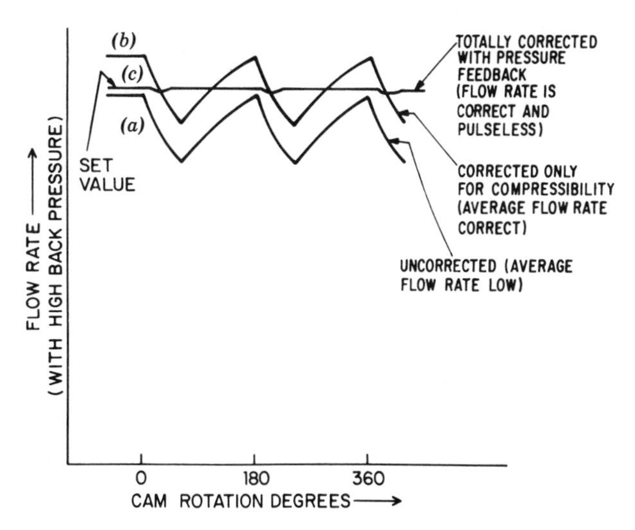

Figure 5.8 Effect of pressure feedback and compressibility correction on pumping. (Adapted with permission of Spectra-Physics, Inc.)

against the column back pressure; mobile-phase pulsations are significant. However, against relatively high column back pressures, the average of this uncorrected flow rate is lower than the desired set value. Output *b* is compensated for solvent compressibility, so the average flow rate now is correct compared to the set value; however, the output is still pulsating. In trace *c* the pump output is corrected with pressure feedback; flow rate is correct and pulsations have been greatly decreased.

A method of positive flow-feedback control uses a "hydraulic capacitor" with a flow restrictor to produce true volumetric flow rate (Hewlett–Packard Model 1080 Series). This system uses twin reciprocating (diaphragm) pumps, where the effective stroke of the pump heads is employed to vary flow rate. As illustrated schematically in Figure 5.9, the hydraulic-capacitor sensors, which double as pulsation dampeners, monitor the movement of the pulsating diaphragms in the pump. The signal generated in the hydraulic capacitor (pulsation damper) is used to control the pump stroke. The signal furnished by the flow controller is generated by integrating the pressure change in the hydraulic capacitor over a period of several strokes (several seconds) during certain time-fractions of the pump duty cycle to obtain accurate flow control. Two pulsation dampeners or flow-measuring systems are used to achieve independent flow control for each of the two pumps. The result is stabilization

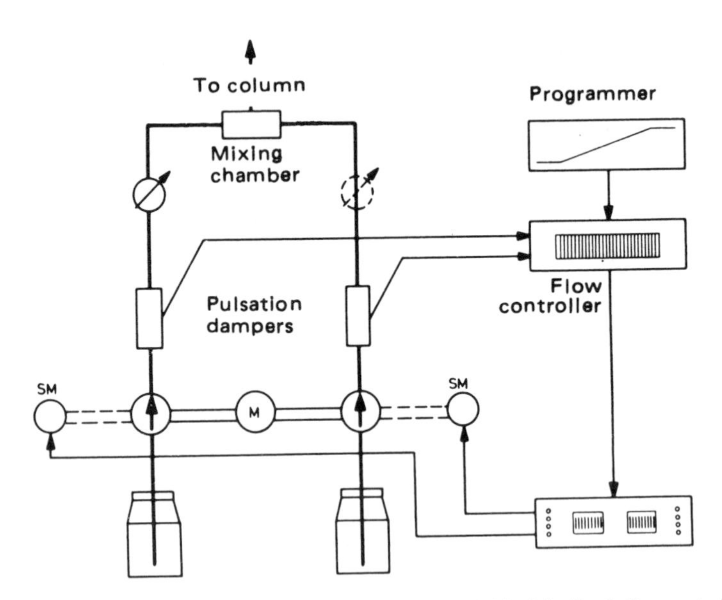

Figure 5.9 Block diagram of solvent-metering system with digitized feedback-flow control. (Reprinted with permission of Hewlett-Packard.)

of mobile-phase output against changes due to temperature, mobile-phase viscosity, and column back pressure. With this system, there is no need for recalibration when the solvent or the column is changed; output is virtually independent of the type of solvent pumped. This is a very useful system for HPSEC, but cost is relatively high, as is generally the case for the more sophisticated pumps.

Diaphragm reciprocating pumps are similar to piston pumps, except that a flexible stainless steel or Teflon diaphragm is in contact with the mobile phase. This diaphragm is actuated by a piston working on an oil cavity which, on each stroke of the piston, flexes the diaphragm to produce a pulsating solvent output.

A general advantage of reciprocating pumps is that solvent delivery is continuous; therefore, there is no restriction on the size of the reservoir that is used or the length of time that a pump is operating. Thus these pumps are particularly useful in equipment for automatic operation (e.g., overnight). A specific advantage of piston-reciprocating pumps is that their internal volume can be made small, and this type is particularly useful for recycle chromatography (Sect. 11.3).

Positive-Displacement Pumps

Positive-displacement pumps are of two forms, the screw-driven syringe and hydraulic-amplifier types. Figure 5.10 schematically shows a single-stroke displacement pump (e.g., Isco, Perkin–Elmer). The system is similar to a large syringe, the plunger being actuated by a screw-feed drive through a gear box, usually actuated by a stepping motor. The rate of solvent delivery by the piston is controlled by the voltage on the motor. This pump provides a flow that is relatively independent of the operating pressure and solvent viscosity. However, minor compressibility of mobile phase (e.g., about 4% at 7000 psi) requires that the solvent in the continually emptying syringe be

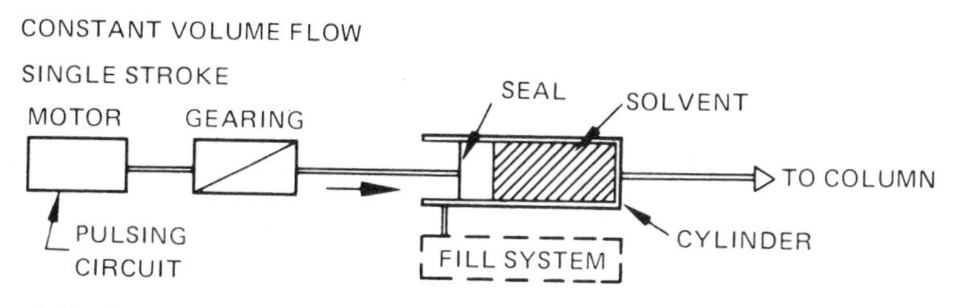

Figure 5.10 Schematic diagram of simple displacement pump.
(Reprinted with permission from Ref. 2.)

maintained at a constant pressure by a variable-restrictor valve before a constant-flow delivery is obtained. The initial equilibration can take several minutes unless the pump is first pressurized to the approximate operating pressure. A disadvantage of the positive-displacement pump is that it has a limited solvent capacity (e.g., 250 ml) and must be stopped for refilling. This type of pump also can be quite expensive, but it is pulseless and convenient to operate.

The second type of positive displacement pump is the hydraulic amplifier, used by Micromeritics Instruments Corp. In this approach a gear pump supplies oil to two ~2.5 ml hydraulic intensifiers which alternately pump mobile phase into the column, one intensifier delivering the mobile phase while the other is refilling. These intensifiers produce mobile-phase pressures which are about nine times the input oil supply pressure. This system uses a pressure-feedback circuit to compensate for change in flow rate during the switchover from one intensifier to another. Thus this pump produces an output which is relatively pulse-free, and it can operate from a limitless solvent supply.

Constant-Pressure Pumps

Gas pressure can be used either to drive the mobile phase directly or to regulate its pressure. In the simplest form, gas pressure is applied to the mobile phase contained in a special reservoir such as a coil of tubing (e.g., DuPont, Varian). Input pressure is limited by the available cylinder gas pressure (generally less than 2500 psi). While this pumping system is the least expensive, it has not been widely used in HPSEC, since it is relatively inconvenient to refill the mobile-phase reservoir, and it is pressure-limited (generally 1500–2500 psi). Moreover, column back pressure *must* be held constant for constant flow, and this is not easily maintained experimentally. An advantage of this system is that the flow produced is pulseless.

The pneumatic amplifier pump (e.g., DuPont) is similar to the simple gas-displacement system, except that pressure amplification is obtained by using a large-area pneumatic gas piston to drive a small-area liquid piston (Fig. 5.11). Since the pressure on the outlet is directly proportional to the ratio of the piston areas, a relatively low inlet gas pressure can be used to create high liquid output pressures. The pneumatic amplifier pump is equipped with a power-return stroke which permits very rapid refilling of the empty piston chamber with mobile phase. This system provides essentially pulseless, continuous pumping, and can achieve relatively high volumetric flow rates. Since this device operates only at constant pressure, flow rate is constant only as long as the column back pressure remains constant. Automatic flow-feedback systems that continually adjust the input air supply are available

Table 5.3 Comparison of Pumps for HPSEC

Type	Resettability	Drift	Short-Term Precision (Including "Noise")	Accuracy	Versatility and Convenience	Serviceability	Cost
Reciprocating							
Single-head, simple drive	+	+	–	+	–	+	Low
Single-head, special drive	+	+	+	+	+	+	Moderate
Dual-head with compressibility correction	++	+	+	+	++	+	High
Dual-head with compressibility correction and electronic pulse damping	++	++	++	+	++	+	High
Dual-head with closed-loop flow control	++	++	++	++	++	+	Very high
Triple-head low-volume	++	++	++	++	++	+	Very high
Positive displacement							
Syringe-type	++	++	++	+	–	+	Moderate to very high
Hydraulic amplifier	++	++	+	+	++	–	Moderate
Pneumatic							
Simple	–	–	+	–	–	++	Low
Pneumatic amplifier	–	+	+	–	++	++	Moderate
Pneumatic amplifier with closed-loop flow control	++	++	++	+	+	+	High

++ Optimum performance.
+ Satisfactory or usable performance.
– Limiting in some applications.

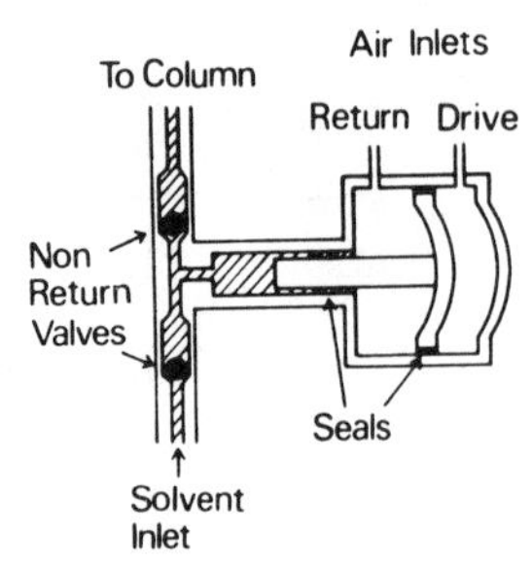

Figure 5.11 Schematic diagram of pneumatic-amplifier pump. (Reprinted with permission from Ref. 3.)

with some of the pneumatic amplifier pumps (DuPont) so that constant flow rate is maintained (Fig. 5.4).

Comparison of Pumps

A comparison of various pump characteristics for HPSEC is given in Table 5.3. In addition to the general qualities previously discussed, others, such as versatility, convenience, serviceability and cost, are also compiled. All the pumping types discussed are usable for HPSEC, but better performance is usually obtained with the more sophisticated solvent-metering systems, which unfortunately are also more expensive.

5.5 Sample Injectors

The method of introducing the sample onto the column can be a significant factor in determining HPSEC performance. As discussed in Section 5.2, the sample should be introduced onto the column in a sufficiently narrow band so that peak broadening from this cause is negligible. Sample injection in HPSEC with its relatively short, high-efficiency columns is of particular concern, since peak volumes are very small. Ideally, the versatile sample injector should introduce sharp plugs of a wide variety of samples into the columns with insignificant band broadening. Injectors should be convenient to use, reproducible, and operable against high back pressures. Some sample types require injection at elevated temperatures to meet solubility requirements.

The simplest form of sample introduction involves syringe injection into the pressurized column through a self-sealing elastomeric septum contained in a low-volume inlet port. This approach is rarely used in GPC but sometimes has utility in simple GFC systems. Reproducibility of sample injection with a syringe is rarely better than 2%, and often much poorer. Syringe injection creates practical problems because of possible pluggage of the column inlet with small particles of the elastomeric septum used. A buildup of these particles increases the column back pressure and may cause unsymmetrical peak shapes and a marked decrease in column efficiency. In addition, it is not generally feasible to make injections through elastomeric septa with high-pressure syringes above about 1500 psi.

Sometimes syringe injection is made using a stop-flow technique, where the pump is first turned off until the column inlet pressure becomes essentially atmospheric. The sample is then injected in the usual fashion, and the pump is turned on. Since diffusion in liquids is very slow, stop-flow injection can be made without affecting column efficiency. However, the stop-flow mode still retains the other disadvantages of syringe injection.

Syringe-septumless injection devices are available in some instruments (e.g., Hewlett-Packard), and these can eliminate the difficulties of sample leakage from syringe injection at higher pressures. This sampling mode has not been widely used in HPSEC, and appears applicable only for systems not requiring elevated temperatures. Advantages of the syringe-septumless injection technique include its high-pressure capability, variable injection volume, wide chemical compatibility, and elimination of the problems associated with elastomer septums. A disadvantage is the limited sample volume range available. If not properly designed, these devices also can cause significant extracolumn band broadening.

The most generally useful sampling device in HPSEC is the microsampling injector valve (e.g., Valco Instruments, Rheodyne). These special valves permit the sample to be introduced reproducibly into pressurized columns without significant interruption of solvent flow, even at higher temperatures. Figure 5.12 shows schematic drawings of a six-port, plug-type valve (e.g., Valco Instruments) in which the sample is contained in an external loop. (Long, narrow loops are preferred over shorter, wider-i.d. loops when large sample volumes are required.) The loop of appropriate volume is filled at low pressure by flushing it thoroughly with the sample solution, using an ordinary syringe ("Insert," Fig. 5.12). A clockwise rotation of the valve rotor places the sample-filled loop into the mobile-phase stream with subsequent injection of the sample onto the top of the column. ("Inject," Fig. 5.12). Other valve types (e.g., Siemans) use sample cavities which consist of annular rings in a sliding rod that can be thrust into the flowing stream.

The particular advantage of valve injection is the rapid, reproducible delivery of large volumes (e.g., up to several milliliters at $\ll 1\%$ error) with pressures to 7000 psi. High-performance valves suitable for HPSEC deliver appropriate sample volumes without significant extracolumn band broadening. These valves are only moderately expensive, delivery volumes are

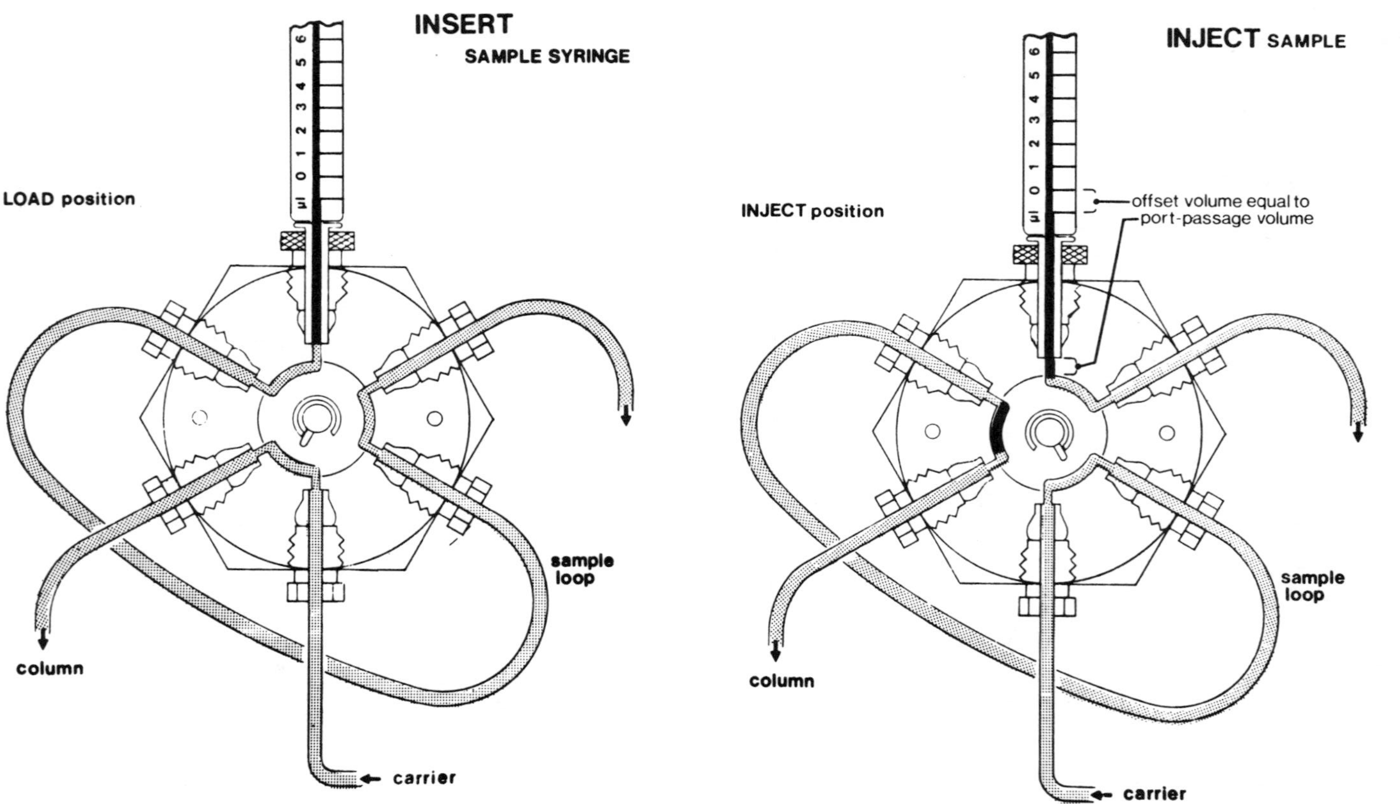

Figure 5.12 Six-port microsampling valve for HPSEC.
(Reprinted with permission of Valco Instruments Co.)

essentially operator-independent, and valves can be obtained in automated versions. One minor disadvantage is that the sample loop must be changed to obtain various sample volumes. (However, with the device shown in Figure 5.12, it is possible to inject variable sample sizes into a given loop, using a specially designed syringe.) A special advantage of valves is that they can be located within a controlled-temperature environment, such as an oven, for use with samples that require dissolution and injection at elevated or controlled temperatures (e.g., up to 150°C).

Low-volume, high-pressure switching valves are also available (e.g., Valco Instruments, Rheodyne, Siemans) for use in special techniques such as recycle (Sect. 11.3). These valves come in a variety of configurations and can be operated at pressures up to 7000 psi; some can be used at elevated temperatures but at lower pressures.

Automatic sampling devices are commercially available so that large numbers of samples may be routinely analyzed without need for operator intervention. These devices have not yet been widely used with HPSEC, but undoubtedly will develop popularity, particularly when coupled with data-handling systems. The device shown schematically in Figure 5.13 allows up to 64 samples contained in small vials to be pressurized consecutively into a sampling valve for injection. Automatic loop flushing is part of the

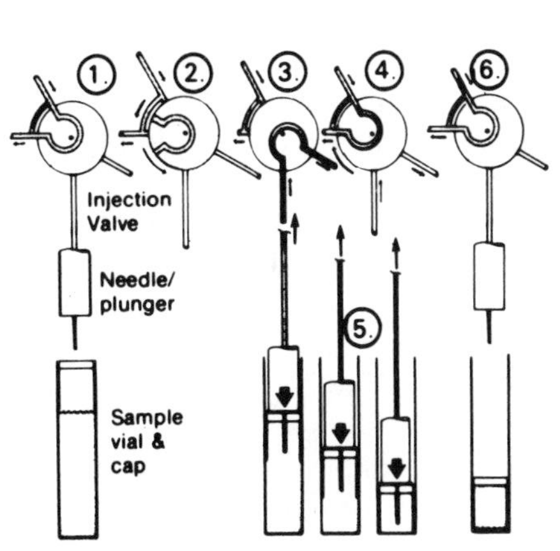

1. Vial is positioned beneath needle for injection.
2. Motorized injection valve rotates to "load" position as solvent diversion valve opens to allow continuing flow of mobile phase around sample loop.
3. With motorized injection valve in "load" position, hollow needle penetrates polyethylene vial cap and displaces sample into injector loop.
4. To make injections, motorized valve rotates to allow reverse displacement of sample on to column.
5. Valve action in 3 & 4 repeat for additional injections.
6. After injections completion, needle is withdrawn.

Figure 5.13 Positive displacement automatic sampler.
(Reprinted with permission of Micromeritics Instruments Corp.)

sampling cycle, and single, duplicate, or triplicate injection can be made with total control of the chromatographic cycle. Devices of this type are now available from several manufacturers (e.g., Waters Associates, DuPont, Micromeritics, Varian).

5.6 Miscellaneous Hardware

Line filters should be used between the pump and the sample injector to prevent particulates from clogging the column inlet. Most commercial instruments use stainless steel frits or filters of 2μm porosity. However, experience has shown that with columns of <10 μm particles, it is advantageous to use 0.5 μm porosity filters. The volume of these devices should not be large to facilitate solvent changeover.

Pressure monitors are required in the apparatus as diagnostic tools in separation, and to indicate system plugging or leaks. Diaphragm or Bourdon-type gauges are less expensive, simpler, and more robust than pressure transducers of the strain-gauge type. However, the latter are generally more precise and have a lower internal dead volume, which facilitates solvent changeover. In addition, they are available with high and low-pressure alarms or cutoff circuits to protect the chromatographic system against the effects of column plugging or solvent leaks.

Pulse dampers are required by certain pumping systems. (The effects of a pulsating mobile phase are discussed in Sect. 5.4.) Many modular and most integrated commercial instruments containing reciprocating pumping systems are equipped with pulse-damping devices. An effective damping system for homemade equipment is a combination of about 5 m of 0.25 mm i.d. capillary tubing and an associated diaphragm or Bourdon-type gauge. The capillary tubing acts as a flow restrictor and the gauge as a capacitor, so that the combination of these two components usually reduces pulsations of simple reciprocating pumps to manageable levels. Pulse-dampening devices increase the volume between the pump and the sample injector, and therefore decrease solvent changeover convenience. Detector output may be affected by mobile-phase pulsation, but these pulses have no effect on column efficiency.

All fittings and connectors between the sample injector and the detector should be designed to be cleanly swept and with a minimum dead volume. Extraneous volumes act as mixing chambers, which significantly contribute to extracolumn band broadening (Sect. 5.2). Comparison of ordinary compression fittings with "zero-dead-volume" fittings for use with columns (Fig. 6.21) shows that low-volume fittings must be used between the sample injector and detector to minimize extracolumn band broadening.

Thermostats are needed to control above-ambient column temperatures.

Use of circulating-air baths is convenient and generally preferred, since $\pm 1°C$ is easily maintained around the column. This usually results in a variation of no more than about $\pm 0.2°C$ in the temperature of the column packing. Column air baths in liquid chromatographs are very similar to those used in gas chromatographs and usually consist of high-velocity air blowers and electronically controlled thermostats. Alternatively, the SEC columns may be jacketed and the temperature controlled by circulating a fluid through the jacket system from a constant-temperature bath. This approach is generally less flexible but is practical for routine analyses.

As discussed in Section 7.2, mobile-phase flow rates must be precise in HPSEC, since retention times are often used to develop sample MW information. Since flow rate variations can result from even minor failure of the pumping system, it is important that techniques be available for carefully monitoring the flow rate during sample analysis. Flow rates either can be determined manually, or they may more conveniently be measured automatically with any one of the several devices described below.

Volumetric measurements of flow rate are most commonly used. The mobile phase simply is collected for a measured time in a calibrated vessel, usually a small volumetric flask. "Flow-tube" methods sometimes are used. Typically, an air bubble is introduced into the detector eluent stream, which passes through a transparent, volume-calibrated tube. The bubble is then timed while it travels between two volume markers on the tube. With this approach the flow rate can be measured quickly with a precision of about 1 %.

In GPC, mobile-phase flow rate is often measured automatically using a syphon-counter assembly such as that shown in Figure 5.14. To obtain the desired precision for HPGPC applications, the syphon volume must be relatively small (e.g., 1 ml), since the total peak volumes are also relatively small. If the column is operated at elevated temperature, the syphon must also be heated in order to [1] properly equate volumetric flows, [2] allow the syphon to empty properly, and, most important, [3] eliminate difficulties with sample precipitation. Each "dump" of the syphon actuates a photoelectric switch which indicates the event on the recorder trace. Note that the syphon in Figure 5.14 has a vapor-bypass tube to eliminate solvent evaporation during mobile-phase collection. Evaporation causes appreciable dump-volume errors with volatile solvents such as tetrahydrofuran (4).

An alternative approach to the syphon-counter method is to sum the volume of the total chromatogram and to calculate an average flow rate over the total analysis. In this manner the more precise, "integrated total volume" is used rather than individual dump volumes. The absolute volume of the syphon dump need not be known; actually arbitrary "dummy" values can be incorporated into the calibration procedure. The total integrated volume can be supplied directly to a computer for appropriate calculations of the MW data (Sect. 10.5).

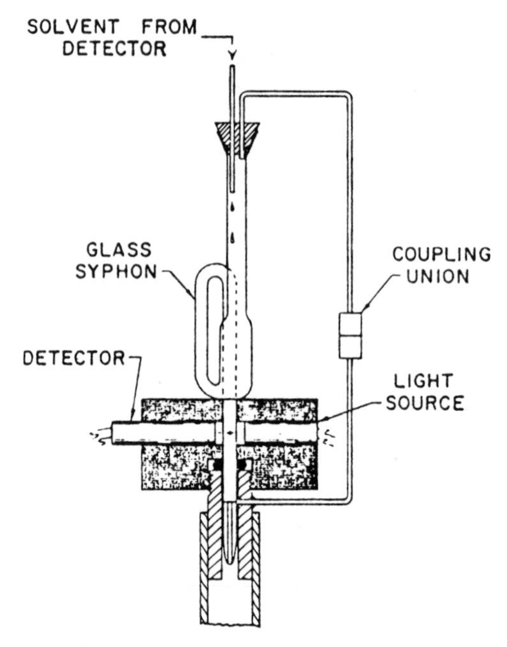

Figure 5.14 Vapor feedback syphon.
(Reprinted with permission from Ref. 4.)

Fraction collectors are normally not used in HPSEC, since the chromatogram requires only a few minutes to develop and manual collection is quite convenient. Built-in sample collection ports are available on the detector outlets of many commercial LC instruments. Fraction collectors may be convenient in the use of larger-diameter preparative SEC columns, because larger volumes of mobile phase are used and the separation is usually slower.

Special data-handling hardware is generally used to obtain precise molecular weight information on synthetic polymers in GPC. Detailed discussions of this topic are provided in Chapters 9 and 10. In all cases a high-speed recording potentiometer is a desirable part of the HPSEC apparatus. Preferred characteristics include pen response of 1 sec or less full scale, a high input impedance (e.g., >1 mΩ), good noise rejection, a floating input, and variable chart speeds (e.g., 10 cm/min to 10 cm/hr).

5.7 Integrated Microprocessor-Controlled Instruments

Most recent LC instruments include a microprocessor to control operating parameters and to serve as a data-handling system. Such equipment generally shows improved versatility, convenience and precision, but at increased cost. An advantage of microprocessor-controlled systems is that all separation parameters are continuously monitored, to ensure control at the level set by the operator. The microprocessor can be used to control pump flow output, composition of the mobile phase, sample injection, column oven temperature, detector output, and other functions, such as data processing and data printout.

Several forms of operation are used for modern integrated microprocessor-controlled SEC instruments. For example, some instruments (e.g., Spectra-Physics, Hewlett-Packard) are accessed by alphanumeric keyboard systems (Fig. 5.2), which are quite flexible but require some operator training. Other instruments (e.g., DuPont Model 850) use interactive operator-access systems, consisting of a series of functional touch switches—specifically controlling each of the operating parameters. Other instruments use keyboard access (e.g., Perkin–Elmer). The Varian Model 5000 Series uses a keyboard system plus an interactive cathode ray tube to assist in building separation programs, monitoring parameters, and diagnosing problems.

Integrated microprocessor-controlled equipment usually has additional safety features to protect continuously against overpressuring the column or the possibility of fire due to leakage of flammable solvents. Some microprocessor-controlled instruments also have the capability of communicating with other computers, permitting the use of very sophisticated data-reduction programs accessible only with more powerful external systems. Most integrated microprocessor-controlled instruments have the capability to use internally stored parameter sets to perform different chromatographic tests automatically. Method files can be held in microprocessor memory and recalled sequentially so that the user can mix different types of runs (e.g., different solvents, flow rates, temperature, etc.) in a single, programmed sequence. With automatic sampling, such instruments can be used for unattended overnight operation. Integrated microprocessor-controlled instruments are relatively expensive and are most suited for scouting or research studies. Less expensive and less complicated equipment is often suitable for more routine, less demanding operations.

5.8 Laboratory Safety

The general aspects of solvent handling in HPSEC are described in Section 8.2. Specifically, HPSEC instruments should be operated in well-ventilated areas, and although a hood is usually not required for instrument operation, it is recommended for preparing samples. Certain commercial LC instruments permit purging the reservoir and column compartment continuously with nitrogen to reduce the possibility of fire. Safety cutoff devices are desirable to protect from hazards that might result from solvent spills or leaks.

B. DETECTORS IN HPSEC

5.9 Introduction

As in other methods of liquid chromatography, the column eluent in SEC is continuously monitored by a sensitive detector. Two types of detectors are available: [1] bulk-property or general detectors measure a change in some overall physical property of the solvent or mobile phase due to the presence of solute, such as refractive index; and [2] solute-property or selective detectors, such as the ultraviolet (UV) photometer, are sensitive only to the sample components. Either of these detector types supplies an output signal that is related to the concentration or weight of solute in the column effluent. While no LC detector is "universal" in both applicability and sensitivity, currently available detectors are not usually a limiting factor in HPSEC applications.

Selection of a detector for HPSEC requires consideration of various performance criteria. Some detector specifications are supplied by the manufacturers, while at times others must be determined in the individual laboratory. Detector response can be defined in terms of the output signal that is provided in response to change in solute concentrations. The absolute detector sensitivity represents that total change in solute concentration which must take place for a full-scale deflection of the detector at maximum sensitivity, with a defined amount of noise. Relative detector sensitivity or sample detection limit is the minimum solute concentration that produces a peak measurably larger than (often taken as twice) the value of the noise. The relative detector sensitivity for a sample normally is dependent on the chromatographic system, since efficient columns produce sharp peaks that can be sensed at lower detection limits. The sensitivity of general detectors (e.g., refractive index) also varies with both the mobile phase and the solute structure.

Detector sensitivity or sample detectability is limited by the noise of the detector, arising from instrument electronics, temperature and line-voltage fluctuations, and from any other source, such as flow changes, pump pulsations, and similar effects that affect the detector output signal. Detector drift, which is a steady up-scale or down-scale movement of the detector baseline, is also important. Drift tends to obscure small peaks, causing errors in peak measurement. It should be noted that often many "apparent" detector noise and drift problems actually are caused by other aspects of the LC system (e.g., temperature variations, solvent impurities, dissolved gases), rather than being inherent detector limitations.

For the most utility (versatility and accuracy), detector output should be linear over a wide concentration range, preferably $>10^3$. This aspect is particularly important when both major and minor components must be determined in a single separation. It is also of particular significance in GPC that the detector responds equally to components of different molecular weight, for accurate calibration and subsequent characterization of polymers (see Section 7.4). Detector response also should be repeatable over extended time periods so that frequent calibrations are not required.

To minimize band broadening, the effective volume of the detector in HPSEC should be no greater than one tenth the volume of the peak(s) of interest (5). With columns of <10 μm particles, peaks of ≤ 100 μl are sometimes of interest; therefore, detectors with effective volumes of 10 μl or less are required for optimum results in this case. The volume of detector cell cavities should be minimum and contain no unswept "dead" corners.

Another source of apparent band spreading is detector response time. High-efficiency HPSEC columns can produce very sharp peaks for monomers, and detectors that have a response time greater than 1 sec can contribute significantly to band spreading. Increasing the electronic filtering can decrease detector noise and improve sample detectability, but increased filtering also increases the response time and may cause measurable broadening of sharp HPSEC peaks.

Some other general aspects of detectors should be mentioned when considering HPSEC applications. Normally, the detector cell outlet is at atmospheric pressure, but the cells should permit detector operation at moderate pressures (e.g., >50–100 psi) so that gas bubbles can be eliminated by imposing back pressure on the cell. To prevent diffusion of air into the mobile phase, fittings on the detector and the remainder of the flow system should be airtight. It is often not realized that air can be aspirated into small holes from which pressurized liquids cannot leak. Improperly connected compression fittings can allow air to leak in and saturate the previously degassed mobile phase. (Compression fittings should be tightened only in the manner recommended by the manufacturer. For instance, $>\frac{1}{16}$-in. Swaglok fittings should be tightened only $1\frac{1}{4}$ turns beyond the hand-tightened state; $\frac{1}{16}$-in. fittings only $\frac{3}{4}$ turn.)

Ease of operation and serviceability also are important in routine applications, especially when the detector must be operated by relatively inexperienced personnel. The usefulness of a particular detector may be limited by its susceptibility to mobile-phase flow rate or temperature changes.

Temperature control can be achieved by immersion of the detector in a liquid bath, by circulating fluid from a thermostatted bath, or by placing the detector in a well-insulated (adiabatic) environment.

Flow-sensitive detectors, or detectors sensitive to temperature change (e.g., refractive index), are particularly susceptible to pulsating pump flow. Apparent changes of flow rate are often actually caused by temperature changes. Since changes in response often result from temperature changes, certain detectors (e.g., refractive index) must contain devices for equilibrating the cell temperature to that of incoming mobile phase. Maximum detector stability is obtained by temperature preequilibration of the mobile phase by passing it through a loop of capillary tubing just prior to (and in close contact with) the cell. Of course, this added tubing increases detector dead volume and can contribute to extracolumn band broadening.

In the following sections we discuss those detectors most useful for HPSEC, including their operation, applicability, and limitations.

5.10 Differential Refractometer

Probably the most widely used detector in HPSEC is the differential refractometer (RI detector), and this often is the detector of choice for GPC. The RI detector continuously measures the difference in refractive index between the mobile phase and the mobile phase containing the sample. Being a bulk-property and general detector, the RI device responds whenever the solute differs in refractive index from the mobile phase by typically >0.05 RI units.

Three different types of RI detectors are available commercially. The operation of the deflection-type RI detector (e.g., Waters, Micromeritics, DuPont) is shown schematically in Figure 5.15. This detector employs a beam of light generated from source A, which is limited by mask B, collimated by lens C, then passed through the sample and reference cells D, which are in series. As the incident light first passes through the prism-cell assembly, it is deflected, reflected from mirror E, and again deflected as it passes back through the two cells. When the composition of the mobile phase changes in the sample cell, the new, slightly different refractive index of the solution causes a deflection in the location of the light beam on the surface of the position-sensitive photodetector. The light beam falling on the photodetector produces an electrical signal proportional to the position of the light and the output signal then is amplified and relayed to a recorder. Deflection refractometers have the advantage of a wide range of linearity, and only one prism is needed for the entire RI range. Sample cells for detectors of this type can be made relatively small (10 μl), but heat exchangers leading

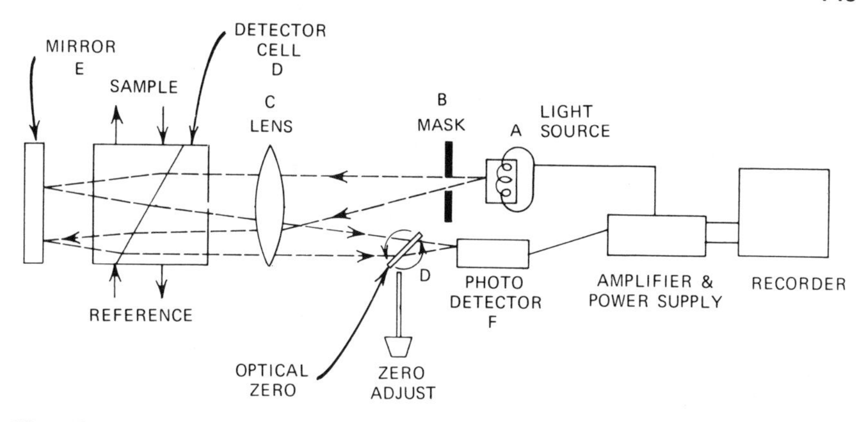

Figure 5.15 Deflection-type differential refractometer detector. (Reprinted with permission of Waters Associates.)

to the sample cell have volumes of 25–70 μl, depending on the manufacturer. Deflection RI detectors appear to be relatively insensitive to the buildup of contaminants on the sample cell windows.

The differential refractometer schematically shown in Figure 5.16 (e.g., LDC) is based on Fresnel's law of reflection, which states that the amount of light reflected at a glass/liquid interface is dependent on the angle of the incident light and the refractive index difference between the two phases. This detector normally operates with the light striking at the glass/liquid interface near the critical angle where the Fresnel relationship becomes linear. In Figure 5.16 light from source SL is vignetted by source mask M1, transmitted through an infrared blocking filter, F, and passed through aperture mask M2. After being collimated by lens L1, the two beams enter the glass prism and are focused on the glass/liquid interfaces in sample and reference cell cavities. (These are formed with perfluoropolymeric gaskets which are clamped between the cell prism and a base plate of stainless steel.) The intensity of the reflected light is a function of the refractive index of the liquid in both cells. Emerging light beams then are focused by lens L2 onto the surface of the dual-element photodetector D for amplification and recording. All the optical components in the projector are mounted on a separate optical bench which can be rotated to allow a change in the incident angle of light.

Fluctuations caused by temperature and flow changes are greatly minimized in both RI detector designs by using a differential measurement, wherein the refractive indices of the reference and sample streams are compared continuously. Since the Fresnel effect occurs at the liquid/glass interface, the cells must remain very clean for proper detector response. Fresnel

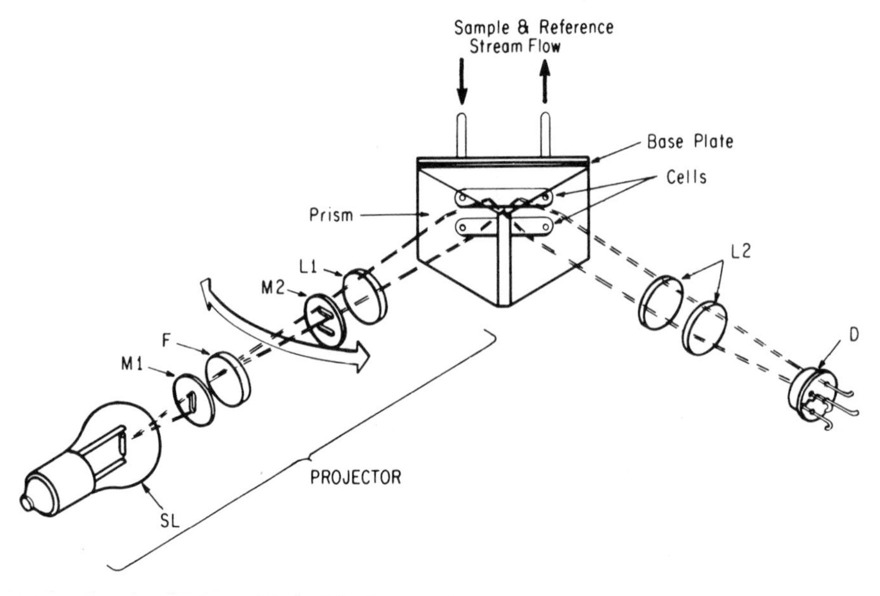

Figure 5.16 Fresnel-type differential refractometer detector.
(Reprinted with permission of DuPont Instrument Products Division.)

RI detector cells can be made very small (e.g., 3 μl), so some workers prefer to use this device with very high performance SEC columns. This refractometer has a restricted range of linearity, and two different prisms must be used to cover the useful RI range. Fortunately, the cell with the lower RI range is useful for most operations.

A new type of refractometer (Optilab) uses the shearing interferometer principle. The difference between the refractive indices of the sample and reference streams produces a difference in optical path length which is measured by the interferometer in fractions of light wavelengths. Sensitivity is claimed by the manufacturer to be an order of magnitude greater than other RI detectors, and under favorable conditions it is possible to detect about 0.1 μg of a monomer.

Differential RI detectors can be made to respond to all solutes by proper selection of mobile phase. For maximum sensitivity the mobile phase and solute should differ in RI as much as possible; however, solubility requirements in GPC often restrict solvent selection. For a series of different solutes and mobile phases, the RI detector may produce positive or negative peaks, depending on the various refractive index differences.

A major weakness of the RI detector is that of modest sensitivity, making it unsuitable for measuring very low solute concentrations. Another limiting feature is the acute sensitivity to temperature change, because total compensation is not obtained even in the differential mode. In practice, it is difficult to control the temperature of the refractometer cell sufficiently to obtain maximum sensitivity. On the other hand, detection is nondestructive, relatively convenient, and reliable. Particular care must be taken when changing mobile phases to ensure complete removal of the previous solvent. Maximum baseline stability is normally obtained by purging the reference cell with mobile phase and closing off this solvent in the cell with compression fittings or a valve. Reference cells should be flushed periodically with the mobile phase to maintain baseline stability. Degassing of the mobile phase may be required for minimum baseline noise, and this appears to be particularly important with aqueous or partially aqueous solvents.

A very wide range of organic and aqueous mobile phases can be used with RI detectors. Table 8.3 lists the refractive indices for some of the common solvents used in HPSEC.

In summary, typical properties of modern RI detectors for HPSEC include: cell volume 5–10 μl (30–70 μl total with heat exchanger); noise level, 10^{-7} refractive-index unit (RIU); linear range, 10^4; drift, $<10^{-6}$ RIU/hr; sensitivity to favorable sample, 5×10^{-7} g/ml; range, 10^{-7}–10^{-3} RIU.

5.11 Ultraviolet Photometers and Spectrophotometers

While detectors based on ultraviolet absorption are widely used in modern LC and GFC, they have more restricted application in GPC. Often a UV-transmitting solvent cannot be found which meets the solubility requirements of polymers. Obviously, solutes must absorb in the ultraviolet to be detected (e.g., UV detection of polyethylene dissolved in 1,2,4-trichlorobenzene would be impossible).

Very sensitive photometers operating over a wide wavelength range now provide surprising utility in terms of both applicability and sensitivity. These devices require only a single double bond or similar chromophore per molecule for solute detection, and in favorable cases a few nanograms of a solute can be sensed.

The high versatility of UV and visible light-absorption detectors is due largely to the availability of many solvents of widely varying solvent power which have high transmittance at the wavelengths required for detecting many solutes. Table 8.3 lists the UV cutoff wavelengths for a number of these useful solvents.

Monochromatic photometric detectors are more often used in GFC than in GPC. Many of the single-wavelength commercial UV detectors exploit the high-energy 254 nm emission from low-pressure mercury lamps, which permits excellent detectability with low-volume cells. Many biologically important compounds (e.g., proteins, enzymes, aromatic amino acids, nucleic acids) absorb strongly at 254 nm, but because of high inherent sensitivity, the modern photometric detector is useful even when the absorption maximum is not at 254 nm.

UV filter photometers operating at wavelengths other than 254 nm are also available. In the example shown in Figure 5.17, light from a highly regulated source passes through sample and reference cells onto silicon photodiodes. The output of these sensors, amplified with differential log converters, yields a signal that is linear with solute concentration. With a medium-pressure mercury vapor lamp, wavelengths of 254, 280, 312, 316, 436, and 546 nm can be selected simply by changing a filter. However, signal-to-noise specifications are still somewhat inferior to those of photometers operating with a low-pressure mercury source at 254 nm, because of the higher energy associated with this source. Relative detector response at two different wavelengths (e.g., 254 and 280 nm) is sometimes used to confirm the identity of a suspected solute.

Photometric detector cells typically have optical paths with dimensions of about 10 mm by 1 mm i.d. and an internal volume of $<10\ \mu l$. Figure 5.18 pictures a tapered cell construction which apparently reduces the undesired refractive-index effects in the conventional *Z*-type cell shown above, leading to improved stability and lower noise for the photometric detector.

Variable-wavelength spectrophotometric detectors are also very useful for HPSEC. Specially designed equipment, now available for use over a very wide wavelength range (e.g., 190–700 nm), employs a monochromator and continuous-spectrum energy source for maximum versatility and convenience. Spectrophotometric detectors may be set to the absorption maximum of a solute for maximum detection sensitivity, or alternatively at a wavelength that provides maximum freedom from possible interferences. An example of the latter in Figure 5.19 shows that detection of a yellow impurity in a polymer extract (separated on an HPSEC column set optimized for separating small molecules) was enhanced by working at 445 nm as compared to 254 nm.

Modern spectrophotometric detectors for chromatography use high-energy sources, relatively low optical resolution (i.e., wide band-pass monochromaters), and stable low-noise electronics to produce signal/noise and linearity responses which are only slightly inferior to those of UV photometric detectors (but at added cost). A schematic drawing of a double-beam spectrophotometric detector is shown in Figure 5.20. Light from a source lamp (deuterium for ultraviolet, tungsten for visible) is focused and reflected through at 250 Hz chopper onto a single-grating monochromater. The selected wavelength passes through a partially transmitting mirror, where

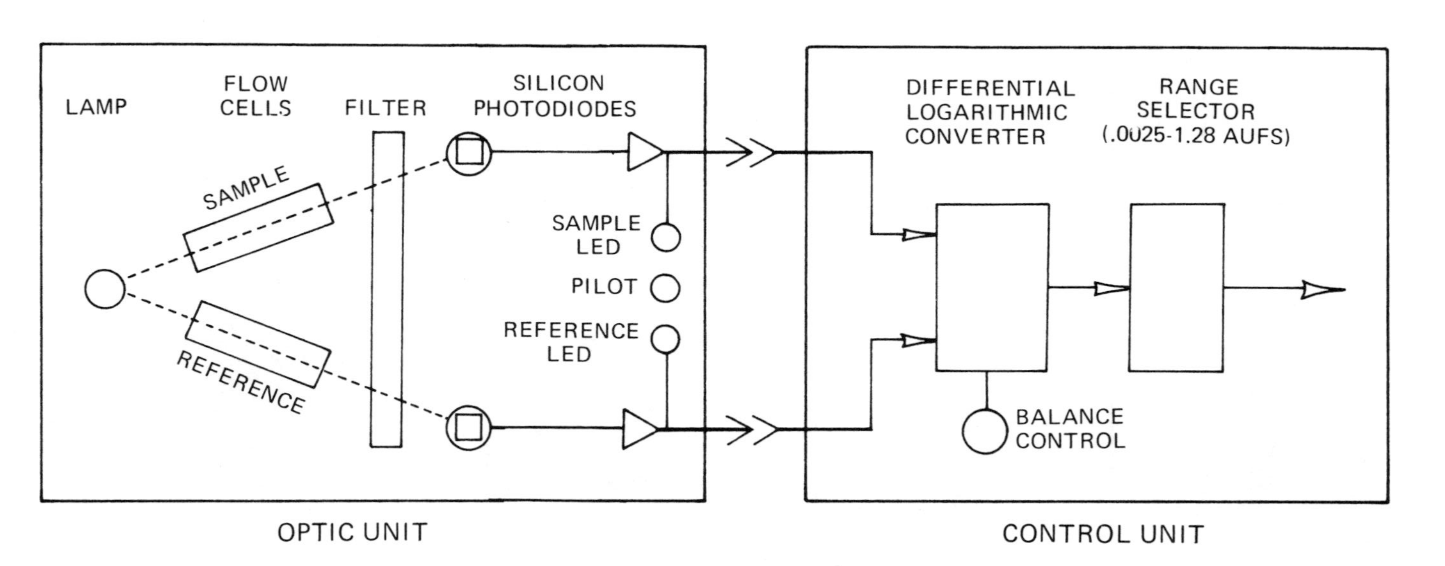

Figure 5.17 Multiwavelength photometric detector.
(Reprinted with permission of Spectra-Physics, Inc.)

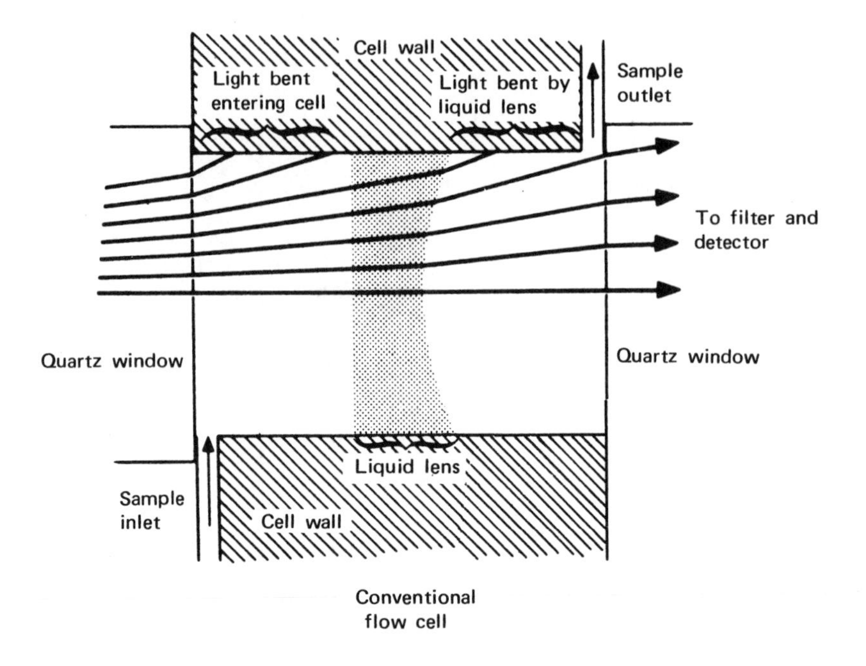

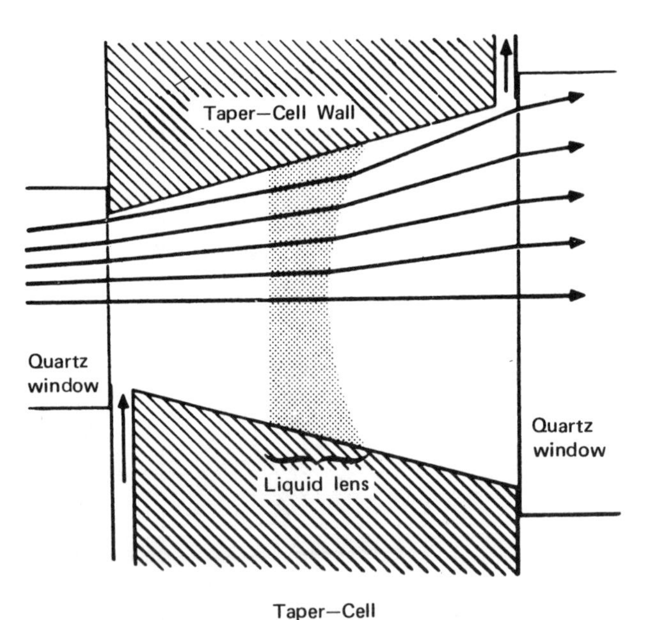

Figure 5.18 Conventional and tapered cells for UV detector.
(Reprinted by permission of Waters Associates.)

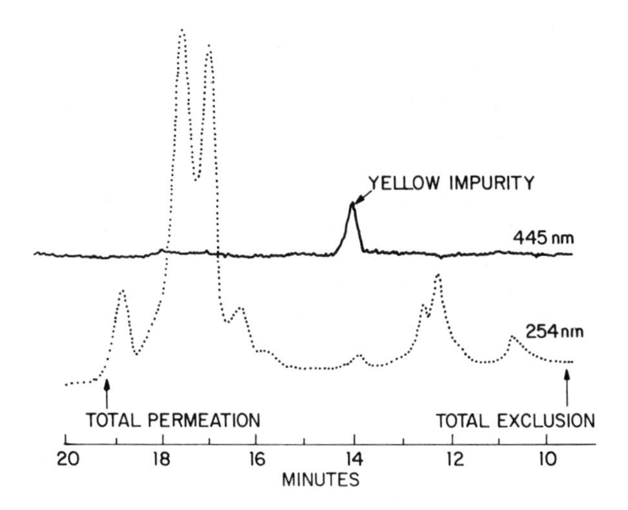

Figure 5.19 Selective detection with spectrophotometer.

Columns, four 25 × 0.62 cm PSM-50S (75 Å, silanized); mobile phase, methanol, 1.25 ml/min at 27°C; detectors: UV photometer, 0.02 AUFS at 254 nm; UV spectrophotometer, 0.02 AUFS at 445 nm; sample, 50 μl methanol extract. (Reprinted with permission from Ref. 6.)

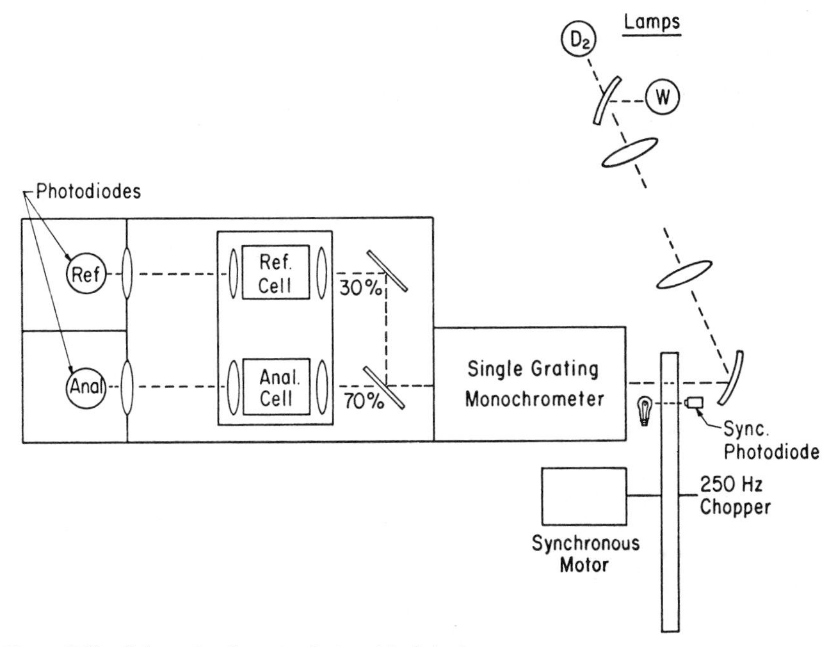

Figure 5.20 Schematic of spectrophotometric detector.

(Reprinted with permission of DuPont Instrument Products Division.)

30 % of its energy is focused onto the reference photodiode. The main energy fraction (70 %) passes through the analytical cell for measurement with the analytical photodiode. Response from this device is linear with concentration as a result of differential logarithmic conversion in the electronics.

UV detectors of all types have high sensitivity (but samples must have some UV absorptivity) and good linearity and can be made with very small cell volumes. They are relatively insensitive to mobile-phase flow rate and temperature changes, and they are very reliable, easy to operate, and attractive for routine operation with relatively inexperienced personnel. UV detectors for HPSEC typically exhibit the following properties: cell volume, 10–20 μl; noise, <0.0001 absorbance unit full-scale (AUFS); linear range, 10^4–10^5; drift, <0.0005 AU/hr; sensitivity to favorable sample, 5×10^{-10} g/ml; absorbance range, 0.005–2.54 for full-scale deflection.

5.12 Other Detectors

Low-angle laser light-scattering (LALLS) photometry has been used to monitor GPC effluents. In conjunction with a concentration detector (e.g., differential refractometer) this device can be used for directly determining the molecular weight distributions of polymers. Figure 5.21 compares the response of refractive index and light-scattering detectors for polystyrene standards. Note that the LALLS detector is sensitive to both sample MW and concentration, while the RI detector senses concentration only. Use of both detectors provides data that can be used to make absolute MW calibrations (7) (see also Sect. 1.4).

Figure 5.22 is a schematic drawing of a low-angle laser light-scattering photometer which can be used in HPSEC. A small helium-neon laser

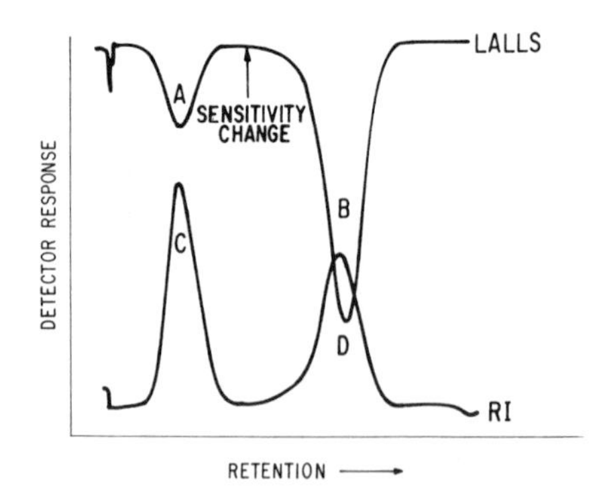

Figure 5.21 Comparison of refractive index and low-angle laser light scattering detectors. Five 4 ft Styragel columns 2–10^6 Å, 10^5 Å, 10^4 Å, and 10^3 Å; mobile phase, chloroform at 1.0 ml/min; room temperature; peaks A and C, 1.8×10^6 MW polystyrene; peaks B and D, 5.1×10^3 MW polystyrene; LALLS, low-angle laser light-scattering, RI, refractive index. (Reprinted with permission from Ref. 7.)

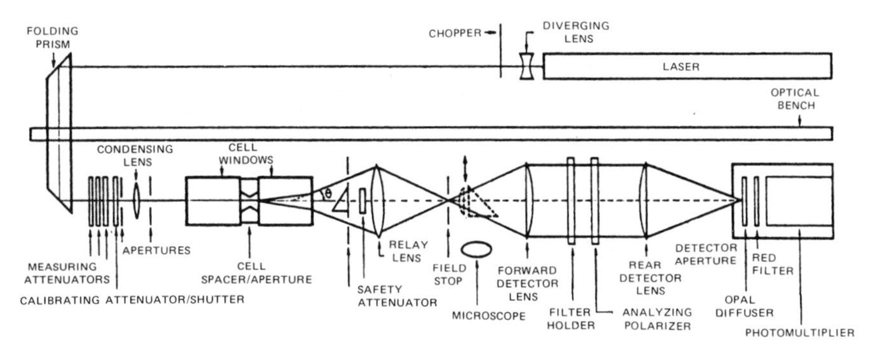

Figure 5.22 Low-angle laser light-scattering photometer.
(Reprinted with permission of Chromatix, Inc.)

operating at 633 nm produces a beam passing through a series of neutral-density attenuators. The incoming light is focused by a condensing lens onto a 100 μm diameter target in the sample cell. The cell is comprised of thick fused-silica windows separated by a perfluoroelastomeric spacer with a hole in the center. The light scattered by the sample through angle θ is collected by a relay lens and is focused onto a field stop. The cell angle through which the scattered light is collected is defined by an annular ring where the included scatter angle is defined by the inner and outer edges of the annulus. Light passing through the field stop is detected by a photomultiplier, or it may be visually inspected with the aid of a microscope. Sample volumes as small as 5 μl may be used with this instrument. Because the detected scattered light arises from an extremely small scattering sample volume within the cell, the problem of scattering from gas bubbles and other contaminants is minimized.

Sample degradation, fluorescence, and absorption difficulties are minimized by using the red 633 nm wavelength of the helium-neon laser. The minimum detectable concentration of a given solute and its molecular weight are inter-related, as illustrated in Figure 5.23. The compounds listed in Figure 5.23 illustrate the range of biologically significant polymers whose average molecular weight is conveniently measured by the light-scattering technique.

Infrared absorption can be used as a selective or general method of detection in GPC, but it usually does not have the characteristics suitable for GFC. Figure 5.24 shows a schematic drawing of an infrared (IR) detector suitable for some GPC applications. Energy from a glowing coil source A passes through rotating chopper B and wheel C which contains three gradient interference filters (2.5–4.5, 4.5–8, and 8–14.5 μm). Any wavelength

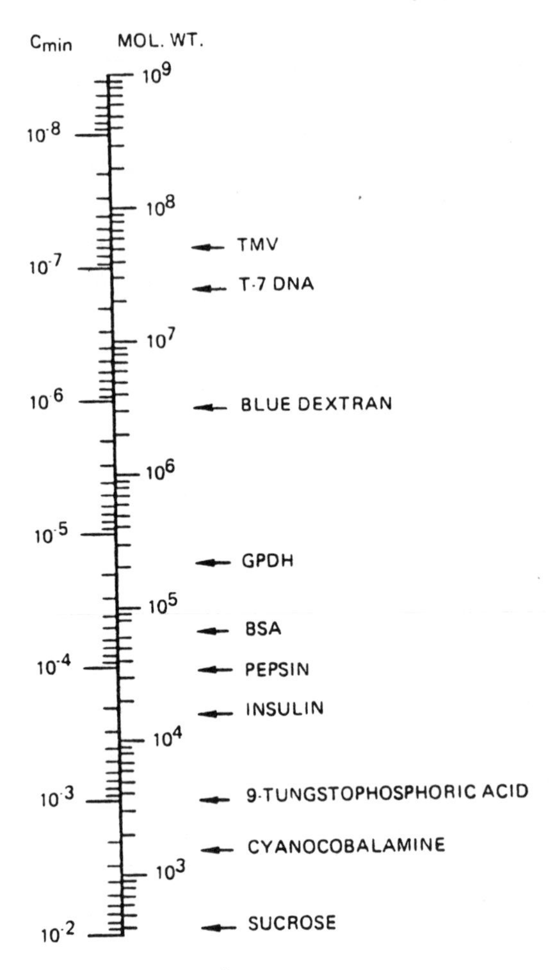

Figure 5.23 Light-scattering nomograph.

Shows concentration of solute in water at which excess of solute Rayleigh ratio (R_θ) equals Rayleigh ratio for pure water as a function of solute MW. (Reprinted with permission of Chromatix, Inc.)

in the 2.5–14.5 μm range may be selected for detection by the proper positioning of the filter wheel. The selected infrared energy passes through slit D and then through the detector cell E, which normally has a 0.2 or 1 mm path length. (This detector cell can be heated for high-temperature operation if desired). A thermoelectric detector F senses the change in energy resulting from sample absorption and the output is amplified for use with a recording potentiometer.

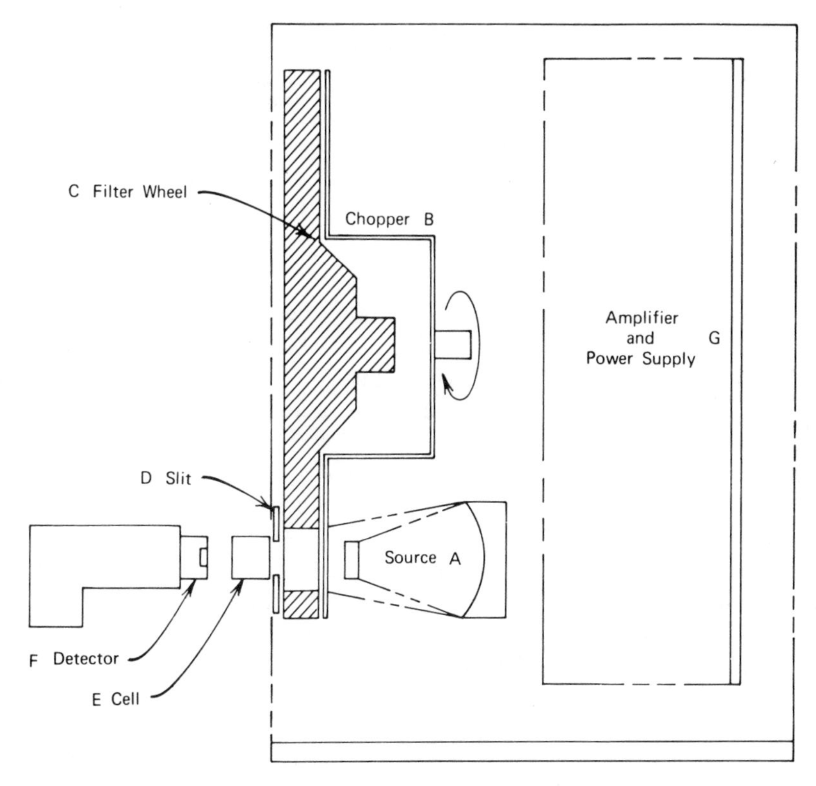

Figure 5.24 Infrared-absorption detector.
(Reprinted with permission from Wilks Instrument Co.)

The sensitivity of the infrared photometric detector generally is comparable to that of the RI detector. Satisfactory sensitivity for a typical organic solute can usually be obtained by monitoring the C—H stretching frequency, if the solvent permits, while carbonyl and other strongly absorbing functional groups often provide somewhat higher sensitivity. Infrared detection is limited to mobile phases which are transparent at the absorption wavelengths to be monitored, as illustrated in Figure 8.1a. This detector can be utilized readily in high-temperature applications (e.g., the 135°C GPC analysis of synthetic polymers such as polyolefins, illustrated in Figures 12.7 and 12.8).

Reaction detectors, useful both as general and selective devices, utilize color-forming reactions between an external reagent and the eluting solute. Monitoring usually is by absorption photometry. An example is the reaction

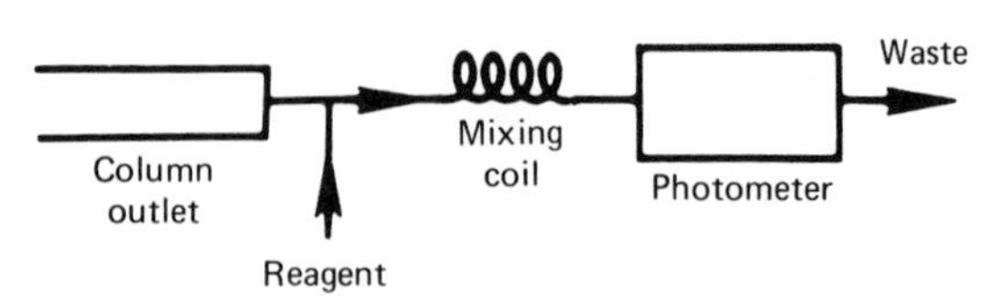

Figure 5.25 Simple postcolumn reaction detector.

of (colorless) amino acids with ninhydrin in a heated reaction coil to form colored products which are sensed with a visible-wavelength photometer. Reaction detectors have special applicability in GFC but have been used in GPC only for detection of end groups. There are large numbers of color-forming reactions available for sensitive, general, or selective detection (see Ref. 9 for over 200 color-forming reactions used in TLC). Actually, any rapid chemical reaction can be used in conjunction with colorimetric, spectrophotometric, fluorescence, or electrometric detection. Reaction with solutes normally is carried out in the column effluent; however, derivatives also can be made prior to chromatographic separation to achieve a similar effect (see Ref. 10 for a detailed discussion).

A simple arrangement of a postcolumn reaction detector is shown in Figure 5.25. This apparatus is applicable for reactions that go essentially to completion at room temperature in a few seconds (e.g., reactions with fluorescamine, Fluoram). For reactions requiring higher temperature and extended reaction times (e.g., ninhydrin, 100°C, 15–20 min) more elaborate arrangements must be used, with increased reaction coil lengths and increased temperature. Because of the larger dead-volume of the reaction detector, resolution can be degraded by extracolumn band broadening; hence careful engineering is required. To reduce band broadening from the detector, the column effluent may be segmented with nitrogen bubbles (11).

The general requirements of a reaction detector are that the reagent and mobile phase must not react in the absence of sample, and the mobile phase must not interfere with the reaction of the sample. Molar absorptivities of 100–1000 for the product are needed for more sensitive solute detection than would be obtained by RI detectors. Degradation of resolution is the most difficult problem to overcome; therefore, the reaction must be chosen carefully and optimized.

Fluorimeters and spectrofluorimeters are quite useful in GFC but are rarely used in GPC applications. In this approach the fluorescent energy emitted from certain UV light-activated solutes is utilized for very sensitive and selective detection. Concentrations less than 10^{-11} g/ml can be measured in this way, and thus picograms of solutes can be detected under optimum

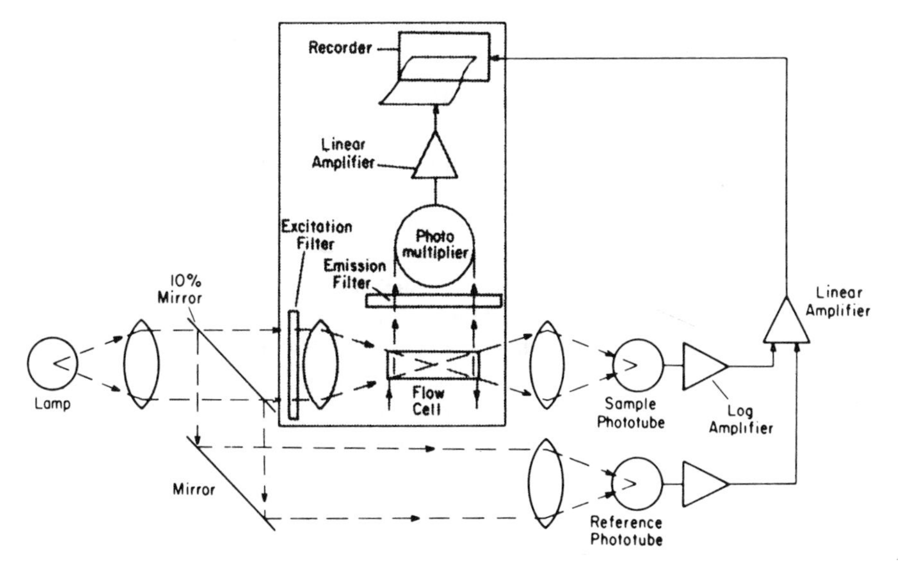

Figure 5.26 Schematic of fluorescence-multiwavelength UV detector.
(Reprinted with permission of DuPont Instrument Products Division.)

chromatographic conditions. These devices are particularly useful for many biologically important compounds since they can be made to fluoresce under proper conditions.

Commercial filter fluorimeters normally utilize a right-angle optical bench configuration. One right-angle fluorescence/multiwavelength UV detector also permits the simultaneous measurement of both solute absorption and fluorescence, as shown in Figure 5.26. In this device, 90 % of the energy from the source is focused with a lens and passes through a wavelength-selective filter which determines the wavelength band for both fluorescence exitation and absorption. This band then passes through the flow cell and is focused on the phototube for the sample, where the intensity for the absorbance measurement is compared with 10 % of the original incident energy. Fluorescence from certain solutes is also collected at a right angle to the special 20 μl cell. It passes through an emission filter to select the desired wavelength and is then detected by a photomultiplier tube.

Spectrofluorimeters using monochromaters to select the desired wavelengths of excitation and emitted energy are especially important in situations where maximum flexibility, sensitivity, and/or selectivity are required, but this advantage is gained only at considerably higher cost than the fixed band-pass detectors (fluorimeters).

Fluorescent derivatives of many nonfluorescing substances can be prepared (10), and this approach is particularly attractive for the selective detection of various classes of compounds for which other sensitive methods are lacking. Undesirable background fluorescence and quenching effects are not a common problem with fluorimetric detectors in high-efficiency systems.

Viscometer detectors have been used to monitor column effluents in gel permeation chromatography. When the viscometer is coupled to a concentration detector (e.g., differential refractometer), two plots are obtained which provide information for determining the MW distribution of the sample. Suitable viscometer detectors for HPSEC are not now available commercially, but designs have been reported in the literature which should be applicable. Figure 5.27 is a schematic of a complete viscometer system having a transducer with $< 10\ \mu l$ cell volume. This viscometer is based on the measurement of the pressure drop through a capillary. The effluent stream

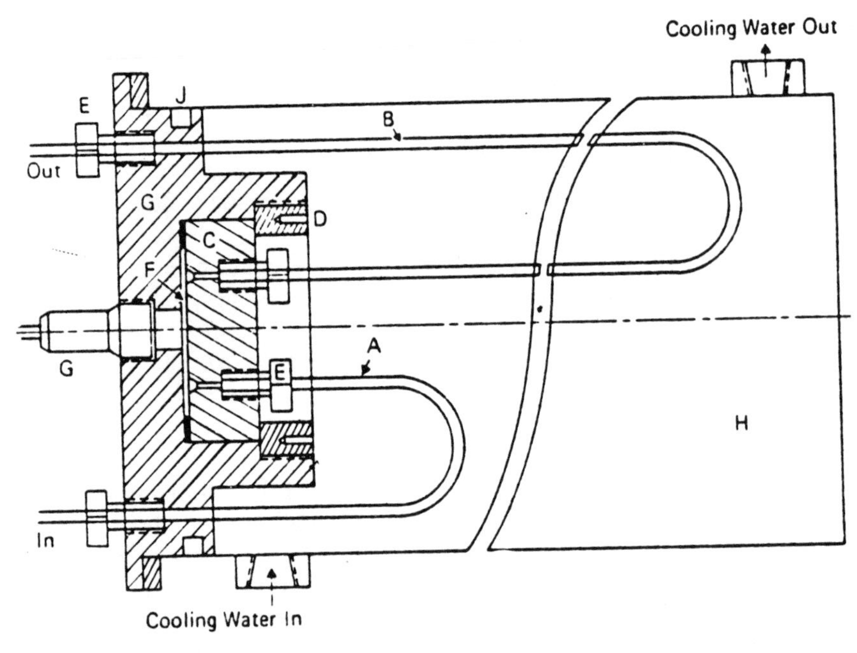

Figure 5.27 Schematic diagram of viscometer detector.

A, Stainless steel inlet connection to the viscometer; B, capillary viscometer; C, pressure cell upper platen; D, upper platen retaining ring; E, Swagelok fitting; F, pressure cell chamber; G, pressure transducer; H, thermostat; J, thermostat O-ring seal. (Reprinted with permission from Ref. 13.)

enters the pressure cell at a 90° angle through a spacer which determines the cell volume, flows axially across a pressure transducer membrane, and exits at a 90° angle through the capillary. For Poiseuille-type flow (i.e., laminar flow through capillaries), the pressure-drop is directly proportional to the viscosity of the fluid.

$$\Delta P = k\eta \tag{5.2}$$

where $k = (8/\pi)F(1/r^4)$. The instrumental constant k depends on the flow rate (F), the capillary length (1), and the radius (r). Consequently, at constant flow rate and for a particular capillary, the ratio of the pressure drop of the sample stream to that of the pure mobile phase ($\Delta P_1/\Delta P_0$) is equal to the ratio of the viscosity of the sample stream to the viscosity of the mobile phase, (η_1/η_0). Since the concentration of the sample solution can be obtained from the chromatogram with the concentration detector (e.g., refractometer), the intrinsic viscosity $[\eta]$ can be calculated from

$$\left[\ln\left(\frac{\Delta P_1}{\Delta P_0}\right)\bigg/ c\right]_{c\to 0} \tag{5.3}$$

where ΔP_0 and ΔP_1 are the pressure drops due to the pure mobile phase and the sample solution, respectively.

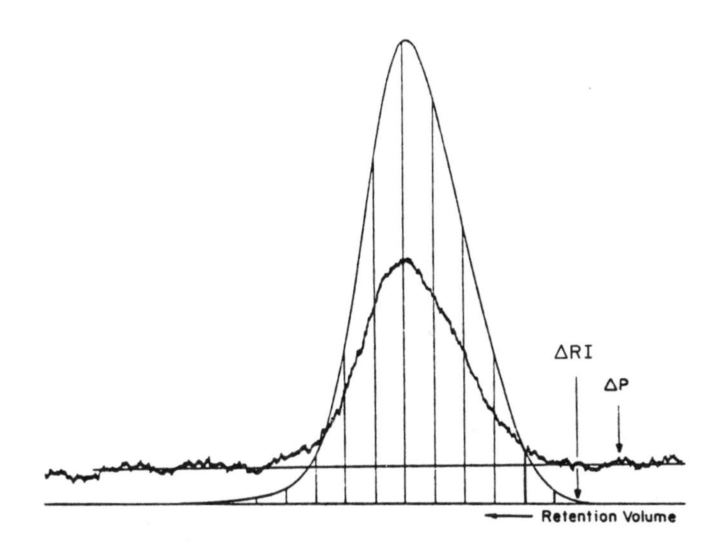

Figure 5.28 Chromatogram with viscometer and refractive-index detectors.
ΔRI, refractive-index detector; ΔP, viscometer detector. (Reprinted with permission from Ref. 13.)

Viscometer detectors require very precise flow rate and temperature control to obtain the baseline stability required for satisfactory operation. The sensitivity of the viscometer appears to be adequate for many GPC analyses, but detection of macromolecules of biological origin in aqueous systems has not been explored. Figure 5.28 shows a chromatogram of a narrow MWD sample with both refractive index (concentration) and viscometer (ΔP) detectors used in data reduction to compute values of $\overline{M}_n$ and $\overline{M}_w$. Data from both detectors can be used to prepare universal SEC calibrations (13).

REFERENCES

1. H. M. McNair and C. D. Chandler, *J. Chromatogr. Sci.*, **14**, 477 (1976).
1a. L. R. Snyder and J. J. Kirkland, *Introduction to Modern Liquid Chromatography*, 2nd ed., Wiley-Interscience, New York, 1979.
2. N. Hadden et al., *Basic Liquid Chromatography*, Varian Aerograph, Walnut Creek, Calif., 1971.
3. J. N. Done and J. H. Knox, *Applications of High-Speed Liquid Chromatography*, Wiley, New York, 1974.
4. W. W. Yau, H. L. Suchan, and C. P. Malone, *J. Polym. Sci., Part A-2*, **6**, 1349 (1968).
5. J. J. Kirkland, W. W. Yau, H. J. Stoklosa, and C. H. Dilks, Jr., *J. Chromatogr. Sci.*, **15**, 303 (1977).
6. J. J. Kirkland and P. E. Antle, *J. Chromatogr. Sci.*, **15**, 137 (1977).
7. A. C. Ouano and W. Kaye, *J. Polym. Sci., Part A-1*, **12**, 1151 (1974).
8. W. Kaye, *Anal. Chem.*, **45**, 221A (1973).
9. E. Stahl, ed., *Thin-Layer Chromatography. A Laboratory Handbook*, Springer-Verlag, New York, 1969.
10. J. F. Lawrence and R. W. Frei, *Chemical Derivatization in Liquid Chromatography*, American Elsevier, New York, 1976.
11. L. R. Snyder, *J. Chromatogr.*, **125**, 287 (1976).
12. R. P. W. Scott and J. G. Lawrence, *J. Chromatogr. Sci.*, **8**, 65 (1970).
13. A. C. Ouano, *J. Polym. Sci., Part A-1*, **10**, 2169 (1972).

Six

THE COLUMN

6.1 Introduction

As discussed in Section 3.3, better column efficiencies and separations are obtained with small particle packings and solutes having high diffusion rates. Plate height is essentially dependent on the square of the particle diameter (d_p^2) but is a linear reciprocal function of the solute diffusion coefficient (D_m). Thus the effect of particle size is most important for macromolecules that diffuse slowly, and the use of columns with very small, totally porous particles is particularly favored in SEC.

In Section 3.1 we have described the band broadening that is inherent in the HPSEC method. What can be done to prepare columns to minimize these band-broadening effects? Eddy diffusion can be reduced by preparing

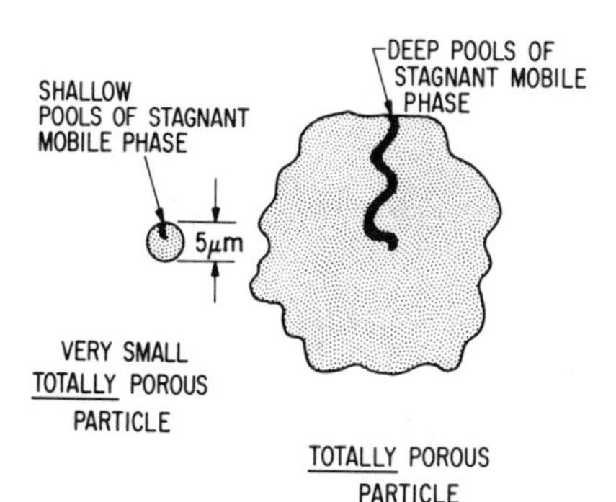

Figure 6.1 Stagnant mobile phase in large and small porous particles.

homogeneously packed beds. This generally is accomplished more readily with spherical particles rather than with irregular particles. Both mobile-phase and stagnant-mobile-phase mass transfer are significantly improved by using very small particles. As suggested in Figure 6.1, movement of solute molecules in and out of stagnant mobile phases is much faster in very small particles than in larger totally porous particles with deeper pools of stagnant mobile phase.

6.2 Column Packings

A variety of porous packing materials is available for HPSEC. Semirigid organic gels and rigid solids are available in spherical or irregular shape, and from these materials must be chosen the packing best suited for the particular application. Optimum performance of an SEC packing material involves high resolution and low column back pressure, combined with good mechanical, chemical, and thermal stability. A combination of these desirable properties allows a column to be used at high resolution with different solvents at relatively high flow rates and elevated temperatures.

Until recently, most SEC analyses of synthetic organic polymers have been made using cross-linked semirigid polystyrene gel packings. Now, however, small rigid inorganic particles (e.g., silica) are available which have several significant experimental advantages over the organic gels. Rigid particles are relatively easily packed into homogeneous columns which are mechanically stable for long times. A much wider range of mobile phases can be used, providing greater versatility and increased convenience in application. The rigid packings equilibrate rapidly with new solvents, so that solvent changeover is rapid. Columns with rigid packings are stable with the high-temperature solvents required for characterizing some synthetic macromolecules, while organic gels of $<10\ \mu m$ particles are often not usable under these conditions. Rigid particles can also be used in aqueous systems for separating high-molecular-weight, water-soluble solutes by gel-filtration chromatography.

A potential disadvantage of the rigid inorganic particles is adsorption or degradation of solutes (e.g., denaturing of proteins). However, siliceous particles can often be easily modified with certain organic functional groups to effectively eliminate these difficulties for most applications (1, 4, 5) (Sect. 6.2).

Since soft particles (e.g., agarose in gel filtration chromatography) collapse at high inlet pressures, they are not utilized in HPSEC. While soft gel packings traditionally have been used for separating high-molecular-weight water-soluble substances, such GFC separations are now being carried out at high pressures using columns of deactivated rigid particles (Sect. 13.3).

There is a variety of commercially available packings for HPSEC, and others are currently under development to improve the versatility, resolution, or speed of the SEC process. Many of these special packings are made in small lots and are relatively expensive. However, this expense is more than compensated by high performance. The many analyses that can be carried out with a single packed column generally make the cost per analysis relatively insignificant. Figure 6.2 shows scanning electron micrographs of some of the packings that have been specially developed for HPSEC.

As discussed in Section 1.1, we have arbitrarily defined HPSEC as a system with about a tenfold greater number of theoretical plates per second than that of conventional SEC systems. In keeping with this, we here further define column packings for HPSEC as those having particle sizes less than about 30 μm that can be used at relatively fast flow rates and high input pressures to accomplish superior separations in a short time. In this section we describe some of the commercially available packings that fit these specifications.

Semirigid Organic Gels

Columns packed with spherical semirigid, highly cross-linked, styrene/divinylbenzene copolymers of about 10 μm diameter are now available in a variety of pore sizes for $\sim$15 min analyses. Table 6.1 lists the characteristics of columns of these small styrene/divinylbenzene gels; bulk packings are not available for some forms. Columns of these packing materials are available

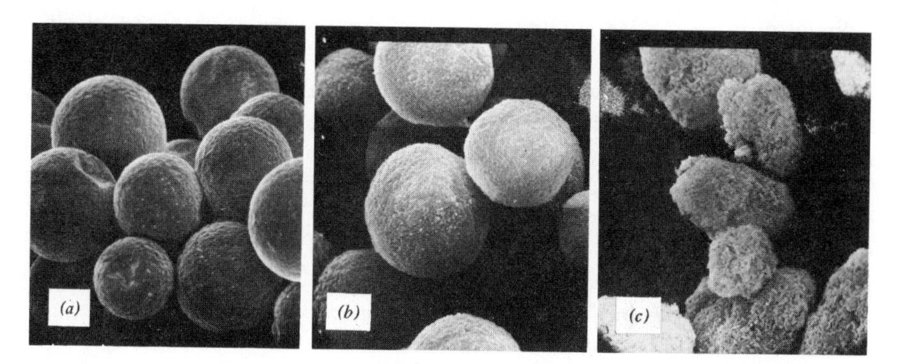

Figure 6.2 Scanning electron photomicrographs of some rigid particles for HPSEC.
(*a*) LiChrospher 500 (silica, 500 Å pores); (*b*) porous silica microspheres (500 Å); (*c*) Poramina-1200A-FF (alumina, 1200 Å). (Figures from Ref. 7.)

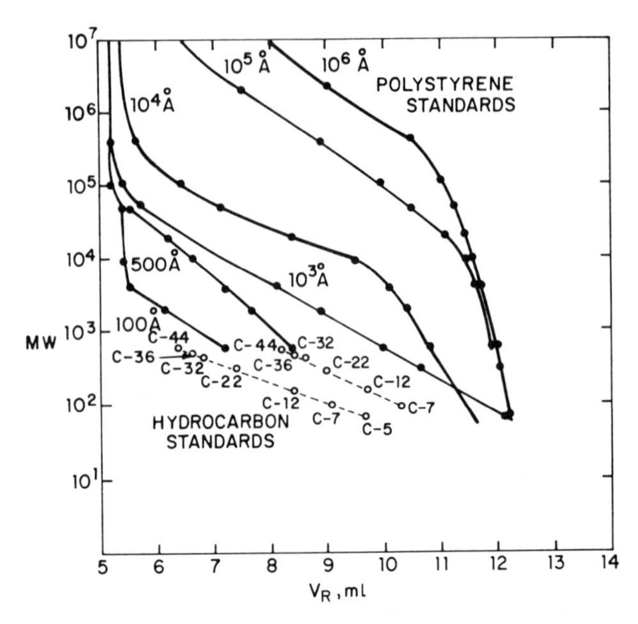

Figure 6.3 **μ-Styragel molecular weight calibration curves.**
(Reprinted courtesy of Waters Training Dept., Waters Assoc.)

with pore sizes for separating a range of about 5×10^1–10^7 MW (polystyrene standards). Figure 6.3 shows typical polystyrene calibration plots reported for μ-Styragel (Waters Associates) columns, and Figure 6.4 shows plots for the similar MicroPak BKG Series and Toyo Soda TSK polystyrene gel columns (Varian Associates).

Small-pore gel (e.g., designated 100 Å) columns are widely used for separating small molecules by HPSEC. Such packings probably have an average pore diameter of ~30 Å when swollen. Because of the unique small-pore gel structure, columns of these packing particles are restricted to a limited range of mobile phases for separating small molecules (see Table 8.4).

Packed columns of larger-pore cross-linked polystyrene gels can be used with appropriate solvents at temperatures up to ~175°C. High-temperature use of small-pore gel columns (i.e., μ-Styragel 100 Å) is more restricted and must be used at temperatures no greater than 150°C. μ-Styragel columns are shipped in a closed condition with a bellows end connector to minimize problems during shipping and handling.

Spherical particles of controlled-pore-size vinylacetate copolymers are also suitable for use in organic solvents. Type OR-PVA gels (Fractogel, E. Merck; EM Laboratories) are used mainly for separating synthetic

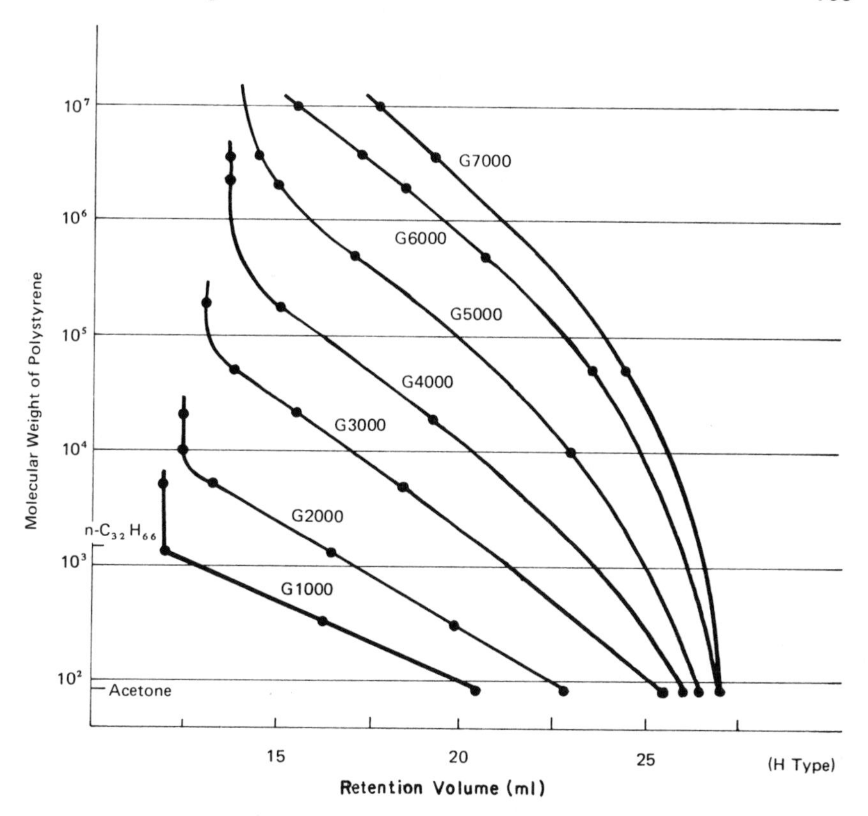

Figure 6.4 Molecular weight calibration curves for TSK gel columns.
Mobile phase, tetrahydrofuran; sample, polystyrenes, hydrocarbons. (Reprinted courtesy of Varian Associates.)

polymers (e.g., polystyrene, phenolformaldehyde). Pore sizes are available to separate materials with an exclusion limit up to about 10^6 MW, but the larger-pore-size particles are too soft for use in HPSEC. These small-pore-size, semirigid PVA gels are stable in organic solvents such as alcohols and acetone, even at elevated temperature ($<100°C$) but are limited to maximum column pressures of about 300–600 psi (20–40 bar). Because the gel particles are relatively large, 30–63 μm (230–450 mesh), chromatographic performance is borderline relative to the small-particle HPSEC packings. Properties of these vinylacetate gels are summarized in Table 6.1. Figure 6.5 shows typical calibration curves for OR-500 and OR-2000, the two packings most suitable for operation at higher flow rates and column inlet pressures up to 300–600 psi.

Table 6.1 Some Semirigid Organic Gels for HPSEC

Type	Column Packing	Particle Size (μm)	Approximate MW Fractionation Range	Approximate Maximum Pressure (psi)	Mobile-Phase Capability	Supplier[a]	Comments
Cross-linked styrene/divinyl benzene copolymer	μ-Styragel 10^6 Å	10 ± 1	$1 \times 10^5 - > 1 \times 10^7$	3000	Organic solvents but not acetone, alcohols, other very polar	1	Available only in 30 × 0.7 cm packed columns. MW range determined with polystyrene, hydrocarbons. Maximum flow rate, 3 ml/min at 0.6 cP. Used in GPC.
	10^5 Å		$1 \times 10^4 - 1 \times 10^7$	3000			
	10^4 Å		$7 \times 10^3 - 2 \times 10^6$	3000			
	10^3 Å		$4 \times 10^2 - 4 \times 10^5$	3000			
	500 Å		$1 \times 10^2 - 8 \times 10^4$	3000			
	100 Å		$<1 \times 10^2 - 3 \times 10^3$	2000			
Cross-linked styrene/divinyl benzene copolymer	MicroPak BKG1000H	9 ± 1	$<1 \times 10^2 - 2 \times 10^3$	NA[b]	Organic solvents but not acetone, alcohols, other very polar	2	Available only in 30 and 50 cm × 0.8 cm packed columns. Guaranteed plate count, >8000 plates/30 cm.
	G2000H		$1 \times 10^2 - 1 \times 10^4$				
	G3000H		$1 \times 10^2 - 8 \times 10^4$				
	G4000H		$5 \times 10^2 - 4 \times 10^5$				
	G5000H		$5 \times 10^3 - 3 \times 10^6$				
	G6000H		$5 \times 10^4 - > 10^7$				
	G7000H		$1 \times 10^5 - > 10^7$				
	GMH		$1 \times 10^2 - > 10^7$				
Cross-linked styrene/divinyl benzene copolymer	μ-Spherogel 10^6 Å	NA	$>1 \times 10^6$	NA	Organic solvents but not acetone, alcohols, other very polar	2a	Available only in 30 × 0.8 cm packed column. Guaranteed plate count, >6000 plates/30 cm.
	10^5 Å		$1 \times 10^5 - 5 \times 10^6$				
	10^4 Å		$1 \times 10^4 - 5 \times 10^5$				
	10^3 Å		$1 \times 10^3 - 5 \times 10^4$				
	5×10^2 Å		500–10,000				
	10^2 Å		100–5000				
	5×10 Å		<2000				

Vinylacetate copolymer	EM Gel Type OR PVA 500 2000	30–63	Up to 1.5×10^3 Up to 8×10^3	300–600	Organic solvents, including alcohols, acetone	3	Larger pore sizes too soft for HPSEC. Used only at low flow rates (e.g., <1 ml/min). MW range determined with polystyrenes and oligo-phenylenes in tetrahydrofuran.
Hydroxylated organic	TSK-Gel G-2000SW 3000SW	10 ± 2	Up to 8×10^4 Up to 2×10^6	NA	Aqueous solvents	2	Available in 60 × 0.75 cm and 60 × 2.0 cm packed columns. Use at <45°C.
Polyacryamide	Bio-Gel P-2	<28	$<1 \times 10^2$–1.8×10^3	200	Aqueous systems, including buffers	4	Used for GFC. Larger pore sizes too soft for HPSEC. Used only at low flow rates (e.g., <1 ml/min).
Polysaccharide	Sephadex G-25	10–40	1×10^2–5×10^3	200	Aqueous systems	5	Used for GFC. Larger pore sizes too soft for HPSEC. Used only at low flow rates (e.g., <1 ml/min).
	Sephadex LH-20	25–100	1×10^2–4×10^3	200	Polar organic solvents, alcohols	5	Hydroxypropylated Sephadex G-25. Used only at low flow rates (e.g., <1 ml/min).
Sulfonated cross-linked styrene/divinylbenzene copolymer	Hydrogel-II -IV -VI	<37	$<1 \times 10^2$–2×10^3 2×10^2–4×10^4 7×10^3–2×10^6	3000	Aqueous systems, with salts	1	On either side of pH 7 may show ionic sorption effects and/or hydrolysis.

[a] Suppliers: (1) Waters Associates, Milford, Mass.; (2) Varian Associates, Palo Alto, Calif.; Toyo Soda Manufacturing Co., Tokyo, Japan; (2a) Altex Scientific, Inc., Berkeley, Calif.; Showa Denko, Tokyo, Japan; (3) E. Merck, Darmstadt, German Federal Republic; EM Laboratories, Inc., Elmsford, N.Y.; (4) Bio-Rad Laboratories, Richmond, Calif.; (5) Pharmacia Fine Chemicals, Inc., Pictaway, N.J.

[b] NA, not available

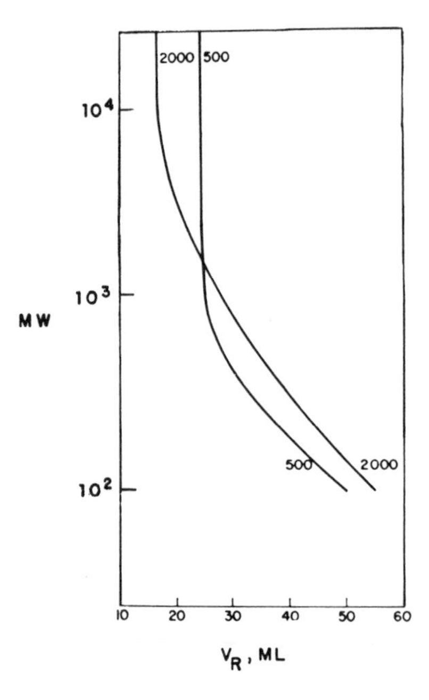

Figure 6.5 Calibration curves for EM Gel-OR. (Reprinted with permission from Ref. 6.)

Porous organic gels that permit the high-efficiency separation of water-soluble solutes are yet not as well developed as packings for use in organic solvents. However, several useful packings are available and a larger selection can be anticipated. The SW type TSK-Gel (Toyo Soda) is a rigid, hydrophilic, spherical gel for use with primarily aqueous solvents (Table 6.1), but other polar solvents are also permitted. Very little information is available on this new packing, but interesting applications have been reported (3a).

Some of the conventional hydrophilic organic gels long used in GFC also can be used in HPSEC with certain restrictions. The Bio-Gel P series of neutral, hydrophilic poly(acrylamide) gels, available from Bio-Rad Laboratories, are made by the copolymerization of acrylamide with the cross-linking agent, *N,N′*-methylene-bisacrylamide. By regulating the concentration of these monomers, a series of covalently bonded gel products of various average pore size is produced. While 10 different Bio-Gel P products are available with peptide or globular protein exclusion limits ranging from 1800 to 400,000 MW, only Bio-Gel P-2 has sufficient mechanical strength to be included in our list of HPSEC packings (Table 6.1). Columns of Bio-Gel P-2 can be used at maximum pressures of about 200 psi for separating water-soluble materials up to 1800 MW. Larger particle sizes are available for preparative applications.

Dextrans cross-linked with epichlorohydrin, Sephadex (Pharmacia Fine Chemicals) are also useful for aqueous-phase separations. Because of high hydroxyl-group concentrations in the polysaccharide chains, Sephadex is strongly hydrophilic and swells in water and electrolyte solutions. While most of these materials are relatively soft, Sephadex G-25 has sufficient strength to be used at pressures to about 150 psi for MW of up to about 5000 (Table 6.1).

Sephadex LH-20 is a semirigid gel prepared by the propylation of Sephadex G-25. This gel is used with polar organic solvents (e.g., alcohols, tetrahydrofuran), but is also compatible with water. It is supplied in a relatively wide particle-size range (25–100 μm); however, narrower particle-size fractions (e.g., 25–38 μm) can be obtained by sizing to improve column performance. Column input pressure is limited to 150–200 psi, which restricts applications in HPSEC.

Hydrogel packings (Waters Associates) listed in Table 6.1 also may be useful in HPGPC. These water-wettable beads are made from lightly sulfonated, cross-linked polystyrene. Because of the surface sulfonic acid groups, ionic sorption effects may occur on either side of pH 7 in aqueous media, thus limiting mobile-phase solute compatibility.

Rigid Inorganic Packings

Siliceous packings for HPSEC are of increasing interest and utility. Table 6.2 summarizes some significant properties of various silica packing materials obtainable in packed columns (only a few can be purchased in bulk). Some particles are untreated, others have been modified by adding organic functionality to reduce adsorption. Columns of all of these packings may be used with both organic and aqueous mobile phases. However, with aqueous systems, appropriate additives may be required to eliminate solute adsorption (Sect. 13.3).

LiChrospher—totally porous spherical silica microparticles with a range of pore sizes—is manufactured by E. Merck. These microparticles have a relatively large internal porosity, which provides both good sample capacity and column efficiencies (4). Typical calibration plots for LiChrospher columns are shown in Figure 6.6.

Gel Type SI is an irregularly shaped, untreated porous silica, also manufactured by E. Merck. It was originally designed for general use in conventional adsorption, gel permeation, and partition chromatography. However, because of its availability in relatively small particle size, it can be utilized in HPSEC for some operations and may be particularly suited for preparative GPC. For analytical HPSEC, LiChrospher is generally preferred because of higher performance. Figure 6.7 shows typical calibration curves for columns of the EM Type SI particles.

Table 6.2 Some Porous Silica Packings for HPSEC

Column Packing	Particle Size (μm)	Approximate MW Fractionation Range (polystyrene)	Mean Pore Diameter (Å)	Specific Porosity (ml/g)	Specific Surface Area (m^2/g)	Approximate Maximum Pressure (psi)	Supplier[a]	Comments[b]
LiChrospher SI-100	5, 10, 20	1×10^3–7×10^4	100	1.2	370	>3000	1[c]	Untreated spherical particles.
SI-500	10	5×10^3–5×10^5	500	0.8	50			
SI-1000	10	1×10^4–3×10^6	1000	0.8	20			
SI-4000	10	5×10^4–7×10^6	4000	<0.8	6			
Zorbax SE-60	6	1×10^2–4×10^4	60	0.6	400	>3000	2	Untreated spherical particles; 25×0.62 cm packed columns only.
-100	10	5×10^3–7×10^4	100	1.2	370			
-500	10	3×10^3–5×10^5	500	0.8	50			
-1000	10	1×10^4–3×10^6	1000	0.8	20			
Zorbax PSM-60	6	1×10^2–4×10^4	60	0.6	400	>3000	2	25×0.62 cm columns. PSM-60 and PSM-1000 combination as bimodal pore size set for linear MW calibration. Untreated and trimethylsilyl-modified spherical silica.
PSM-500	6	1×10^4–5×10^5	350	0.6	60			
PSM-1000	6	3×10^4–2×10^6	750	0.6	15			

μ-Bondagel-E-1000	10 ± 1	5×10^4–2×10^6	1000	NA[d]	NA	> 3000	3	Chemically modified with ether groups. 30 × 0.4 cm packed columns only.
-500		5×10^3–5×10^5	500					
-300		3×10^3–1×10^5	300					
-125		2×10^3–2×10^4	125					
-linear		2×10^3–2×10^6	—					
μ-Porasil-60-GPC-60 Å	10	1×10^2–1×10^4	60	NA	500	> 3000	3	Untreated spherical particles; primarily for small molecules.
EM Type SI-200	10–40	2×10^2–5×10^4	200	0.8	120–170	NA	1	Untreated irregular particles.
-500		5×10^3–4×10^5	500	0.8	35–65			
-1000		1×10^4–1×10^6	1000	0.7	10–20			
Glycophase-G/GPC-40	5–10	1×10^3–8×10^3	60	> 0.12		NA	4	Irregular particles, chemically modified with glyceryl groups. Primarily used for HPGFC.
-100		1×10^3–3×10^4	100	> 1.0				
-250		2.5×10^3–1.3×10^5	250					
-550		1×10^4–3.5×10^5	550					
-1500		1×10^5–1×10	1500					
-2500		2×10^5–1.5×10^6	2500					

[a] Suppliers: (1) E. Merck, Darmstadt, German Federal Republic; EM Laboratories, Inc., Elmsford, N.Y.; (2) E. I. du Pont de Nemours & Co., Instrument Products Division, Wilmington, Del.; (3) Waters Associates, Milford, Mass.; (4) Pierce Chemical Co., Rockford, Ill.

[b] All packings compatible with both organic and aqueous solvents.

[c] Also offered in Chromegapore 30 × 0.46 cm packed columns by E.S. Industries, Marlton, N.J., in 100, 500, and 1000 Å sizes.

[d] NA, not available.

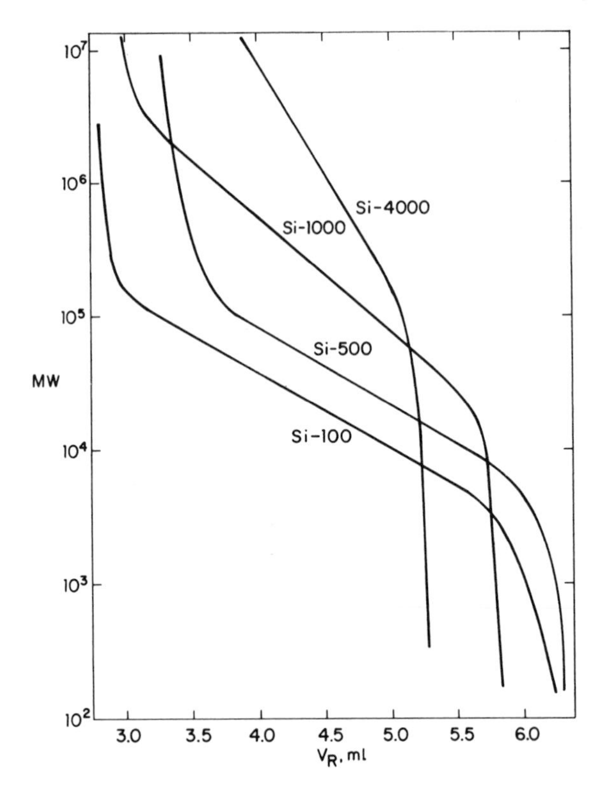

Figure 6.6 Calibration curves for LiChrospher columns

Columns, 0.62 × 25 cm; mobile phase, tetrahydrofuran; flow rate, 2.0 ml/min; temperature, 22°C; detector, UV, 254 nm; sample, 25 μl of 0.25% polystyrenes. (Data from Ref. 7.)

Columns of very high efficiency silica SE packings are available from E.I. du Pont de Nemours and Co. These spherical particles are untreated porous silica with a range of narrow-distribution pore sizes. DuPont SE-60 porous silica microspheres have been especially designed for separating small molecules (<20,000 MW) by HPSEC (1, 18). Figure 6.8 shows typical polystyrene calibration curves for DuPont SE packed columns, which can separate throughout the 10^2–10^7 MW range. Another series of these high-efficiency particles (Zorbax PSM) have been developed for specific applications in SEC. For example, the Zorbax PSM-60, PSM-1000 combination with a bimodal pore-size distribution is used to obtain linear calibration curves for determining the molecular weight distribution of polymers (Sect. 8.5). The latter particles are available in both untreated or silanized form, for use in aqueous and organic mobile phases, respectively.

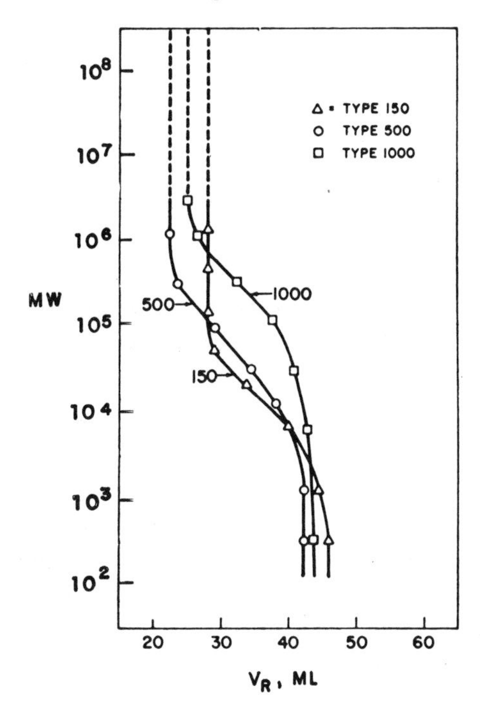

Figure 6.7 Calibration curves for EM Gel Type SI porous silica.
(Reprinted with permission from Ref. 6.)

All the untreated siliceous packings listed in Table 6.2 can be used with HPGPC at low and high temperatures and with HPGFC using the proper aqueous mobile phase. If adsorption is a problem, it can usually be reduced or eliminated by adding certain components to the mobile phase as discussed in Chapter 13.

Siliceous particles which have been surface-modified by adding organic functionality to reduce adsorptive properties have recently created great interest. μ-Bondagel (Waters Associates) is a porous silica to which an ether functionality has been chemically bonded. This material is offered in four pore-size columns, plus an E-linear column which is a blend of pore sizes. The manufacturer states that these packings are compatible with solvents of all polarity from *n*-pentane to buffered aqueous mobile phases; however, there is the usual pH limitation of ~2–8. Figure 6.9 shows polystyrene calibration curves for individual μ-Bondagel columns, while Figure 6.10 gives polystyrene calibràtion curves for two connected series of μ-Bondagel columns, one for an *E*-(125, 300, 500, and 1000) series and one for the E-linear column series. A chromatogram of polystyrene standards separated with the μ-Bondagel E-linear column set is shown in Figure 6.11. This chromatogram

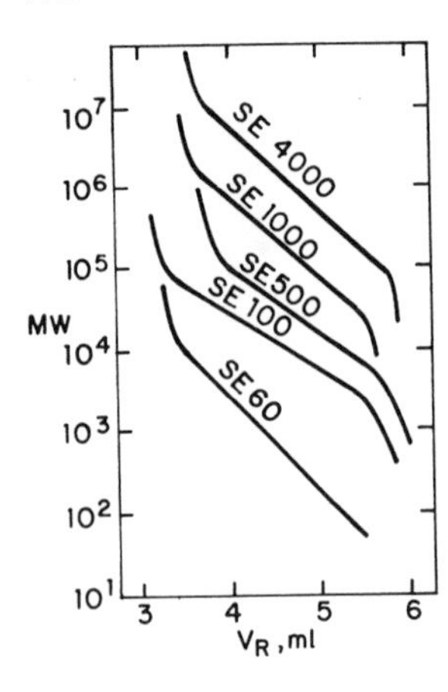

Figure 6.8 Calibration curves for DuPont SE silica columns. Columns, 0.62 × 25 cm; mobile phase, tetrahydrofuran; flow rate, 2.0 ml/min; temperature, ambient; detector, UV, 254 nm; sample, 25 μl polystyrene standard solutions. (Data from Ref. 7.)

shows the high-resolution capability of the column set within a linear molecular weight calibration range of about $2\frac{1}{2}$ decades for polystyrene. Because of the small fractionation volume associated with this 0.4 × 30 cm column set, it is important that the flow rate be accurately controlled or monitored during molecular weight determination (see Sect. 7.2).

Because of the compatibility of the ether-modified μ-Bondagel with water, these columns also can be used for high-efficiency gel-filtration chromato-

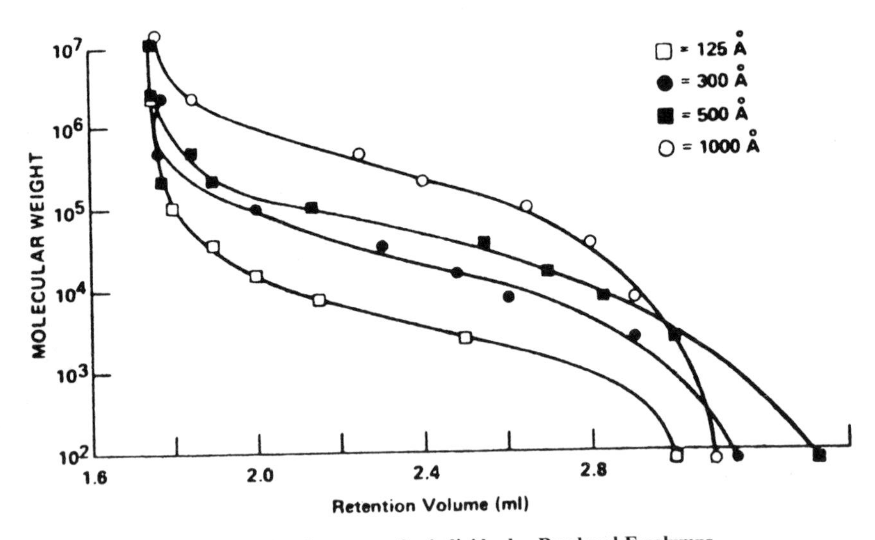

Figure 6.9 Polystyrene calibration curves for individual μ-Bondagel E columns. Columns, 30 × 0.4 cm; mobile phase, toluene. (Reprinted courtesy of Waters Training Dept., Waters Associates.)

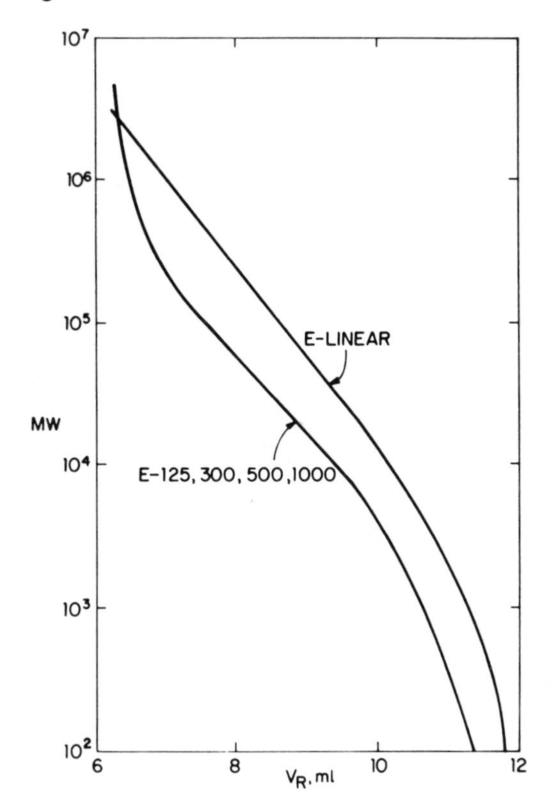

Figure 6.10 Polystyrene calibration curves for μ-Bondagel E series.
Columns, 120 × 0.40 cm; mobile phase, dichloromethane; flow rates and pressures: 0.5 ml/min and 650 psi for E-125, 300, 500, 1000 series, 2.0 ml/min and 2500 psi for E-linear series. (Data courtesy of Waters Training Dept., Waters Associates.)

graphy. Figure 6.12 shows dextran calibration curves with water as the mobile phase.

Glycophase-G/GPC packings (Corning Biological Products Department; Pierce Chemical Company) and SynChropak GPC (The Anspec Co., Ann Arbor, Mich.) are irregular siliceous particles which have been modified with a hydrophilic glycol. These materials, which were designed especially for use in aqueous and other highly polar mobile phases, have thus far found greatest application in HPGFC. As with the other siliceous packings, they are stable throughout the range pH 2–8 and can be used at high pressures without fear of particle degradation.

Small-particle porous aluminas such as Poramina (2, 3) have been proposed as stationary phases for HPSEC but are generally not available.

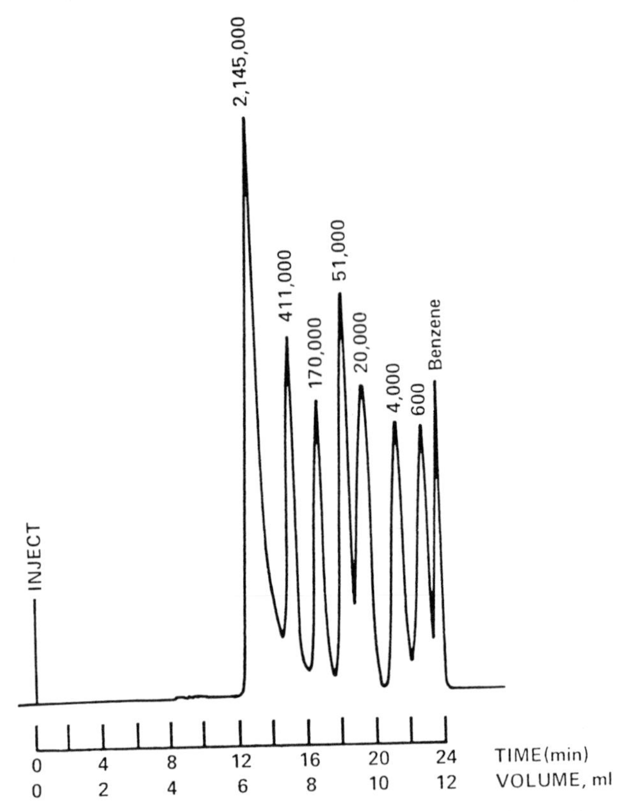

Figure 6.11 Separation of polystyrene standards with μ-Bondagel-E linear column series. Columns, four 30 × 0.4 cm; mobile phase, dichloromethane, 0.5 ml/min; pressure 650 psi; velocity, 0.08 cm/sec. (Courtesy of Waters Training Dept., Waters Associates.)

Packings for Preparative SEC

Some column packings useful for preparative-scale size-exclusion chromatography are summarized in Table 6.3. In general, these materials consist of larger particles of the same packings used for high-performance analytical applications. The materials listed in Table 6.3 are particularly suitable (e.g., lower cost and ready availability) for use in the longer and larger-diameter columns needed for preparative applications. Resolution equivalent to that obtained in analytical applications with columns of smaller particles can also be obtained with these larger preparative packings, but significantly longer separation times are required (see Sect. 11.2). The packings listed in Table 6.3 can also be used for analysis, but with performance that does not meet the analysis time standards we previously proposed for HPSEC columns (Sect. 6.1).

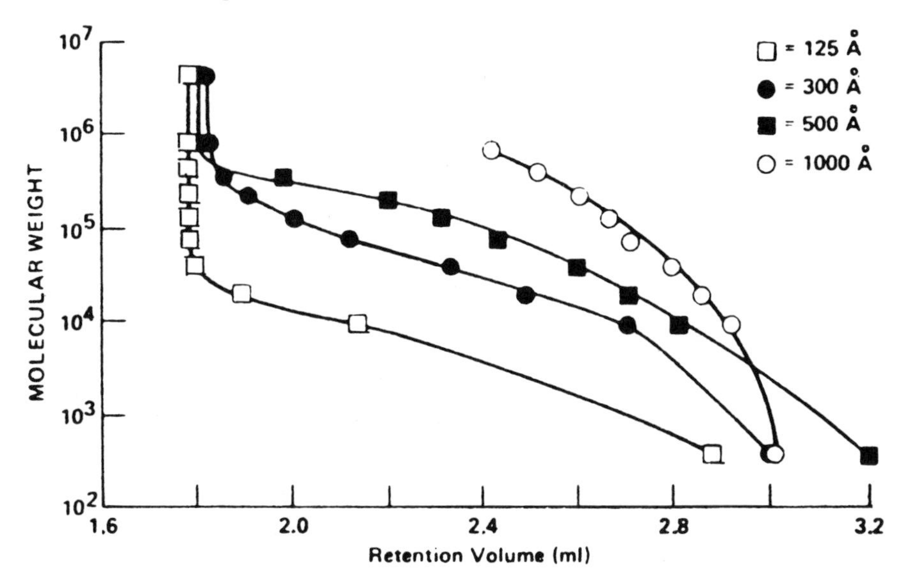

Figure 6.12 Calibration for dextrans in water with μ-Bondagel-E columns. Columns, 30 × 0.4 cm. (Courtesy of Waters Training Dept., Waters Associates.)

Columns of Styragel, cross-linked styrene/divinylbenzene copolymers (Waters Associates) have long been used for GPC analyses and are still valuable for applications in which large sample sizes are needed or if only low-pressure pumps are available. Packed 120 × 0.7 cm Styragel columns are available in a variety of pore sizes (Table 6.3), and typical calibrations are illustrated in Figure 6.13. Styragel columns are compatible with most organic solvents except acetone and alcohols. The swelling of Styragel packings with small pores demonstrates their high sensitivity to solvent polarity, so care must be exercised when using highly polar solvents (e.g., alcohols, dimethylformamide). These packings are stable to about 150°C.

Larger particle sizes of most of the other organic gels listed in Table 6.1 are also available for preparative applications.

Porous silica beads manufactured by Rhone-Poulenc (Fr.) are distributed by Waters Associates in the United States under the name Porasil and by other suppliers worldwide as Spherosil. These untreated siliceous particles can be used for preparative work with either organic or aqueous mobile phases at elevated temperatures. Their properties are the same as those previously described for the untreated silica particles in Table 6.2. Porasils are also offered in a deactivated form in the Porasil-X series. The deactivation is a chemical treatment that decreases, but does not eliminate, all adsorption

Table 6.3 Some Packings for Preparative SEC

Type	Column Packing	Particle Size (μm)	Approximate MW Fractionation Range	Mean Pore Diameter (Å)	Specific Porosity (ml/g)	Specific Surface (m^2/g)	Approximate Maximum Pressure (psi)	Mobile-Phase Capability	Supplier[a]	Comments
Cross-linked styrene/divinyl benzene copolymer	Styragel 10^7 Å	37–75	5×10^5–5×10^8	—	—	—	600	Organic solvents, but not acetone, alcohols, water	1	Spherical particles available only in 122 × 0.7 cm packed columns. Stable to temperature in excess of 150°C.
	10^6 Å		1×10^5–5×10^7							
	10^5 Å		5×10^4–2×10^6							
	10^4 Å		1×10^3–7×10^5							
	10^3 Å		5×10^2–5×10^4							
	10^2 Å		1×10^2–8×10^3							
Porous silica	Porasil A	37–75	NA–4×10^4	< 100		450	> 3000	Organic and aqueous solvents	1	Spherical particles, untreated. Also available in Porasil-X deactivated series.
	B		NA–2×10^5	150		200				
	C		NA–4×10^5	300		75				
	D		NA–1×10^6	600		35				
	E		NA–1.5×10^6	1200		22				
	F		NA–4×10^6	> 1500		15				
Porous silica	Spherosil XOA-400	< 40	NA–4×10^4	80		400	> 3000	Organic and aqueous solvents	2	Spherical particles, untreated.
	-200		NA–2×10^5	150		185				
	-075		NA–4×10^5	300		100				
	-030		NA–1×10^6	600		50				
	-015		NA–1.5×10^6	1250		25				
	-005		NA–$>4 \times 10^6$	3000		10				
Porous silica	Vit-X-[b]-388	36–44	2×10^3–1.3×10^4	84	0.5	NA	> 3000	Organic and aqueous solvents	3[c]	Irregular particles, chemically deactivated.
	-648		4×10^3–2×10^4	101	0.6					
	-1-068		6×10^3–7×10^4	171	0.8					
	-1-195		9×10^3–1.5×10^5	210	0.8					
	-5-150		1.2×10^4–3×10^5	321	1.2					
	-15-300		3×10^4–1.2×10^6	660	0.8					
	-40-550		4×10^4–2×10^6	1206	0.8					
	-120-120		9×10^4–1×10^7	1933	0.9					

[a] Suppliers: (1) Waters Associates, Milford, Mass.; (2) Supelco, Inc., Bellefonte, Pa., and other worldwide supply houses; (3) Perkin-Elmer Corp., Norwalk, Conn.
[b] Similar untreated and glyceryl-modified controlled-pore porosity glass available in 37–74 μm size from Electro-Nucleonics, Inc., Fairfield, N.J.
[c] Also available in 50 × 0.26 and 50 × 0.46 packed columns for analytical studies.

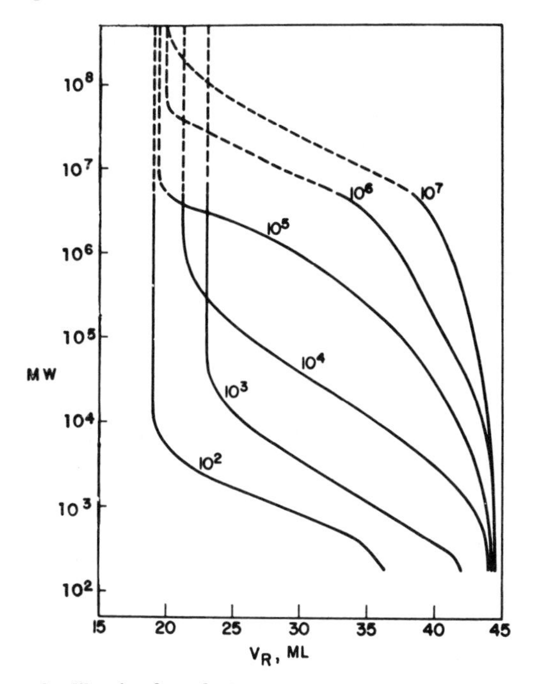

Figure 6.13 Styragel calibration for polystyrene.
Columns, 122 × 0.7 cm; mobile phase, tetrahydrofuran; flow rate, 1.00 ml/min. (Reprinted with permission from Ref. 6.)

problems. Figure 6.14 contains calibration plots for polystyrene obtained on columns of untreated Porasil.

Vit-X packings (Perkin-Elmer Corp.) are irregular porous silica particles, deactivated to reduce adsorptive properties. Bulk packing can be used to make larger-diameter preparative columns, and 50 × 0.26 cm and 50 × 0.46 cm packed columns are commercially available for analytical use.

Modifying Siliceous Particles

Silanol groups on the surface of untreated porous silica packings can cause problems in SEC by adsorbing the solute. This biases the desired size-exclusion mechanism, and the desired relationship between retention volume and molecular size (or molecular weight) will not be obtained. Mixed retention usually is evidenced by tailing chromatographic peaks, lowered column efficiency, and, in extreme cases, retention beyond the total permeation volume. Fortunately, silica surfaces can be altered by adding certain organic functional groups to effectively eliminate this disadvantage for most

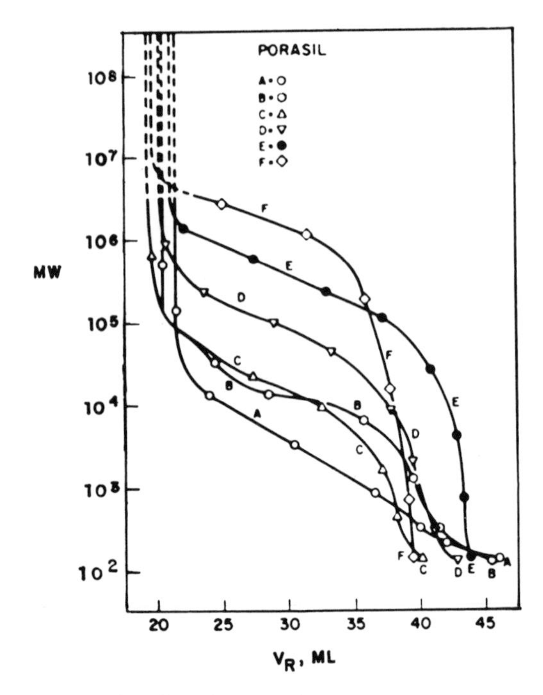

Figure 6.14 Porasil calibrations for polystyrene.
(Reprinted with permission from Ref. 6.)

applications. By proper selection of the organic functionality, the surface of particles can be altered for both organic and aqueous mobile-phase applications.

One approach used to eliminate solute adsorption in organic solvents is to maximize conversion of surface silanol groups on silica to their trimethylsilyl derivatives (1, 4, 8). This hydrocarbon-modified packing can be prepared by refluxing a large molar excess of a short-chain chlorosilane (e.g., chlorotrimethylsilane) with the siliceous support, the surface of which has been previously fully hydrolyzed (e.g., heating at 90–100°C at pH 9 in aqueous systems for several days). The reactions proceed according to

$$\begin{matrix} -\overset{|}{\text{Si}} \\ | \\ -\underset{|}{\text{Si}} \end{matrix} \!\!>\!\text{O} \xrightarrow[\text{H}_2\text{O, pH 9}]{\Delta} -\overset{|}{\underset{|}{\text{Si}}}-\text{OH}$$

$$-\overset{|}{\underset{|}{\text{Si}}}-\text{OH} + \text{ClSi(CH}_3)_3 \xrightarrow[-\text{HCl}]{\Delta} -\overset{|}{\underset{|}{\text{Si}}}-\text{O}-\text{Si(CH}_3)_3$$

Unreacted silanol groups (which constitute about one-half of the total silanol concentration) become shielded by an "umbrella" of tightly packed trimethylsilyl organic groups. As long as the trimethylsilyl groups are at a sufficiently high concentration on the surface (>3.5 μmoles/m^2), the residual silanol groups remain unavailable for unwanted adsorptive interactions. Reaction of surface Si-OH groups with chlorosilane reagents to high yields is promoted by [1] using a large excess of reactant, [2] conducting the reaction in the neat liquid reactant or in a dry solvent, [3] mechanically removing the volatile reaction product during the reaction (e.g., volatilization) (9), or [4] by using an appropriate acid acceptor such as pyridine (10). The following procedure is typical of the reactions that are used to cover the surface of various porous silicas with trimethylsilyl groups.

> The silica particles must first be fully hydrolyzed by heating with a fivefold volume excess of concentrated nitric acid on a steam bath for 4–6 hr (alternatively, at pH 9 in NH_4OH in closed polyethylene or Teflon container for several days at 90–100°C). The particles are repeatedly washed by slurrying with distilled water until neutral. Any fines that are present are discarded during this procedure. To ensure complete removal of acid, the washed packing is boiled in distilled water for about 15 min. The washed particles are filtered, air-dried, and heated in a circulating air oven at 200°C or in a vacuum at 150°C for several hours until dry. This carefully dried solid is refluxed for at least 16 hr in approximately a 10-volume excess of *dry* toluene (e.g., dried over Type 4A molecular sieve) containing a 2 M excess of redistilled chlorotrimethylsilane (based on ~ 4 μmoles/m^2 of reactable silanol groups) and a 0.1 mole ratio of pyridine. Alternatively, *dry* pyridine may be used as the solvent instead of toluene. During the reaction the system should be protected against moisture. The reacted solids are isolated, washed thoroughly with dry toluene and absolute methanol successively, air-dried, and heated at 150°C for 4 hr.
>
> For highest possible concentration of trimethylsilyl groups on the surface, these trimethylsilyl-treated particles should be refluxed in 1 : 1 methanol/0.01 *M* HCl for 30 min, dried at 150°C for 4 hr, refluxed again with chlorotrimethylsilane for at least 4 hr, and treated as described above. This resilanization procedure should result in a concentration of 3.5–4 μmoles/m^2 of trimethylsilyl groups as measured by carbon analysis. Trimethylsilation of both small-pore (60 Å) and large-pore (750 Å) porous silica microspheres causes no significant change in the molecular weight calibration plots for untreated particles, as indicated in Figure 6.15.

If desired, untreated silica packing may be silanized by *in situ* reaction with chlorotrimethylsilane (18). This approach is useful to resilanize a set of

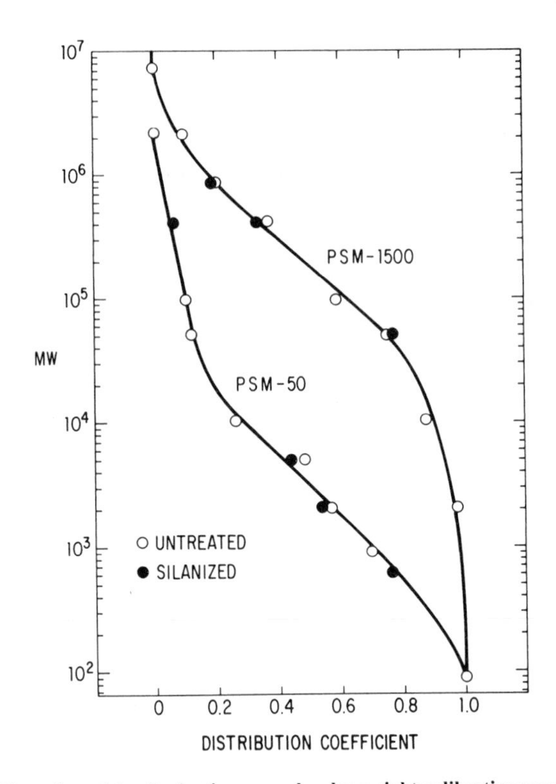

Figure 6.15 Effect of particle silanization on molecular weight calibration curves. Polystyrene standards; mobile phase, tetrahydrofuran, 22°C, flow rate, 2.5 ml/min; pressure 925 psi; UV detector at 254 nm; sample, 25 μl, 0.25%; 60 cm set of porous silica microsphere columns 60–3500 Å. (Reprinted with permission from Ref. 1.)

columns that have become somewhat adsorbing because of loss of deactivating bonded organic groups, but is less convenient than the general procedure described above for silanizing larger quantities of bulk packings.

6.3 Column-Packing Methods

Particle Technology

As indicated in Section 3.3, particle size is an important factor in the preparation of efficient SEC columns. Both plate height and column permeability decrease approximately as the square of the particle diameter. Thus, when using smaller particles to gain higher column resolution, higher column

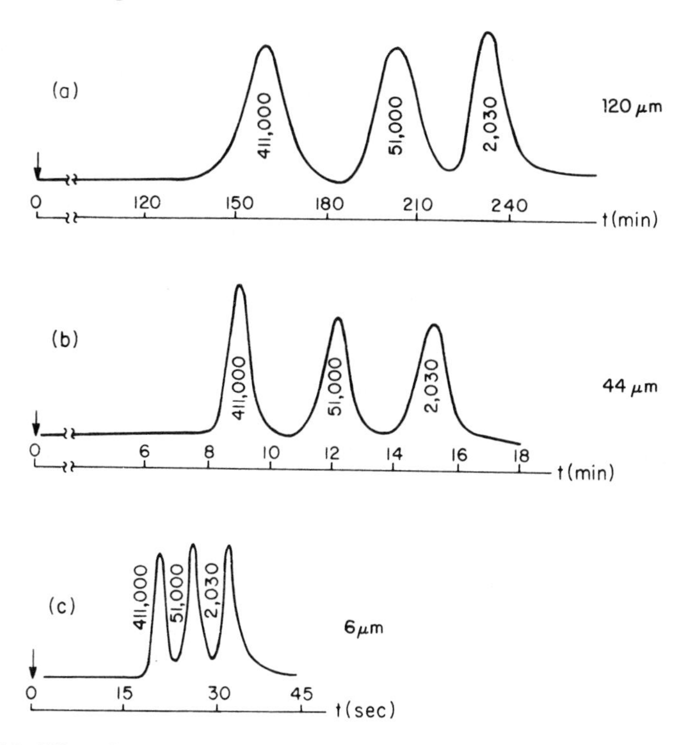

Figure 6.16 Effect of particle size on separation time.
(*a*) $d_p = 120\ \mu m$: one column each; 4 ft × 0.305 in. of CPG 10-2000, -1250, -700, -370, -240; mobile phase, toluene; flow rate, 1.0 ml/min. (*b*) $d_p = 44\ \mu m$: one column each; 50 × 0.26 cm of CPG 10-700, -350, -125; mobile phase, tetrahydrofuran; flow rate, 0.5 ml/min. (*c*) $d_p = 6\ \mu m$: one column, 50 × 0.2 cm porous silica microspheres; mobile phase, tetrahydrofuran; flow rate, 1.0 ml/min. (Reprinted with permission from Ref. 11.)

inlet pressures are required. On balance, columns of smaller particles ($<20\ \mu m$) are worth the increased cost and higher column pressures to gain increased resolution or decreased analysis time. The effect of particle size on analysis time is shown in Figure 6.16. A column of 6 μm particles demonstrates resolution equivalent to that of a column with 120 μm particles, but in about $\frac{1}{400}$ of the analysis time.

A disadvantage of using small-particle ($<20\ \mu m$) columns is that they require a carefully constructed HPSEC apparatus to reduce potential extracolumn effects (see Sect. 5.1b). Also, columns of small particles are more difficult to pack homogeneously. As result, a compromise in particle size often is made in HPSEC—particles are chosen small enough to produce good column efficiency but large enough to be conveniently packed into columns and maintained in practical use. Columns of 7–10 μm particles appear to be a reasonable compromise, but 5–20 μm particles of narrow size range have been satisfactorily used for HPSEC.

When available, spherical particles are preferred over irregular ones, because reproducibly packed, high-efficiency columns are more easily prepared from the former. In addition, columns packed with spherical particles tend to have higher permeability for the same apparent average particle size. Irregularly shaped particles with a wide particle-size range cause particular difficulties in packing homogeneous column beds.

For best results the particle-size range in a given packing should be relatively narrow, for example ±20%, as illustrated for the porous silica microspheres in Figure 6.17. A wider range of particle sizes makes packing homogeneous columns more difficult because of the tendency of the particles to size across the column during the packing process. Segregation of particles

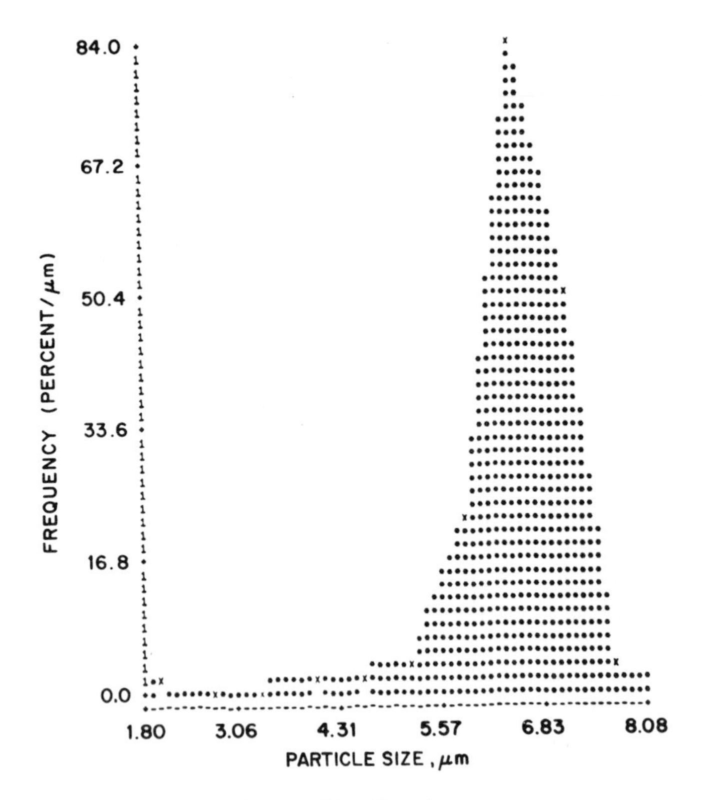

Figure 6.17 Particle-size range of porous silica microspheres. Size range weighted by volume; Quantimet analysis.

during the packing procedure produces a distribution of flow velocities across the column. This difference in the packing structure at the column wall and at the center of the column will cause band distortion. Particle agglomeration during packing makes these flow effects more serious. Inhomogeneous beds result in lower column plate counts and column permeabilities, as compared to homogeneous bed, narrow size-range packings of the same average size. Thus optimum column packings for HPSEC appear to involve spherical particles with a narrow size range.

Good mechanical stability of particles is required for the preparation of high-efficiency columns, particularly by the high-pressure slurrying-packing technique described below. The shear force imposed on particles during this packing process is relatively large. If particles are not sufficiently strong, they will fragment or compress, resulting in significantly reduced column performances and greatly increased column back pressure. The higher strength of rigid siliceous packings permits higher mobile-phase pressures and velocities during packing and subsequent use (i.e., higher column input pressures), as compared to organic gels. The excellent mechanical rigidity of these particles also allows rapid changing of mobile phase without the changes in swelling associated with organic gels.

Basis of Column-Packing Techniques

There is no one "best" column-packing method for all packings, since the optimum procedure is determined by the particle size and the nature of the material. The prime goal is to pack the column uniformly without channels or particle sizing within the column. Rigid solids and semirigid, hard gels generally are packed as densely as possible without fracturing. Columns of rigid solids may be made by dry-packing or slurry-packing techniques, depending on particle size; dry packing is normally used with particles of $>20\ \mu m$. Any packing technique that rapidly establishes a dense, stable structure is applicable. Several techniques (described below) work satisfactorily, and good columns usually can be produced with a little care.

As previously discussed, higher column efficiency results from more homogeneous structures in the packed bed. The desired homogeneous bed structure is not obtained by the classical technique of simply pouring particles into the column until full; with use, a void will usually develop at the column inlet, and a serious decrease in column performance will result. The purpose of all column-packing techniques is to produce a compact, homogeneous, and stable packing structure.

Dry-packing techniques have been used in high-performance liquid chromatography for some time (12). However, columns of high efficiency are difficult to achieve as the particle size is decreased, because small particles have very high surface energies relative to their mass. Small particles tend to form larger aggregates, producing an effect analogous to that obtained in packing a column with a very wide particle-size range. The incidence of particle agglomeration increases as the particle diameter decreases, and totally porous particles smaller than ~20 μm cannot be dry-packed easily into homogeneous beds.

Since particles smaller than about 20 μm are difficult to form into high-efficiency columns by dry-packing techniques, high-pressure "wet"-packing or slurrying techniques are normally used for this purpose. A suitable liquid is utilized to wet the particles, reduce surface energy, and eliminate aggregation. With the proper procedure the small packing particles can be dispersed homogeneously in the liquid by vigorous agitation. However, with wet packing, particle sizing can become a problem in obtaining highly efficient columns, since after initial dispersion, the particles begin to size-fractionate by gravitational sedimentation. Spherical particles settle with increasing velocity until the gravitational force is balanced by frictional resistance. The effect of particle sedimentation is shown in Equation 6.1:

$$v_t = \frac{2gr^2(\rho - \rho_0)}{9\eta} \tag{6.1}$$

where v_t is the terminal velocity of the particle, g the gravitational acceleration, r the radius of the particle, (assumed to be spherical), ρ the particle density, ρ_0 the density of the liquid, and η the fluid viscosity. Equation 6.1 indicates that large particles settle faster than smaller particles by a factor proportional to the square of the particle radius. Table 6.4 shows the relative settling times of particles with various diameters (13). The dependence of settling velocity upon particle size means that the wider the distribution of sizes, the more rapid an initially homogeneous slurry of particles will become heterogeneous, because of particle segregation. This effect is one of the reasons why a narrow particle-size range (e.g., ≤20%) is desirable.

As suggested in Equation 6.1, segregation of particles by sedimentation decreases as the density of the suspending fluid approaches the density of the particles. Therefore, it is sometimes desirable to use a suspending fluid having a density equal to the density of the particles (14, 15). Since particles of ~5 μm

Table 6.4 Relative Settling Rates of Spherical Particles

Particle diameter (μm)	50	40	30	20	10	5
Approximate U.S. standard wet mesh	280	360	480	720	—	—
Approximate relative settling time (s)[a]	1	1.6	3	6	25	100

[a] Required to settle a given distance for particles of equal density.

settle very slowly, they seldom require balanced-density solvents for the column preparation. On the other hand, 10 μm particles settle four times faster, and the balanced-density slurrying technique often is required for producing columns of highest performance from these materials.

The polarity or charge on a particle may also inhibit adequate dispersion without aggregation, and the suspending fluid should be chosen carefully. A small test sample of the packing particles can be suspended in the proposed liquid and observed by optical microscopy. Media that maintain discrete particles with no aggregation are desired. For example, slurries of negatively charged silica particles have been stabilized by using 0.001 *M* ammonium hydroxide (16). The dilute ammonia solution apparently places a positive charge on the particles and reduces their tendency to agglomerate by hydrogen bonding.

It is important that the suspension liquid thoroughly wet the packing. Polar silica or hydrophilic organic-modified silicas require relatively polar liquids, while most low-surface-energy packings, such as trimethylsilyl-modified silicas, need relatively nonpolar media. Table 6.5 lists the properties of fluids that have been used for high-pressure slurry-packing techniques.

Equation 6.1 further suggests that fractionation by sedimentation decreases as the viscosity of the suspending fluid increases. Thus an alternative to the use of the balanced-density technique is to employ viscous solvents as the slurrying agents (17). However, high viscosity increases the resistance to flow through the column so that the packing procedure takes longer or requires impractically higher pressures. Results published to date indicate that the viscous-slurry technique is less convenient and generally less capable of producing efficient columns than methods using low-viscosity slurrying liquids.

Packing Rigid Solids

High-Pressure Slurry-Packing Techniques. Columns of <20 μm rigid particles (e.g., silica) are best prepared by a slurry-packing procedure. Particle segregation is minimized by rapidly forcing a homogeneous slurry into the column with sufficient force (speed) to make the resulting bed compact and stable. The homogeneity and compactness of the formed column bed are increased by using very high flow rates to pump the slurry into the column. Apparently, the force exerted upon the particles by the high-velocity solvent causes a tightly packed bed structure to form. The long-term mechanical stability of the packed bed is markedly enhanced by a series of sudden repressurizations or "slamming" processes at high pressures to further consolidate the packing (9). Columns formed by this approach have been used for many months without evidence of the packed bed settling.

Table 6.5 Properties of Some Slurry-Packing Solvents

	Density, ρ (g/ml)	Viscosity, η (cP, 20°C)
Diiodomethane (methylene iodide)	3.3	2.9
1,1,2,2-Tetrabromoethane[a]	3.0	—
Dibromomethane (methylene bromide)	2.5	1.0
Iodomethane (methyl iodide)	2.3	0.5
Tetrachloroethylene (perchloroethylene)[a]	1.6	0.9
Carbon tetrachloride	1.6	1.0
Chloroform	1.5	0.6
Trichloroethylene	1.5	0.6
Bromoethane (ethyl bromide)	1.5	0.4
Dichloromethane (methylene chloride)	1.3	0.4
Water	1.0	1.0
Pyridine	1.0	0.9
Tetrahydrofuran	0.9	0.5
n-Butanol	0.8	3.0
n-Propanol	0.8	2.3
Ethanol	0.8	1.2
Methanol	0.8	0.6
Cyclohexane	0.8	1.0
n-Heptane	0.7	0.4
Isooctane	0.7	0.5

[a] Most halogenated solvents are somewhat toxic, but this one is particularly toxic.

A typical apparatus for preparing columns by the slurry-packing technique is shown in Figure 6.18. This apparatus uses a constant-pressure, pneumatic-amplifier pump to force the slurry rapidly into the empty column. The velocity of the slurry is proportional to the pressure used, and this velocity decreases as the packed bed is formed. Best column performance is obtained when the slurry mixture is forced into the empty column at the highest possible velocity. Thus the slurry should be introduced into the column at the highest pressure allowed by the compression fittings that connect the column to the slurry-packing apparatus. Since very high initial velocities of the slurry may fracture weak particles and cause heterogeneities in the packing structure, strong particles are required for best results.

Constant-flow (metering) pumps have also been used to prepare columns of small particles by the slurry-packing procedure. Here, the pressure on the

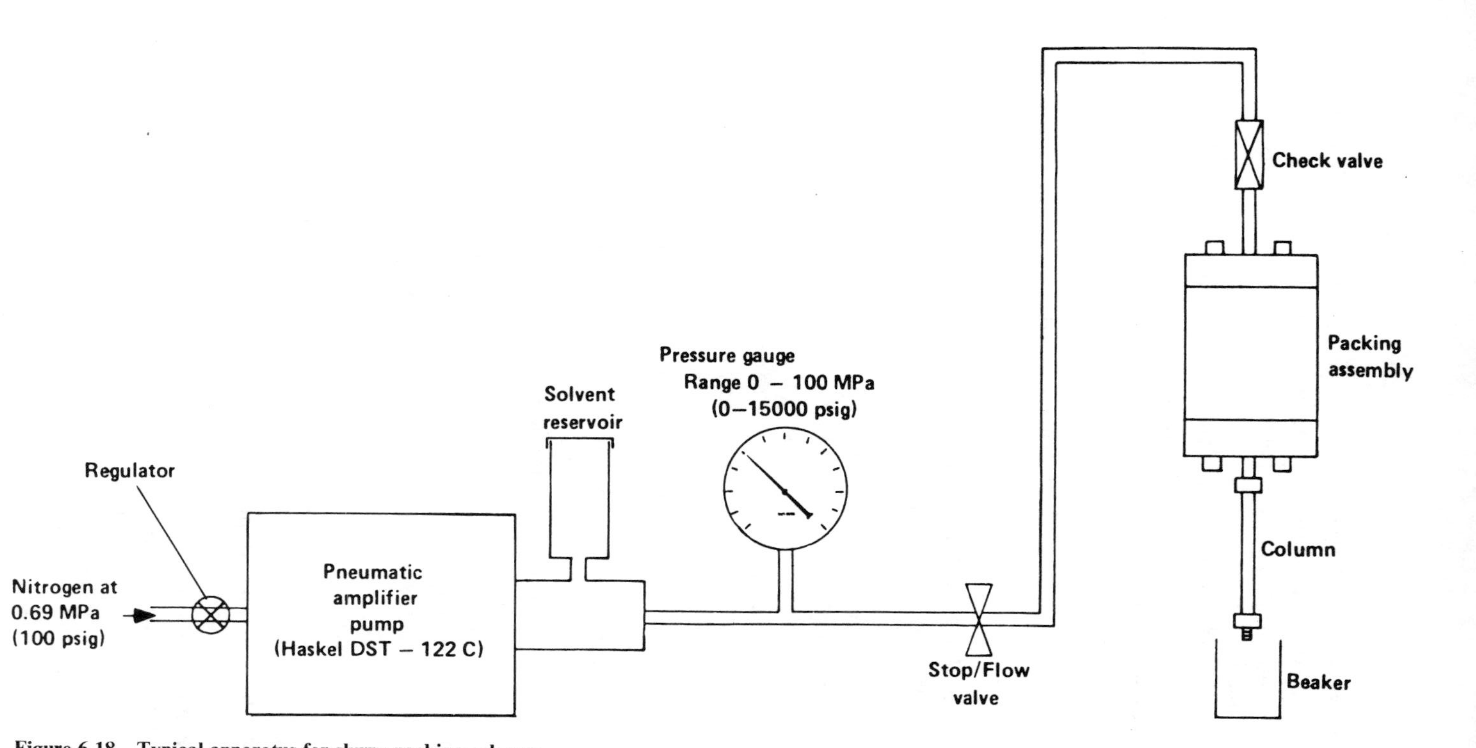

Figure 6.18 Typical apparatus for slurry-packing columns.
(Reprinted by permission from Ref. 23.)

slurry increases from an initially low value as the packed bed is formed. However, constant-volume pumps sometimes do not provide high-enough volume flow rates to prepare compact beds, and the use of constant-pressure pumps is generally recommended.

Slurries of 1–30% (by weight) of the packing have been used, and 5–15% solids generally appear most satisfactory. The slurry does not have to be pumped into the column with the same medium used to prepare it. Generally, a less dense fluid is preferred, and hexane is useful for this operation. It is desirable that the slurry be degassed and dispersed by vigorous shaking (e.g., ultrasonic vibration while evacuating the container containing the slurry).

Typical Slurry-Packing Procedure. A detailed procedure for preparing a column by a typical slurry-packing procedure follows:

> The column must be cleaned before packing. The inner walls are first degreased by washing with dichloromethane, acetone, and water, in turn. The interior of the tubing is then scrubbed with a hot detergent solution (0.5%), using a long pipe cleaner or a lint-free cloth soaked in the detergent solution. The latter may be tied to a fine wire or a nylon line (avoid scratching the inside of the tube!). The column blank is rinsed with pure water followed by absolute methanol, and dried with pure nitrogen or air.
>
> The appropriate porous disc or screen (Sect. 6.5) is then fitted to the outlet of the tubing and retained with an ordinary (not low-volume) outlet fitting, to allow maximum flow rate from the column during the packing operation. (This larger-bore outlet fitting must be replaced with an appropriate low-volume or "zero-volume" fitting before actually using the column for separations.)
>
> The column blank, interfaced with a short length (3–5 cm) of identical connector tubing (not shown in figure), is attached to the slurry-packing apparatus (e.g., that shown in Figure 6.18). This connector tubing or "precolumn" has been found to be useful in the column packing operation to direct the slurry into the column blank, and it also contains sufficient packing to ensure that the column is fully filled at the completion of the packing procedure.
>
> The empty column and precolumn initially are filled with a liquid more dense than that used to make the slurry, and the outlet of the column is temporarily closed to retain this liquid. The packing slurry (described below) is added rapidly to the reservoir, which is immediately attached to the pump. Meanwhile, the pump has been prepressurized with the pressurizing liquid up to the high-pressure shutoff valve at the

highest pressure allowed by the column-blank compression fittings (usually 5000–12,000 psi). The column outlet is opened and the reservoir-column assembly is then pressurized by suddenly opening the shutoff valve for several minutes. The high-pressure shutoff valve is closed and the pump shut down.

The mechanical stability of the packed bed is enhanced by bed consolidation or "slamming" (9). Experience has shown that methanol is desirable for this process, and it is also useful for purging the slurry solvent and conditioning the packing for subsequent chromatography. The consolidation process consists of pressurizing the pump again, with the high-pressure liquid shutoff valve closed, to the highest pressure allowed by the column blank-compression fitting assembly. The shutoff valve is then suddenly opened and 50–100 ml of fluid is allowed to pass through the column. A series of these sudden repressurizations consolidates the packing. Final stabilization is indicated by a constant flow of liquid through the packed bed as the repressurization process is continued at a constant input pressure (constant permeability).

A variety of liquids may be employed to prepare the packing slurry. For untreated silicas of <7 μm with a very narrow particle-size range, single solvents such as methanol, isopropanol, and 0.01 *M* ammonium hydroxide have been used. However, a 1:1 mixture of chloroform and methanol or trichloroethylene and ethanol has been found generally satisfactory for 5–10 μm particles for both unmodified silica and alumina (1, 13). These same organic mixtures, or a 1:3 mixture of isooctane and chloroform, have been used satisfactorily for trimethylsilyl-modified silica.

If the particles have a relatively wide particle-size distribution, a balanced-density technique is generally needed. The ratio of the solvents used for the suspending mixture is adjusted to provide a density equal to that of the particles, and the particles should stay suspended for at least 30 min without evidence of settling. Slurries made with tetrabromoethane/perchloroethylene, tetrabromoethane/tetrahydrofuran, and methyl iodide are all useful for both untreated and trimethylsilyl-modified silicas. For preparing columns of hydrophilic organic modified silicas (e.g., γ-glycidoxypropylsilane-treated) slurrying solvents of chloroform/methanol (1:1), tetrabromoethane/tetrahydrofuran, or *n*-propanol have been used successfully.

To prepare the packing slurry for use with the apparatus shown in Figure 6.18, a 10–15% excess of packing is weighed into a small glass container. (The amounts of packing needed for columns of various sizes are indicated in Table 6.6.) Degassed solvent is added to the container to produce the desired concentration (usually 5–15% by weight), and the contents are mixed vigorously in an ultrasonic bath for about 5 min. This mixture is then

Table 6.6 Approximate Weight (g) of Packing Required for 25 cm Columns

Column i.d. (cm)	LiChrospher[a]	Porous Silica Microspheres (1)[b]
0.21	0.5	0.7
0.32	1.1	1.5
0.46	2.3	3.2
0.62	4.0	5.5
0.78	5.1	8.5
0.85	7.5	11.0

Taken from Reference 7.
[a] Specific porosity approximately 0.8 cm^3/g.
[b] Specific porosity approximately 0.6 cm^3/g.

added to the reservoir and the column packing procedure is implemented immediately.

The systems just described above are only illustrative of those used successfully in the down-flow slurry-packing approach. Experience by many workers has indicated that other liquids and techniques may often be substituted in similar slurry-packing operations without loss of final column efficiency.

A somewhat different approach for preparing slurry-packed columns is used with an up-flow slurry-packing reservoir such as that shown in Figure 6.19 (24, 25). The pumping system of any high-pressure liquid chromatograph is used up to 6000 psi to pressurize the reservoir and provide the flow to move the slurry into the column blank. The packing is dispersed at 1–10% (by weight) concentration in a suitable liquid (e.g., isopropanol), and the slurry is stirred constantly to prevent particle sedimentation while forming the column bed. The continuous stirring homogenizes the slurry so that each increment entering the column contains the same particle-size distribution. Experience with this approach has been limited, but initial results appear approximately equivalent to those obtained for the down-flow procedure previously described.

In the high-pressure slurry packing of columns, certain precautions should be taken to assure operator safety. Although liquids have relatively low compressibility (store very little energy), the column-filling operation can present some dangers. First, in no instance should an air pocket be allowed to form in the high pressure system during slurry packing so that a sudden

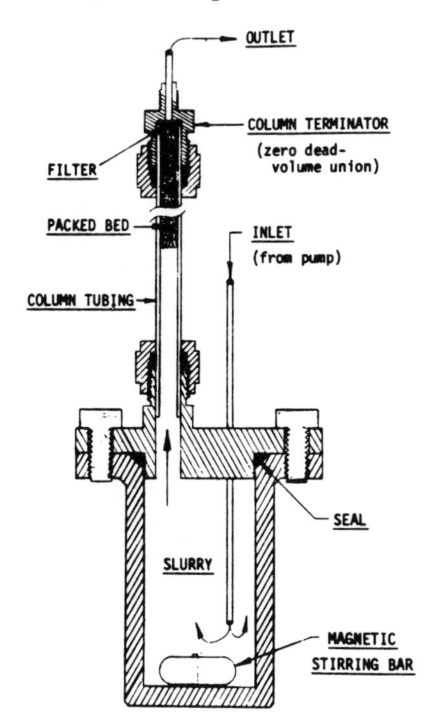

Figure 6.19 Stirred reservoir for slurry-packing columns.
(Courtesy of Micromeritics Instrument Corp.)

equipment failure could result in a sudden release of compressed gas. Second, excessive hydraulic forces can deform the slurry reservoir and compression fittings, causing leakage of solvents, which in certain cases can be fairly toxic (e.g., tetrabromoethane). For these reasons, wearing safety glasses (or shields) should be required for this operation. Further, it is much preferred that the column-filling operation be carried out in a well-ventilated hood behind shielding, to protect from sudden spurts of liquids, and to reduce exposure to toxic vapors.

Dry-Packing Technique. As mentioned in the previous section, dry-packing techniques generally cannot be used with rigid particles of less than 20–30 μm, but should be used for preparing preparative columns of large, rigid particles. The "tap-fill" method for dry-packing columns of rigid solids has been widely used for many years (12). The clean empty column with a porous plug-end fitting attached is held vertically and a small increment of packing (equivalent to about 3–5 mm of bed) is added through a funnel. The column is tapped firmly on the floor or bench top about 2–3 times a second (for

80–100 times total per sample increment) while lightly rapping the side and rotating the column slowly. The column should then be *very gently* tapped vertically for about 20 sec without rapping the side. Another increment of packing is added and the procedure repeated. After the column is filled, very gentle tapping should be continued for 3–5 min (no rapping on the side). The end face of the column tubing is then carefully cleaned off and the inlet fitting attached without disturbing the packing or creating a void. This procedure requires 20–30 min, depending on experience.

For the production of high-efficiency columns, mechanical vibration techniques such as those commonly used to prepare gas chromatographic columns should *not* be used during this dry-packing process. Lateral vibration techniques cause the particles to segregate by size across the column, resulting in poor column efficiency.

Commercially available column packing machines such as that shown in Figure 6.20 automatically add packing and tap the columns which are held vertically. A motor-driven cam causes the column to bounce up and down with the proper force. Vibration of the assembly by this motion is sufficient to cause a steady flow of the dry packing from the feeding reservoir into the column blank through an inlet funnel.

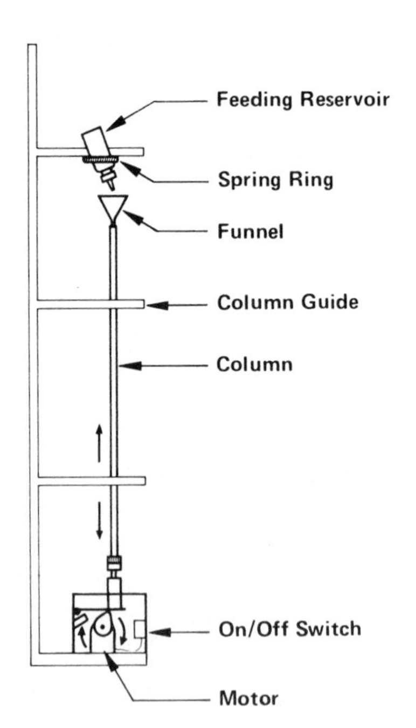

Figure 6.20 Automatic dry column packing machine. (Reprinted with permission from Ref. 13.)

Packing of Semirigid Gels

Columns of semirigid organic gels are generally packed by the balanced-density slurry-packing technique described above, with some modification. Since these polymeric gels are less dense than silica, the actual slurry-packing procedure is carried out with lower-density solvents (e.g., acetone/perchloroethylene mixtures) and at lower pressures. However, the same apparatus illustrated in Figure 6.18 can be used. In the column packing process, pressures and flow rates are limited according to the specific type of particles, as specified by the manufacturer. While no specific limits can be given, packing pressures for the semirigid gels rarely exceed 3000 psi. Generally, these packings must also first be swollen for several hours in the slurrying medium before packing. One commercial supplier of columns of 10 μm cross-linked styrene/divinylbenzene copolymer particles utilizes toluene for the slurry-packing process.

The softer Type OR-PVA, poly(vinyl acetate), gels are usually packed by gravity sedimentation. The gels are first allowed to swell in a suitable organic solvent (e.g., benzene) for at least 5 hr. The swollen gel is then stirred and packed into the appropriate column by introducing the slurry in several increments while allowing the particles to settle. It is important that the gel beds settle continuously. The bed will be more uniform, and better separations will be obtained if the column is rotated slowly about its axis while it is being filled in this manner. Further compaction of this column bed at the operating pressure is desired before use.

Columns of soft gels cannot be packed from dry particles, nor can the high-pressure slurry-packing process be used, since most of the gels compress and become impermeable even at relatively low pressures. Therefore, soft gels are not useful for HPSEC and are not discussed further in this book.

6.4 Column Technology

Column Materials

Resolution in HPSEC is especially dependent on the packed column and its associated hardware. Materials of construction are selected to withstand both the pressures and the chemical action of the mobile phase; most columns are made from stainless steel tubing. However, heavy-wall glass columns are sometimes used for special problems (e.g., halogen corrosion of stainless steel), and glass column blanks that will withstand pressures up to about 600 psi are commercially available (e.g., Laboratory Data Control).

Stainless steel column tubes with special inside walls are usually selected for HPSEC. Because the walls influence the homogeneity of the packed bed produced by the high-pressure slurry-packing technique, the inside diameter of the tubing should be regular and have a mirror finish. Satisfactory column blanks can be made from materials such as Type 316L Superpressure Tubing (Superior Tube Co.) or LiChroma tubing (Handy and Harman Tube Co.). The superiority of mirror-finished tubing is observed only with high-performance packings and equipment; less efficient SEC systems are not influenced significantly by different types of column blanks.

Column Dimensions, Configuration

Straight sections of column in lengths ranging from 15 to 50 cm are recommended for efficient packing, the optimum length depending on the size of the packing particles. Results suggest that maximum column length for 0.62 cm i.d. columns prepared by the down-flow high-pressure slurry technique should be:

Particle Size (μm)	Maximum Column Length (cm)
10–12	50
7–8	25
5–6	10–15

With these lengths the expected plate count can be obtained with the indicated particle sizes (18). Attempts to pack longer columns than those suggested usually result in significantly lower plate count than anticipated.

The end fittings and connectors for columns should be designed to minimized dead volume and unwanted "pockets" which can act as miniature mixing vessels and contribute to extra-column band broadening. Figure 6.21 compares the "zero" dead-volume compression fittings with ordinary commercial connectors. Low-dead-volume or zero-dead-volume fittings with small-diameter passages are now used by most instrument manufacturers.

Porous nickel or stainless steel frits (or plugs) are normally used in the ends of columns to retain the packing. Typically, these frits are ~1.5 mm thick with pores less than one-half the size of the packing particles. These frits (or plugs) must be homogeneous to ensure uniform flow and a minimum of unwanted peak broadening. Small extra column effects due to the column end fittings have been obtained using 1.5 μm porosity, 0.025 cm thick stainless steel screens to retain the column packings (19).

It appears that the optimum internal diameter of HPSEC columns is a complex function of particle size, the ratio of column length to internal

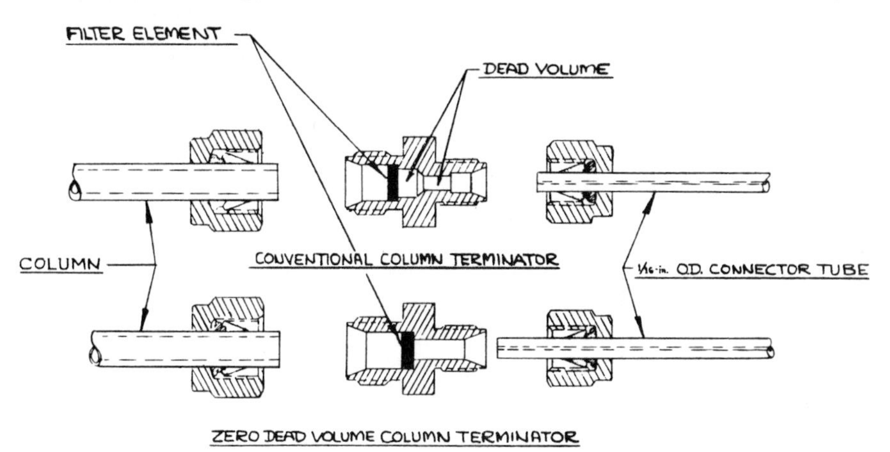

Figure 6.21 Comparison of conventional and "zero-dead-volume" column terminators. (Reprinted with permission of Micromeritics Instrument Corp.)

diameter, sample-injection technique, and other effects. The very small radial dispersion of a solute injected at a point on the inlet of a microparticulate column results in the solute never reaching the column walls before it emerges from the column, as illustrated in Figure 6.22. Columns in which this occurs are called *infinite-diameter*, because they behave as if they were of very large diameter and, under this condition, demonstrate highest efficiency (26, 27). There also is evidence that it is advantageous to avoid allowing the solute to reach the walls of columns, where axial peak dispersion is more rapid than in the middle of the column (27). However, HPSEC columns are often not operated in the infinite-diameter mode. In many cases both the mobile phase and solute are introduced (e.g., from a microinjector) via a small capillary onto the center of the column inlet, and in an unfavorable situation this can result in the sample being largely dispersed (sometimes unevenly) over the column inlet cross section (20). Under these conditions, an "infinite-diameter" column cannot exist, since a portion of the sample approaches the more highly dispersive wall area rather than remaining in a tight sphere at the column center. To approach the generally desired infinite-diameter column mode for analysis, allowances should be made for a perturbed sample-injection profile and a finite sample volume. Therefore, columns must generally be wider than might be theoretically predicted for infinite-diameter columns. It has been suggested that for 5 μm particles, the minimum internal diameter for full column performance should be 0.5 cm and the maximum length about 10 cm (28).

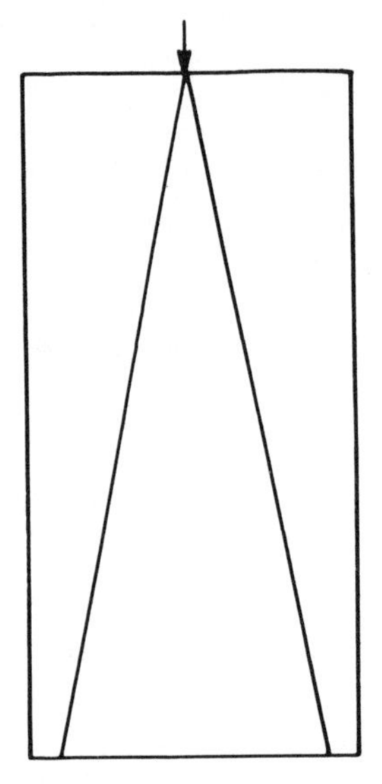

Figure 6.22 Schematic of flow pattern for "infinite-diameter" column.

Sample injected into center of column inlet radially disperses, but never reaches the column walls.

Complete experimental definition of the effect of column internal diameter on resolution in HPSEC has not been reported. However, a study made with 10 cm columns of 6 μm porous-silica microspheres indicate that highest plate count and best peak symmetry were obtained with 0.78 cm i.d. columns (1). On balance it appears that HPSEC columns with internal diameters larger than 0.5 cm are desirable. Internal diameters of 0.6–0.8 cm appear to be a good compromise between column performance and the amount of mobile phase and column packing required (see Fig. 7.14).

Column Performance

Even though conventional SEC is generally considered to be a low-resolution technique, use of columns with particles ≤20 μm substantially increases resolution. "Good" HPSEC columns exhibit reduced plate heights ($h = H/d_p$) $\simeq$ 2.0–3.5 for a totally permeating monomer such as toluene (e.g., 7000–10,000 theoretical plates for a 25 cm column of 10 μm particles). For well-packed columns, particle silanization appears to have little or no effect on efficiency. For example, no differences in plate height were found between columns of untreated or trimethylsilyl-modified porous silica microspheres of about 7 μm (1).

Specifications for column performance are not yet standardized. Most commercial suppliers do provide data on minimum plate count for their products, determined by internally prescribed operating conditions. For example, Waters Associates currently specifies a minimum plate count of 12,000 plates per meter for 100 Å μ-Styragel and 9000 for the other μ-Styragel columns; minimum column efficiency specification for μ-Bondagel-E columns is 4500 plates/30 cm, measured at the total permeation volume. These specifications are normally established with a low-viscosity solvent (~0.5 cP) and a low-molecular-weight monomer (~100). Plate count standards for higher-molecular-weight materials and higher-viscosity mobile phases have not been standardized. However, plate counts of >1200 have been reported for proteins on 30 cm columns of hydrophilic organic-modified silica particles (10 μm) using aqueous mobile phases (21).

As discussed in Section 4.2, the packing resolution factor, R^*_{sp}, is the more useful parameter for evaluating and comparing single columns (see also Ref. 22). An example of the use of R^*_{sp} is shown in Table 4.2. Other results not shown in Table 4.2 have also verified the utility of R^*_{sp} for comparing various packings using columns of different dimensions.

Manufacturers are now beginning to supply much-needed specifications on peak tailing (or peak asymmetry) for columns. For example, μ-Styragel and μ-Bondagel columns are specified with a peak asymmetry factor of 2.4 or less, hand-calculated by the technique shown in Figure 6.23. A more precise and accurate measure of peak tailing involves determining with a computer (22) the τ/σ ratio (a measure of peak skew; Sect. 3.5), where σ is the standard deviation of the Gaussian portion of the peak and τ the time constant of an exponential peak tailing (20). This ratio can then be used to

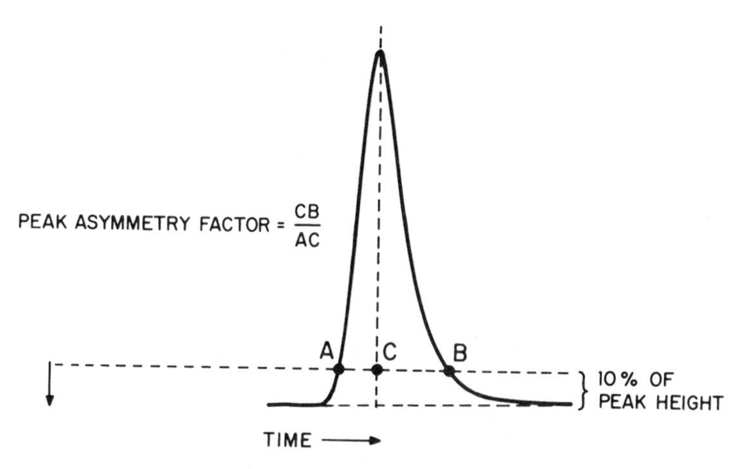

Figure 6.23 Calculation of empirical peak asymmetry factor. (Reprinted with permission from Ref. 22.)

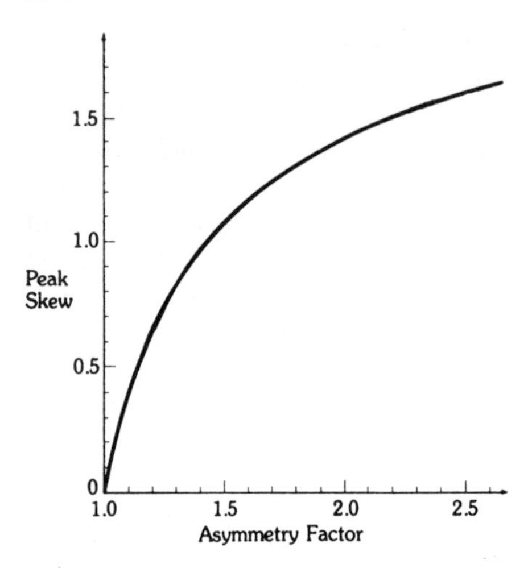

Figure 6.24 Relation of peak asymmetry to peak skew.

(Reprinted with permission of DuPont Instrument Products Division.)

calculate the mathematical peak skew (Eq. 3.39). The graphical relationship between peak skew value γ' and the peak asymmetry factor is given in Figure 6.24. A peak skew of <0.7 (corresponding to a peak asymmetry factor of <1.25) is desirable to ensure accurate column performance and GPC molecular weight data (22).

The pressure drop or permeability of a newly packed or purchased column should be checked before use. Column specific permeability is a function of mobile phase velocity, v, as given by the well-known relationship

$$K^{\circ} = \frac{v\eta L}{\Delta P} \tag{6.2}$$

where K° is the column permeability, ΔP the pressure drop across the column, η the viscosity of the mobile phase, L the length of the column, and v the velocity of the mobile phase. To check the pressure drop of a column in practice, Equation 6.3 is more easily used:

$$\Delta P = \frac{1.2 \times 10^4 L^2 \eta}{t_0 d_p^2 f} \tag{6.3}$$

where P is in psi, L in cm, η in cP, d_p in μm, t_0 in s (measured with totally excluded solute), and f assumes the value of 1 or 2 for irregular porous or spherical porous particles, respectively. A pressure drop significantly different from that predicted by Equation 6.3 indicates that the column may be poorly packed, or that the outlet frits or frittings may be partially plugged by particles. In either situation the problem should be rectified before attempting a separation. Column pressure drop (or permeability) should also be checked periodically during use, to ensure that changes have not occurred as a result of plugging or for some other reason.

A new column, whether purchased or packed in the laboratory, should always be tested before use to ensure that it satisfies the intended need and meets the proper specifications. Performance can be compared to that of columns previously used or to published data. Also, this testing provides reference performance data for future comparison. In the event that the performance of a column becomes doubtful, information on the initial pressure drop and plate count is important. Comparison of column performance should only be made when pertinent variables are maintained constant: flow rate, mobile phase, solute, temperature, and apparatus.

Techniques with Columns of Small Particles

The use of columns with small particles requires certain precautions to maintain column efficiency, capacity, and permeability (9). Good laboratory practice dictates that carefully filtered mobile phases should always be used, since columns of less than 10 μm particles are readily plugged by minute particulates (see also Sect. 8.7). Water is a particularly important source of contamination of this type. Buildup of particles at the column inlet can be eliminated by employing in-line solvent filters (e.g., 0.5 μm porosity) and by prefiltering samples before chromatographing them (Sect. 8.7).

To maintain initial performance, highly efficient columns for HPSEC should be used at pressures well below (about one-half) that employed in the slurry packing operation, since otherwise a void can develop at the inlet. While degraded performance caused by a void at the inlet can sometimes be improved by adding glass microbeads or more of the same packing, experience has shown that the initial performance of a good column rarely can be reestablished with such techniques. Vibration and extremes of temperature should also be avoided. In addition, while no change in column performance occurs when columns are used and stored with many neat organic mobile phases, aqueous mobile phases should be used only in the pH range 2–8 to prevent degradation of the packing. It is unwise to allow aqueous systems with the pH at the extremes of this range to stand in a column for long periods of time. Of particular importance when using aqueous mobile phases (e.g., buffer solutions) is bacterial growth, which can form within the column and cause pluggage. It is suggested that the column be purged with absolute methanol (or another pure, inert organic solvent) before storage for long periods. The drying out of columns of semirigid gels must be avoided. Closing both ends of all columns with compression-fitting caps is desirable before storage.

In some situations, adsorbed sample residues can affect retention-time reproducibility and column efficiency, or cause non-Gaussian peaks. In unusual situations such as this, it may be necessary to purge the column periodically with a solvent strong enough to remove tightly held components that have collected at the inlet. This condition generally does not occur when the proper choice of the combination of mobile phase and stationary phase has been made.

Because of the relatively high cost of HPSEC column packings, the question often arises whether it is feasible to recover and reuse materials from columns whose performance has deteriorated. Although it is sometimes possible to reconstitute materials by solvent extraction and repack satisfactory columns, this attempt is not generally recommended. Actually, it is often difficult (and usually impossible) to return packings fully to their original state, since used particles can be irreparably changed by use. For example, the structure of a silica packing can be significantly altered by dissolution in an aqueous buffer, or polystyrene gel particles can be partially collapsed as a result of solvent or salt ionic-strength changes, and so on. Thus it is strongly recommended that fresh column packing be utilized whenever possible to prepare columns that are more likely to be reproducible and of higher performance.

REFERENCES

1. J. J. Kirkland, *J. Chromatogr.*, **125**, 231 (1976).
2. S. Sato and Y. Otaka, *Chem. Econ. Eng. Rev.*, **3**, 40 (1971).
3. S. Sato, Y. Otaka, N. Baba, and H. Iwasaki, *Jap. Anal.*, **22**, 673 (1973).

3a. Toyo Soda Manufacturing Co., Ltd., Technical Data Bulletin, TSK-Gel SW Type Column, 1978.

4. K. K. Unger, R. Kern, M. C. Ninou, and K. F. Krebs, *J. Chromatogr.*, **99**, 435 (1974).
5. F. E. Regnier and R. Noel, *J. Chromatogr. Sci.*, **14**, 316 (1976).
6. W. A. Dark and R. J. Limpert, *J. Chromatogr. Sci.*, **11**, 114 (1973).
7. J. J. Kirkland, E. I. DuPont de Nemours & Co., unpublished studies, 1975.
8. K. K. Unger and P. Ringe, *J. Chromatogr. Sci.*, **9**, 463 (1971).
9. J. J. Kirkland, *Chromatographia*, **8**, 661 (1975).
10. I. Halasz and I. Sebestian, *Chromatographia*, **7**, 371 (1974).
11. E. P. Otocka, *Acc. Chem. Res.*, **6**, 348 (1973).
12. L. R. Snyder and J. J. Kirkland, *Introduction to Modern Liquid Chromatography*, Wiley-Interscience, New York, 1974, Chapt. 6.
13. S. Bakalyar, J. Yuen, and R. H. Henry, Spectra-Physics Chromatography Technical Bulletin 114–76, 1976.
14. J. J. Kirkland, *J. Chromatogr. Sci.*, **9**, 206 (1971).
15. R. E. Majors, *Anal. Chem.*, **44**, 1722 (1972).

16. J. J. Kirkland, *J. Chromatogr. Sci.*, **10**, 593 (1972).
17. J. Asshauer and I. Halasz, *J. Chromatogr. Sci.*, **12**, 139 (1974).
18. J. J. Kirkland and P. E. Antle, *J. Chromatogr. Sci.*, **15**, 137 (1977).
19. J. J. Kirkland, in *Gas Chromatography, 1972, Montreux*, S. G. Perry, ed., Applied Science Publishers, Barking, Essex, England, p. 39, 1973.
20. J. J. Kirkland, W. W. Yau, H. J. Stoklosa, and C. H. Dilks, Jr., *J. Chromatogr. Sci.*, **15**, 303 (1977).
21. K. K. Unger and N. P. Becker, Pittsburgh Conference on Analytical Chemistry and Applied Spectroscopy, March 1, 1977, paper 171.
22. W. W. Yau, J. J. Kirkland, D. D. Bly, and H. J. Stoklosa, *J. Chromatogr.*, **125**, 219 (1976).
23. T. J. N. Webber and E. H. McKerrell, *J. Chromatogr.*, **122**, 243 (1976).
24. P. A. Bristow, *J. Chromatogr.*, **131**, 57 (1977).
25. H. P. Keller, F. Erni, H. R. Lindner, and R. W. Frei, *Anal. Chem.*, **49**, 1958 (1977).
26. D. S. Horne, J. H. Knox, and L. McLaren, *Sep. Sci.*, **1**, 531 (1966).
27. J. H. Knox, G. R. Laird, and P. A. Raven, *J. Chromatogr.*, **122**, 129 (1976).
28. J. H. Knox, *J. Chromatogr. Sci.*, **15**, 353 (1977).

Seven

OPERATING VARIABLES

7.1 Introduction

Some considerations of the separating variables in SEC are different from those normally encountered in the other LC methods. In SEC the substrate (the porous packing) primarily determines retention and resolution. As a result, the mobile phase is chosen primarily for sample solubility, and secondarily to eliminate unwanted solute or substrate effects. With large molecules, diffusion is slow, and the need to maintain high column efficiency by using very small particles and low viscosity solvents is an important consideration. In this chapter we discuss the general aspects of mobile phase, substrate, and sampling effects in HPSEC. We also bring together information from other chapters and summarize the adjustment of these variables for optimum operation. Chapter 8 presents the experimental laboratory procedures involving these variables.

7.2 Solvent Effects

Sample Solubility

The dissolution process for polymers is, in some respects, rather different from that for low-molecular-weight substances. Polymer dissolution is preceded by swelling of the bulk solid phase, which is typically a slow process. In a second stage the molecules of the swollen polymer phase disentangle and enter into solution. The rate-controlling step is the solvent diffusion rate within the swollen polymer. Warming the mixture will reduce solvent viscosity and may speed up diffusion.

Polymer dissolution rate is affected by the crystallinity of the sample. In the case of highly crystalline polymers (e.g., linear polyethylene), the first step toward dissolution is to melt the crystalline region by heating to facilitate solvent permeation into the polymer. While the physical state of the polymer sample affects the rate of dissolution, the chemical structure of the polymer dictates the total solubility. Generally, polymer/solvent structural similarity favors enhanced polymer solubility. Dissolution occurs because the net attractive forces between the solvent and the solute outweigh those between pairs of solvent and solute molecules, respectively. As a general rule solubility decreases with increasing solute MW.

Solvents that provide rapid dissolution rates usually are small, compact molecules. However, solvents promoting high rates of solution do not necessarily possess the favorable thermodynamic properties that control the quantity of polymer that is dissolved. Some liquids exhibit both qualities and will dissolve polymers quickly to relatively high concentrations.

The dissolution process occurs if the free energy of mixing, ΔG, is negative:

$$\Delta G = \Delta H - T\,\Delta S \tag{7.1}$$

The entropy of mixing, ΔS, is positive for polymer solutions. If the enthalpy of mixing, ΔH, is negative (meaning there is a net positive attraction favoring solvent/solute pairs), then dissolution will occur at any temperature. When the polymer is in a poor solvent (e.g., ΔH is positive), the thermodynamics of dissolution depends on the temperature. Dissolution can occur at high temperatures because the negative $T\,\Delta S$ term in Equation 7.1 becomes dominant. Conversely, as temperature decreases, the $T\,\Delta S$ term is less important and the solvent becomes thermodynamically poorer (i.e., ΔG becomes less negative). Finally, with some systems a consolute temperature, T_c, is reached below which polymer and solvent are no longer miscible in all proportions and the mixture separates into two phases.

The dependence of T_c on polymer MW can be expressed by (1)

$$\frac{1}{T_c} \simeq \frac{1}{\theta}\left(1 + \frac{C}{M^{1/2}}\right) \tag{7.2}$$

where C is a constant and θ is the Flory theta temperature for the specific polymer/solvent system (1–3). (Both θ and T_c are in Kelvin units.) The Flory theta temperature is the critical miscibility temperature at "infinite" molecular weight. By convention a solvent whose theta temperature is close to

Table 7.1 Theta-Temperature Data for Polymer/Solvent Systems

Polymer	Solvent	Θ (°K)
Polystyrene	Octadecanol	474
Polystyrene	Cyclohexanol	358.4
Polystyrene	Cyclohexane	307.2
Polystyrene	Ethylcyclohexane	343.2
Polyethylene	Nitrobenzene	503
Polyisobutene	Diisobutyl ketone	331.1
Poly(methyl methacrylate)	Heptanone-4	305
Polydimethylsiloxane	Phenetole	358
Polydimethylsiloxane	Butanone	298.2
Cellulose tricaprylate	Dimethylformamide	413
Cellulose tricaprylate	3-Phenylpropanol-1	323
Poly(acrylic acid)	Dioxane	302.2
Polymethacrylonitrile	Butanone	279

Taken from Reference 4.

room temperature is called a "poor" solvent, since the polymer solution is then close to precipitation. Similarly, a "good" solvent is one whose theta temperature is well below room temperature. Table 7.1 shows the theta temperatures of some polymer/solvent systems. Also by convention, theta solvents are designated as those in which the theta temperature is approximately room temperature, or specified for whatever temperature is to be employed. In a thermodynamically good solvent where polymer/solvent interactions are favored, a polymer coil in solution is more extended than in a poor solvent. The radius of gyration of a polymer molecule (i.e., R_g; see Sect. 2.4) in solution is expanded by a factor α times that in the theta condition. Above the theta temperature the following relationship holds (1):

$$(\alpha^5 - \alpha^3) \propto \left(1 - \frac{\theta}{T}\right) M^{1/2} \tag{7.3}$$

At $T = \theta$, $\alpha = 1$, but α becomes larger with increasing T (°K); at the same time, α becomes more molecular-weight-dependent. At large α values, $\alpha^5 \propto M^{1/2}$, or the limiting molecular weight dependence of $\alpha \propto M^{0.1}$. These relationships explain why R_g increases with temperature in a poor solvent, as illustrated in Table 2.3. (The opposite effect that R_g decreases with increasing temperature can occur in a good solvent). The predicted molecular weight dependence on the expansion factor α is in agreement with observations (1, 2) that

$$R_g \propto M^{0.5-0.6} \tag{7.4}$$

and

$$[\eta] \propto M^{0.5-0.8} \tag{7.5}$$

In SEC experiments, good solvents help to avoid possible packing surface adsorption effects. Near θ conditions, the solvent is less effective in preventing polymer solute molecules from adsorbing to the packing surface.

Another measure of polymer/solvent interaction is provided by the solubility parameter. When only dispersion forces are involved, the enthalpy of mixing can be calculated by (5)

$$\Delta H = \phi_s \phi_m (\delta_s - \delta_m)^2 \tag{7.6}$$

where ϕ_s and ϕ_m are the volume fractions of solvent and macromolecule, and δ_s and δ_m are the solubility parameters defined by

$$\delta = \left(\frac{\Delta E}{V}\right)^{1/2} \tag{7.7}$$

where $\Delta E/V$ is the energy of vaporization per unit volume, or the cohesive energy density. For ΔH to be minimal, δ_s and δ_m should be similar. This solubility parameter approach was originally devised for small molecules, but it has been successfully applied to predict polymer solubility (6). Equation 7.6 predicts that $\Delta H = 0$ if $\delta_s = \delta_m$, so that two substances with equal solubility parameters should be mutually soluble because of the negative entropy factor. This is in accordance with the general rule that chemical and structural similarity favors solubility. As the difference between δ_s and δ_m increases, the tendency toward dissolution decreases. As with the theta-temperature approach, the solubility parameter method is used to study polymers in poor solvents (i.e., positive ΔH). For cases where ΔH is negative, the polymer is so easily solubilized that evaluation of solubility has no real meaning. Polymer and solvent are miscible in any proportion at any temperature.

Whereas δ_s can be measured directly, δ_m cannot. Polymers do not vaporize, and therefore δ_m has to be measured indirectly through swelling or viscosity experiments using solvents of known δ_s. Typical values of δ_m and δ_s for common polymers and solvents are shown in Tables 7.2 and 7.3. The values of "δ calc." in the last column of Table 7.2 were obtained by summing contributions to δ from structural groups in the polymer (6). The validity of using structural group summation to estimate δ_m is illustrated by the good agreement shown for experimental and calculated values of δ for polymers. Table 7.3 shows

Table 7.2 Experimental and Calculated Values of δ_m for Some Polymers

Polymer	δ_m, exp range ($J^{1/2}/cm^{3/2}$), From	To	δ_m, calc. ($J^{1/2}/cm^{3/2}$)
Polyethylene	15.8	17.1	16.0
Polypropylene	16.8	18.8	17.0
Polyisobutylene	16.0	16.6	16.4
Polystyrene	17.4	19.0	19.1
Poly(vinyl chloride)	19.2	22.1	19.7
Poly(vinyl bromide)	19.4	—	20.3
Poly(vinylidene chloride)	20.3	25.0	20.6
Poly(tetrafluoroethylene)	12.7	—	11.7
Poly(chlorotrifluoroethylene)	14.7	16.2	15.7
Poly(vinyl alcohol)	25.8	29.1	—
Poly(vinyl acetate)	19.1	22.6	19.6
Poly(vinyl propionate)	18.0	—	18.8
Poly(methyl acrylate)	19.9	21.3	19.9
Poly(ethyl acrylate)	18.8	19.2	19.2
Poly(propyl acrylate)	18.5	—	18.7
Poly(butyl acrylate)	18.0	18.6	18.3
Poly(isobutyl acrylate)	17.8	22.5	18.7
Poly(2,2,3,3,4,4,4-heptafluorobutyl acrylate)	13.7	—	15.8
Poly(methyl methacrylate)	18.6	26.2	19.0
Poly(ethyl methacrylate)	18.2	18.7	18.6
Poly(butyl methacrylate)	17.8	18.4	17.9
Poly(isobutyl methacrylate)	16.8	21.5	18.3
Poly(*t*-butyl methacrylate)	17.0	—	18.0
Poly(benzyl methacrylate)	20.1	20.5	19.3
Poly(ethoxyethyl methacrylate)	18.4	20.3	18.6
Polyacrylonitrile	25.6	31.5	25.7
Polymethacrylonitrile	21.9	—	22.8
Poly(α-cyanomethyl acrylate)	28.7	29.7	23.8
Polybutadiene	16.6	17.6	17.5
Polyisoprene	16.2	20.5	17.4
Polychloroprene	16.8	18.9	19.2
Polyformaldehyde	20.9	22.5	20.5
Poly(tetramethylene oxide)	17.0	17.5	17.6
Poly(propylene oxide)	15.4	20.3	18.9
Polyepichlorohydrin	19.2	—	20.1
Poly(ethylene sulfide)	18.4	19.2	18.9
Poly(styrene sulfide)	19.0	—	19.6
Poly(ethylene terephthalate)	19.9	21.9	20.5
Poly(8-aminocaprylic acid)	26.0	—	25.7
Poly(hexamethylene adipamide)	27.8	—	28.0

[a] Conversion factors: $J^{1/2}/cm^{3/2} = 0.49\ cal^{1/2}/cm^{3/2}$.

Taken from Reference 6.

Table 7.3 Solubility Parameter and Hydrogen-Bonding Tendency of Solvents[a]

δ_S	Poorly Hydrogen-Bonded	Moderately Hydrogen-Bonded	Strongly Hydrogen-Bonded
			ethylene glycol
		ethylene carbonate	
30			
			methanol
		butyrolactone	
28			
		propylene carbonate	
			ethanol
26			
	nitromethane		
		DMF	formic acid
			n-propanol
		acetonitrile	
24			
			isopropanol
		HMPT	
		NMP	
	nitroethane		m-cresol
		DMA	
22			
		TMU	
		dioxane	
		acetone	
	tetrachloroethane		
20			
		tetrahydrofuran	
	chlorobenzene	cyclohexanone	
	Tetralin	methyl acetate	
	chloroform	methyl ethyl ketone	
	benzene	ethyl acetate	
	toluene		
18	p-xylene		
	carbon tetrachloride		
	n-butyl chloride	butyl acetate	
	cyclohexane		
16			
	heptane	diethyl ether	

Taken from Ref. 6.

[a] DMA, dimethylacetamide; DMF, dimethylformamide; HMPA, hexamethylphosphoramide; NMP, *N*-methylpyrrolidone; TMU, tetramethyl urea.

solvents in three groups representing different hydrogen-bonding tendencies. The arrangement of Table 7.3 is convenient to use, since mutual solubility only occurs if the degree of hydrogen bonding between the polymer and solvent is approximately matched. The values of δ in Tables 7.2 and 7.3 have the dimensions $J^{1/2}/cm^{3/2}$, where J = joules. Values of δ expressed in $cal^{1/2}/cm^{3/2}$ have become familiar quantities for many investigators.

Conversion of $cal^{1/2}/cm^{3/2}$ to $J^{1/2}/cm^{3/2}$ requires multiplication by a factor of 2.046.

The effect of δ_h (hydrogen bonding) is significantly different from those of δ_d (dispersion) and δ_p (polar forces). Therefore, it is advantageous to classify the ability of solvents to dissolve solutes by separating the contributions of δ_h, δ_d, and δ_p. Utilization of the parameter $\delta_v = (\delta_d^2 + \delta_p^2)^{1/2}$,

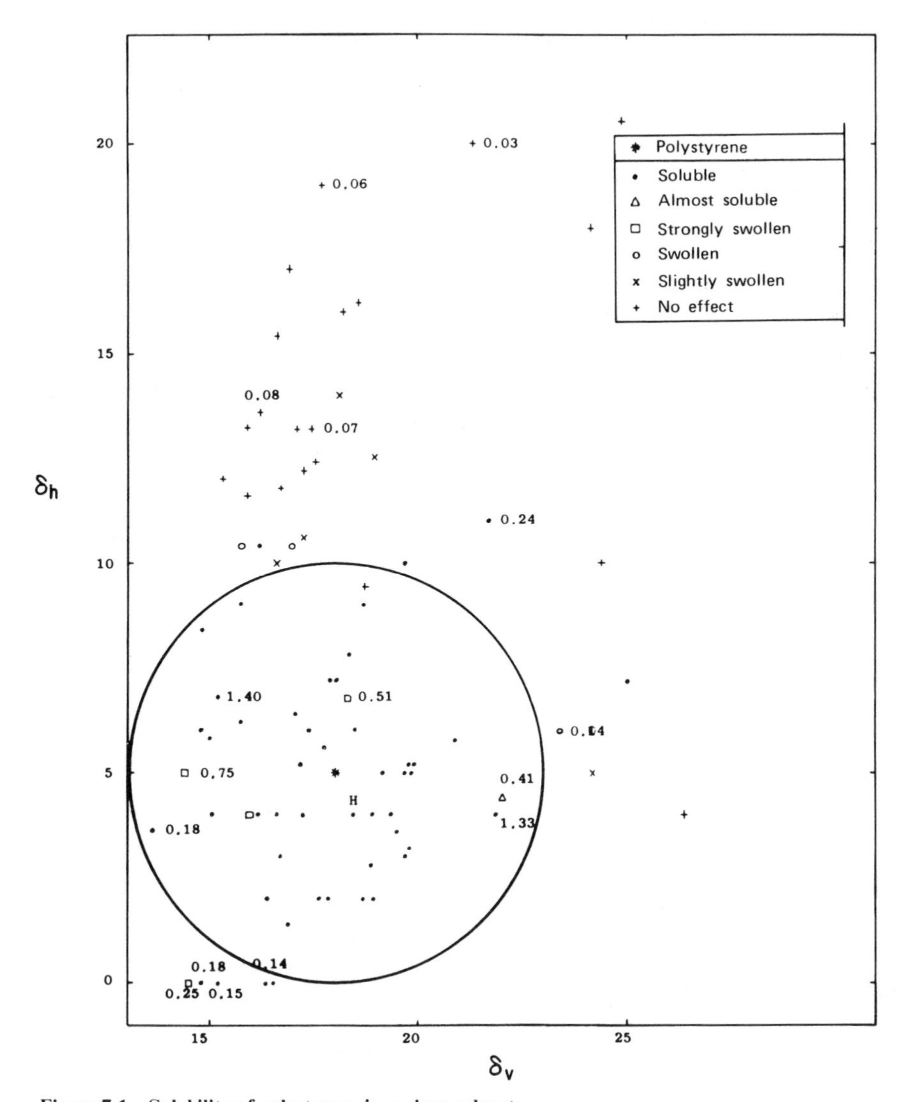

Figure 7.1 Solubility of polystyrene in various solvents.
Numbers: volume polystyrene/volume solvent. (Reprinted with permission from Ref. 6.)

leads to a degree-of-solubility diagram involving δ_v versus δ_h, as illustrated in Figure 7.1 for the interaction between polystyrene and a number of solvents (7). The majority of the points for good solvents fall in a single region, which can be defined approximately by a circle. The center of the circle is indicated by the symbol * in Figure 7.1 and has the coordinate values $\delta_v = 18$, $\delta_h = 5$. Thus in the (δ_h, δ_v) diagram the degree of solubility (volume of polymer per volume of solvent) can be indicated by a number. The solubility region can be approximately defined by a circle with a radius of about 5 units of δ. It can be seen that the solubility generally decreases as the distance from the center increases. As a general rule, polystyrene is soluble in solvents for which

$$\sqrt{(\delta_v - 18)^2 + (\delta_h - 5)^2} < 5 \tag{7.8}$$

The (δ_h, δ_v) diagram is an efficient way to represent polymer/solvent interactions.

Solubility limits of a given polymer are closely related to the Flory theta temperatures of the polymer in various solvents, as shown in the (δ_h, δ_v) diagram for polystyrene in Figure 7.2. Thus the closer the data points are to the center of the circle in Figure 7.2, the better is the solubility of the system and the lower are the values of the theta temperature.

Typical values of δ_d, δ_p, and δ_h for common polymers and solvents are listed in Tables 7.4 and 7.5.

Until now our discussion of polymer solubility has been limited to the use of a single solvent. In *mixed solvents* the solubility relationships become

Table 7.4 Specified Solubility Parameters for Some Polymers[a]

Polymer	δ	δ_d	δ_p	δ_h
Polyisobutylene	17.6	16.0	2.0	7.2
Polystyrene	20.1	17.6	6.1	4.1
Poly(vinyl chloride)	22.5	19.2	9.2	7.2
Poly(vinyl acetate)	23.1	19.0	10.2	8.2
Poly(methyl methacrylate)	23.1	18.8	10.2	8.6
Poly(ethyl methacrylate)	22.1	18.8	10.8	4.3
Polybutadiene	18.8	18.0	5.1	2.5
Polyisoprene	18.0	17.4	3.1	3.1

Taken from Reference 6.
[a] All δ units in $(J/cm^3)^{1/2}$.

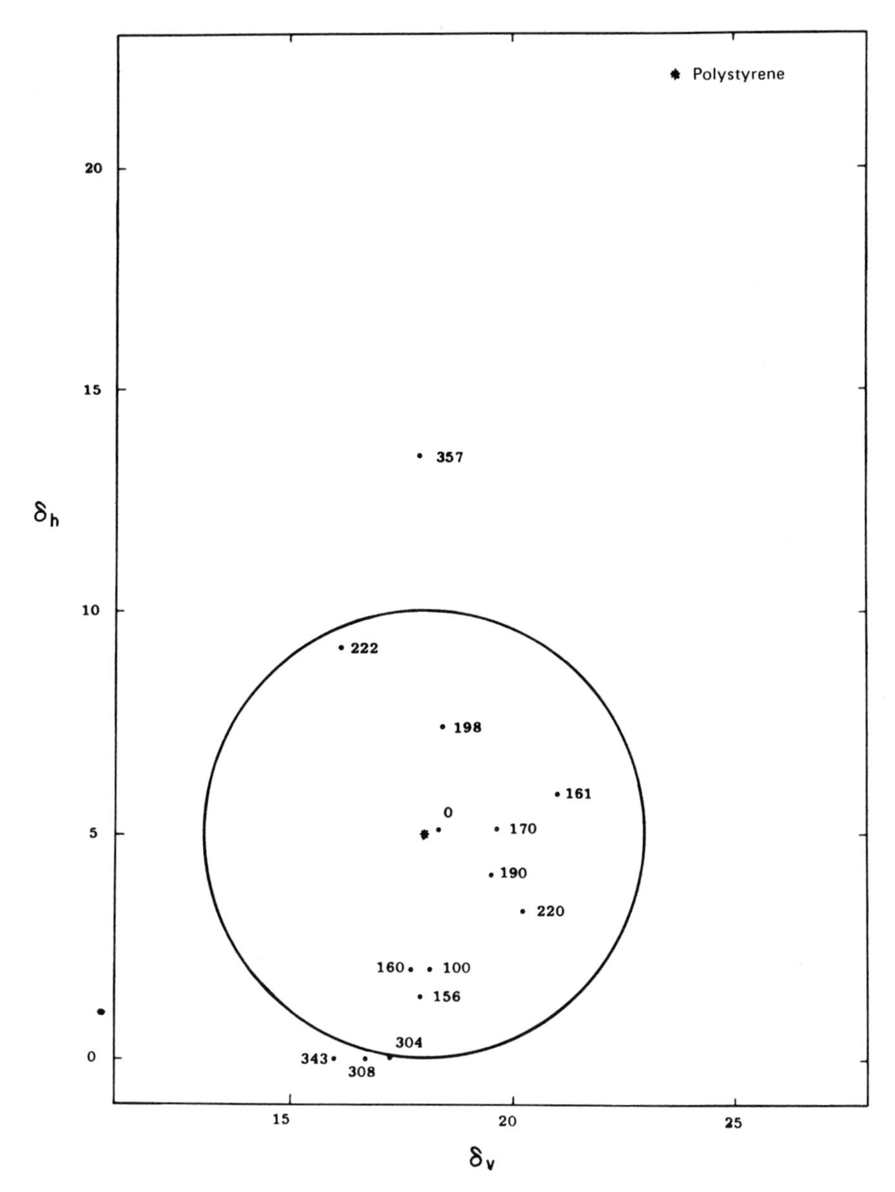

Figure 7.2 Flory temperature of polystyrene in various solvents.
Numbers represent theta temperatures. (Reprinted with permission from Ref. 6.)

Table 7.5 Three-Dimensional Solubility Parameters of Solvents[a]

Solvent	δ_o	δ_d	δ_p	δ_h	Solvent	δ_o	δ_d	δ_p	δ_h
Acetic acid	10.5	7.10	3.9	6.6	Diethylene glycol monobutyl ether	8.96	7.80	3.4	5.2
Acetic anhydride	10.30	7.50	5.4	4.7	Diethylene glycol monomethyl ether	10.72	7.90	3.8	6.2
Acetone	9.77	7.58	5.1	3.4	Diethyl ether	7.62	7.05	1.4	2.5
Acetonitrile	11.9	7.50	8.8	3.0	Diethylsulfide	8.46	8.25	1.5	1.0
Acetophenone	9.68	8.55	4.2	1.8	Diisobutyl ketone	8.17	7.77	1.8	2.0
Aniline	11.04	9.53	2.5	5.0	Dimethylformamide	12.14	8.52	6.7	5.5
Benzaldehyde	10.40	9.15	4.2	2.6	Dimethylsulfoxide	12.93	9.00	8.0	5.0
Benzene	9.15	8.95	0.5	1.0	Dioxane	10.0	9.30	0.9	3.6
α-Bromonaphthalene	10.25	9.94	1.5	2.0	Dipropylamine	7.79	7.50	0.7	2.0
1,3-Butanediol	14.14	8.10	4.9	10.5	Dipropylene glycol	15.52	7.77	9.9	9.0
n-Butanol	11.30	7.81	2.8	7.7	Ethanol	12.92	7.73	4.3	9.5
2-Butoxyethanol	10.25	7.76	3.1	5.9	Ethanolamine	15.48	8.35	7.6	10.4
n-Butyl acetate	8.46	7.67	1.8	3.1	Ethyl acetate	9.10	7.44	2.6	4.5
n-Butyl lactate	9.68	7.65	3.2	5.0	Ethylbenzene	8.80	8.70	0.3	0.7
Butyric acid	9.2(?)	7.30	2.0	5.2	2-Ethylbutanol	10.38	7.70	2.1	6.6
γ-Butyrolactone	12.78	9.26	8.1	3.6	Ethylene chloride	9.76	9.20	2.6	2.0
Butyronitrile	9.96	7.50	6.1	2.5	Ethylene glycol	16.30	8.25	5.4	12.7
Carbon disulfide	9.97	9.97	0	0	Ethylene glycol monoethyl ether	11.88	7.85	4.5	7.0
Carbon tetrachloride	8.65	8.65	0	0	Ethylene glycol monoethyl ether acetate	9.60	7.78	2.3	5.2
Chlorobenzene	9.57	9.28	2.1	1.0	Ethylene glycol monomethyl ether	12.06	7.90	4.5	8.0
1-Chlorobutane	8.46	7.95	2.7	1.0	2-Ethylhexanol	9.85	7.78	1.6	5.8
Chloroform	9.21	8.65	1.5	2.8					

m-Cresol	11.11	8.82	2.5	6.3	Ethyl lactate	10.5	7.80	3.7	6.1
Cyclohexane	8.18	8.18	0	0	Formamide	17.8	8.4	12.8	9.3
Cyclohexanol	10.95	8.50	2.0	6.6	Formic acid	12.15	7.0	5.8	8.1
Cyclohexanone	9.88	8.65	4.1	2.5	Furan	9.09	8.70	0.9	2.6
Cyclohexylamine	9.05	8.45	1.5	3.2	Glycerol	21.1	8.46	—	—
Cyclohexylchloride	8.99	8.50	2.7	1.0	Hexane	7.24	7.23	0	0
Diacetone alcohol	10.18	7.65	4.0	5.3	Isoamyl acetate	8.32	7.45	1.5	3.4
o-Dichlorobenzene	9.98	9.35	3.1	1.6	Isobutyl isobutyrate	8.04	7.38	1.4	2.9
2,2-Dichlorodiethyl ether	10.33	9.20	4.4	1.5	Isophorone	9.71	8.10	4.0	3.6
Diethyl amine	7.96	7.30	1.1	3.0	Mesityl oxide	9.20	7.97	3.5	3.0
Diethylene glycol	14.60	7.86	7.2	10.0	Methanol	14.28	7.42	6.0	10.9
Methylal	8.52	7.35	0.9	4.2	*n*-Propanol	11.97	7.75	3.3	8.5
Methylene chloride	9.93	8.91	3.1	3.0	Propylene carbonate	13.30	9.83	8.8	2.0
Methyl ethyl ketone	9.27	7.77	4.4	2.5	Propylene glycol	14.80	8.24	4.6	11.4
Methyl isoamyl ketone	8.55	7.80	2.8	2.0	Pyridine	10.61	9.25	4.3	2.9
Methyl isobutyl carbinol	9.72	7.47	1.6	6.0	Styrene	9.30	9.07	0.5	2.0
Methyl isobutyl ketone	8.57	7.49	3.0	2.0	Tetrahydrofuran	9.52	8.22	2.8	3.9
Morpholine	10.52	9.20	2.4	4.5	Tetralin	9.50	9.35	1.0	1.4
Nitrobenzene	10.62	8.60	6.0	2.0	Toluene	8.91	8.82	0.7	1.0
Nitroethane	11.09	7.80	7.6	2.2	1,1,1-Trichloroethane	8.57	8.25	2.1	1.0
Nitromethane	12.30	7.70	9.2	2.5	Trichloroethylene	9.28	8.78	1.5	2.6
2-Nitropropane	10.02	7.90	5.9	2.0	Water	23.5	6.0	15.3	16.7
Pentanol-1	10.61	7.81	2.2	6.8	Xylene	8.80	8.65	0.5	1.5

Taken from Reference 8.

[a] All δ units in $(cal/cm^3)^{1/2}$.

Table 7.6 Dissolution of Polymers in Mixtures of Two Nonsolvents for the Polymer

Polymer	Nonsolvent 1	Nonsolvent 2
Polystyrene	Acetone	Nonane
Polystyrene	Methyl acetate	Nonane
Polystyrene	Phenol	Acetone
Buna S	Methyl acetate	Pentane
Buna N	Dimethyl malonate	*p*-Cymene
Gel rubber	Benzene	Butanol
Poly(vinyl acetate)	Water	Ethanol
Poly(vinyl acetate)	Ethanol	Carbon tetrachloride
Poly(methyl methacrylate)	Propanol	Water
Poly(vinyl alcohol) with 30% acetyl groups	Methanol	Water
Poly(vinyl isobutyral)	Benzene	Ethanol
Poly(vinyl chloride)	Acetone	Carbon disulfide
Poly(vinyl chloride)	Nitromethane	Trichloroethylene
Polychloroprene	Acetone	Hexane
Cellulose tribenzyl ether	Ethanol	Carbon tetrachloride

Taken from Reference 9.

more complicated, but nevertheless can be formulated entirely from considerations of intermolecular forces. In Table 7.6 some examples are compiled for the solubility of polymers in mixtures of two liquids, where neither liquid by itself is a solvent for that polymer. These solubility phenomena can be explained as follows (9).

Polystyrene does not dissolve in acetone because the intermolecular association between the acetone molecules through polar forces is too strong. Polystyrene is also insoluble in nonane, since the intermolecular dispersion forces between the polystyrene molecules (because of high π-electron polarizability) are stronger than the intermolecular forces between polystyrene and nonane. However, in the mixture of acetone and nonane, polystyrene dissolves easily at 20°C to form a homogeneous solution, because the intermolecular acetone associations that inhibit strong solvation are broken up by the nonane molecules. The single acetone molecules that result are more able to solvate polystyrene molecules through polar forces to an extent sufficient for solution.

Similar arguments define the solubility behavior of poly(vinyl chloride) in acetone/carbon disulfide, poly(vinyl chloride) in nitromethane/trichloroethylene, polychloroprene in acetone/hexane, and Buna S in methyl acetate/pentane. In systems of poly(vinyl acetate) + ethanol/carbon tetrachloride and (cellulose-tribenzylether) + ethanol/carbon tetrachloride, dissociated alcohol molecules hydrogen bond to the polymer oxygen-containing groups. With phenol alone, the intermolecular H bonding is so strong that polystyrene does not dissolve. However, if the phenolic hydroxyl groups are made partially inactive for intermolecular association through complexation by the addition of acetone, then polymer dissolution is possible.

Propanol is a nonsolvent for poly(methyl methacrylate) because of steric hindrance. However, if the propanol contains water, assocations form between propanol and water and the "solubility arm" of the alcohol is lengthened. In this way the macromolecules are solvated adequately. A similar situation is also true for poly(vinyl acetate) + ethanol/water and other such systems.

With random copolymers solubility, in general, is greater than that of the respective homopolymers, because the irregular arrangements of groups responsible for the intermolecular forces in the copolymer cause a lower degree of interaction between adjacent molecules. Therefore, random copolymers can be solvated more easily than the corresponding homopolymers (9).

So far we have dealt primarily with dissolving synthetic polymers in organic solvents. However, dissolution of biopolymers in aqueous solvents can also be troublesome at times, particularly if the material is labile. Since the hydrogen bonding of water molecules largely dictates the mode and extent of solvation, the various solubility forces discussed above are not generally operable in aqueous systems. However, biopolymer solubility can be significantly altered by adjusting salt concentration and pH levels. For example, at low salt concentrations (e.g., $<0.01\ M\ PO_4^{-3}$), a strong increase in the solubility of nonapeptides occurs with increasing salt concentration (salting-in effect) (10). At higher salt concentrations (e.g., $>0.05\ M$), this increase reaches a maximum and then decreases (salting-out effect). Thus ionic strength or salt concentration significantly affects solute conformation and size through solvation effects (11). As with synthetic polymers, if possible biopolymers should also be dissolved in the mobile phase used for the separation. However, provided that denaturization or precipitation is not a problem, sometimes it is feasible to dissolve a sample at the ionic strength required to obtain a high sample concentration, then chromatograph this solution in a mobile phase at another ionic strength for optimum separation.

The adjustment of pH for sample dissolution is limited by solute stability. Some solutes (e.g., enzymes) often are unstable outside pH 5–9. Also, some solutes (e.g., DNA polymers) are less soluble at low pH. In the latter case it may be desirable to raise the solution to pH 9 with dilute sodium or ammonium hydroxide solution, and then carefully neutralize after the polymer has dissolved.

Sometimes it is difficult to wet a freeze-dried polymer due to gel formation at the surface. Agitation on a vortex mixer is useful in these cases, but ultrasonic dissolution is prohibited for higher-molecular-weight polymers (i.e., $\gtrsim 10^5$) because of possible shear degradation (Sect. 7.3). Heating some biopolymers at 40–50°C for 5–10 min to increase dissolution rate is permitted. However, solutions of very unstable compounds may have to be kept cool at all times (5°C for best protection), and chromatographed as soon as possible to ensure minimum decomposition.

Other Solvent Effects

Unlike other LC methods, the mobile phase in SEC is not varied to control resolution. Rather, the mobile phase is limited to the solvents that can dissolve the sample macromolecule. If permitted, a solvent of low viscosity at the temperature of separation (due to improved mass transfer; see Sect. 3.4) is preferred to ensure high column plate count. To maintain high resolution, mobile phases that have boiling points only about 25–50°C higher than the column temperature should be used. In such a case, the viscosity of the mobile phase will usually be <1 cP and solute diffusion rates will be relatively high. Of course, in the case of difficultly soluble samples, the solvent must be selected primarily to provide sufficient solubility, and viscosity considerations are then secondary.

A particular mobile phase has also to be selected on the basis of its compatibility with the solute detector. Thus if a differential refractometer is to be used, the refractive index of the mobile phase should be as different as possible from that of the sample. Or the mobile phase must have a much lower absorption than the solute at the wavelength of detection with a UV or IR photometric detector (Sect. 5.2).

An occasional problem in SEC is molecular association of the sample. For example, in some solvents ionic surfactants form micelles which produce asymmetrical bands and retentions that change with sample size. However, by using a more dilute sample solution or a mobile phase that is a better solvent for the sample, this type of association can usually be eliminated.

Controlled sample association can be advantageous for certain separations. For instance, carboxylic acids dimerize in nonbasic solvents such as benzene, but are solvated with more basic solvents such as tetrahydrofuran and remain monomeric. The retention of a carboxylic acid varies with its apparent molecular size and decreases in nonbasic solvents. Association between solvent and sample molecules (e.g., certain alcohols plus tetrahydrofuran) can also be used to control certain separations. As shown in Figure 7.3,

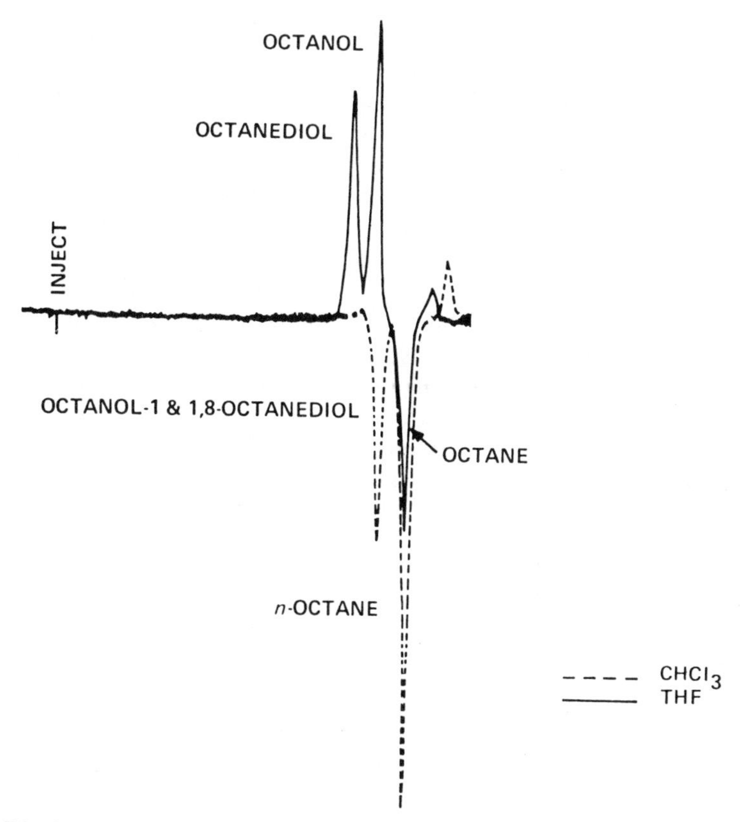

Figure 7.3 Sample/solvent association effects.
Column, four 30 × 0.78 cm μ-Styragel 100 Å; mobile phase, chloroform or THF; flow rate, 2.0 ml/min; sample, 1,8-octanediol, 1-octanol, *n*-octane; detector, refractive index. (Reprinted with permission from Waters Associates.)

1,8-octanediol and 1-octanol coelute in chloroform, but with tetrahydrofuran complete resolution of these solutes is obtained. Presumably, bonding with tetrahydrofuran increases the size of the difunctional alcohol relative to that of the monofunctional, resulting in improved resolution. Changes in retention due to bonding with tetrahydrofuran are usually small. However, by adding a larger associating solvent to the mobile phase, the molecular size of a particular solute may sometimes be artificially increased through association, to carry out an advantageous separation (e.g., association of benzoic acid with aniline in *N*,*N*-dimethylformamide mobile phase).

The effect of the solvent on the various SEC packings must be considered. For example, the cross-linked polystyrene gels for GPC can tolerate a moderate range of organic solvents, but acetone, alcohols, and other highly polar solvents cannot be used (Sect. 8.5). Aqueous systems outside the pH range of about 2–8 degrade siliceous packings. Strong bases, such as NaOH and tetramethylammonium hydroxide, should be avoided but organic amines (e.g., triethylamine) are well tolerated (12). Salts in general appear to hasten the degradation of silica packings, particularly at temperatures above ambient, the effect being a direct function of pH and ionic strength. As discussed in more detail in Sections 6.2 and 8.5, solvents that cause collapse of organic gels for GPC must be avoided. In GFC, and in GPC with certain very polar solvents such as dimethylformamide, a salt must often be added to the solvent to reduce solute adsorptive effects and maintain a constant ionic strength. However, large changes in salt concentration can cause an organic gel to collapse. Salt concentration also affects molecular size in some cases, as discussed in Sections 12.7 and 13.3. The manufacturer's literature should be consulted for each packing to determine which mobile phases are allowed. Many commonly used solvents for HPSEC are listed in Table 8.3, together with most of the properties of interest.

Flow Rate Effects

Flow rate level has a significant influence on the efficiency and the resolution of an HPSEC column. Figure 7.4 shows the effect of mobile-phase velocity on plate height for a monomer and a series of polystyrene standards with a set of porous silica microphere columns. For lower-molecular-weight solutes, this set of columns exhibits a relatively small increase in plate height with increased mobile phase velocity. This is characteristic of the relatively rapid solute equilibration associated with small ($<10\ \mu m$) porous particles. In this case, mobile-phase velocity can be substantially increased without significant sacrifice in the resolving power because of the excellent mass transfer characteristics. The shape of the plate height versus velocity plot for the totally permeating solute toluene in Figure 7.4 agrees with plate height theory (Sect. 3.2). The plate height minimum, at a mobile phase velocity of $\simeq 0.15$ cm/s, represents about 25,000 theoretical plates for this high-efficiency system. With this column set a 97,200 MW polystyrene standard exhibited an apparent plate count of 10,000 at the same mobile phase velocity.

The polymer standards in Figure 7.4 show the anticipated increase in plate height with increasing velocity; however, the less steep slope of the 390,000 and 37,000 MW polystyrene plots at low velocities does not agree with theory. The most plausible explanation for this artifact is peak broadening as a result of partial fractionation of these polymer standards by the high-efficiency system. This trend is particularly apparent for the 3600 MW polystyrene (dashed line in Fig. 7.4), which shows an anomalous, almost

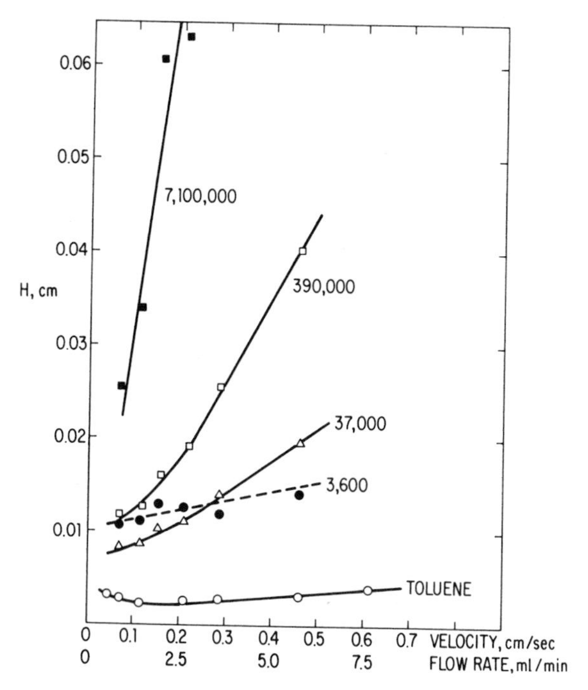

Figure 7.4 Effect of mobile-phase velocity and flow rate on column plate height. Columns, porous silica microspheres: 10 cm, 60 Å; 15 cm, 125 Å; 10 cm, 300 Å, 10 cm, 750 Å; 15 cm, 3500 Å (60 cm total, all columns 0.78 cm i.d.); mobile phase, tetrahydrofuran, 22°C; sample volume, 25 μl, polystyrene standard solutions. (Reprinted with permission from Ref. 13.)

flat plate height versus velocity plot, presumably because of significant molecular weight fractionation. These data indicate that because of partial fractionation of polymer molecular sizes present in the standards, measured plate height (and σ values) may be significantly larger than the true values for these materials. This effect often causes the apparent resolution of HPSEC columns to be less than it actually is.

The practical effect of mobile-phase velocity on column efficiency is illustrated by the separation of small molecules in Figure 7.5. When optimum resolution is required and high separation speed is of secondary importance, SEC columns are normally operated at mobile-phase velocities of 0.2 cm/s or less (measured with a totally excluded solute). Typically, flow rates of <2 ml/min are used with a 0.78 cm i.d. column, and 1 ml/min is often preferred as a compromise between resolution and speed.

Very high molecular weight polystyrenes are degraded at high mobile-phase velocities, probably as a result of the mechanical shear of the large

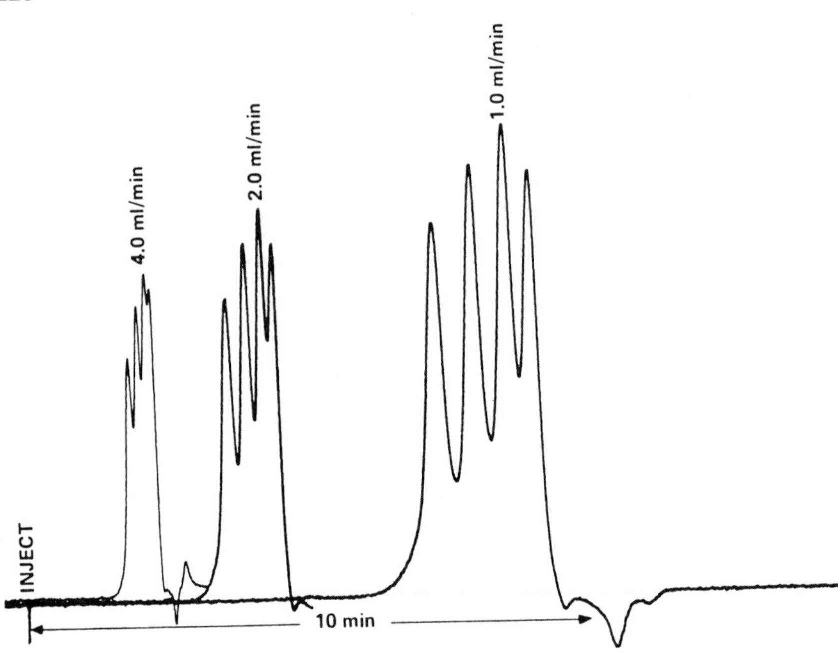

Figure 7.5 Effect of mobile velocity on separation of small molecules.
Column, 30 × 0.78 cm μ-Styragel 100 Å; mobile phase, tetrahydrofuran; sample, dioctylphthalate, dibutylphthalate, diethylphthalate, and dimethylphthalate; detector, UV. (Reprinted with permission from Waters Associates.)

polymer chain under these conditions (13, 14). Table 7.7 shows that the relative retention of a 7.1×10^6 MW polystyrene standard apparently increases (i.e., the molecular weight decreases) at successively higher mobile-phase velocities on a column of ~8 μm porous silica microspheres. This is in contrast to the constant relative retention for lower-molecular-weight polystyrenes under the same conditions. Collection of the 7.1×10^6 MW polystyrene chromatographed at about 0.5 cm/s, followed by reinjection into the column at a mobile phase velocity of 0.1 cm/s, yielded a chromatogram for a material of lower molecular weight and broader molecular weight distribution than the original, substantiating the fact that the starting polymer sample had undergone degradation during the first separation. This degradation at higher mobile-phase velocities is also in accord with observations that polystyrene of $>2 \times 10^6$ MW is degraded by agitation of solutions in an ultrasonic bath. These phenomena suggest that a mobile-phase velocity no greater than about 0.1 cm/s should be used when characterizing very high molecular weight polystyrenes. This effect may also

Table 7.7 Effect of Mobile-Phase Retention on Relative Retention and Shear of Polystyrene Standards

Flow Rate (ml/min)	Approximate Velocity (cm/s)[a]	V_R/V_R^O (PS/Toluene)			
		7,100,000	390,000	37,000	3600
0.50	0.05	0.567	0.680	0.796	0.910
0.76	0.07	0.569	0.678	0.794	0.910
1.28	0.11	0.580	0.679	0.795	0.909
1.78	0.15	0.592	0.681	0.796	0.909
2.50	0.21	0.609	0.683	0.796	0.910
3.39	0.29	[b]	0.684	0.799	0.909
5.26	0.46	[b]	0.680	0.793	0.912
7.27	0.61	[b]	0.684	0.794	0.917

Taken from Reference 13.
[a] Based on 7.1 mm PS peak, not totally excluded.
[b] Not calculated; badly overlapping with 390,000 PS peak.

be expected for polymers other than polystyrene. It should also be noted that shear degradation can occur in sample filtration (see Sect. 8.7).

In HPGPC, instrumental flow rate fluctuations can cause large errors in calculated sample $\overline{M}_w$ and $\overline{M}_n$ with systems in which the separations are monitored according to retention time rather than retention volume. In high-performance systems, solute retention volumes are relatively small, and errors associated with flow rate variations can be significant. An accurate, constant flow rate is required not only during a given experiment, but across the entire time span, from the calibration runs to the sample analyses.

The influence of flow rate fluctuations on values of $\overline{M}_w$ and $\overline{M}_n$ has been determined in a simulation study. Figure 7.6*a* illustrates the effect of a random variation of flow rate about a fixed level from time zero (sample injection) to completion of the GPC analysis. This effect could arise from random flow rate fluctuations arising from the pump (e.g., uneven check-valve action). If during the analysis a 5% random variation in flow rate occurs, the $\overline{M}_n$ for the test polymer ($\overline{M}_n = 112{,}900$; $\overline{M}_w = 278{,}100$) can vary from the expected $\overline{M}_n$ by 7000 (2σ) and $\overline{M}_w$ by 11,200 (2σ). However, if enough runs are made, the mean values of $\overline{M}_n$ and $\overline{M}_w$ approach the expected values.

The effect of a gradual change in flow rate during the entire time a sample elutes from the columns is shown in Figure 7.6*b*. Flow rate variations of this type could arise when the viscosity of a polymeric sample is high, and the viscosity of the total liquid system decreases as sample emerges from the

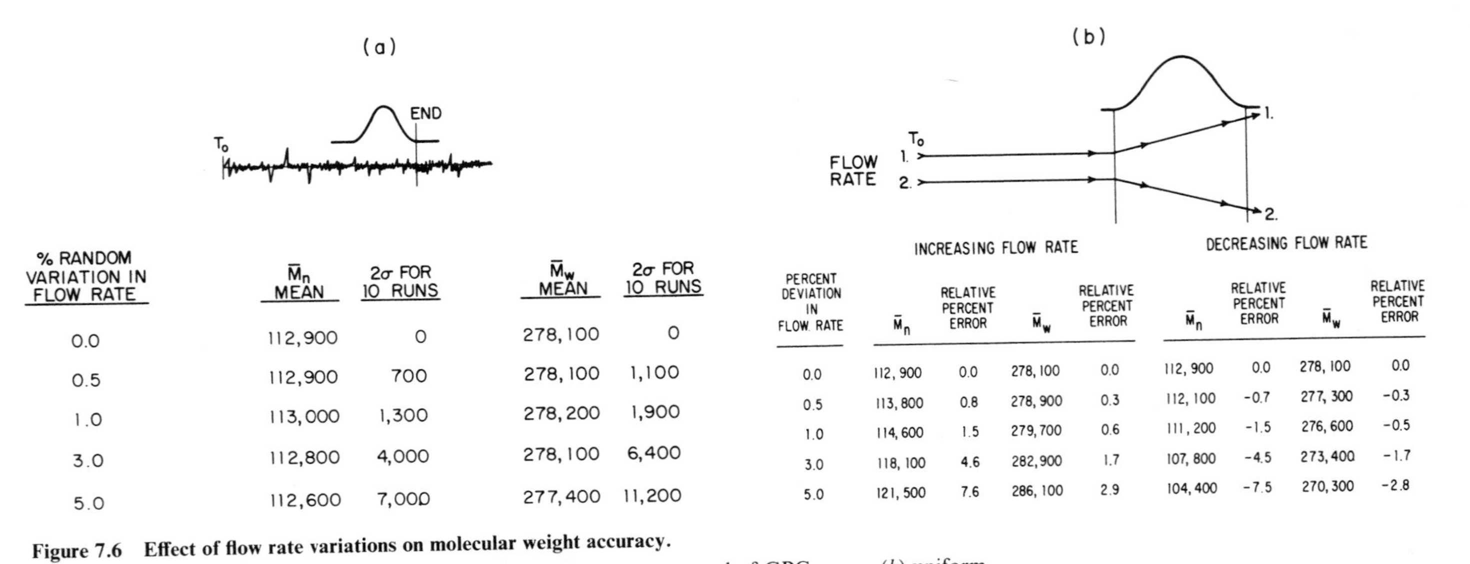

% RANDOM VARIATION IN FLOW RATE	$\bar{M}_n$ MEAN	2σ FOR 10 RUNS	$\bar{M}_w$ MEAN	2σ FOR 10 RUNS
0.0	112,900	0	278,100	0
0.5	112,900	700	278,100	1,100
1.0	113,000	1,300	278,200	1,900
3.0	112,800	4,000	278,100	6,400
5.0	112,600	7,000	277,400	11,200

PERCENT DEVIATION IN FLOW RATE	INCREASING FLOW RATE				DECREASING FLOW RATE			
	$\bar{M}_n$	RELATIVE PERCENT ERROR	$\bar{M}_w$	RELATIVE PERCENT ERROR	$\bar{M}_n$	RELATIVE PERCENT ERROR	$\bar{M}_w$	RELATIVE PERCENT ERROR
0.0	112,900	0.0	278,100	0.0	112,900	0.0	278,100	0.0
0.5	113,800	0.8	278,900	0.3	112,100	-0.7	277,300	-0.3
1.0	114,600	1.5	279,700	0.6	111,200	-1.5	276,600	-0.5
3.0	118,100	4.6	282,900	1.7	107,800	-4.5	273,400	-1.7
5.0	121,500	7.6	286,100	2.9	104,400	-7.5	270,300	-2.8

Figure 7.6 Effect of flow rate variations on molecular weight accuracy.
(*a*) Random variation of flow rate about a fixed level from time zero to end of GPC curve; (*b*) uniform change in flow rate across the GPC curve. (Reprinted with permission from Ref. 15.)

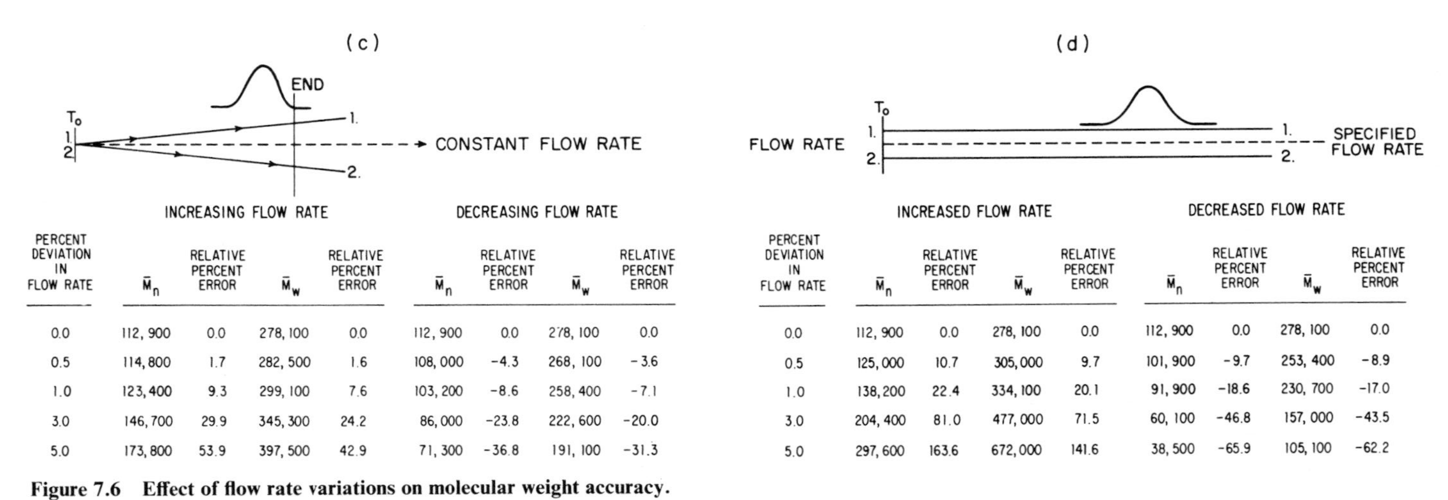

PERCENT DEVIATION IN FLOW RATE	INCREASING FLOW RATE $\bar{M}_n$	RELATIVE PERCENT ERROR	$\bar{M}_w$	RELATIVE PERCENT ERROR	DECREASING FLOW RATE $\bar{M}_n$	RELATIVE PERCENT ERROR	$\bar{M}_w$	RELATIVE PERCENT ERROR
0.0	112, 900	0.0	278, 100	0.0	112, 900	0.0	278, 100	0.0
0.5	114, 800	1.7	282, 500	1.6	108, 000	-4.3	268, 100	-3.6
1.0	123, 400	9.3	299, 100	7.6	103, 200	-8.6	258, 400	-7.1
3.0	146, 700	29.9	345, 300	24.2	86, 000	-23.8	222, 600	-20.0
5.0	173, 800	53.9	397, 500	42.9	71, 300	-36.8	191, 100	-31.3

PERCENT DEVIATION IN FLOW RATE	INCREASED FLOW RATE $\bar{M}_n$	RELATIVE PERCENT ERROR	$\bar{M}_w$	RELATIVE PERCENT ERROR	DECREASED FLOW RATE $\bar{M}_n$	RELATIVE PERCENT ERROR	$\bar{M}_w$	RELATIVE PERCENT ERROR
0.0	112, 900	0.0	278, 100	0.0	112, 900	0.0	278, 100	0.0
0.5	125, 000	10.7	305, 000	9.7	101, 900	-9.7	253, 400	-8.9
1.0	138, 200	22.4	334, 100	20.1	91, 900	-18.6	230, 700	-17.0
3.0	204, 400	81.0	477, 000	71.5	60, 100	-46.8	157, 000	-43.5
5.0	297, 600	163.6	672, 000	141.6	38, 500	-65.9	105, 100	-62.2

Figure 7.6 Effect of flow rate variations on molecular weight accuracy.
(*c*) Uniform change in flow rate from time zero to end of GPC curve; (*d*) fixed flow rate deviation from specified value. (Reprinted with permission from Ref. 15.)

column. This kind of flow rate variation can also accompany the technique of staggered injections, in which a new sample is injected at the head of the column while the previous sample is emerging at the outlet. The data in Figure 7.6*b* show that a 5% change in flow rate of this type would cause an 8% error in $\overline{M}_n$ and a 3% error in $\overline{M}_w$. However, a 5% change is unusual with high-performance systems, and with a more realistic 1% variation, only about 1–2% change in the computed molecular weight occurs.

As indicated in Figure 7.6*c*, long-term constant drift in flow rate causes much more serious errors. Such flow rate variation may result from reciprocating-pump check valves that become dirty or clogged, from variations in solvent metering systems that are temperature dependent, or from temperature drifts in the column. If the flow rate changes by 5% between the time of sample injection and that of completion of the run, there is a 37% or 54% change in $\overline{M}_n$ and a 31% or 43% change in $\overline{M}_w$, depending on whether the flow rate is increasing or decreasing during the course of the experiment. Even at only 1% flow rate drift, errors amount to 7–10% in $\overline{M}_w$, which can be important in many SEC applications.

As illustrated in Figure 7.6*d*, the most significant molecular weight errors are caused by poor flow rate repeatability, defined as the ability of the pump to deliver a specified volume per unit time on a day-to-day basis under the same set of operating conditions. For this particular example, a flow rate of 2.00 ml/min was specified and the calibration was made by establishing a relationship between retention time and molecular weight. The sample was analyzed in the same manner, except that it was computed for a new flow rate, which had been changed by simulation. (In this case there was no additional random fluctuation or drift assumed in the flow rate.) The data for a 1% flow rate change between the time of the calibration experiment and the time of analysis show a corresponding relative error of 22% in $\overline{M}_n$ for the increased flow rate, and −19% for the decreased flow rate, with similar results for $\overline{M}_w$. Such errors generally are intolerable, and a flow rate repeatability of much better than 1% (e.g., 2.00 ml/min ± 0.02 ml/min) is desired for precise molecular weight analysis in HPGPC when chromatograms are monitored on a time basis.

The results of the preceding study can be used to predict specifications for the solvent delivery systems in HPGPC. Of particular importance is the requirement that the repeatability of the solvent delivery system on a long-term basis must be better than about 0.3% for errors of ≤6% in $\overline{M}_n$ and $\overline{M}_w$. Long-term flow rate drift (increase or decrease) is also serious, because molecular weight errors of about 6% are caused by flow rate variations of somewhat less than 1% for a 20 min analysis. Random fluctuation or short-term instability of the pumping system appears to be less critical in GPC (unless it seriously affects detector baseline stability), and random short-term flow rate variations of 1–4% apparently can be tolerated for all but the most critical applications. These conclusions assume that flow rate variation has no effect on detector response or the separation mechanism.

Since design and fabrication of sophisticated pumps with a solvent delivery of ±0.3% is difficult and expensive, other approaches are suggested. One technique is to utilize real-time flow rate measurements with data acquisition software to correct for flow rate changes (16) (see also Sect. 10.5). Another approach that appears promising for certain systems is to use oligomeric internal standards to compensate for flow rate variation (17).

Retention volumes of macromolecules injected at low concentrations in small volumes are essentially independent of flow rate for practical HPSEC systems, as shown in Figure 7.7. In this case, flow rates of 0.1 to 12.5 ml/min were used with 0.85 cm i.d. columns. The rate of mass transfer associated with permeation into the pores is fast compared to the SEC experiment, so solute distribution between phases is not affected by flow rate (see also Sect. 2.3). No change in retention volume occurs for either large molecules (polystyrene MW = 411,000; diffusion coefficient, $D_s = 22 \times 10^{-8}$ cm²/s) or small molecules (acetonitrile, MW = 41; $D_s = 1 \times 10^{-5}$ cm²/s). This

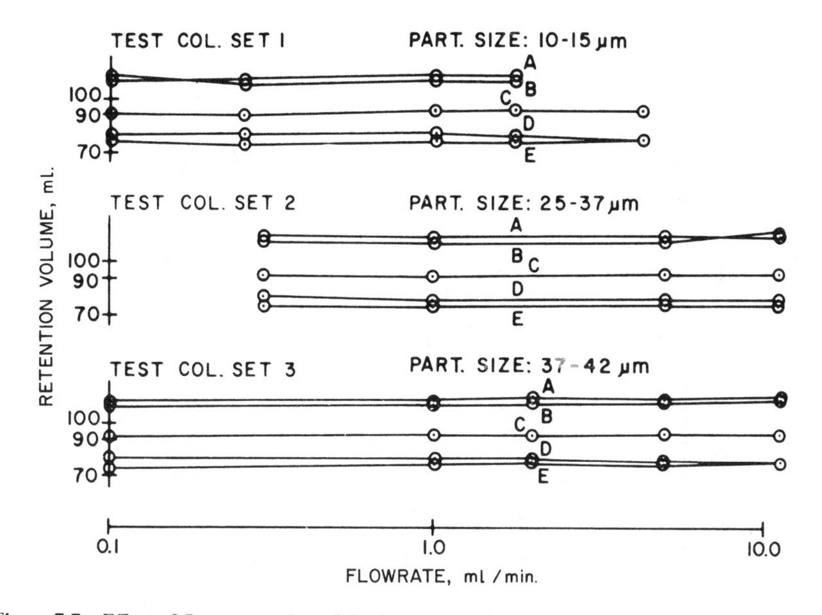

Figure 7.7 Effect of flow rate and particle size on retention volume.
(A) 0.05% CH_3CN; (B) 0.5% CH_3CN; (C) 0.05% and 0.5% polystyrene MW 19,850; (D) 0.5% polystyrene MW 411,000; (E) 0.05% polystyrene MW 411,000 (all tests under same operating conditions). (Reprinted with permission from Ref. 18.)

effect is observed even with columns packed with larger particles (37–42 μm) which have much wider interstitial spaces and deeper pore networks. It appears that most macromolecules have time to equilibrate fully with the column porous network at flow rates normally employed for SEC. Note also in Table 7.7 that the relative retention volumes of the lower-molecular-weight polystyrenes (i.e., 390,000, 37,000, and 3,600) are independent of mobile-phase flow rate. The small variation in elution volume with changes in flow rate that have been reported (19,20) can be attributed to the apparent shifting of peak maximum because of changes in peak shape at high flow rates, or error in peak retention (or flow rate) measurements (e.g., because of solvent evaporation problems in open syphon counters).

Temperature Effects

In HPSEC, temperature is increased to enhance sample solubility or to improve column efficiency by decreasing solvent viscosity. However, for

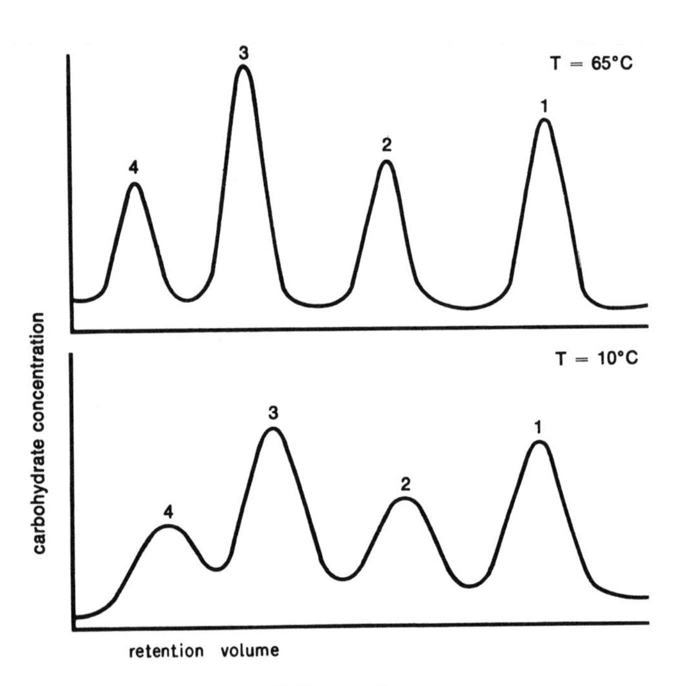

Figure 7.8 Effect of temperature on a GFC separation.
Column, 100 × 2.5 cm Bio-Gel P-2; mobile phase, water; sample, 20 μl of a solution containing 320 μg glucose (1), 240 μg lactose (2), 360 μg raffinose (3), and 200 μg stachyose (4); detector, RI. (Reprinted with permission from Ref. 21.)

Table 7.8 Effect of Temperature on Column Efficiency[a]

MW (Polystyrene Standards)	Plate Count			
	297°K	323°K	373°K	417°K
411,000 (excluded)	2601	2315	2100	2427
173,000	130	141	230	235
98,000	151	212	235	298
51,000	220	242	329	329
19,000	363	389	460	465
5,000	1126	1163	1214	1238
600	3125	3571	3677	3968
106 (ethylbenzene)	6757	6803	6898	6945

Taken from Reference 22.
[a] Column 25 × 0.4 cm porous silica 70–600 Å, 10 μm; mobile phase, tetrahydrofuran; velocity, 0.1 cm/s; detector, UV at 254 nm.

convenience, many GPC separations are carried out at room temperature. The characterization of higher-MW polyolefins (and polyamides in some solvents) requires temperatures greater than 100°C, since these materials are difficult to dissolve at lower temperatures (Sect. 12.4). If the sample is sufficiently stable, GFC also can be employed at higher temperatures (e.g., 50–70°C), to improve column efficiency and resolution. Figure 7.8 shows the influence of temperature on the separation of some carbohydrates by GFC, and Table 7.8 shows the effect of temperature on the plate count for a series of polystyrene standards with a silica microparticle column.

While temperature can have a significant effect on column resolution, its influence on the slope and position of the molecular weight calibration curve is relatively minor as long as the macromolecule is dissolved in a good solvent (Sect. 2.3). Figure 7.9 shows the dependence of retention volume on column temperature for polystyrenes chromatographed on a set of Bioglas columns with 1,2,4-trichlorobenzene. A slight shift toward smaller retention volumes is observed for this system with increasing temperature.

While adsorption of the solute on the column packing plays no part in an ideal SEC separation, in actual practice, limited adsorption does occur at times. Adsorption is more noticeable with lower-molecular-weight solutes, since through permeation they are exposed to a much higher surface area in the packing. Therefore, decrease of adsorption with increasing temperature can be expected in some situations with low-molecular-weight materials.

Operation of HPSEC columns at high temperatures also can have a significant effect on some detector outputs, as discussed in Section 5.2. Thus devices

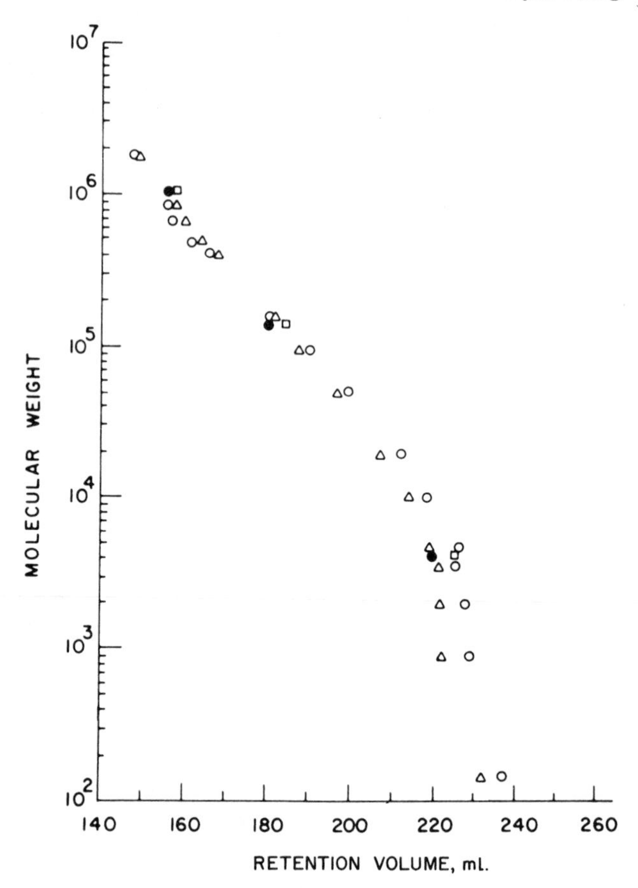

Figure 7.9 Effect of temperature upon SEC retention calibration.
Retention volume/molecular weight relationships for polystyrene and polyisobutene eluted from treated Bioglas columns with 1,2,4-trichlorobenzene: ○, polystyrenes, 25°C, △, polystyrenes, 150°C; □, polyisobutenes, 25°C; ●, polyisobutenes, 150°C. (Reprinted with permission from Ref. 23.)

based on the bulk solution properties (e.g., refractive index and viscosity) must be carefully thermostated to maintain the required baseline stability.

7.3 Substrate Effects

As discussed in Chapter 2, pore structure of the substrate (the packing material) largely determines retention in SEC. Consequently, the first step in an SEC separation is the selection of an appropriate packing to cover the expected molecular-size range of the sample (Chapts. 6 and 8).

The physical characteristics of the substrate have considerable effect on the operation of the HPSEC system. The general effect of particle size on column performance was discussed in Sections 3.3 and 6.4. Figure 7.10 further illustrates how particle size affects the performance of columns operated at increasing mobile-phase velocities. The plate count advantage of the smaller particles is particularly evident at the higher mobile-phase velocities needed for very high speed HPSEC separations. At the present time HPSEC columns of 5–10 μm particles are commonly offered by suppliers. A narrow particle-size distribution (e.g., not more than $\pm 20\%$ of the average size) is desired to produce high-efficiency columns with the best permeability (Sect. 6.4), while the internal pore volume or the specific porosity of a packing should be as large as possible for optimum resolution and highest molecular weight accuracy (Sect. 2.5). The method of preparation largely determines particle porosity. Porosity can be measured accurately by mercury intrusion for rigid particles, but not for organic gels. As discussed in Section 8.5, pore-volume constancy for the various pore-size packings is particularly important to obtain wide linearity in the molecular weight calibration by the bimodal pore-size approach. Pore volume for each pore size used in the column set generally should vary by no more than about 20% and preferably less than 10% to allow useful linearity of the molecular weight calibration (25, 26). Packing pore-size distribution and pore geometry are also important SEC variables that can significantly influence resolution (Chapt. 4).

Improper selection of the packing pore size can lead to false double peaks in SEC. When part of a polymer sample is totally excluded (i.e., has a molecular weight greater than the exclusion limit), that part elutes at the total exclusion volume. Therefore, the presence of a sharp polymer band at the total exclusion volume should always be considered suspect in molecular weight interpretation. In such cases, the sample should be rechromatographed on a column with a larger exclusion limit if accurate molecular weight is required. If the sharp band at the exclusion limit from the first separation is eliminated, the false exclusion peak in the first experiment is identified. However, if the initial band is still apparent in the second separation with the larger exclusion limit but no longer at the total exclusion volume, a true bimodal molecular weight distribution exists.

Retention of the solute by mechanisms other than size exclusion greatly complicates the process of obtaining sample molecular weight information from SEC chromatograms. Adsorption or "matrix" effects involving a form of partition or adsorption can be superimposed on size exclusion, resulting in excessive retention, so that the desired relationship between retention

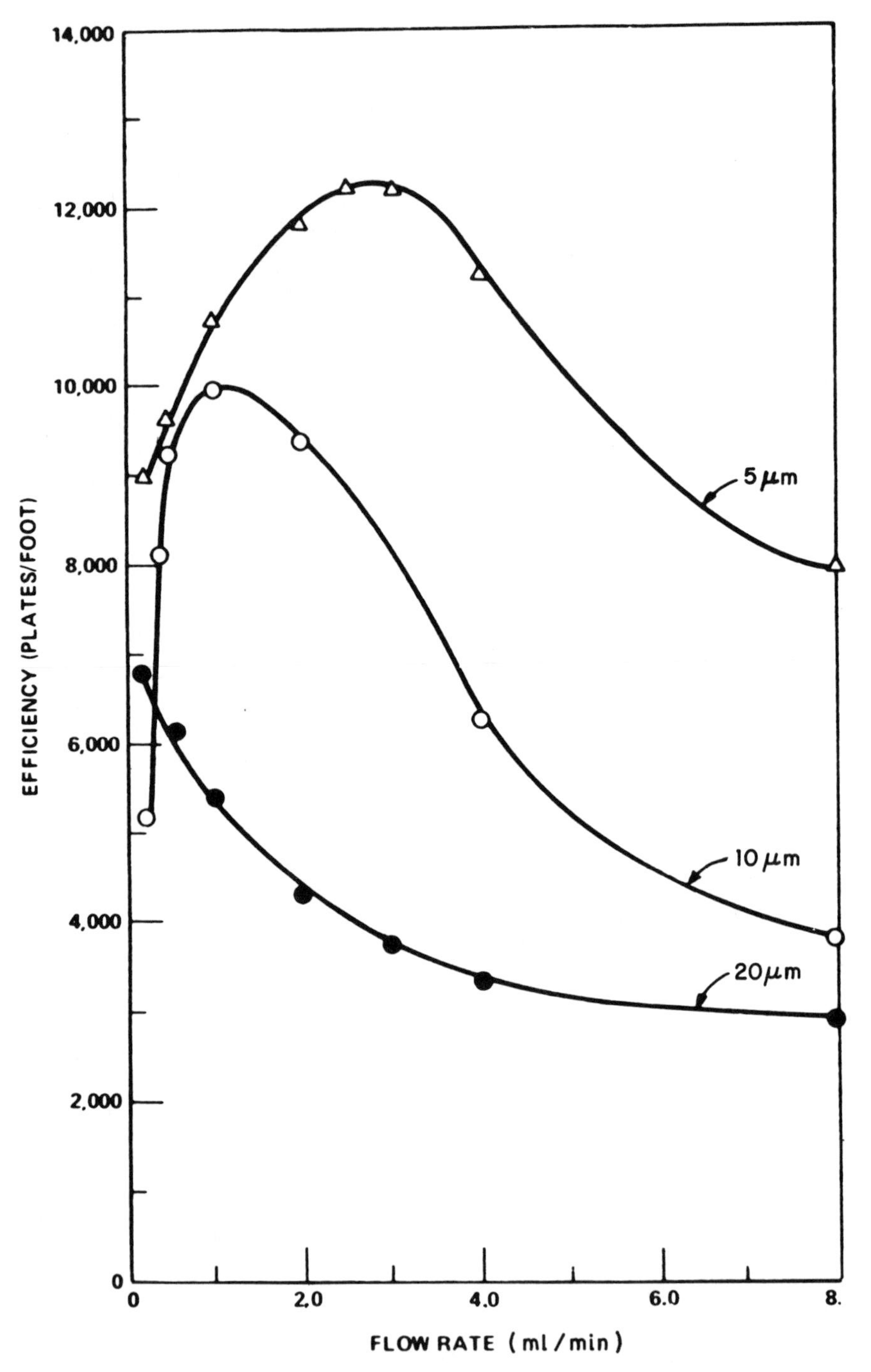

Figure 7.10 Effect of particle size on column plate count as a function of flowrate. Columns, 30 × 0.85 cm μ-Styragel. (Reprinted with permission from Ref. 24.)

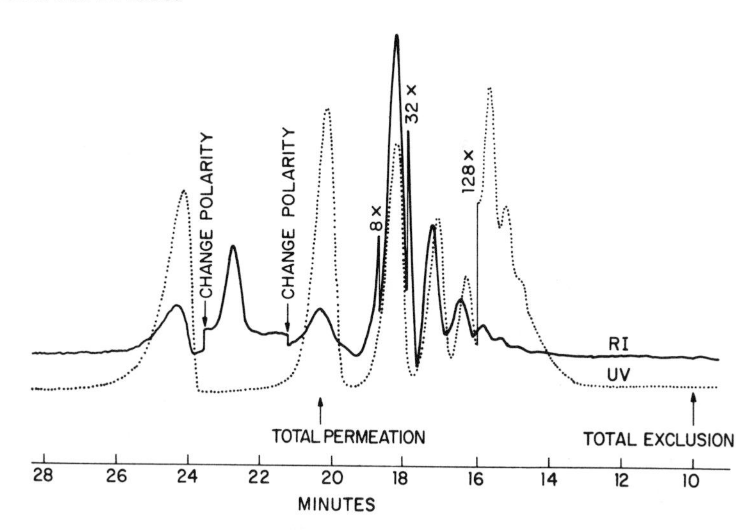

Figure 7.11 Separation of poly(ethylene terephthalate) oligomers.
Column, 100 × 0.62 cm porous silica microspheres, 60 Å (silanized); mobile phase, 2:8 *m*-cresol: chloroform at 22°C; flow rate, 1.18 ml/min; detectors: UV, 0.32 AUFS (initial) at 300 nm; RI, 8X (initial); sample, 50 μl, 25 mg/ml in mobile phase. (Changes in detector polarity and sensitivity noted.) (Reprinted with permission from Ref. 27.)

volume and molecular size is not obtained. As discussed in Section 6.2, exhaustive reaction of silica particles with chlorotrimethylsilane reduces the adsorption tendency of silica columns for GPC. In GFC, the modification of siliceous surfaces is made with hydrophilic groups (e.g., ether or diol) to avoid unwanted interactions between the solute and substrate (Sect. 13.2). In some instances, an improper combination of solute and mobile phase results in excessive retention. In Figure 7.11 two peaks are shown eluting after the total permeation volume for a poly(ethylene terephthalate) sample in a *m*-cresol/chloroform mobile phase, indicating unwanted interaction of the solutes and substrate. In this example a better selection of the mobile phase (e.g., hexafluoroisopropanol) would have eliminated the extra retention problem. Similar effects are also observed with columns of organic gels.

Adsorption or matrix effects can often be minimized by utilizing the most polar mobile phase permitted by sample solubility. For example, water is an effective mobile phase with unmodified siliceous particles for some separations. At pH > 4, the acidic SiOH groups on silica are ionized and can function as ion-exchange sites. Aqueous phases containing buffers or salts can be used effectively to eliminate undesired ion-exchange interaction of certain solutes (e.g., proteins) with unmodified siliceous surfaces (Sect. 13.3). Where permitted by the sample properties, working at pH < 4 often will also eliminate unwanted ion-exchange effects. In GPC highly polar solvents such as hexafluoroisopropanol and dimethylsulfoxide have proven to be useful in eliminating the adsorption of polar polymers to surfaces of unmodified silica particles. Additional discussion of this effect is given in Section 13.3.

The mobile phase can affect the surface or pore characteristics of organic gel columns, and these columns are limited to a few solvents (see Sects. 6.2 and 8.5). Improper mobile phases can collapse organic gel structures and destroy the column. On the other hand, columns of rigid particles (e.g., silica) can be used with a wider range of solvents without harm. Different solvents can also cause substantial changes in the calibration curves for

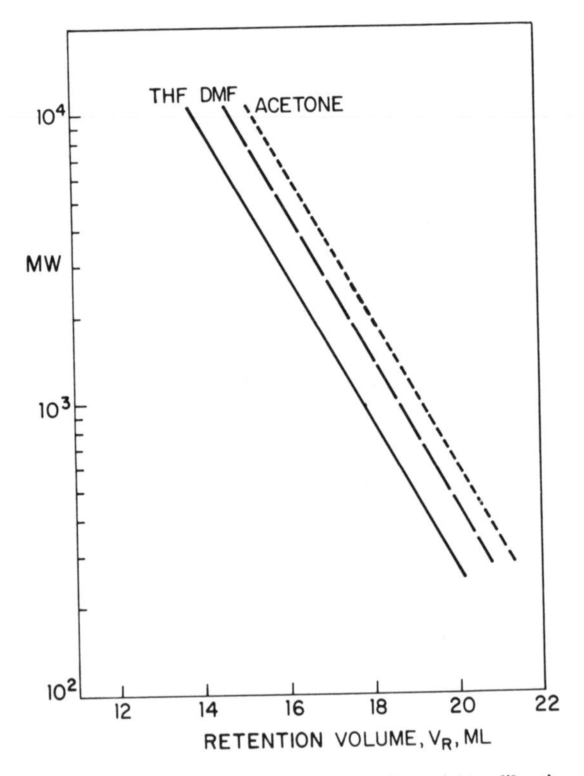

Figure 7.12 Effect of mobile phase on polystyrene molecular weight calibration. Column, 100 × 0.62 cm porous silica microspheres, 60 Å (silanized); calibrations in solvents: acetone, dimethylformamide, tetrahydrofuran; flow rate 1.5 ml/min; sample, 25 μl; detector, UV with THF, RI with DMF and acetone. (Reprinted with permission from Ref. 27.)

organic gel columns because of swelling differences. On the other hand, changes in the solvent result in only minor variations in the calibration curves for columns of rigid particles. Figure 7.12 shows the effect of various mobile phases on polystyrene calibration curves obtained on porous silica microsphere columns optimized for small-molecule HPSEC. The slopes of the calibration curves are constant, and the small changes in the intercept are probably due to changes in the hydrodynamic volume of the polymer standards with different solvents.

The stability of the substrate can also be influenced by other aspects of the mobile-phase environment. For instance, with silica-based particles the mobile phase must be in the pH 2–8 range, or particle degradation can occur. With columns of organic gels or organic-modified rigid particles, particularly with silica-based particles with glycol/ether or ether-bonded organic-modifying films that may form peroxides, it is desirable to exclude oxygen from the mobile phase to improve long-term column stability.

The operating pressures used for the separation can have an effect on the HPSEC stationary phase. Columns of strong, rigid particles (e.g., silica) can often be used at pressures up to ~5000 psi if properly packed (13). However, as discussed in Section 6.2, pressures are more limited for columns of organic gels, because the column bed can be compressed and resolution destroyed. The stability of various silica packings to pressure is largely a function of the particle shape, the pore structure, and the pore volume. Particles with a very high internal volume may be mechanically less stable to high packing pressures and high mobile-phase velocities, but definitive information on these effects is lacking.

7.4 Sample Effects

Sample Volume

Some sample injection variables lead to significant band broadening in HPSEC, which ultimately reduces resolution and causes inaccurate molecular weight measurements. Extracolumn variance caused by the volume of injected sample is accurately predictable whether syringe or valve sampling is used. For plug sampling, $\frac{1}{12}V_{inj}^2$ (where V_{inj} is the injection volume) is added to the peak variance (28, 29). Band broadening due to sample injection is independent of flow rate; it is part of the extracolumn effect that differs from the band broadening associated with the detector, tubing, and connectors, whose variances are flow-rate-dependent. Thus the band broadening caused by injected sample volume is not separable from

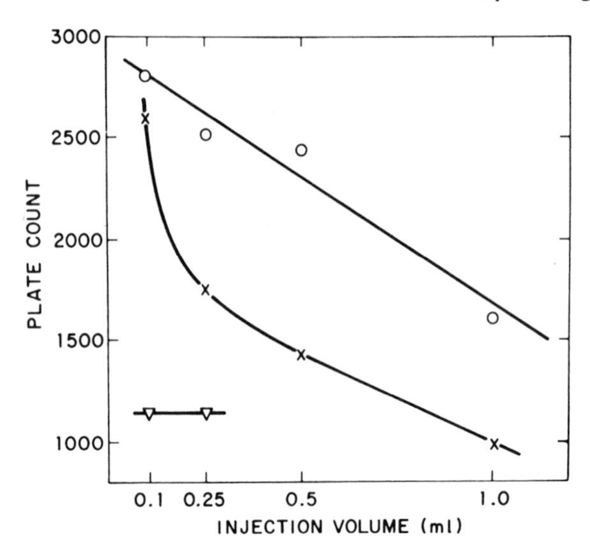

Figure 7.13 Effect of injection volume on column plate count for polymers.
Column, 30 × 0.85 cm each of μ-Styragel 10^6, 10^5, 10^4, and 10^3 Å; mobile phase, toluene; detector, RI; solute, 200,000 MW polystyrene: ○, 0.1% concentration at 1 ml/min; ×, 0.4% concentration at 1 ml/min; ▽, 0.4% concentration at 4 ml/min. (Redrawn from data in Ref. 20.)

the eddy-diffusion term in the chromatographic plate height equation (Sect. 3.2).

As a practical guide in HPSEC analysis, sample volumes should generally be limited to one-third or less of the baseline volume of a monomer peak measured with a very small sample. Under these conditions no more than about 8% loss in column resolution occurs as a result of injection volume band broadening. Thus at a flow rate of 2 ml/min (or 0.03 ml/s), with a 20-sec monomer baseline width, the maximum allowable sample volume V_s would be $\frac{1}{3}(20 \times 0.03)$, or 0.2 ml. Figure 7.13 shows the effect of injection volume on column plate count for a μ-Styragel column series. For greater column capacity, such as that required in preparative studies or for better detection, larger sample volumes can sometimes be used with only a modest decrease in overall column resolution.

Sample volumes also affect apparent solute retention volumes. Generally, one-half of the injected sample volume is contributed to the measured retention volume (30). In view of this effect, it is important to use a constant injection volume during calibration and analysis.

For a constant sample volume, improved performance can be expected for larger-diameter columns; sample injection volume variance σ^2_{inj} is

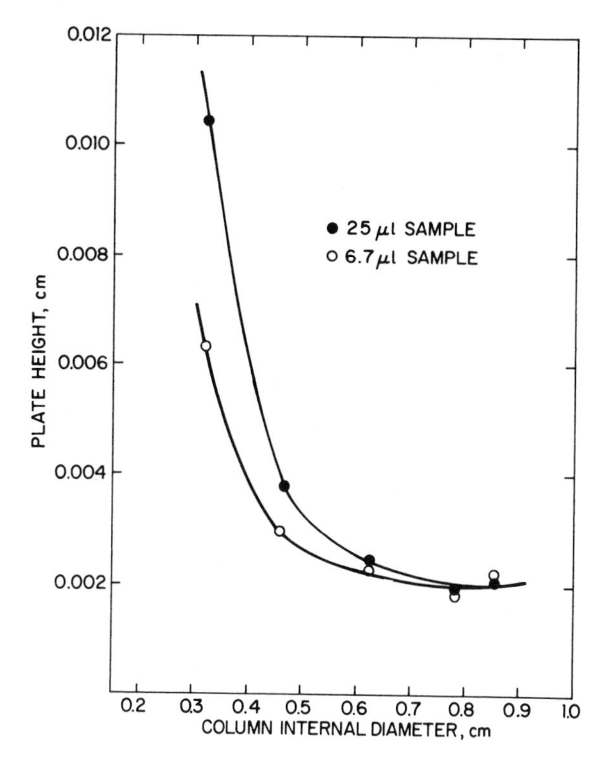

Figure 7.14 Effect of column internal diameter on plate height.
Columns, 10 cm porous silica microspheres, 300 Å, 6 μm; mobile phase, tetrahydrofuran; velocity, 0.25 cm/s; UV detector, 0.2 AUFS; sample, 1 mg/ml toluene in tetrahydrofuran. (Redrawn from data in Ref. 13.)

fixed, but peak volume variance σ_v^2 increases with increasing column diameter. Thus plate height decreases because of the reduced contribution of the sample volume. Figure 7.14 illustrates the effect of increasing internal diameter on the plate count of short (10 cm) columns for two different sample volumes. For column internal diameters of 0.6–0.8 cm and lengths of 50–100 cm, sample volumes larger than about 200 μl are not recommended in practice, and 25–50 μl samples are typical for many applications.

Sample Weight or Concentration

Contrary to what is found with the other LC methods, the retention volumes of polymers have been observed to increase with increased sample concentration. Figure 7.15 shows that with a set of μ-Styragel columns, the magnitude

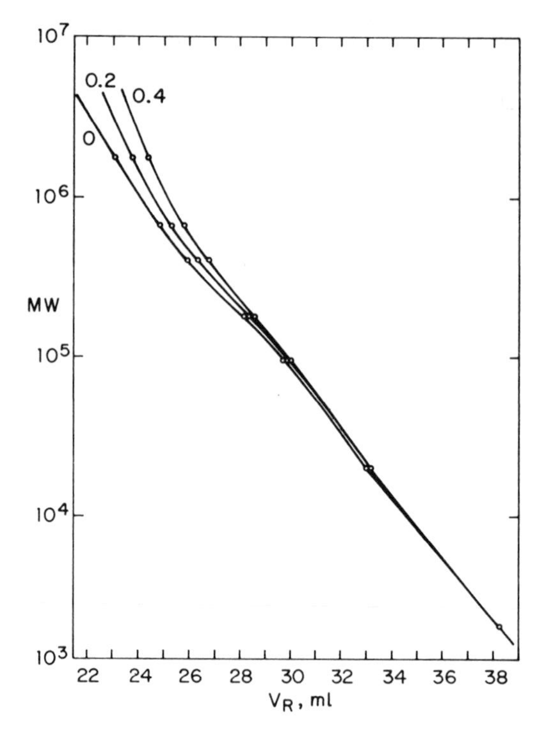

Figure 7.15 Effect of solute concentration on molecular weight calibration.
Same conditions as for Fig. 7.13, except flow rate, 2 ml/min; sample, 250 μl. Numbers on curves refer to concentration in % ("0" represents extrapolated zero concentration). (Reprinted with permission frcm Ref. 20.)

of concentration dependence on retention volume V_R is larger for higher-molecular-weight polystyrenes than for lower-molecular-weight standards. The same trend is also illustrated by the data in Table 7.9 for a set of porous silica columns, with the higher-molecular-weight standards again showing larger increases in V_R with increasing polymer concentration. This type of column overloading appears to be due to a change in the effective dimensions (i.e., radius of gyration) of the macromolecular coils with concentration, the effect increasing with polymer molecular weight and decreasing as the solvent approaches the Flory theta condition (35). Column overloading effects are only a function of the weight of sample injected (V_{inj} × concentration). To obtain highest accuracy in determining molecular weights, the concentration dependence of retention volumes should be minimized by working at the lowest possible constant sample concentration consistent with the required detector signal/noise ratio.

Table 7.9 Dependence of Retention Volume on Concentration

Polystyrene $M_w \times 10^3$	Concentration (wt. %)	V_R (counts)
2610	0.8	18.2
	0.4	18.0
	0.2	17.9
	0.1	17.8
	0.05	17.7
	0.025	17.6
867	0.4	22.8
	0.2	22.3
	0.1	21.8
	0.05	21.7
	0.025	21.6
498	0.8	25.8
	0.4	25.2
	0.2	24.7
	0.1	24.4
	0.05	24.2
	0.025	24.1
200	0.4	29.2
	0.2	28.9
	0.1	28.9
	0.05	28.9
	0.025	28.9

Taken from Reference 31.

Sample size in HPSEC is also limited at times by sample viscosity. At high sample concentrations, increased solution viscosity can cause significant band broadening due to viscous streaming or "viscous fingering" on the trailing side of the solute band. To prevent problems of this kind, a rough guide is that an injected sample solution should have a viscosity no greater than twice that of the mobile phase (32). For high-molecular-weight polymers, concentrations of $\leq 0.1\%$ are often required to eliminate the undesirable effects of concentration on both molecular coil dimensions and sample viscosity.

Another type of column overload which can be significant in the HPSEC analysis of small molecules is shown in Figures 7.16 and 7.17 for low-molecular-weight polymers and monomers. Both retention volume and plate

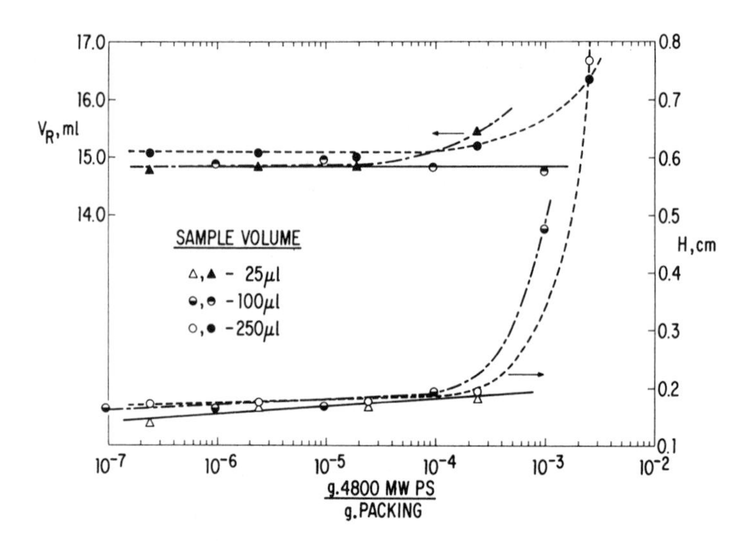

Figure 7.16 Effect of polymer sample size on column performance.

Column, 100 × 0.62 cm porous silica microspheres, 47 Å (silanized); mobile phase, tetrahydrofuran, 22°C; flow rate, 1.5 ml/min; detector, UV, 254 nm; solute, 4800 MW polystyrene in tetrahydrofuran. Upper curve, effect of sample loading on retention volume; lower curve, effect of sample loading on plate height. (Reprinted with permission from Ref. 27.)

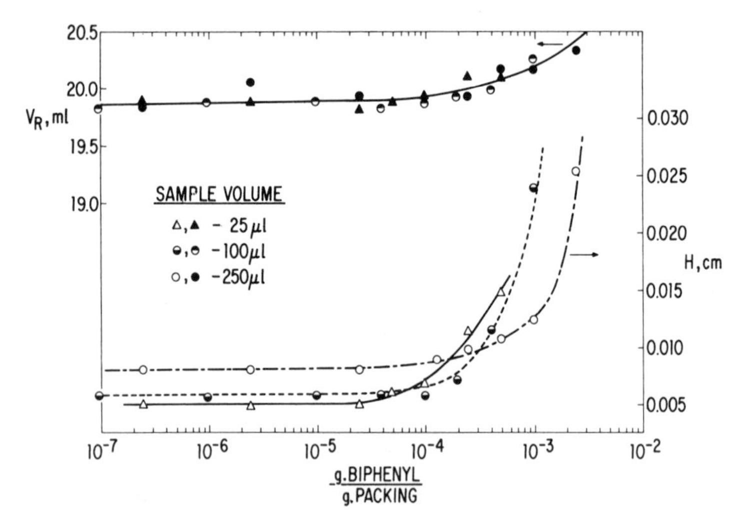

Figure 7.17 Effect of monomer sample size on column performance.

Conditions, same as for Figure 7.16, except solute biphenyl in tetrahydrofuran. Upper curve, effect of sample loading on retention volume; lower curve, effect of sample loading on plate height. (Reprinted with permission from Ref. 27.)

height increase sharply with sample size when sample sizes are relatively large. Since changes in molecular coil dimension and viscosity effects should not be significant with these low-molecular-weight solutes, a mechanism different from that just described for large macromolecules is probably involved. It is likely that increased plate height H and V_R at high sample load result from variations in the distribution coefficient K (influenced by the concentration level) as the solute proceeds through the column. Initial high solute concentrations at the top of the column locally increase K. As the solute band proceeds through the column and becomes more dilute, K decreases and finally reaches a constant value. Retention volumes and plate heights are only constant when K is invariant during the separation. For the particular systems in Figures 7.16 and 7.17, sample loads of $\leq 10^{-4}$–10^{-3} g/g packing for a single component can be injected without significant change in H or V_R. Larger loadings can be tolerated for preparative separations, but at some sacrifice in column plate count and resolution. The data in Figures 7.16 and 7.17 also show increases in H and V_R with increasing sample volume (25–250 μl), which is in keeping with the discussion on the effect of sample volume in the previous section.

The refractive index of a polymer solution depends on the concentration of the solute, but solute MW also can have a slight influence. With a differential refractometer detector, the difference between the refractive index of the polymer solution and that of pure solvent is assumed to be linear with concentration, and the concentration variation of the specific refractive index increment dn/dc is assumed negligible for the low concentrations normally used. However, dn/dc can change with molecular weight in the low-molecular-weight range. As illustrated in Figure 7.18, a significant dependence of refractive index on molecular weight is observed up to ~20,000 MW for polystyrene in toluene. (The refractive index does not become essentially constant until about 300,000 MW. In methyl ethyl ketone, the refractive index/molecular weight relationship is not yet constant at 1,800,000 MW.) As a result, GPC chromatograms that are obtained with a differential refractometer do not exactly represent a linear weight concentration/retention volume relationship (33). Molecular weight errors from this source may be significant for samples with low-molecular-weight fractions.

For some addition polymers a linear relationship exists between dn/dc and the reciprocal of molecular weight. If the dn/dc of the monomer and/or oligomers is known, as well as the limiting value for the macromolecule, this relationship may be approximated by (34)

$$\frac{dn}{dc} = a + \frac{b}{\text{MW}} \tag{7.9}$$

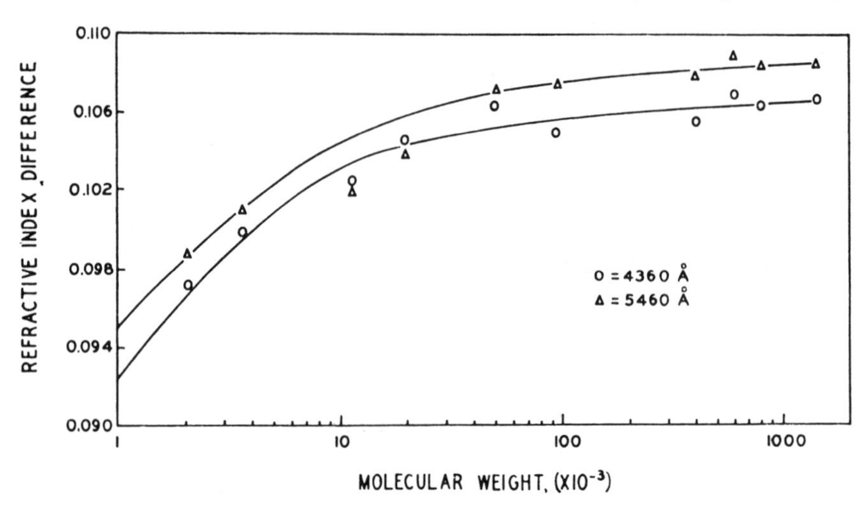

Figure 7.18 Effect of molecular weight on refractive index difference.
Solvent, toluene; samples, polystyrene standards; temperature, 25 ± 0.1°C; measurements with 436 and 546 nm mercury lines. (Reprinted with permission from Ref. 33.)

where a and b are constants. When dn/dc is unknown, it may be estimated by

$$\frac{dn}{dc} = \frac{n_2 - n_1}{d_2} \tag{7.10}$$

where n_1 is the refractive index of the solvent and n_2 and d_2 are the refractive index and density of the polymer, respectively. This refractive index correction is only required for samples having a high population of low-molecular-weight materials. The value of dn/dc is also useful for estimating the refractometer detector sensitivity required for the desired response of a particular polymer/solvent pair in a GPC analysis.

REFERENCES

1. F. W. Billmeyer, Jr., *Textbook of Polymer Science*, 2nd ed., Wiley-Interscience, New York, 1971, Chapt. 2.
2. P. W. Allen, in *Techniques of Polymer Characterization*, P. W. Allen, ed., Academic Press, New York, 1959, Chapt. 1.
3. P. J. Flory, *J. Chem. Phys.*, **10**, 51 (1942).
4. H. Morawetz, *Macromolecules in Solution*, 2nd ed., Wiley, New York, 1975.
5. J. Hildebrand and R. Scott, *The Solubility of Non-Electrolytes*, 3rd ed., Reinhold, New York, 1949.

6. D. W. Van Krevelen, *Properties of Polymers*, American Elsevier, New York, 1976, Chapter 7.
7. E. B. Bagley, T. P. Nelson, and J. M. Scigliano, *J. Paint Technol.*, **43**, 35 (1971).
8. J. Brandrup and E. H. Immergut, eds., *Polymer Handbook*, 2nd ed., Wiley-Interscience, New York, 1975, Sect. IV, pp. 348–349.
9. O. Fuchs, *Fortschr. Chem. Forsch.*, **11**, 74 (1968).
10. K. Krummen and R. W. Frei, *J. Chromatogr.*, **132**, 27 (1977).
11. J. Steinhardt and J. A. Reynolds, *Multiple Equilibria in Proteins*, Academic Press, York, 1969, Chapt. 4.
12. J. Wehrli, *J. Chromatogr.*, **149**, 199 (1978).
13. J. J. Kirkland, *J. Chromatogr.*, **125**, 231 (1976).
14. E. L. Slogowski, L. J. Fetters, and D. McIntyre, *Macromolecules*, **7**, 394 (1974).
15. D. D. Bly, H. J. Stoklosa, J. J. Kirkland, and W. W. Yau, *Anal. Chem.*, **47**, 1810 (1975).
16. D. D. Bly, H. J. Stoklosa, and W. W. Yau, *Anal. Chem.*, **48**, 1256 (1976).
17. G. N. Patel, *J. Appl. Polym. Sci.*, **18**, 3537 (1974).
18. J. N. Little, J. L. Waters, K. J. Bombaugh, and W. J. Pauplis, *J. Polym. Sci., Part A-2*, **7**, 1775 (1969).
19. W. W. Yau, H. L. Suchan, and C. P. Malone, *J. Polym. Sci., Part A-2*, **6**, 1567 (1968).
20. S. Mori, *J. Appl. Polym. Sci.*, **21**, 1921 (1977).
21. G. Trenel et al., *FEBS Lett.*, **2**, 74 (1968).
22. K. Unger and R. Kern, *J. Chromatogr.*, **122**, 345 (1976).
23. A. R. Cooper and A. R. Bruzzone, *J. Polym. Sci., Part A-2*, **11**, 1423 (1973).
24. R. V. Vivilecchia, B. G. Lightbody, N. Z. Thimot, and H. M. Quinn, *J. Chromatogr. Sci.*, **15**, 424 (1977).
25. J. J. Kirkland and W. W. Yau, U.S. patent applied for.
26. W. W. Yau, C. R. Ginnard, and J. J. Kirkland, *J. Chromatogr.*, **149**, 465, 1978.
27. J. J. Kirkland and P. E. Antle, *J. Chromatogr. Sci.*, **15**, 137 (1977).
28. J. C. Sternberg, in *Advances in Chromatography, Vol.* **2**, J. C. Giddings, and R. A. Keller, ed., Marcel Dekker, New York, 1966.
29. J. J. Kirkland, W. W. Yau, H. J. Stoklosa, and C. H. Dilks, Jr., *J. Chromatogr. Sci.*, **15**, 303 (1977).
30. K. A. Boni, F. A. Sliemers, and P. B. Stickney, *J. Polym. Sci., Part A-2*, **6**, 1567 (1968).
31. J. Janca, *J. Chromatogr.*, **134**, 263 (1977).
32. L. R. Snyder and J. J. Kirkland, *Introduction to Modern Liquid Chromatography*, Wiley, New York, 1974, Chapt. 10.
33. E. M. Barrall, II, M. J. R. Cantow, and J. F. Johnson, *J. Appl. Polym. Sci.*, **12**, 1373 (1968).
34. J. M. Evans, *Polym. Eng. Sci.*, **13**, 406 (1973).
35. T. Bleha, D. Bakos, and B. Berek, *Polymer*, **18**, 897 (1977).

Eight

LABORATORY TECHNIQUES

8.1 Introduction

This chapter provides the reader with a checklist of options and guidelines for optimizing HPSEC experiments. Besides "how-to" information, typical values for operating parameters are provided as well as experimental suggestions and caveats.

Prior to running an experiment, the problem objectives must be understood clearly. A qualitative survey scan may be sufficient to establish that there are several components in a sample or that molecular-weight-distribution differences exist between two samples. Such survey scans can probably be made without close attention to detail or to optimizing resolution. On the other hand, if small differences in molecular weight distribution must be discerned or absolute molecular weight values are required, careful attention must be given to experimental details to optimize resolution and obtain accurate data. It should also be determined if the experiment is to be a single run or a repetitive analysis. Generally, a large number of analyses justify the cost of more sophistication and automation. Preparative SEC experiments require yet another approach.

Table 8.1 Solvent/Polymer Combinations Compiled from Conventional GPC Investigations[a]

	o-Dichlorobenzene	Benzene or Toluene	Methylene Chloride	Tetrahydrofuran	Chloroform	Dimethylformamide	1,2,4-Trichlorobenzene	Water	m-Cresol
Acenaphthylene MMA copolymer				X		X		N	
Acenaphthylene styrene-acrylic				X		X		N	
Acrylic butadiene styrene				X		X		N	
Acrylics		X		X		X		N	U
Acrylonitrile butadiene rubber		X				X	X	N	
Alkyd resins		X		X	X				
Alkyl resins				X	X				
Antioxidants for polymers		X		X					
Asphalt	X	X	X	X	X	X	X	N	X
Butene-1	X	X	N	U	N		X	N	
Butyl rubber	X	X		U			X	N	
Carbowaxes	X			U	X	X	X		U
Cellulose acetate				X		X			
Cellulose nitrate				X		A		N	
cis-Polybutadiene	X	X		U			X	N	
Coal-tar pitch	X	X	X	X	X	X	X	N	X
Dextrans								X	
Dialkyl phthalate	X	X		X	X		X	N	
Dimethylpolysiloxanes	X	X	X	U	U		X	N	
Drying oils	X	X	X	X	X	X			

	o-Dichlorobenzene	Benzene or Toluene	Methylene Chloride	Tetrahydrofuran	Chloroform	Dimethylformamide	1,2,4-Trichlorobenzene	Water	m-Cresol
Polybutadiene acrylic acid acrylonitrile polymer		X				X		N	
Polycaprolactam							N	N	X
Polycarbonates	X		X	X			X	N	
Polyelectrolytes				X				X (salt water)	
Polyesters	X	X		X	X		X	N	U
Polyethers		X		X		X		N	
Polyethylene, branched	X	N	N	N	N	N	X	N	
Polyethylene, linear		N	N	N	N	N	X	N	
Polyethylene oxide				X		X			
Polyethylene terephthalate		X				X		N	
Polyglycols	X	X	X	X		X	X		X
Polyisobutylene		X		X				N	
Polyisobutylene copolymers		X		U				N	
Polyisoprene		X					X	N	
Polynuclear aromatics	X	X	X	X	X	X	X	N	
Polyols				X		X			
Polyphenylene oxide							X		

Epichlorohydran				U						Polypropylene							X	N	
Epoxy resins (not cross-linked)		X		X	X					Polystyrene	X	X	X	X	X	X	X	N	X
Ethyl acrylates	X	X		U		X		N	X	Polysulfonates				X					
Ethylene propylene polymer	X						X	N		Polysulfone				X					
Ethylene vinyl acetate		U	U				X	N		Polyurethane			X	U		X		N	
Fatty acid derivatives	X	X		X	X		X			Poly(vinyl acetate)	X			X		X			X
Fatty acids	X			X	X		X			Poly(vinyl acetate) copolymers	X			X					X
Furfurylalcohol	X			X	X		X			Poly(vinyl alcohol)								X	
Glycerides	X			X			X			Poly(vinyl butyral)				X		X		N	
Isocynates		X		X	X	X				Poly(vinyl chloride)		X		X			R	N	
Lignin sulfonates								X		Poly(vinyl fluoride)						X			
Lipids			X	X						Poly(vinyl methyl) ethers				X		X			
Lube oils	X	X	X	X	X	X	X	N	X	Propylene-butene-1 copolymers	X	X					X	N	
Melamines	X			X			X	N		Rubber, natural	X	X					U	N	
Methyl methacrylate		X		X		X		N	X	Silicones	X	X		U	X		X		
Methyl methacrylate/styrene	X	X		X		X		N		Styrene/acrylonitrile			U	X		X		N	
Neoprene		X						N		Styrene/butadiene rubber	X	X		X			X	N	
Nonionic surfactants				X	X					Styrene/isoprene		X		X			X	N	
Nylon (46,66, etc.)							N	N	X	Trifluorostyrene				X					
Phenol formaldehyde resins				X			X			Urethane prepolymers		X		X		X			
Phenolic novalacs				X	X			N		Vinylchloride/vinyl acetate/vinyl									
Plasticizers (esters)	X	X	X	X		X	X	N	X	acetate/maleic acid terpolymer						X			
Polyalkylene glycols	X	X		X	X					Vinylidene fluoride/hexafluoro-									
Polybutadiene	X	X		X			X	N		propylene (Viton)				U					
Polybutadiene acrylic		X				X		N		Waxes (hydrocarbons)	X	X	X	X		X	X	N	X

Taken from Reference 6.

[a] Samples containing gels will not completely dissolve. In some cases, a low-molecular-weight polymer will dissolve in a given solvent but a higher-molecular-weight polymer of the same composition will not dissolve. X, suitable; N, not suitable; U, generally suitable; A, adsorption can occur; R, soluble but no RI difference; *, requires temperature of at least 125°C; a blank indicates that no data are available.

Table 8.2 Other Solvent/Polymer Combinations, *dn/dc*, and K and a Values Compiled for Use in Conventional GPC

Polymer	Solvent	Temperature (°C)	*dn/dc*, 5890 Å	Monomer	Monomer *dn/dc* 5890 Å	K × 10^4	a	K × 10^4 (for polystyrene)	a (for polystyrene)	Molecular Weight Range Applicable × 10^{-3}
Polystyrene	THF	25	0.198	Styrene	0.156	1.60	0.706	—	—	>3
Polystyrene	THF	23	—	—	—	68.0	0.766	—	—	50–1000
Polystyrene (comb.)	THF	23	—	—	—	2.2	0.56	1.68	0.69	150–11,200
Polystyrene (star)	THF	23	—	—	—	0.35	0.74	1.68	0.69	150–600
Poly(vinyl chloride)	THF	23	0.106	Vinyl chloride	(0.007)	1.63	0.766	1.60	0.706	20–170
	—	—	0.115	—	—	1.50	0.77	1.22	0.72	20–105
	—	—	—	—	—	1.60	0.77	1.17	0.725	10–1000
	—	—	—	—	—	7.2	0.61	1.68	0.69	27–100
Poly(methyl methacrylate)	THF	23	0.0865	Methyl methacrylate	0.008	0.93	0.72	1.68	0.69	170–1300
	—	—	—	—	—	1.28	0.69	1.41	0.70	150–1200
	—	—	—	—	—	21.1	0.406	1.60	0.706	<31
	—	—	—	—	—	1.04	0.697	1.60	0.706	>31
Polycarbonate	THF	25	0.181	—	—	3.99	0.77	NV	—	—
Polycarbonate	THF	25	—	—	—	4.9	0.67	NV	—	7–77
Polydioxalane	THF	25	—	3-Dioxalane	0.010	0.937	0.874	NV	—	—
Poly(vinyl acetate)	THF	25	0.058	Vinyl acetate	0.010	3.5	0.63	1.17	0.725	10–1000
Poly(vinyl bromide)	THF	20	0.112	Vinyl bromide	0.027	1.59	0.64	NV	—	—
Poly(vinyl ferrocene)	THF	30	—	—	—	0.72	0.72	NV	—	—
Polyisoprene	THF	25	0.156	Isoprene	0.021	1.77	0.735	1.25	0.717	40–500
Natural rubber [poly(*cis*-isoprene)]	THF	25	0.125	Isoprene	0.021	1.09	0.79	1.0	0.70	10–1000
Butyl rubber	THF	25	0.112	Isobutylene	0.021	0.85	0.75	1.25	0.707	4–4000
(isobutene coisoprene)	—	—	—	Tetraisobutylene	0.054	—	—	—	—	—
Poly(1,2-butadiene)	THF	20	—	1,2-butadiene	0.023	M_n(PB) = 0.167M_n(PS)		—	—	9–25
Poly(1,4-butadiene)	THF	40	0.132	1,3-butadiene	0.039	5.78	0.67	1.9	0.68	10–100
Poly(1,4-butadiene)	THF	25	—	—	—	76.0	0.44	1.68	0.69	270–550
Poly/1,4-butadiene) (*cis/trans* ≃ 0.8)	THF	25	—	—	—	—	—	—	—	—
8% vinyl	THF	25	—	—	—	4.57	0.693	1.25	0.717	80–1100

20 % vinyl	THF	25	—	—	—	4.51	0.693	1.25	0.717	20–200
52 % vinyl	THF	25	—	—	—	4.28	0.693	1.25	0.717	20–200
73 % vinyl	THF	25	—	—	—	4.03	0.693	1.25	0.717	20–200
Polybutadiene (20 % *cis*/20 % vinyl)	THF	25	—	—	—	2.36	0.75	1.2	0.71	3–6
SBR (25 % styrene)	THF	40	0.179	—	—	3.18	0.70	1.9	0.68	70–1000
SBR (25 % styrene)	THF	25	—	—	—	4.1	0.693	1.25	0.717	24–40
SBR 1507	THF	30	—	—	—	3.0	0.70	1.25	0.707	10–1000
SBR 1808	THF	30	—	—	—	5.4	0.65	1.25	0.707	10–1000
Cellulose nitrate	THF	25	—	—	—	25.0	1.0	68.0	0.766	95–2300
Cellulose trinitrate	THF	25	—	—	—	3.21	0.83	NV	—	DP 60–6000
Amylose acetate	THF	25	—	—	—	108.0	0.70	NV	—	20–500
Amylose butyrate	THF	25	—	—	—	111.0	0.70	NV	—	20–500
Amylose propionate	THF	25	—	—	—	248.0	0.61	NV	—	20–500
Polystyrene	ODCB	135	0.068	Styrene	0.025	1.38	0.70	—	—	2–900
Polyethylene	ODCB	135	−0.090	*n*-hexatriacontane	−0.141	4.77	0.70	4.57	0.606	6–700
Polyethylene	ODCB	135	HDPE-0.069	*n*-octacosane	−0.152	5.046	0.693	1.51	0.693	10–1000
Polyethylene	ODCB	138	LDPE-0.104	*n*-eicosane	−0.170	5.06	0.70	1.38	0.70	0.2–200
Polybutadiene (hydrogenated)	ODCB	135	−0.017	—	—	2.7	0.746	1.51	0.693	10–500
Polypropylene	ODCB	135	—	—	—	1.30	0.78	0.736	0.75	28–460
Poly(dimethyl siloxane)	ODCB	138	0.078	—	—	3.83	0.57	1.38	0.70	25–300
Poly(dimethyl siloxane)	ODCB	87	—	—	—	8.19	0.50	1.00	0.73	20–800
Polystyrene	*m*-Cresol	135	—	—	—	2.02	0.65	—	—	4–2000
Poly(ethylene terephthalate)	*m*-Cresol	135	—	—	—	1.75	0.81	2.02	0.65	2.7–32
Poly(ethylene terephthalate)	*m*-Cresol	135	—	—	—	2.0	0.90	2.02	0.65	0.45–0.80
Poly(ethylene terephthalate)	*m*-Cresol	25	−0.073	—	—	0.077	0.96	NV	—	—
Nylon 66	*m*-Cresol	130	−0.016	—	—	0.40	1.00	0.846	0.715	8–24
Nylon 6	*m*-Cresol	25	—	—	—	32	0.62	NV	—	0.5–5
Nylon 610	*m*-Cresol	25	—	—	—	1.35	0.96	NV	—	8–24
Nylon 66	*m*-Cresol	25	—	—	—	$(0.15 + 3.53 \times 10^{-4} M)$	0.79	NV	—	0.15–50
Nylon 6	OCP	90	−0.012	—	—	6.2	0.64	1.2	0.70	10–1000
Poly(ethylene terephthalate)	OCP	—	−0.070	—	—	3.0	0.77	NV	—	1–300

Taken from Reference 8.
NV = not verified

8.2 Solvent Selection and Preparation

Solvent selection for HPSEC involves a number of considerations, including convenience, sample type, column packing, operating variables, safety, and purity. These are discussed now in terms of how they affect the SEC process.

Convenience

The most convenient solvent is that already in the instrument. The greatest expenditure of time in preparing HPSEC experiments often occurs in solvent changeover and in obtaining detector baseline stability with a new solvent. If a solvent change can be avoided, analyses of different sample types can be continued without delay.

Sample Type

For samples of any type, the solvent used for the mobile phase in HPSEC must satisfy the following criteria:

1. The solvent must completely dissolve the sample; if not, the sample may be partially fractionated by the solvent according to molecular weight, crystallinity, or composition.
2. The solvent must permit adequate detection of solute in the eluent; typically with a differential refractometer, the solvent refractive index (RI) must differ from the sample RI by ± 0.05 unit or more, while with a UV detector, the solvent should transmit more than 10% of the incident energy at the chosen wavelength.
3. The solvent must not degrade the sample during dissolution and use; if degradation is suspected for polymer solutions, the viscosity can be measured several times over a few hours. A constant viscosity is reasonably good assurance of sample solution stability, since large molecules in a sample are statistically more susceptible to chemical degradation and most influence the solution viscosity.
4. The solvent also must not corrode any of the components of the chromatograph.

Tables 8.1 and 8.2 show solvents that have been used in conventional GPC for various sample types. These lists can provide guidelines for HPSEC solvent selection. Some important physical properties, safety aspects, and restrictions on the use of these solvents are provided in Table 8.3. Additional information for dissolving synthetic polymers is found in Section 7.2 and in Reference 1.

Effect on Column Packing

To be effective with organic-gel-type packings (e.g., μ-Styragel), the solvent (mobile phase) must swell the packing. Thus aqueous and certain organic solvents cannot be used with these organic packings. Tables 8.4 and 8.5 list the solvents most commonly used with μ-Styragel column packings.

Most solvents can be used with silica packings because of the rigid, permanent nature of the particle and pore structure of these kinds of column-packing materials. Figure 8.1 shows that aqueous solvents with silica packings should be maintained at pH < 8, since at higher pH, silica is slowly dissolved. A pH < 2 also should be avoided if possible.

The solvent should have strong affinity for the packing to avoid sample partitioning or adsorption that will bias the size-exclusion mechanism. Unwanted solute retention can be recognized if the retention volume (V_R) of any portion of the sample exceeds the total permeating volume obtained for an inert single compound or monomer (see Sect. 8.8).

Operation

With organic-gel column packings, the boiling point of the solvent should be 25–50°C above the column operating temperature and usually above the maximum temperature used to dissolve the sample. Operating the column close to the mobile-phase boiling point may cause outgassing (bubble formation), which in turn can upset column bed structures and interfere

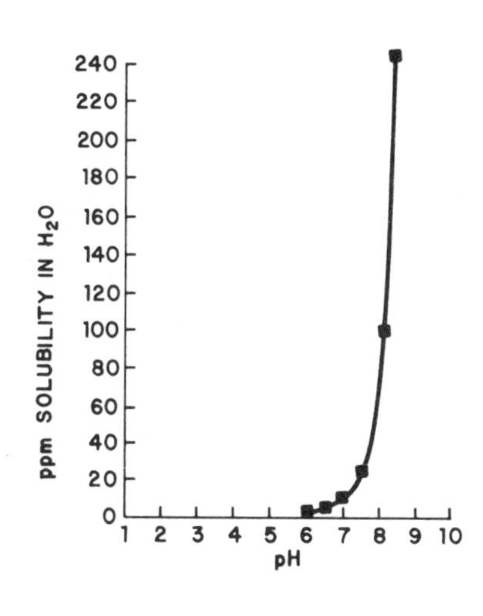

Figure 8.1 Silica solubility curve.
(Reprinted with permission from Ref. (6))

SOLVENT/FUNCTIONAL GROUP CORRELATION CHART

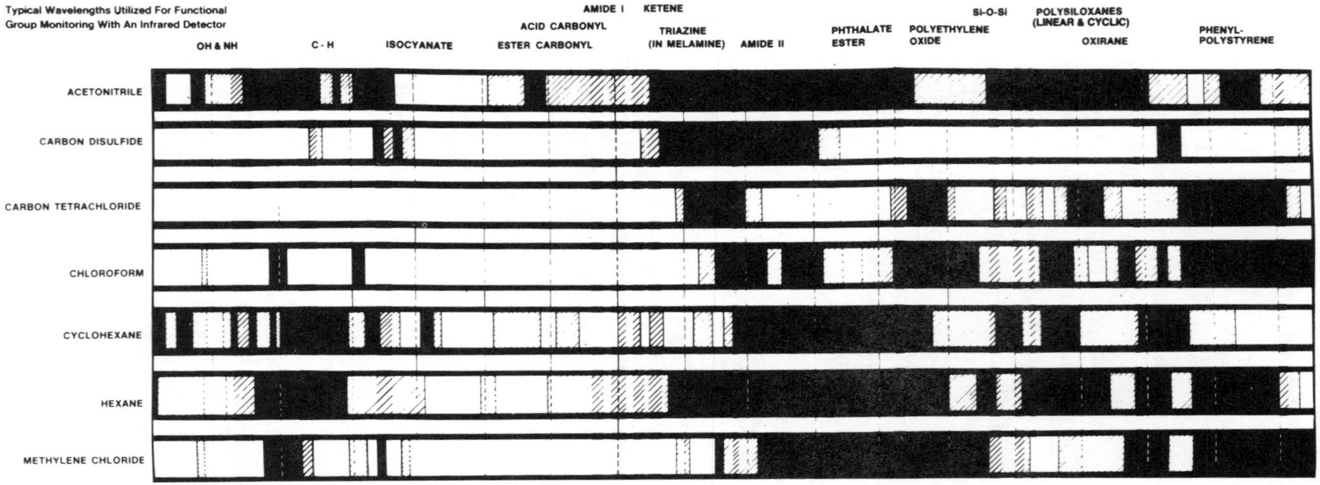

Typical Wavelengths Utilized For Functional
Group Monitoring With An Infrared Detector

OH & NH
C - H
ISOCYANATE
ESTER CARBONYL
ACID CARBONYL
AMIDE I
KETENE
TRIAZINE
(IN MELAMINE)
AMIDE II
PHTHALATE
ESTER
POLYETHYLENE
OXIDE
Si-O-Si
POLYSILOXANES
(LINEAR & CYCLIC)
OXIRANE
PHENYL-
POLYSTYRENE

ACETONITRILE
CARBON DISULFIDE
CARBON TETRACHLORIDE
CHLOROFORM
CYCLOHEXANE
HEXANE
METHYLENE CHLORIDE

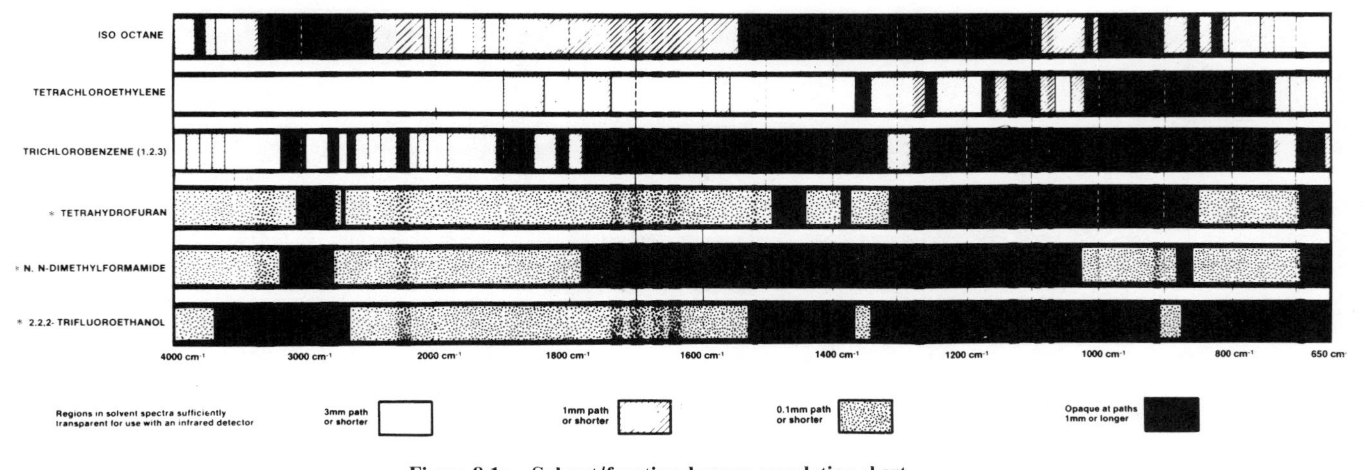

Figure 8.1a Solvent/functional group correlation chart.
(Reprinted with permission from Ref. 9.)

Table 8.3 Properties of Solvents Commonly Used in Gel Permeation Chromatography[a]

Solvents	Melting Point (°C)	Boiling Point (°C)	Density at 20°C	UV Cutoff (nm)	Viscosity at 20°C (cP)	Refractive Index at 20°C	Flash Point (°C)	Oral LD_{50} in Rat (mg/kg)	TLV^R in Rat (ppm)	Irritant to Skin and eye	Toxicity
Tetrahydrofuran[b]	−65	66	0.8892	220	0.55	1.4072	14	3,000	200	Mild	Slight
1,2,4-Trichlorobenzene[c]	17	213	1.4634	307	$1.89^{25°C}$	1.5717	99	756	5	Moderate	Slight
o-Dichlorobenzene[c]	−19	180	1.3048	294	1.26	1.5515	66	500	50	Moderate	Moderate
Toluene	−95	110.6	0.8669	285	0.59	1.4969	4	5,000	100	Moderate	Slight
N,N′-Dimethyl formamide[c]	−61	153	0.9445	275	0.90	1.4294	58	3,500	10•	Moderate	Slight
Methylene chloride (dichloromethane)	−97	40.1	1.3266	245	0.44	1.4237	None	2,136	200[f]	Severe	Slight
Ethylene dichloride (dichloroethane)	−36	84	1.235	230	0.84	1.4443	13	680	50	Slight	Slight
N-Methylpyrrolidone[d]	−24	202	1.027	262	1.65	1.47	95.4	7,000	Not set	Moderate	Very low
m-Cresol	12	202.8	1.034	302	20.8	1.544	94	242	5	Severe	Moderate
Benzene	5.5	80.1	0.8790	280	0.652	1.5011	27	3,800	1[g]	Mild	Slight
Dimethylsulfoxide	18	189	1.014	260	2.24	1.4770	95.0	20,000	Not set	None	Very low
Perchloroethylene	−19	121	1.622	290		1.505	None	5,000	100[h]	Severe	Very low
o-Chlorophenol	7	175.6	1.241		4.11	1.5473^{40}	None	670	Not set	Severe	Slight
Carbon tetrachloride	−23	76.8	1.589	265	0.969	1.4630	None	1,770	10[h]	Moderate	Slight
Water	0	100.0	1.00		1.00	1.33	None	—	—	—	—
Trifluoroethanol		73.6	1.382	190	2.00	1.2910	40.6	240	Not set	Mild	Moderate
Chloroform[e]	−64	61.7	1.483	245	0.58	1.4457	None	2.000	10[f]	Moderate	Slight
Hexafluoroisopropanol	−3.4	58.2	1.59	190	$0.642^{25°C}$	1.2752	—	1,040	Not set	Very severe	Slight

[a] The toxicity data were collected and evaluated by Dr. Clifford L. Dickinson Jr. of the Haskell Laboratory for Toxicology and Industrial Medicine, Central Research and Development Department, E. I. du Pont de Nemours, Inc., Newark, Del. The LD_{50} data refer to the dosage (mg chemical/kg body weight) which kills 50% of rats tested in time. A somewhat arbitrary correspondence with toxicity used in the table is: <50 mg/kg, highly toxic; 50–500 mg/kg, moderately toxic; 500–5000 mg/kg, slightly toxic; 5000 mg/kg, very low toxicity. The TLV^R is the *threshold limit value* for airborne contaminants set by the American Conference of Governmental and Industrial Hygienists.

[b] Generally contains butylated hydroxytoluene at a few hundredths of a percent as stabilizer.

[c] These solvents are usually used at 135°C. The use of an antioxidant is recommended: Santonox R (Monsanto) 1.5 g/gal.

[d] Quite hydroscopic. Relatively large amounts of water (several percent) may drastically affect fractionation.

[e] Ordinarily contains 0.75% ethanol as stabilizer.

[f] Proposed new level.

[g] Recently proposed level because of carcinogenicity.

[h] Value may be lowered due to possibility of carcinogenicity.

with detection. Outgassing normally does not interfere with the bed structure of well-packed porous silica columns.

The solvent should have a low viscosity (e.g., <1 cP) for maximum separation efficiency and minimum operating pressures. High pressure may lead to collapse or shear degradation of organic packings but has little effect on porous silica column packings. Thus it is more convenient to work with the rigid material columns. Table 8.4 indicates the relationship of solvent viscosity to utility for μ-Styragel.

If the universal calibration concept is to be used, a solvent should be chosen for which literature values of **K** and **a** are known (Sects. 9.2 and 10.7). Values of **K** and **a** for several polymer/solvent combinations are listed in Table 8.2, while Table 10.2 is a detailed list of **K** and **a** values for various polymers in tetrahydrofuran, a common and very useful GPC mobile phase (Sect. 12.3).

Table 8.4 Most Commonly Used Solvents for Use with μ-Styragel

Solvents	Viscosity at 20°C (cP)	Recommended Maximum Flow[a] (ml/min)
N,N'-Dimethylformamide[b]	0.924	2.5
Ethylene chloride	0.84	2.8
p-Dioxane	1.439	1.6
Trichloroethane	1.20	2.0
Cyclohexane[b]	0.98	2.4
Carbon tetrachloride	0.969	2.4
Toluene	0.59	3.0
1,1,1-Trifluoroethanol[b]	1.996	1.1
Benzene	0.652	3.0
1,1,1,3,3,3-Hexfluoroisopropanol[b]	0.642	1.2
Chloroform	0.58	3.0
Hexane[b]	0.326	3.0
Xylene	0.81	2.9
Tetrahydrofuran	0.55	3.0
Methylene chloride	0.44	3.0

Taken from Reference 6.

[a] Maximum flow should not exceed 3.0 ml/min when using 100 Å μ-Styragel columns.

[b] These solvents should not be used with 100 Å or 500 Å μ-Styragel columns.

Safety

Some safety considerations for various solvents employed in HPSEC are listed in Table 8.3. Generally, all solvents should be treated as dangerous and should be used only in well-ventilated areas. The use of gloves (e.g., neoprene) and eye protection is recommended. Various acid solvents, phenols, and cresols cause skin burns. *N*-methylpyrrolidone, *N*,*N'*-dimethylformamide, and dimethyl sulfoxide facilitate the transport of other chemicals through the skin, making them potentially more toxic than expected. Hexafluoroisopropanol is an extremely potent solvent for the cornea of the eye. Other solvents may be carcinogenic; for example, benzene, perchloroethylene, carbon tetrachloride, and chloroform have recently been implicated and should only be used in chemical hoods. Reference 2 may be consulted for additional information on solvent toxicity and handling.

Table 8.5 Other Solvents for Use or Avoidance with μ-Styragel

Satisfactory	Marginal	Damaging
Benzene	Cycloheptane	Acetone
Cyclopentane	Diethylamine	Benzyl alcohol
Carbon tetrachloride	Methylene chloride	Butyl acetate
Chloroform	Diethyl ether	Carbon disulfide
p-Dioxane	Ethyl acetate	*m*-Cresol
Diethylbenzene	Methyl ethyl ketone	*o*-Chlorophenol
Divinylbenzene	Triethylamine	Dimethylformamide
Pyridine	90% Tetrahydrofuran/methanol	Dimethylacetamide
Toluene		*n*-Dodecane
Tetrahydrofuran		50% Tetrahydrofuran/ methanol
Trichlorobenzene		Acetic acid
Tetrachloroethane		Acetic anhydride
Tetrahydropyran		Acetonitrile
Tetrahydrothiophene		Cyclohexane
p-Xylene		Hexafluoroisopropanol
		Heptane
		Hexane
		Isooctane
		Trifluoroethanol
		Alcohols
		Water

Taken from Reference 6.

Solvent Purification and Modification

Solvents or mobile phases used in HPSEC should be of high purity. The objects of using pure solvents are [1] to avoid suspended particulates that may abrade the solvent pumping system or cause plugging of small-particle columns, [2] to avoid impurities that may generate baseline noise, and [3] to avoid impurities that are concentrated by evaporation in preparative work. Solvents are purified for HPSEC use by distillation, degassing, and filtration. At times it may be necessary to add certain oxidation or corrosion inhibitors to the solvent.

Distillation may be required for some solvents to eliminate impurities that cause baseline drift (e.g., with differential refractometers) and unwanted adsorption with UV detectors. As suggested by the data in Figure 8.1, distillation is required after solvent purification by silica adsorption because many solvents (or contaminating water) contain small amounts of dissolved silica which interfere in subsequent operations.

Degassing may be required to remove dissolved gases that can nucleate to form bubbles in the detector. Degassing is accomplished by warming the solvent under vacuum, or by purging with helium.

If the solvent is not distilled, filtration is essential to protect the high-performance column and pump. If the solvent contains a noticeable number of particles, it should be prefiltered through a coarse, sintered glass funnel. Solvents should be final-filtered under vacuum through a 1 μm membrane filter (e.g., Fluoropore for organic solvents and Mitex for aqueous solvents, Millipore Company). Filtered solvent should be stored under an inert gas in a container rinsed with the filtered solvent. The equipment should also be protected by a sintered-metal filter (0.5 μm) inserted in line between the solvent reservoir and the pump.

After obtaining a distilled, degassed, and filtered solvent, it is desirable in some cases to add certain inhibitors. With ethers such as tetrahydrofuran, a peroxide inhibitor such as BHT (butylated hydroxytoluene) should be added, especially for large-volume solvent use and storage. Whether an inhibitor is used or not depends on the purpose of the experiment. Usually it is not desirable to have an inhibitor present in preparatory work, since it interferes with the characterization of the fractions. Examples where use of inhibitors may be desirable include [1] chloroform, where ethanol normally is used as a photolysis inhibitor; [2] 1,2,4-trichlorobenzene, where Santonox-R or another equivalent antioxidant is added to protect the sample, solvent, and packing from high-temperature oxidation effects; and [3] aqueous buffers, where for long-time use, antimicrobial agents such as 0.02% NaN_3 should often be used. Generally, columns should not be stored in phosphate buffers, as they encourage microbial growth.

8.3 Selection and Use of Standard Reference Materials

Standard reference materials are used both to evaluate system performance and to calibrate column retention in terms of specific molecular weights. Usually, pure compounds, monomers, or narrow-molecular-weight-distribution polymer standards are used for evaluating column performance. For calculating accurate sample molecular weights and molecular weight distributions it is necessary to calibrate SEC retention with known molecular weight polymers of the same molecular type as the sample (Sect. 9.3). The number of useful standards is very limited relative to the large variety of molecular structures requiring analysis. Table 8.6 lists the sources of many of the currently commercially available macromolecular standards for GFC and GPC.

If standards of the sample type of interest are not available, one of several alternative approaches may be employed for MW calibration. The universal calibration approach uses polystyrene standards for organic solvents and dextrans or sulfonated polystyrenes for aqueous solvents. While often applicable, the universal calibration method cannot be used with every system (Sect. 10.7) and values for the constants **K** and **a** are required. In

Table 8.6 Standard Reference Materials for Calibration Use in HPSEC

Polymer Type	Nominal Molecular Weights Available, $M \times 10^{-3}$	Source Code
Narrow-MWD polystyrenes	(a) 2, 4, 50, 100, 200, 400, 1200, 2500, 4000	1
	(b) 30, 3000, 10,000	5
	(c) 2, 4, 10, 20, 50, 100, 200, 400, 600, 900, 2000	4
	(d) ($\overline{M}_w/\overline{M}_n \leq 1.1$) 171; ($\overline{M}_w/\overline{M}_n \leq 2.1$) 136	6
	(e) ($\overline{M}_w/\overline{M}_n = 1.1$), $\overline{M}_w = 4100, 7100$	8
	(f) 120, 240, 320, 460, 550, 760, 770, 1340	9
	(g) Long list available; see catalog	10
Polypropylene glycol	(a) 0.8, 1.2, 2, 4	1
Polypeptides	(a) Varied	2
	(b) Varied	3
Proteins (various types)	(a) 1.4, 12.4, 17.8, 25, 45, 67, 160, 480	4
	(b) Kit 1: 14.5, 18, 32, 45, 68; kit 2: 13.7, 25, 45, 158, 2000	5
Dextrans	(a) Varied	15
Polyisoprene	(a) ($\overline{M}_w/\overline{M}_n < 1.2$)	1

Table 8.6 *Continued*

Polymer Type	Nominal Molecular Weights Available, $M \times 10^{-3}$	Source Code
Polybutadiene	(a) ($\overline{M}_w/\overline{M}_n < 1.1$) 1, 3	1
	(b) 1, 2, 3	5
	(c) (Carboxy-terminated) 2, 3	5
	(d) (Narrow MWD) 16, 135, 206, 286	7
	(e) (Hydrogenated PBD) 14, 126, 158	7
Poly(dimethyl siloxane)	(a) 77, 168, 610, 10,000	5
	(b) 47.2 and (kit: 30.8, 57.2, 150)	12
Polyethylene	(a) ($\overline{M}_w/\overline{M}_n < 1.1$) 0.7, 1, 2	5
	(b) (known MWD, $\overline{M}_w/\overline{M}_n = 2.9$) 18.3	6
	(c) (Narrow MWD) 11.4, 28.9, 100.5	6
	(d) 7.1, 17.9, 27.7, 35.7, 32 (broad MWD), 41.7	10
Poly(methyl methacrylate)	(a) 106	5
	(b) 33.2	12
Polyether epoxy resins	(a) Oligomers; unit separation for $\overline{M}_n$ calibration in low-MW range	13
Poly(vinyl chloride)	(a) ($\overline{M}_w/\overline{M}_n \simeq 2.5$ for all) 25.5, 41, 54	10
	(b) 37.4	
Linear hydrocarbons	(a) $C_{28}H_{58} = 394.77$, $C_{36}H_{74} = 506.99$	11
Various polymers (usually only one MW of a given type available)	(a) Approx. or exact MW values may be given; some kits available	2, 3, 5, 11, 12, 14

[1] Waters Associates, Inc., Maple Street, Milford, Mass. 01757.
[2] Research Plus, Inc., Denville, N.J. 07834.
[3] Gallard-Schlessinger, Inc., 584 Mineola Ave., Carle Place, N. Y. 11514.
[4] Schwarz/Mann, Division of Becton, Dickinson & Co., Orangeburg, N.Y. 10962.
[5] Polysciences, Inc., Paul Valley, Industrial Park, Warrington, Pa. 18976.
[6] National Bureau of Standards, Office of Standard Reference Materials, Washington, D.C. 20234.
[7] Phillips Petroleum Company, Bartlesville, Okla. 74003.
[8] Duke Standards, 445 Sherman Ave., Palo Alto, Calif. 94306.
[9] Toyo Soda Mfg. Co. Ltd., Toso Bldg., Tokyo, Japan.
[10] Pressure Chemical Co., 3419 Smallman Street, Pittsburgh, Pa. 15201.
[11] Aldrich Chemical Co., Inc., 940 W. St. Paul Ave., Milwaukee, Wis. 53233
[12] (SP^2) Scientific Polymer Products, Inc., 99 Commercial St., Webster, N.Y. 14580.
[13] Shell Chemical Company, One Shell Plaza, Houston, Tex. 77002.
[14] Monomer-Polymer Laboratories, Division of Haven Industries, Inc., 5000 Langdon St., Philadelphia, Pa. 19124.
[15] Pharmacia Fine Chemicals A.B., Uppsala 1, Rapsg, Sweden.

another approach, one or two of the sample materials of interest can be characterized by light scattering and osmometry. The characterized sample then becomes the standard required for the single broad standard calibration method (Sects. 1.4 and 9.3).

Sometimes polymer standards that are molecularly similar to the sample can be used for calibration. By using this calibration approximation, the true values of MW and MWD for the samples will not be obtained, but relative comparison will be valid, especially if the sample molecular weights are not extremely varied.

Of various alternatives, the creation of standards from selected samples by light scattering and osmotic pressure measurements normally provides the most reliable calibration approach. However, characterization is time consuming and expensive. Service laboratories (3–5) can be employed to perform the needed measurements if in-house equipment or expertise is unavailable. By using the $\overline{M}_w$ and $\overline{M}_n$ values obtained by characterization of a single sample, calibration is possible with the Hamielec, GPCV2, and GPCV3 methods discussed in Chapter 9.

A light-scattering detector can also be used as an *in situ* molecular weight detector, eliminating the need for calibration (Sect. 5.12). This approach is not fully developed nor in general use, but it should see increased application because it reduces the need for many standard reference materials.

8.4 Detector Selection

A number of detectors are available for use in HPSEC and their general characteristics are described in Chapter 5. Frequently, the detector to be used is the one readily available on the instrument. Occasionally, however, solvent or sample type dictates the detector that must be used for a given analysis. (Sect. 8.2.) Three common methods of detection in HPSEC are: refractive index, UV absorption, and infrared absorption. To facilitate use of these techniques we have included in Table 8.3 solvent RI and UV cutoff data for commonly used solvents; IR transmitting "windows" are shown in Figure 8.1a.

MW-sensitive detectors such as light scattering and viscosity can be used for SEC calibration where standards do not exist. However, they are expensive and are not in general use. These detectors must be used in series with a concentration-sensitive detector to derive the desired MW information (Sect. 5.12). Occasionally, it is desirable to use other combinations of detectors for composition studies. An example showing the serial use of UV and RI detectors is discussed in Section 12.9.

8.5 Column Selection and Handling

The general aspects of columns and column packings are discussed in detail in Chapter 6. Column selection in HPSEC is based largely on the required molecular weight range of separation and the nature of the sample/solvent combination. Other considerations of convenience sometimes influence the selection: Are the needed columns readily available? Must several columns be coupled? Have the columns been in storage for some length of time, and could problems exist because of this?

Optimum Single Pore-Size Separations

The choice of column packing depends on the purpose of the separation. Pore size is normally the most significant parameter to be considered, since it dictates the range of molecular weight separation. Columns of the smallest single pore size (e.g., 40–60 Å) are desirable for separating mixtures of small molecules (< 5000 MW) (Sect. 11.1). For separating individual components in samples of 5000–500,000 MW (e.g., proteins) columns of a single intermediate pore size (e.g., 200–500 Å) are normally used.

To select the proper pore size for optimizing a given separation, the individual column calibration curves (MW versus K_{SEC} or V_R) are first obtained. The chromatographic peaks of two or more of the standards used in developing the calibration curves are then compared. The method is illustrated in Figure 8.2, where the calibration curves are real data for porous silica microsphere (PSM) columns, but the curves for the standards, although typical, are hypothetical. The distribution coefficient K_{SEC} is defined by Equations 2.7 and 2.8. For column packing 1 and the two standards in the example (MW = 10^4 and 10^5, respectively), the separation is poor (small ΔV_1) and on a nonlinear portion of the calibration curve. The same is true for column packing 3 (nonlinear and small ΔV_3), while the separation is greatest (largest ΔV) and on a linear portion of the calibration with column packing 2.

Once the pore size is selected, resolution can be improved by coupling together two or more columns of the same pore size. These individual columns should have very similar internal pore volumes, as discussed below. Figure 8.3 shows the effect of coupling two columns of the same pore size. Notice that while the range of separation remains the same (log MW axis), the volume over which the separation is made (V_R axis) is doubled and the slope of the calibration curve C_2 is doubled by doubling the pore volume. The column coupling technique is discussed further in Section 8.6.

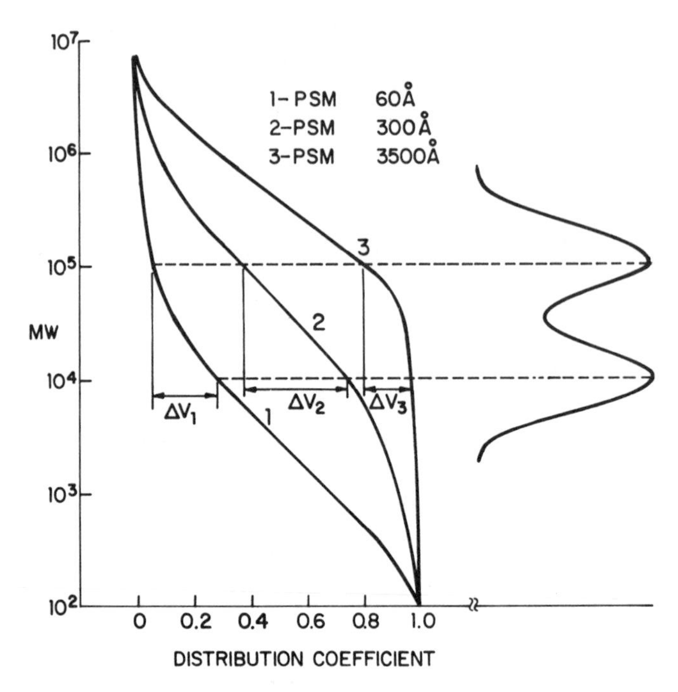

Figure 8.2 Selection of single-pore-size column packing for maximum resolution of two adjacent bands.

(See text for discussion.)

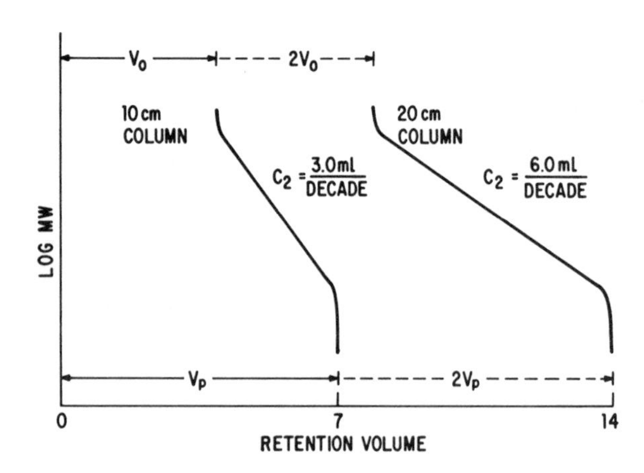

Figure 8.3 Pore volume determines the separation capacity (slope of MW calibration curve).

Bimodal Pore-Size Separations: Optimum Linearity and Range

For separating sample components that extend over more than two decades of molecular weight (e.g., broad-MWD polymers) a set of bimodal pore sizes is optimum. Knowledge of the individual column calibration curves provides a useful guide for choosing which pair of pore sizes to use. For the broadest range of separation and maximum linearity, the individual column calibration curves should be adjacent but nonoverlapping. Also, the pore volumes of each size should be equal for best linearity. Optimized range and linearity yield the most accurate calculated molecular weight values and permit the most accurate visual comparison of polymer chromatograms.

To better understand how bimodal coupling is easily accomplished in practice, Figure 8.4 should be consulted. Here the linear portions of the calibration curves for two types of column packings in individual columns are presented. The vertical axis is the slope of the calibration, C_2, which is

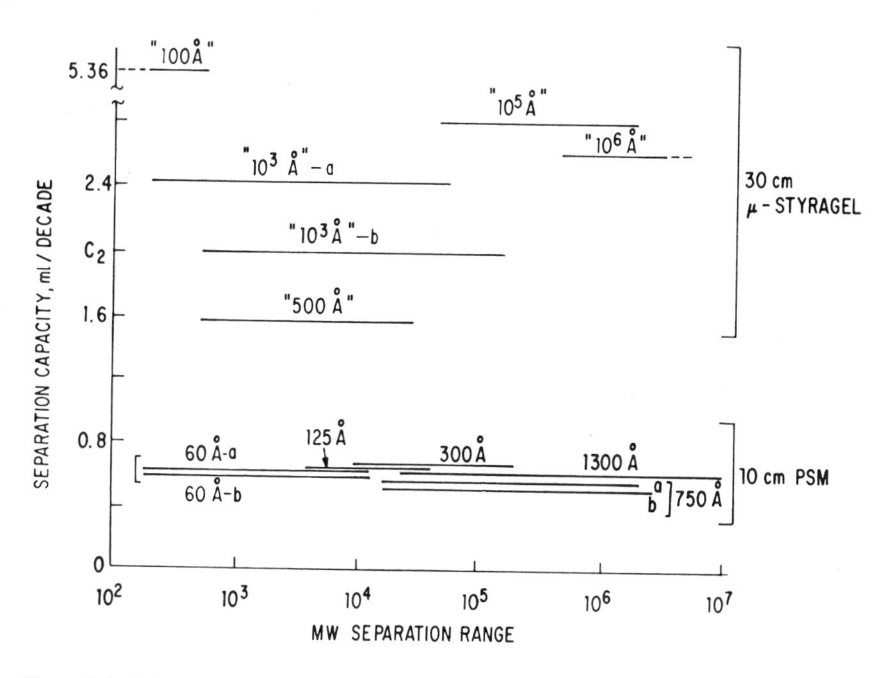

Figure 8.4 Selection of bimodal pore-size column packing for maximum range and linearity. Data for PSM (porous silica microsphere column packing); 10 × 0.78 cm; tetrahydrofuran, 22°C; flow rate, 2.5 ml/min; UV detector, 254 nm; sample, 25 μl. Data for μ-Styragel, 30 × 0.78 cm, tetrahydrofuran, 23°C; flow rate, 1 ml/min; RI detector, sample, 100 μl. (Reprinted with permission from Ref. 11.)

directly related to pore volume, and the horizontal axis is the log MW range which is separated by each pore size (in this case, for polystyrene standards in THF). As mentioned above, maximum calibration linearity and fit are obtained by coupling columns of equal pore volume (equal C_2 values) and adjacent separation ranges. Recall also that pore volumes (C_2 values) are directly additive (Sect. 4.5). Using these concepts we can now select a set of columns for wide linear MW calibration. For example (Fig. 8.4), to obtain a linear range of separation from about 500–1,000,000 MW with μ-Styragel, two columns of 500 Å (yielding a slope C_2 of $2 \times 1.6 = 3.2$) must be coupled with one column of 10^5 Å ($C_2 \simeq 3$). To obtain a linear molecular weight separation range of 1000–50,000, three columns of 500 Å ($C_2 = 3 \times 1.6 = 4.8$) should be coupled with one column of 100 Å ($C_2 = 5.4$). Although the linearity of the latter system is not perfect, it is the best available with the columns of this set.

Based on the data of Figure 8.4, it can be seen that arranging the porous silica microsphere (PSM) columns for bimodal operation is easier and more accurate because the internal pore volume (measured by C_2) and the separating range are much more nearly uniform for all pore sizes. For example, either 60 Å column (*a* or *b*) could be coupled with either 750 Å column to obtain a linear separation covering more than four decades (200–2,000,000 MW). A set of two columns each of 60 Å and 750 Å could conveniently be used to increase the resolution further. The calibration data in Figure 8.5 illustrate the results of proper bimodal-pore-size column arrangement.

Other Column Selection Guidelines

Sometimes column selection must be based on the solvent used for dissolving and separating the sample, because the solvent and column packing must be compatible. For example, organic-gel packings can only be used with organic solvents which swell them. A list of those solvents which do not adequately swell μ-Styragel is presented in Table 8.5. Rigid silica-based packings modified by grafting organic groups on the surface are available for organic and aqueous solvent use (Sect. 6.2).

An example involving column packing and solvent decisions in GPC is found in the analysis of poly(ethylene oxide). While water is an excellent solvent for this polymer, organic column packings such as μ-Styragel cannot be used because they are not swollen in water. On the other hand, with silica-based packings, the poly(ethylene oxide) adsorbs from water onto the packing surface. Therefore, either a compatible organic solvent must be used with the organic-gel packing, or else the water must be modified with

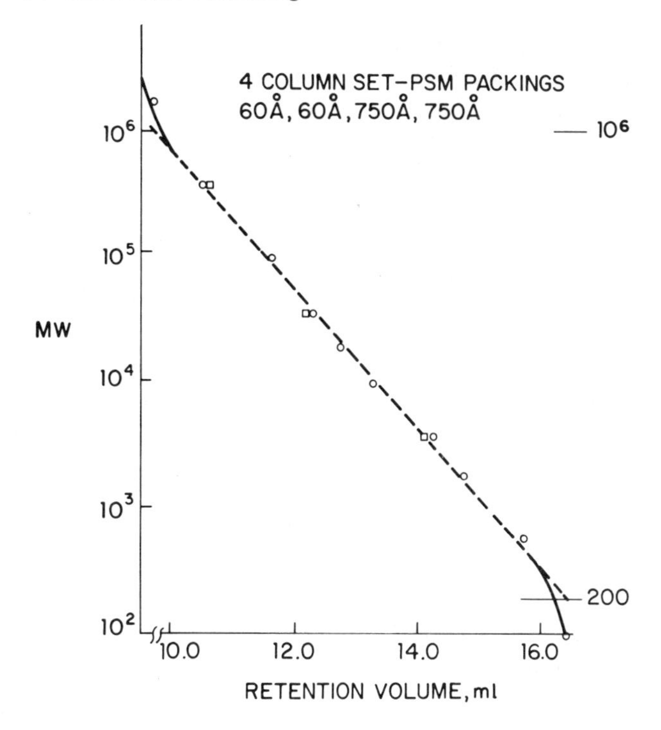

Figure 8.5 Broad-range, linear calibration for polystyrene standards using the bimodal-pore-size selection principle.

Four 10 × 0.78 cm PSM packings, 60 Å, 60 Å, 750 Å, 750 Å; mobile phase, tetrahydrofuran, 22°C; flow rate, 2.5 ml/min; UV detector, 254 nm; sample, 25 μl. (Reprinted with permission from Ref. 11.)

salt to eliminate adsorption on unmodified rigid silica packings (Sect. 13.3). Alternatively, an appropriate, surface-modified porous silica packing may be used with either solvent system. In our laboratories it has been found to be convenient to analyze poly(ethylene oxide) in *m*-cresol using the same organic-gel columns used for nylon and polyester analyses. While *m*-cresol is not an optimum solvent for poly(ethylene oxide) or for SEC, the sample is fully soluble and it is convenient to use this approach because a solvent change need not be made.

Column Handling

It is important to know the performance characteristics of new columns (Sect. 6.4). The separation range and performance of each new column should be checked by calibration under laboratory operating conditions. Particle and pore-size specifications provided by suppliers vary, and equipment variables (e.g., injection valve, connector tubing, and detector) all affect the reported values. Column performance should be tested at typical flow rates and pressures. A flow rate of 1–3 ml/min (e.g., 0.1–0.3 cm/s) is normally useful for operation. The pressure at which the column has been packed and the vendor's specifications on flow rate should not be exceeded in the SEC experiment. (See, for example, Table 8.7, specifying solvent viscosity/ maximum flow rate specifications for μ-Styragel columns.) If the column is found to be plugged during test, it should be returned to the vendor for exchange. High pressure should not be imposed on a column set suddenly, as this may disrupt the packing uniformity and may actually deform organic gel packings. Especially with columns of organic gels, startup flow rates and pressures should be imposed gradually.

Columns should be stored at constant temperature in a solvent that is inert to the packing. Columns of silica packings are less susceptible to

Table 8.7 Maximum Flow Rate and Physical Properties for Solvents Commonly Used with μ-Styragel GPC Columns

Solvent	Maximum Flow Rate[a] (ml/min)	Viscosity at 20°C (cP)
N,*N*′-Dimethylformamide[b]	2.5	0.92
Ethylene dichloride	2.8	0.84
p-Dioxane	1.6	1.44
Trichloroethane	2.0	1.20
Carbon tetrachloride	2.4	0.97
Toluene	3.0	0.59
1,1,1-Trifluoroethanol[b]	1.1	2.00
Benzene	3.0	0.65
1,1,1,3,3,3-Hexfluoroisopropanol[c]	1.2	0.64
Chloroform	3.0	0.58
Xylene	2.9	0.81
Tetrahydrofuran	3.0	0.55
Methylene chloride	3.0	0.44

[a] Flow rate valid for one column or a set of columns.
[b] Cannot be used with 100 or 500 Å μ-Styragel columns.
[c] Cannot be used with 100 Å μ-Styragel columns, not guaranteed with 500 Å μ-Styragel columns.
Taken from Reference 6.

Table 8.8 Solvent/Column Compatibility for Storage of μ-Styragel Columns

Solvent	Compatible with μ-Styragel for Storage	
	Yes	No
N,N′-Dimethylformamide[a]	×	
Ethylene dichloride	×	
p-Dioxane	×	
Carbon tetrachloride	×	
Toluene	×	
Benzene	×	
Xylene	×	
Methylene chloride	×	
Trichloroethane		×
1,1,1-Trifluoroethanol		×
1,1,1,3,3,3-Hexfluoroisopropanol		×
Chloroform		×
Tetrahydrofuran		×

Taken from Reference 6.

[a] Should not be used to store 100 or 500 Å μ-Styragel columns.

change on storage than are columns of organic-gel packings. (See Table 8.8 for conditions for storing μ-Styragel columns.) With organic-gel columns, the use of a bellows is recommended to compensate for swelling and temperature changes that might occur during storage. Silica-based columns stored with aqueous solvents should be at pH 2–8 and include an antimicrobial agent (e.g., 0.02% sodium azide).

8.6 Chromatographic Design Considerations

Various chromatographic compromises usually are required for a given set of analyses, and these depend on the objectives of the HPSEC experiment. Efficiency and resolution, flow rate, pressure, time, temperature, column-packing particle size, cost, viscosity, sample concentration, and so on, must all be considered and suitably adjusted to meet the required conditions. For example, the number of theoretical plates, N, decreases with an increase in flow rate or an increase in particle size. In addition, N increases linearly with L so that maximum efficiency usually is obtained with several small particle-size columns coupled together and used at low mobile-phase velocity. Typically, flow rates of 1–3 ml/min are used with 5–8 mm i.d. columns packed with 10–30 μm particles.

Peak variances for individual columns are additive when the columns are coupled in a set:

$$\sigma_t^2 = \sigma_1^2 + \sigma_2^2 + \cdots + \sigma_n^2 \tag{8.1}$$

where σ_t^2 is the observed variance for the connected column set and σ_1^2, σ_2^2, and σ_n^2 are the variances for the individual columns 1 to n. If the test solute peaks are Gaussian and the retention times, t_R, are approximately equal for each column ($t_{R1} \simeq t_{Rn}$), then

$$N_t = \frac{n^2}{\sum (1/N_i)} \tag{8.2}$$

where N_t is the calculated total plate count, n is the number of columns in series, and N_i is the plate count for each of the individual columns. As shown in Table 8.9, when connecting a set of columns of $\sim$7 μm porous-silica microspheres, a total plate count of 20,260 was calculated from Equation 8.2 for the test solute, toluene, compared to 23,890 plates actually measured at a flow rate of 2.5 ml/min.

It is important that matched columns always be used to assemble column sets. The variance relationship in Equation 8.1 predicts that if a low-efficiency column is connected to a high-efficiency column, the result is a

Table 8.9 Connected Columns for High-Speed GPC

Column Design	Column Length (cm)	t_R, Toluene (min)	N	σ^2
PSM-50S	10	1.27	2970	5.43×10^{-4}
PSM-300S	15	1.92	3425	10.76×10^{-4}
PSM-800S	10	1.26	5400	2.94×10^{-4}
PSM-1500S	10	1.26	3130	5.07×10^{-4}
PSM-4000S	15	1.92	8145	4.53×10^{-4}
				$\sigma_t^2 = 28.73 \times 10^{-4}$
				$N_t = 20{,}260$
				$N_{obs} = 23{,}890$

Taken from Reference 10.

total column set of poor efficiency. Therefore, a low plate count column never should be included with a high plate count column set. Generally, connecting columns of different diameter or of different packing types should be avoided. In coupling columns there has been some debate over which column to place first in the series—the one with the large or the one with the small pores. No definitive experiments have been published for the recommended bimodal pore systems. However, it is best to be consistent, and probably the small porosity columns should be placed first (i.e., close to the injection port of the sample). This arrangement effectively allows the large solute molecules to be diluted in the mobile phase with little or no retention as the sample passes through the small-pore-size column. The large molecules are then separated in the second column at a lower local concentration.

Increasing the column length by coupling increases the mobile-phase back pressure and the time required for analysis. At some point adding more columns is not practical because the pressure required for useful mobile-phase flow rates will exceed the maximum safe operating pressure of the system (typically 5000 psi).

Measurements of column efficiency by plate count indicate only the extent of deleterious band spreading, but they do not provide information about the effectiveness of peak separation. The extent of separation of different molecular sizes is provided by the calibration curve (V_R versus molar volume for small molecules, or V_R versus MW for macromolecules). Resolution accounts for both the extent of separation and column efficiency and is thus a measure of the useful separation. Maximum resolution in SEC is usually obtained with a single pore-size column packing which separates 1.5–2 decades of molecular weight. If the sample encompasses a larger molecular weight range than this, two pore-size distributions should be coupled to obtain maximum accuracy and range (approximately 4–5 decades of molecular weight, Sect. 8.5).

To determine SEC column resolution, two very narrow MWD polymers, or two compounds of differing molecular weights, are injected sequentially into the chromatograph and their peak positions (V_R) and baseline peak widths (W_b) evaluated for determining the resolution via

$$R_{sp} = \frac{0.58}{D_2 \sigma} \tag{8.3}$$

Derivation of Equation 8.3 and theoretical concepts are covered in Section 4.2, as are typical performance values. If at any time the value of σD_2 becomes excessively large, either the columns have low plate count (large σ values) or the column set is made up of improper pore sizes (large D_2).

The hardware used for HPSEC which must be considered in experimental design has been discussed in Chapter 5 and will not be elaborated upon further here. The equipment should be made ready for operation and a stable baseline obtained. If the pumping system does not deliver mobile phase at an assured constant flow rate ($<0.5\%$ error), a syphon assembly similar to that shown in Figure 5.14 is recommended for retention volume monitoring. This syphon prevents solvent evaporation and very precisely measures the total liquid volume of mobile phase used in the experiment (Sect. 10.5). The syphon should be coupled to an electronic counter which keeps track of the dumps. The need for accurate, constant-flow-rate monitoring, and the effect of flow rate variation on molecular weights calculated from GPC experiments are discussed in Section 7.2.

For preparative work, larger column capacities are needed and it is often more economical to operate with lower-efficiency, lower-cost, larger-diameter particles (30–50 μm) (Sect. 11.2). In terms of the column packing technology, it takes more sophisticated equipment and procedures to pack the smaller particles (10 μm) into highly efficient columns than for the large particles, which sometimes can even be hand-packed. These procedures are discussed in Section 6.3.

8.7 Making the Separation

This section discusses preferred procedures for sample preparation, injection, and obtaining the chromatogram.

Dissolving the Sample and Standards

Before dissolution, it must be ascertained that the sample to be analyzed is accurately representative of the whole by grinding and thoroughly mixing the larger bulk materials. The sample may also need to be dried (e.g., nylon picks up several weight percent of water at room temperature and 50% relative humidity), and for some synthetic polymers, melt quenching is desirable for facilitating dissolution by reducing the crystallinity.

The sample must be fully dissolved so that it is not fractionated by differential solubility based on molecular weight, crystallinity, or composition. To eliminate spurious peaks and to provide optimum baseline stability, the solvent used for dissolving both the calibration standards and the sample should be the same mobile phase that is in the chromatograph. The dissolving solvent chosen should be based on the various criteria described in Section 8.2.

Most samples can be dissolved at room temperature by gentle agitation with a magnetic stirring bar or laboratory shaker. Ultrasonic devices should *not* be used for dissolution of macromolecules since shear degradation may result. Workers using ultrasonic devices to disrupt biological cells should take special note of this caution, since macromolecular materials such as high-molecular-weight proteins can be degraded with this procedure. In sample dissolution, care should be taken not to let the sample adhere to the neck of the flask, where it may not be dissolved. For example, volumetric flasks should be filled to only one-third capacity during the initial dissolution step. With some synthetic polymers such as polyacrylonitrile it is advantageous to add ice-cold solvent to the polymer sample and then to warm gradually with stirring to accomplish dissolution. This permits small pieces of polymer sample to dissolve before they swell and congeal into one large intractable mass. In other cases it is necessary to heat the mixture for complete dissolution. For example, there are no known solvents that will break up the crystalline bond forces of high-molecular-weight polyethylene at room temperature.

To inspect the sample solution for complete dissolution, it is sometimes useful to direct a narrow beam of light through the solution flask in a dark box or darkroom to find undissolved or suspended particles. The viscosity of a macromolecular sample solution can be monitored as a function of time to indicate completeness of dissolution, and the NMR technique also has been used, by correlating proton band narrowing with the completeness of dissolution. For guidelines with regard to the total sample mass and concentration to be injected, the reader is referred to Section 7.4 and Table 8.10, which relates concentrations and molecular weight for use with μ-Styragel columns.

Sample Solution Filtration

After the sample is dissolved, the resulting solution should be filtered gently through membrane filters with pore sizes of not less than about 0.4 μm

Table 8.10 Molecular Weight versus Sample Concentration for μ-Styragel Columns

Molecular Weight Range	Maximum Sample Concentration (%)
Up to 20,000	0.25
34,000–200,000	0.10
400,000–2,000,000	0.05
2,000,000+	0.01

Taken from Reference 6.

and not greater than 1 μm. Smaller filter pores may remove or shear-degrade high-molecular-weight dissolved polymer in the sample, while larger-pore-size filters may permit passage of undissolved particles or gel which will subsequently plug the column or its end fittings. A convenient device for filtering organic solutions is a stainless steel Swinney syringe adapter (Millipore Company) containing Fluoropore filters or comparable materials. A plastic Swinnex filter holder with Mitex membranes is useful for aqueous work. Partial or complete plugging of the pores of the membrane during filtration, observed as excessive pressure drop, can reveal the presence of unsuspected gel in the sample solution, even though the solution may appear clear. If this happens, some high-MW soluble portions of the polymer may also be removed and the sample will no longer be representative.

With some polymer samples the solvent type itself can have a striking effect on filtration. Depending on solvation effects, polymer chains may be expanded to a greater or lesser degree (i.e., their hydrodynamic volume may be larger or smaller). The solvent effect is particularly important in the case of elastomers, where there frequently exists a small fraction of very high molecular weight, soluble molecules as well as insoluble gel. If a good solvent is used (causing more expanded chain conformations), some of the very high molecular weight soluble material may be retained on the filter along with the true gel. On the other hand, if a less solvating but still adequate solvent is used, the polymer molecules will uncoil to a lesser degree. Consequently, the larger molecules, which could have been filtered out using the good solvent, will pass through the filter with the poor solvent. Thus the resulting chromatogram is more nearly representative of the total sample.

If any plugging of the filter pores by sample solution is indicated, chromatographic results obtained with this solution are of doubtful validity. All such observations should be considered as part of the HPSEC analysis. Even when the filtration is proceeding normally, the "edges" of the dissolution container should be checked for undissolved material. At this point the operator must decide whether or not to continue the analysis, depending on the degree of final sample dissolution.

Sample Injection

In general for HPSEC, sample injection with a valve is recommended (Sect. 5.5). The valve provides efficiency equal to that of syringe injection, permits continuity of flow in the high-pressure system, and eliminates septum fragments which in syringe injection may contaminate the mobile phase with particles caused by needle punctures.

Prior to sample injection, a silver paper disc filter (Swinney) or equivalent should be placed in line between the filling syringe and the sampling valve. Then the valve should be rinsed with 3–4 volumes of solvent via the syringe attached to its inlet (e.g., by a Luer Lok fitting). Air bubbles should always be purged from the end of the syringe before it is used. The valve is then purged with 3–4 volumes of sample solution before final filling. Alternatively, the rinse and sample solutions can be drawn back through the valve from a reservoir. Frequently, this approach helps to prevent or eliminate bubbles during filling.

For HPSEC analyses, the sample concentration should be low to facilitate filtration and avoid overloading the column. However, the solution must contain sufficient solute for detection, and the total weight of sample injected can be controlled by sample volume. Concentration limits for polymers analyzed on μ-Styragel columns are given in Table 8.10. In general, injection of 50–300 μl solutions of approximately 0.1% (w/v) sample is useful; however, the injection volume should be less than one-third the peak-width volume for a totally permeating peak (Sect. 7.4). Preliminary runs may be required to optimize the sample concentrations. Highly accurate sample concentrations are generally not required for a determination of molecular weight or molecular weight distribution, but nominal accuracy is required. Concentration variations of no more than $\pm 0.05\%$ (w/v) should occur in a sample series. For preparative isolations it may be desirable to overload the columns or to use macroparticle column packings to increase the mass of material that can be injected (Sect. 11.2).

Baseline Stability

A stable detector baseline is required for a successful experiment, but little information has been published to specify the stability needed. For accurate quantitative results it is necessary to have a flat, linear baseline prior to the emergence of the chromatographic peak and a return to that baseline after the peak has emerged. Therefore, the desired baseline may be specified in terms of acceptable percentage drift over the course of the sample and calibration experiments and in terms of long- and short-term signal/noise (S/N) ratios during the course of the experiment. Baselines that exhibit noise and drift of $\leq 0.5\%$ of the height of the peak of interest usually will provide satisfactory data. Variations caused by temperature fluctuation, air bubbles, leaks, flow variation, solvent inhomogenities, bleed from an old adsorbed sample, and electronic noise all may contribute to baseline instability. The actual construction of the baseline is discussed in the next section.

Obtaining and Using a Chromatogram Baseline

If a macromolecular sample is chromatographed, it will result in a display on the strip-chart recorder similar to that in Figure 8.6. (Many additional

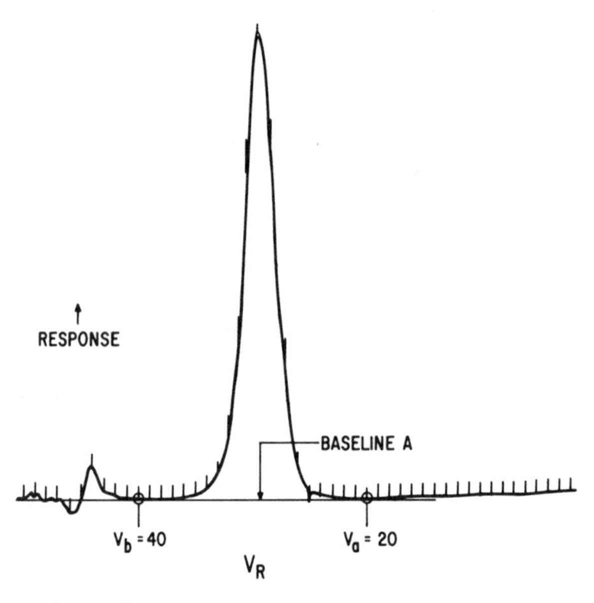

Figure 8.6 Typical chromatogram which illustrates a good baseline.
(Reprinted with permission from Ref. 7.)

chromatograms for GPC and GFC are shown in Chapters 12 and 13.) Since this chromatogram is used to compute a sample MW and MWD, it becomes especially important to accurately specify the start and end points of the curve envelope. In Figure 8.6 retention volumes V_a and V_b correspond to the beginning and end of the polymer chromatographic envelope, respectively. To establish the baseline, a straight line is drawn across the base of the chromatogram, as shown.

The definition of V_a, the low-retention-volume or high-molecular-end of the chromatogram, is normally straightforward, since the baseline is usually flat and is not influenced by impurities at this point. Defining V_b frequently is more difficult and depends on the separation of the polymer peak from low-molecular-weight materials and on the reestablishment of a stable flat baseline following elution of the peak of interest. With baseline resolution of all peaks, the choice of V_b is obvious. For example, Figure 8.6 shows adequate separation between polymer and impurity peaks for establishing a good baseline. However, in Figure 8.7, the chromatogram for polystyrene overlaps with oligomer and impurity peaks. Here there is a serious problem in establishing the correct baseline and the start and end-point limits for the chromatogram. Sometimes (as in Fig. 8.7) there is no

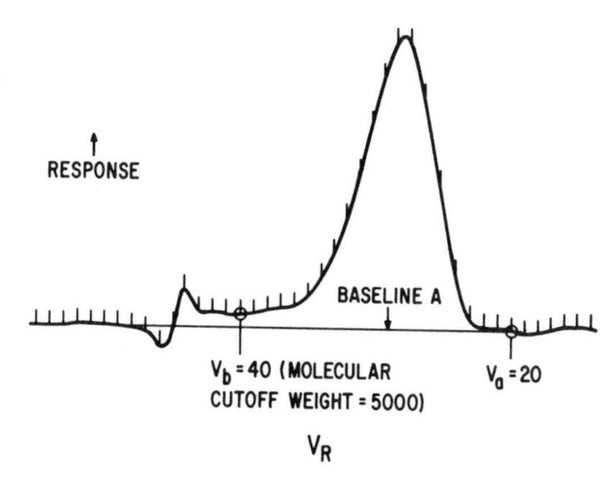

Figure 8.7 Typical chromatogram which illustrates a poor baseline. (Reprinted with permission from Ref. 7.)

choice in setting the limits except to be arbitrary in determining those components which adequately represent the sample.

The arbitrary selection of curve limits is currently one of the most restrictive aspects of accurate molecular weight analysis by GPC. The high sensitivity of the calculated MW values to the ends of the chromatogram ("curve cutting") is shown in Figures 8.8 and 8.9. Figure 8.8 depicts a typical chromatogram for a broad-MWD polystyrene sample with arbitrary curve limits for the low-molecular-weight end, set in 0.5 min increments from 17.0 to 18.5 min. Figure 8.9 shows the calibration curves obtained using these limits with the single broad-standard calibration method (Chapt. 9).

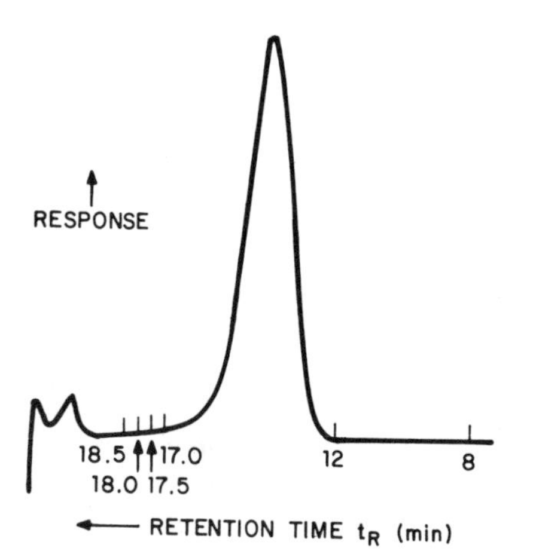

Figure 8.8 Chromatogram of polystyrene with various possible termination limits. Columns, 30 × 0.78 cm μ-Styragel, $10^2 + 10^3 + 10^5 + 10^6$ Å; tetrahydrofuran, 25°C; flow rate, 1 ml/min; RI detector; sample, Dow B8 polystyrene, $\overline{M}_w = 279{,}000$, $\overline{M}_n = 113{,}000$, 200 μl, 0.25% concentration.

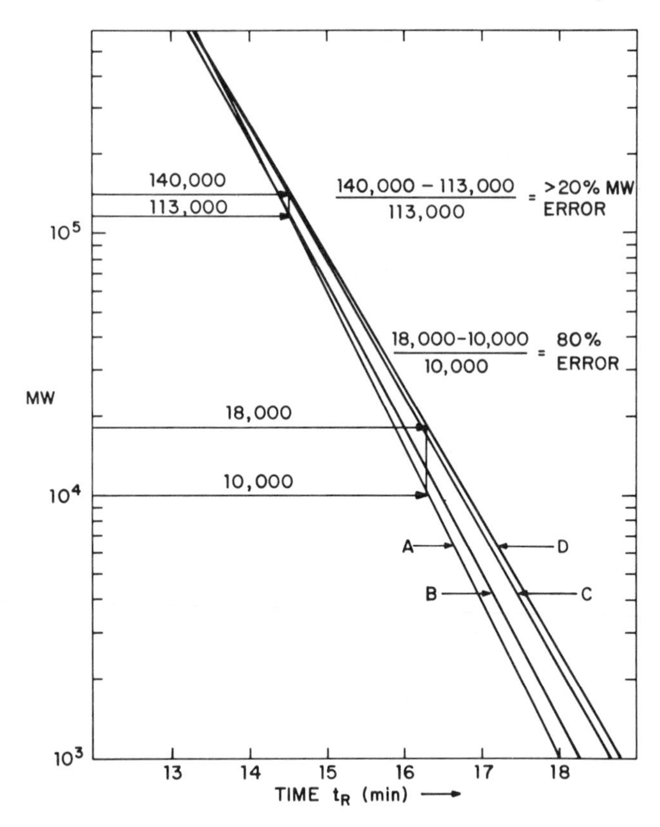

Figure 8.9 Calibration curves from Dow B-8 polystyrene of Figure 8.8 using various termination limits.

Calibration made using single broad standard method (Chapt. 9). Peak end: A, 17.0 min; B, 17.5 min; C, 18.0 min; D, 18.5 min.

The difference between calibration curve A and curve D indicates a calculated discrepancy in $\overline{M}_n$ of about 20% for this sample if it were calibrated with the 17.0 min cut point (Fig. 8.8) but then calculated as an unknown with an 18.5 min cut point. If the same calibration set were used for a sample of much lower $\overline{M}_n$ (e.g., ~10,000) the error could be as large as 80%. These data illustrate the importance of obtaining good reproducible HPSEC curves and baselines, of using reproducible curve cut points, and the value of using standards for calibration that are similar in MW and MWD (elution profile) to the unknowns to be analyzed. Sometimes the cut-point problem can be eliminated or minimized by additional resolution, that is, by using different pore sizes or added column length to separate out confusing peaks.

8.8 Troubleshooting

Experimental problems that may be encountered during the process of developing and making a separation and methods for overcoming some of these problems are now discussed.

Too High Pressure

If the pressure on the column is too high for acceptable operation (e.g., exceeds pressure specifications), one or more of several correction steps may be employed. If excessive back pressure is caused by high sample viscosity, the solution may be diluted, but detector sensitivity will have to be increased. Raising the temperature will reduce viscosity of the mobile phase and, if applicable, may help to redissolve precipitated or adsorbed material in the columns or on the end fittings. Pressure can be reduced by removing some of the fractionating columns from the system if the loss of resolution can be tolerated. An increase in column pressure is often indicative of partial plugging of the column inlet fitting assembly. Rigid particle columns that have been well prepared can sometimes be back-flushed to remove clogging particles from the column inlet, but this approach should only be used as a last resort. Back flushing of high-performance gel columns generally should not be attempted.

Column Plugging

If the flow of mobile phase stops completely, it is necessary to locate the source of the plug. Pump failure may be responsible for the stopped flow, and this possibility should be checked first. Plugged connector tubing or filters should be disconnected from the system and replaced. If the plugging has occurred in the columns or the column end fittings, eliminating the problem without loss of column efficiency should be attempted. Slowly raising the temperature of the columns may permit restoration of flow if a precipitated phase is clogging the inlet frit. If it becomes necessary to replace an inlet, the column should be equilibrated to room temperature and atmospheric pressure and mounted in a vise in a vertical position. The fitting should be removed carefully (do *not* touch the column, as the packing may be forced out by thermal expansion) and the end frit cleaned or replaced. The end fitting should be cleaned, checked for obstruction, and refitted onto the column. If some packing is unavoidably lost, additional packing can be added to the top of the column as a thick slurry with a spatula. Such repairs should only be used as a last resort, and the efficiency of a repaired column should always be checked. Experience has shown that the success rate is low for reestablishing the original plate count of high-performance columns. However, if lower-performance columns can be tolerated, such repairs can be useful.

Air Bubbles and Leaks

Obvious solvent leaks should be repaired immediately. If air bubbles are observed in the detector, it is necessary to isolate parts of the system to find their source. Suspected fittings should be replaced with new fittings, since leaks may be caused by repeated tightening (i.e., by deformation; air can diffuse into the mobile phase even against high pressure). Small compression fittings are a particular problem in this regard. Bubbles from solvent outgassing are best prevented by degassing prior to use.

Poor Resolution

Poor initial column resolution can be improved by [1] reducing the flow rate of the mobile phase, [2] diluting the sample if the column is overloaded, [3] raising the column temperature, [4] adding column length, or [5] substituting more efficient columns. If the performance of an initially good column significantly deteriorates and cannot be restored as described above, it should be replaced with a good column.

Low Solute Recovery

Low solute recovery means that not all of the sample has eluted from the column, either because of adsorption or degradation of the sample on the column. Suspect analyses can be checked by collecting and rechromatographing the total sample. Peak areas should be equal for the original and rerun chromatograms. Alternatively, the sample peak can be collected, characterized by UV absorption or refractive index, and compared to known concentrations of the sample solution. If irreversible column adsorption has occurred, an alternative mobile phase or column packing may be required. Successful elution of the sample from the column may also be attempted by changing the temperature, ionic strength, pH, or solvent polarity (see especially Sect. 13.3).

Constancy of Separation

Periodic comparison of calibration runs is useful for column systems that have been used for a long time. If plate count N is slowly changing, the column packing or packing structure is probably deteriorating. If sudden changes in N are observed, checks for plugging, leaks, pressure changes, and so on, should be made. If the changing of solvents and columns is common laboratory practice, one should ensure that the proper columns are in the instrument. The presence or absence of commonly used additives which could affect sample/substrate interactions should also be checked.

Peak Shape

If sudden or unexpected changes occur in sample peak shapes, the experiment should be suspect. Standard samples should be run and the peak shapes compared to those obtained previously. The user should always be on the lookout for surface effects that may add bias to the size-exclusion mechanism. Such effects can be noted by observing whether the retention volume for a particular solute has increased, or in extreme cases is greater than the retention volume for total permeation, or whether curve shapes occur other than those expected. These surface effects can be eliminated by changing polarity, pH, or ionic strength of the solvent, or by changing the packing.

If negative peaks are obtained, the detector polarity or recorder zero can be reversed for visual convenience. If fused peaks are obtained for discrete species, more resolution is needed. As discussed in Chapter 10 under curve summation techniques, it is not possible to obtain distinct peaks for each molecular species present in a high-molecular-weight polymer.

REFERENCES

1. J. Brandrup and E. H. Immergut, eds., *Polymer Handbook*, Wiley, New York, 1972.
2. N. I. Sax, *Dangerous Properties of Industrial Materials*, Reinhold, New York (1957).
3. Springborn Testing Institute, Enfield, Conn. 06802.
4. Arro Labs, Inc., Caton Farm Road, Joliet, Ill. 60434.
5. Rubber and Plastics Research Association, Shawbury, Shrewsbury SY4 4NR, England.
6. Courtesy of Waters Training Division, Waters Associates, Inc., Milford, Mass. 01757.
7. *ASTM* D3536-76, Standard Method of Test for Molecular Weight Averages and Molecular Weight Distribution of Polystyrene by Liquid Exclusion Chromatography (Gel Permeation Chromatography—GPC), ASTM, 1916 Race St., Philadelphia, Pa. 19103.
8. J. M. Evans, *Polym. Eng. Sci.*, **13**, 401 (1973).
9. Wilks Scientific Corp, Box 449, South Norwalk, Conn. 06856.
10. J. J. Kirkland, *J. Chromatogr.*, **125**, 231 (1976).
11. W. W. Yau, C. R. Ginnard, and J. J. Kirkland, *J. Chromatogr.*, **149**, 465 (1978).

Nine

CALIBRATION

9.1 Introduction

At the end of an HPSEC experiment, polymer molecules of different sizes are separated and their concentrations detected as a function of the retention volume V_R. In HPSEC, V_R increases with decreasing size of the eluting solute molecule. For HPGFC of biopolymers, solutes of distinctive sizes can appear as separate peaks in the chromatogram. On the other hand, broad chromatograms are often observed in HPGPC, because most synthetic polymer samples have broad molecular weight distributions (MWD). An HPSEC elution curve can be looked at as a profile of the molecular size distribution of the sample. Since many physical and chemical properties of polymers vary with molecular size, the raw-data sample elution curves are useful for relative sample comparisons. However, relative sample comparison is valid only for data obtained under the same experimental conditions, because the profile of an HPSEC elution curve is a function not only of the sample molecular size distribution but also of the specific columns and instrumentation used in the experiment. Only when the elution curves are transformed into MWD curves for the polymer samples can the HPSEC data from different instruments be compared and treated quantitatively. Unlike the elution curves, the MWD of a polymer sample is an intrinsic polymer property that determines the end-use properties of the polymer.

The HPSEC elution curves contain the MWD information of the sample, and the task is to extract this MWD information from the elution curves with accuracy and precision. The objective is to remove from the HPSEC elution curves the influences that result from specific features of the particular experiment, but not the features of the intrinsic MWD of the sample. The first step in extracting molecular-weight information from HPSEC is to establish a calibration relating the retention volume to the molecular weight (MW) of the polymer sample. The molecular weight rather than the size of the polymer molecules is used to describe the calibration, because the actual size of polymer molecules in solution can change with temperature and solvent, but the molecular weight of the polymer chain is directly proportional to its contour length, a more intrinsic property of the polymer. Like the elution curves, HPSEC calibration curves should be reported in terms of retention volume V_R, not retention time t_R, to minimize the effect of flow rate changes. In theory (Sect. 2.1), V_R is a more fundamental parameter than is t_R for describing peak retention.

The molecular weight calibration in HPSEC is an experimental approach that is valid only for the particular polymer/solvent system and the experiment in question. To describe the relationship between the polymer molecular weight and the particular retention volume, one must have specific knowledge of the experimental conditions. HPSEC retention is determined by the relative sizes of the solute macromolecules and the sizes of the pores in the column packing. Different pore sizes of column packings can change the extent that macromolecules are excluded from the packing pore structures and thus can also affect retention. Therefore, the MW-V_R calibration relationship holds only for specified HPSEC columns as well as only for specified polymer/solvent systems.

HPSEC calibration should be repeated often to compensate for column deterioration. While small fluctuations in temperature or flow rate may not upset the calibration, changing columns, solvent, or the nature of the polymer samples will require recalibration of the HPSEC experiment. Frequently, the calibration curve is described as a property of the column sets. This is not true because the calibration also depends on the polymer/solvent system and the HPSEC instrumentation, including all external tubing, connectors, and detector volume.

The term "molecular size" is difficult to define quantitatively, as there are many size parameters that can be used to describe molecules or particles of different shapes and conformations. For a rigid spherical particle, the radius or diameter is the obvious size parameter to use. However, even for a simple shape such as a rigid cylindrical rod, there exist two characteristic geometric dimensions, the radius and the length of the rod. Neither of these two dimensions alone can uniquely define the size of a rod. For small organic molecules, the long axis of the molecules may still be used in the first approximation for rough size comparisons of solute molecules. An example of this is the calibration curve shown in Figure 9.1, where the length of the long axis of

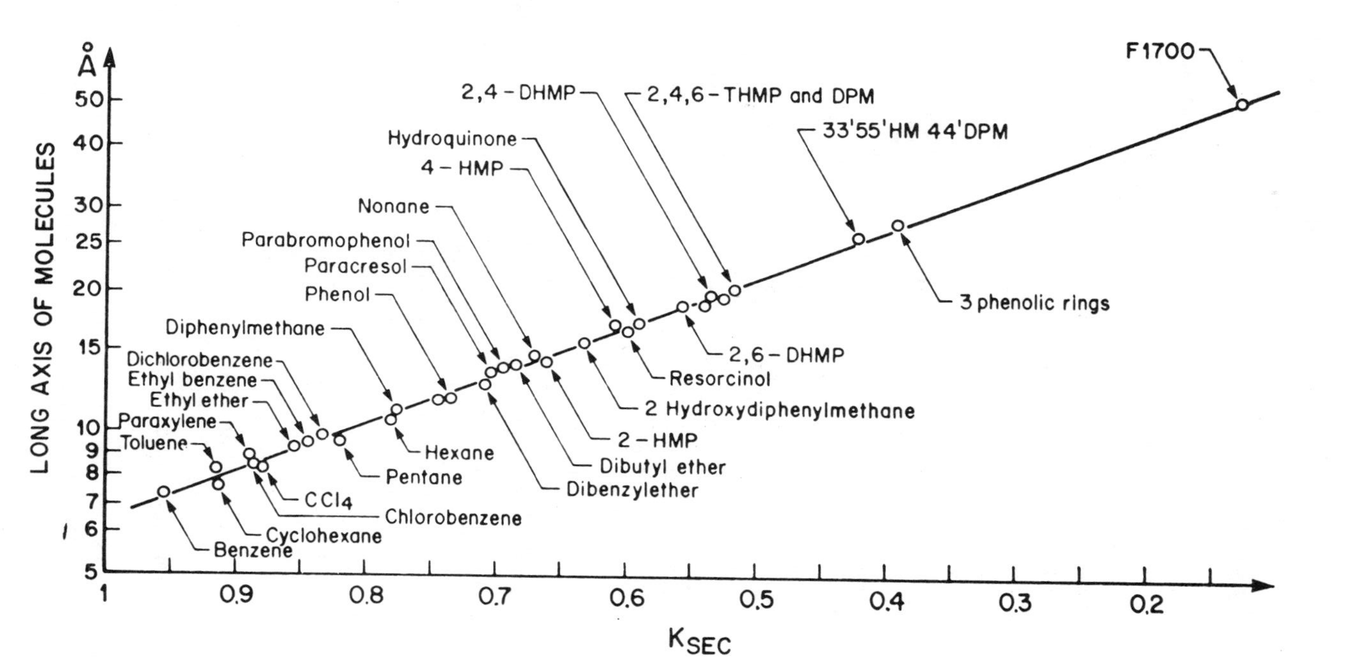

Figure 9.1 Calibration curve for small molecules.
(Reprinted with permission from Ref. 1.)

the molecules measured in angstroms is plotted against the SEC distribution coefficient K_{SEC}. For small molecules, the molar volume or the ratio of molecular weight to density has also been used as the GPC calibration dimension (2).

For macromolecules there are conformations ranging from random coils to rigid spheres and rods (Sect. 2.5). To use a unique dimension to define the size of these molecules is not possible. Therefore, there cannot be a general way of relating molecular size to molecular weight for macromolecules. The best approximation to describe macromolecular size is the use of the radius of gyration (R_g) of the macromolecules. The use of R_g makes it possible to compare HPSEC retention characteristics of macromolecules of different shapes.

The relationship between MW and R_g depends on solute/solvent interactions and chemical structures, even for macromolecules of similar conformations. For example, if we compare polystyrene [$(CH_2{-}CH\phi)_n$] to polyethylene [$(CH_2{-}CH_2)_n$], which are both in the random-coil conformation, one can see that the heavy benzene ring [ϕ] does not increase the polymer chain length of polystyrene as compared to that of a polyethylene molecule having a chain backbone of the same number of carbon atoms. In other words, for the same chain length, a polystyrene molecule will have a much larger molecular weight than a polyethylene molecule. Or, for the same molecular weight, a polystyrene molecule will have a much smaller R_g than polyethylene. A similar situation is found with low-molecular-weight compounds. For example, the R_g of cyclohexane is smaller than that of *n*-hexane because of its cyclic nature, even though the two compounds have nearly equal molecular weight. The use of R_g provides the common denominator for comparing sizes of solute molecules of different chemical structures and shapes. For macromolecules in solution, the degree of coil or rod comformation can change depending on temperature and polymer/solvent interactions. Such conformational changes can affect the polymer R_g and its HPSEC retention volume.

Since there are many polymer/solvent systems involved in HPSEC as well as many types of columns of different pore sizes, the best way to obtain accurate calibration is by experiment rather than by theoretical calculations. Errors in the molecular weight calibration affect the accuracy of the molecular weight measurement by HPSEC. These errors are minimized by obtaining the experimental calibration curve under the same conditions as the samples.

The next two sections describe various ways in which one can calibrate the HPSEC experiment so that a quantitative transformation of the HPSEC elution curve into a MWD curve can be made. The actual data transformation is described in Section 10.3.

9.2 Calibration with Narrow-MWD Standards

Since narrow-MWD standards are available only for polystyrene, peak position calibration cannot be used for molecular weight analyses of polymers in general. The importance of polystyrene peak position calibration is to check and optimize column resolution and separation conditions. As with peak position calibration, the universal calibration is valuable for fundamental studies of HPSEC separation mechanisms and operating variables. The knowledge of these calibration concepts is important for a full appreciation of the fundamentals of HPSEC calibration.

Peak Position Calibration

The calibration curve for a set of HPSEC columns can be established experimentally by relating the peak retention volume to molecular weight for a series of narrow-MWD standards. Polymer fractions of narrow MWD elute as sharp peaks with their retention volumes varying with differences in molecular weight. When there exist a sufficient number of narrow fractions of the same polymer type (about 10 different molecular weight fractions), one can accurately determine the calibration curve of the particular HPSEC columns and polymer/solvent system by the peak position method. The peak position calibration curves for columns of several pore sizes are illustrated in Figure 9.2. The peak position calibration curve is sometimes referred to as the true calibration curve because of its experimental accuracy.

While peak position calibration has its significance in defining the HPSEC calibration concept, it has rather limited use in practice. The problem is the lack of narrow standards. Polystyrene is the only polymer for which a series of commercial standards of different molecular weight and narrow MWD is available. (The standards are made by an anionic polymerization process which gives them the narrow MWD characteristics.) Molecular weight standards of very narrow MWD are generally not available for other synthetic polymers. The preparation of narrow standards by fractionating whole polymers into large-scale fractions is rarely done in practice because of high cost.

Only for biopolymers are there truly monodisperse polymers where all molecules in the sample are of the same molecular weight. In fact, calibration curves in GFC are commonly obtained by the peak position calibration method. An example of the peak position calibration curve in HPGFC is shown in Figure 9.3 for a series of proteins of different molecular weight. The accuracy of this calibration depends on whether the proteins all have a similar conformation and chemical structure, so that molecular size and molecular weight are uniquely related.

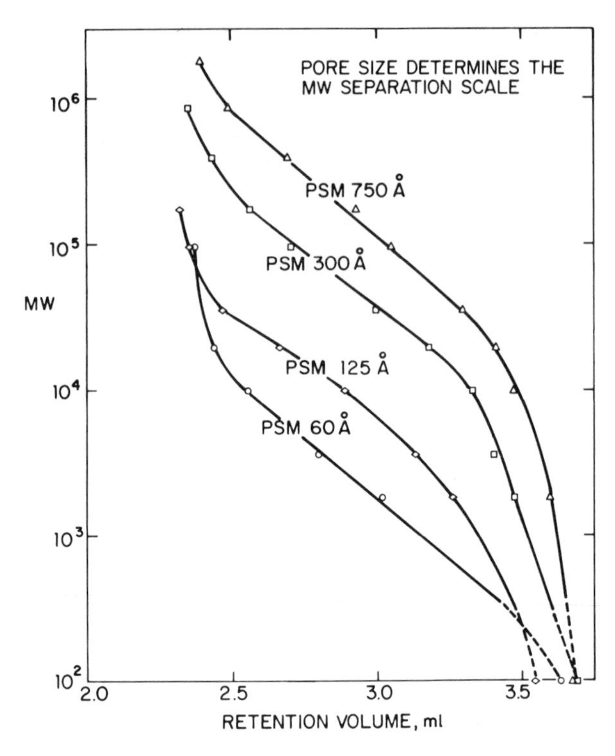

Figure 9.2 Molecular weight calibration range as a function of pore size.
Columns, 10 × 0.78 cm each porous silica microspheres; mobile phase, tetrahydrofuran, 22°C; flow rate, 2.5 ml/min; UV detector, 254 nm; sample, 25 μl solutions of polystyrene standards. (Reprinted with permission from Ref. 3.)

As discussed in Chapter 1, a major application of HPSEC is to study the MWD of synthetic polymers. Before other more practical calibration methods became available, GPC users learned about column characteristics only through the polystyrene peak position calibration curve. Therefore, a common practice was developed in interpreting the polymer MWD based on equivalent polystyrene MW units by means of the polystyrene calibration curve. However, for an unknown polymer, the MW values calculated by using the polystyrene calibration curve are just an arbitrary numerical ranking of sample molecular weight. The absolute MW values of the unknown polymers can be quite different.

The problems associated with the equivalent polystyrene MW approach led to the development of more accurate calibration methods. The universal calibration curve concept discussed next was evolved as a rigorous way to

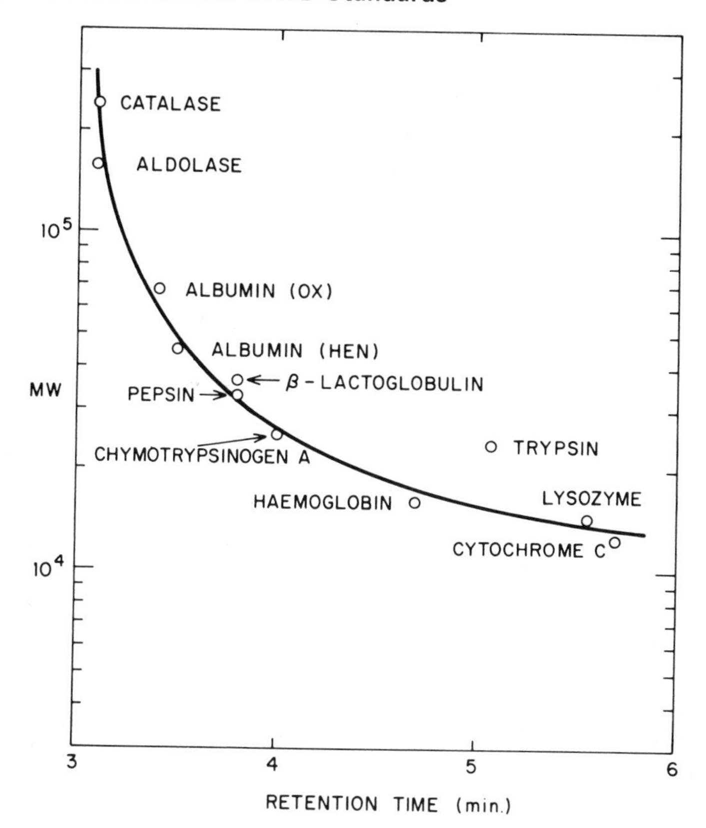

Figure 9.3 Protein peak position calibration curve.
(Data reproduced with permission from Ref. 4.)

transform the polystyrene peak position calibration curve to a suitable calibration curve for characterizing many types of polymers.

Universal Calibration

The universal calibration method introduced by Benoit et al. (5) utilizes the concept of the hydrodynamic volume of polymer molecules. The hydrodynamic volume can be expressed in terms of the product of the molecular weight M and the intrinsic viscosity $[\eta]$ of the polymer sample. In general GPC calibration curves for polymers of different types merge into a single plot when the calibration data are plotted as $\log [\eta]M$ as illustrated in Figure 9.4, instead of on the usual $\log M$ scale.

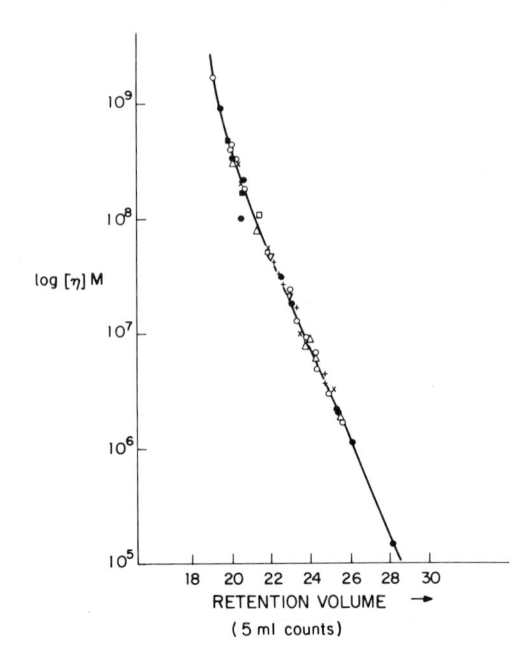

Figure 9.4 Universal calibration plot.
Solvent: THF. ●, PS; ○, PS (comb.); +, PS (star); △, heterograft copolymer; ×, poly-(methacrylate); ⬡, poly(vinyl chloride); ▽, graft copolymer PS/PMMA; ■, Poly(phenyl siloxane); □, polybutadiene; (Reprinted with permission from Ref. 5.)

Intrinsic viscosity $[\eta]$ is an experimental quantity derived from the measured viscosity of the polymer solution. The value of $[\eta]$ for a linear polymer in a specific solvent is related to the polymer molecular weight through the empirical Mark-Houwink equation,

$$[\eta] = \mathbf{K}M^{\mathbf{a}} \tag{9.1}$$

where values for the Mark-Houwink constants **K** and **a** vary with polymer type, solvent, and temperature. They are tabulated for a wide variety of polymer/solvent/temperature combinations in Tables 8.2 and 10.2 and in Reference 6. Experimentally, intrinsic viscosity is defined as

$$[\eta] = \lim_{c \to 0} \frac{\eta_{sp}}{c} \tag{9.2a}$$

or

$$[\eta] = \lim_{c \to 0} \frac{\ln \eta_{rel}}{c} \tag{9.2b}$$

where η_{rel} (relative viscosity) $= \eta_{solution}/\eta_{solvent}$, η_{sp} (specific viscosity) $= \eta_{rel} - 1$, and c is the weight concentration of the polymer solution. Accurate values of $[\eta]$ can be measured by extrapolation of measured viscosities to zero concentration. At high dilution ($\lim_{c\to 0}$), the measured viscosities in Equations 9.2a and 9.2b approach the same limiting value, $[\eta]$. Approximate values of $[\eta]$ can be obtained from a single viscosity measurement at low sample solution concentration (7, 8).

The product $[\eta]M$ of a polymer chain in solution is directly proportional to the hydrodynamic volume V_h of an equivalent sphere (9):

$$[\eta] \propto \frac{V_h}{M} \tag{9.3}$$

On the other hand, V_h (or $[\eta]M$) is expected to be proportional to R_g raised to the third power according to the statistical theories of polymer solutions (Eq. 2.22 and References 10 and 11). Therefore, the concept of universal calibration is consistent with Casassa's SEC theory based on solute R_g considerations (Sect. 2.4 and Ref. 12). Figure 9.4 substantiates that the radius of gyration R_g is the basic parameter for HPSEC size separation. When separation occurs strictly by SEC involving only entropy changes, the theory predicts that polymers of different chemical structures, whether branched or not, will elute at the same retention volume from any given SEC columns provided that the polymers have the same value of $[\eta]M$, V_h, or R_g. Although available literature data generally have supported this universal calibration concept, there are still significant deviations between experiment and theory.

The universal calibration approach is potentially useful for studying polymer branching, as discussed in Chapter 12. For polymer samples having nonuniformity in long-chain branching, copolymer composition, blends or mixtures, the polymer flowing through the detector cell may be polydispersed in molecular weight. The cell may contain a mixture of polymer molecules having the same hydrodynamic volume but of different molecular weights. To account for the polydispersity of the polymer in the detector cell, the correct universal calibration parameter should be $[\eta]\overline{M}_n$ (13).

The universal calibration method is conceptually sound, but its use is still rather limited. The accuracy and precision of the method have not been adequately evaluated. Although the method is applicable to calibration curves obtained with broad-MWD standards as well, it is used most often with polystyrene peak position calibration curves. The steps for using the universal calibration approach are outlined in Section 10.7. Needed are values for the Mark-Houwink constants **K** and **a** for the polystyrene and the unknown polymer in the particular HPSEC solvent. Accurate **K** and **a** values are difficult to obtain experimentally. Usually, there will not be a sufficient number of polymer standards of known molecular weight to allow accurate determination of the molecular weight dependence of $[\eta]$. Imprecision in the viscosity measurements needed to determine **K** and **a** is also a problem. How much these accuracy and precision errors in **K** and **a** can propagate errors in the final molecular weight determination by the universal HPSEC calibration approach is a question not yet answered. Until these considerations are resolved, the broad-MWD standard calibration methods described below will continue to be the preferred approach to GPC calibration.

9.3 Calibration with Broad-MWD Standards

There are two different ways of using broad-MWD polymer standards for GPC calibration. The integral-MWD method utilizes the complete MWD curve of the polymer standard. The linear calibration methods use only the average MW values of the polymer standard but assume a linear approximation of the GPC calibration curve. Both approaches are valid and useful at times, depending on SEC conditions. The linear calibration methods are more versatile for analyzing polymers of different types. Polymer standards are more readily available for these methods, because the average MW values are more readily attainable than the complete polymer MWD.

The accuracy of the broad-standards calibration methods varies, depending on the accuracy of the available MWD information on the standards and the accuracy of the linear-calibration approximation. When the separation columns are purposely selected to assure linearity in calibration (Sect. 9.6), the use of linear calibration methods is definitely recommended over integral-MWD methods.

Unlike those from narrow-standards methods, the calibration curves obtained by broad-standards methods are affected by instrumental peak broadening in the HPSEC experiment. Without corrections, this calibration error can cause errors in the molecular weight analyses of polymer samples. Proper account of the effect of peak broadening on calibration is provided in GPCV2 and GPCV3 calibration methods.

Integral-MWD Method

The integral-MWD calibration method requires that the complete MWD of the broad polymer standard be known. For a known polymer MWD curve, there is a unique correspondence between molecular weight and the weight fraction of the polymer below a given molecular weight. Similarly, there is a unique correspondence between the retention volume V_R and the weight fraction or the fractional area under the observed GPC elution curve. The GPC calibration curve for integral MWD method is obtained by matching those MW and V_R values which correspond to the same value of sample weight fraction on the MWD and GPC elution curves.

Initially, complete MWD information for the broad polymer standards required for this calibration approach was obtained experimentally. A broad-MWD polymer sample intended for a calibration standard was fractionated by solvent extraction or column fractionation and the molecular weight determined for each of the fractions. The MWD information obtained in this way for the whole polymer, now the calibration standard, was then used to establish the SEC calibration curve. This experimental polymer fractionation approach has not generally been followed in practice because of the tedious nature and the questionable molecular weight precision of the experimental fractionation and the molecular weights of the characterized fractions.

A practical alternative approach suggested by Weiss (14) makes use of theoretical polymer MWD and average molecular weight values to provide the needed MWD information for the integral-MWD calibration. This approach has been applied to both water-soluble polymers (15) and organic-soluble polymers (e.g., nylon 66) (16). We will now use the illustration in Reference 16 to explain the integral-MWD calibration method in more detail.

Many polymers follow predictable MWD curve shapes which depend on the type and condition of polymerization. For example, anionic polymerizations often follow a Poisson distribution; certain vinyl polymerizations (at low conversion and with termination via radical coupling) yield a Shultz-Zimm distribution (Fig. 1.5); and condensation polymers give a Flory "most probable distribution" if prepared under equilibrium conditions (Chapt. 1 and Ref. 10). The Flory MWD, accepted as the idealized MWD function for nylon 66 polymers and for polyamides and polyesters in general, is now utilized to illustrate the use of the integral-MWD method. Consider the most probable distribution function,

$$W_X = (1 - p)^2(X)(p^{X-1}) \tag{9.4}$$

where W_X is the weight fraction of polymer with X repeat units of molecular weight M_0 and p is the extent of reaction with $p = (\overline{M}_n - M_0)/\overline{M}_n$. As the value for p approaches unity (as is usually the case for high-molecular-weight polymers), the value for polydispersity ($\overline{M}_w/\overline{M}_n$) approaches 2 for polymers of this theoretical MWD. Equation 9.4 is plotted in Figure 9.5, where the total area under the curve is unity. With a known $\overline{M}_n$ or $\overline{M}_w$ value and the M_0 value for the repeat-unit MW, the complete MWD of a particular sample of a condensation polymer can be predicted according to Equation 9.4. This sample can now be used as the standard for the integral-MWD calibration. The shaded area a in Figure 9.5 represents the weight fraction

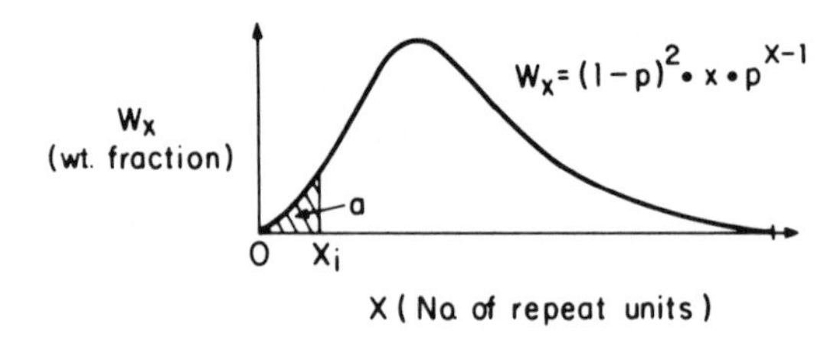

Figure 9.5 Flory most probable distribution function.
(Reprinted with permission from Ref. 16.)

of the molecules that have MW values less than X_iM_0 in this polymer standard.

Now consider the GPC curve represented for a polymer standard, such as in Figure 9.6. The detected concentration response is normalized so that the area under the curve equals unity. Since small molecules elute last in GPC, the low-molecular-weight fractions in the shaded area *a* of Figure 9.5 can be identified with the shaded area at the long retention time in Figure 9.6. For the same fractional area of *a*, a point V_i can be defined on the GPC elution curve. Without column dispersion, all molecules with molecular weight less than X_iM_0 should elute after V_i. This gives an unique pairing between V_i and the MW of X_iM_0. A series of these volume/MW pairs can then be generated to produce the final GPC calibration curve.

The presumed advantage of the integral method is that it makes no assumptions regarding the calibration curve shape and thus permits an accurate search for nonlinear calibration curves. However, this advantage often does not occur in practice, owing to the rather poor precision of the method.

An example of a calibration curve obtained by this method using a nylon 66 standard is shown in Figure 9.7. The curve is compared with the effective

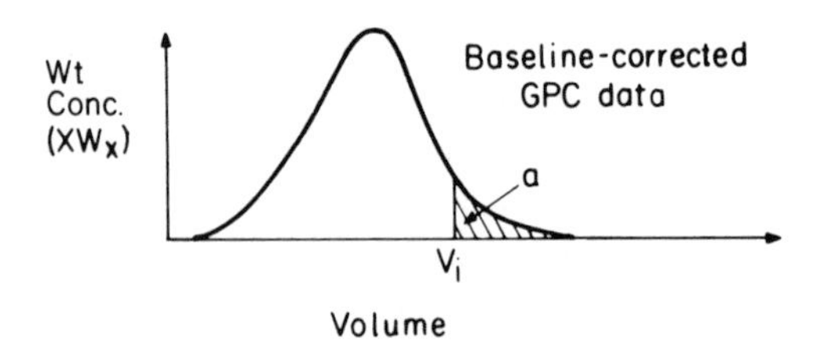

Figure 9.6 GPC elution-curve sketch.
(Reprinted with permission from Ref. 16.)

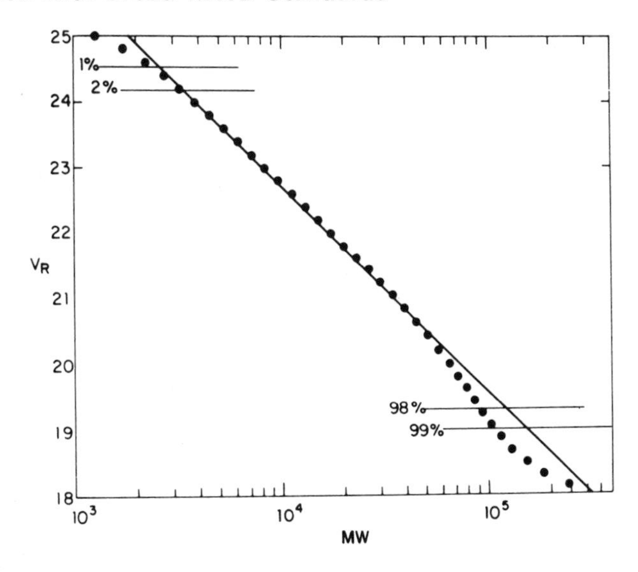

Figure 9.7 Broad-MWD calibration curves.

Calibrations by most probable distribution method (●●●) and by Hamielec's linear calibration method (——) for nylon 66 sample. The retention volumes that correspond to the cummulative elution peak areas of 1, 2, 98, and 99% are marked in the figure. (Reprinted with permission from Ref. 16.)

linear calibration line from the Hamielec method, which is discussed in the next section. Differences between the calibration curves for these two methods occur at the extremes of very low (<2%) or very high (>98%) weight fractions of the calibration standard. At these extremes, the calibration points for the integral method are not very reliable, because they can be greatly affected by column dispersion, the choice of baseline, and the choice of the integration limits (i.e., the beginning and the end of the GPC elution-curve data points used in the calibration computations). Because the extrapolation of the calibration curve is questionable at the extreme molecular weights, the useful calibration range of the integral method is limited to the narrow-molecular-weight region, where the curve agrees with the linear calibration range. As a consequence, the integral calibration method is no more versatile or accurate than the linear calibration approaches.

To use the integral-MWD method of calibration, one must know that the polymer standard can be represented closely by the theoretical MWD function, meaning that there cannot be a mechanism or kinetic bias during the polymerization of the polymer standard. Unfortunately, this restriction can be prohibitive for many commonly occurring polymer MWD problems.

Linear Calibration Methods

The linear calibration methods using broad polymer standards are discussed here beginning with the original form developed by Hamielec (17), followed by the discussions of the improved versions (GPCV2, V3) of Yau et al (18, 22), and others (19–21).

The *Hamielec method* (17) consists of a search for an "effective" linear calibration $M_H(V)$ line having the form of Equation 9.5. The aim is to find the right values for the "effective" calibration constants D'_1 and D'_2 in Equation 9.5, so that the computed $\overline{M}_w$ and $\overline{M}_n$ values according to Equations 9.6 and 9.7 are in agreement with the known MW values for the polymer standard.

$$M_H(V) = D'_1 e^{-D'_2 V} \tag{9.5}$$

$$\overline{M}_w = \sum_V F(V) M_H(V) \tag{9.6}$$

$$\overline{M}_n = \frac{1}{[\sum_V F(V)/M_H(V)]} \tag{9.7}$$

The SEC retention volume is represented by V and the SEC elution curve is represented by $F(V)$. Since the experimental chromatogram $F(V)$ in Equations 9.6 and 9.7 is affected by instrumental peak broadening, the Hamielec method provides an "effective" calibration curve of the SEC experiment, not the true curve.

An alternative expression for the effective linear calibration line is

$$V = C'_1 - C'_2 \log M_H(V) \tag{9.8}$$

where the calibration constants C'_1 and C'_2 are the intercept and the slope of the calibration line as they appear in the usual logarithmic calibration plot. The two sets of calibration constants are related according to the following equalities:

$$C'_1 = \frac{\ln D'_1}{D'_2}, \qquad C'_2 = \frac{\ln 10}{D'_2} \tag{9.9}$$

or

$$D'_1 = 10^{(C'_1/C'_2)}, \qquad D'_2 = \frac{\ln 10}{C'_2} \tag{9.10}$$

The essential element of the computer program for the Hamielec method is a trial-and-error search routine that iteratively adjusts the values of C'_1 and C'_2 (or D'_1 and D'_2) until Equations 9.6 and 9.7 are satisfied by the known values of $\overline{M}_w$ and $\overline{M}_n$ of the standard. The desired calibration curve is defined by the final values of the calibration constants. The effective $M_H(V)$ calibration curve so obtained can then be used for calculating the molecular weight averages for unknown samples from their experimental SEC elution curves.

The Hamielec method offers a truly practical way of obtaining GPC calibration curves that are specific to polymer type. The method needs only one broad-MWD standard of the same structure as the unknown samples. This can usually be provided either by commercial standards (Sect. 8.3) or by converting one of the samples into a working standard through independent determinations of its values of $\overline{M}_w$ and $\overline{M}_n$ using, for example, light-scattering and osmotic pressure techniques (Sect. 1.4).

Although a single broad standard is often used, the Hamielec method can also be used with two different molecular weight standards, as long as there are two average MW values known. Mathematically, two known MW values in any combination of $\overline{M}_n$ and $\overline{M}_w$ are all that are needed to solve for the two unknown calibration constants. The precision of the method increases with the difference between the two MW values used in calibration. For two standards, the two GPC elution curves, or $F(V)$ curves, are used in the search for the effective calibration constants D'_1 and D'_2 (or C'_1 and C'_2) from Equations 9.6 and 9.7. The calibration constants can be found by use of a trial-and-error computer algorithm, as in the case of a single broad standard.

In addition to dependence on the accuracy of the linear calibration approximation and on the experimental $\overline{M}_w$ and $\overline{M}_n$ values of the standard, the Hamielec method has two other weaknesses: [1] the physical significance of the effective calibration curve is not defined, and [2] the calculated MW values are accurate only for samples having an MWD (or GPC elution) curve similar to that of the standard.

Unlike the true calibration curve, the effective calibration curve is not unique to the specific GPC column and polymer/solvent system, but varies as a function of the column efficiency and the MW and MWD of the standard. The sketch in Figure 9.8 is helpful in explaining the properties of the calibration curves obtained by the Hamielec method. Besides the Hamielec calibration line $M_H(V)$ and the experimental GPC elution curve $F(V)$ of the broad standard, the sketch in Figure 9.8 also shows the true calibration curve $M_t(V)$ and the hypothetical dispersion-free elution curve $W(V)$. As discussed in Section 9.2, the true calibration curve $M_t(V)$ can be obtained experimentally by peak position calibration if there are narrow fractions of known MW available for the polymer of interest. The linear approximation of the true calibration curve is described here by

$$M_t(V) = D_1 e^{-D_2 V} \tag{9.11}$$

where D_1 and D_2 are the true calibration constants in contrast to the "effective" calibration parameters D'_1 and D'_2 in Equation 9.5.

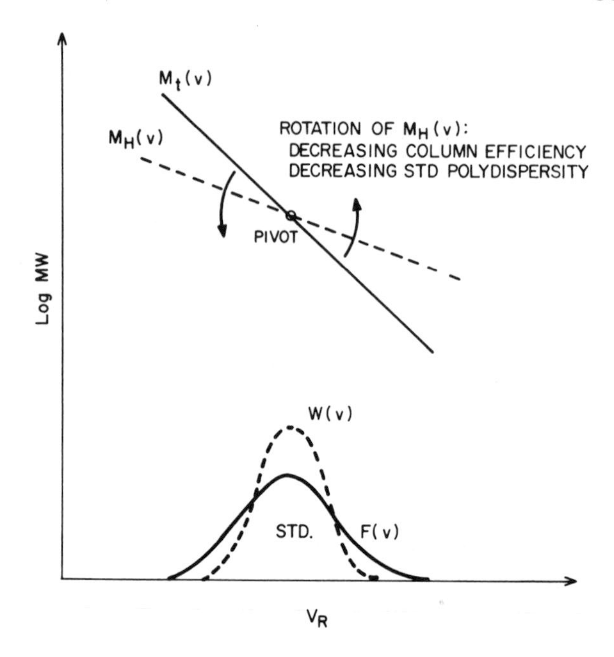

Figure 9.8 Effect of column dispersion and standard polydispersity on rotation of the Hamielec effective linear calibration line.

As illustrated in Figure 9.8, column band dispersion causes the experimental $F(V)$ to be broader than $W(V)$ and makes the apparent $\overline{M}_w$ value too high and the $\overline{M}_n$ value too low if the $M_t(V)$ calibration is used in the computation. To compensate for molecular weight errors when forcing a fit to the known $\overline{M}_w$ and $\overline{M}_n$ of the standard, the required effective calibration curve $M_H(V)$ will have to be rotated away from $M_t(V)$ in a counterclockwise direction. This rotation of the $M_H(V)$ line increases with increasing column dispersion, or decreasing column efficiency, as the $F(V)$ curve becomes increasingly different from $W(V)$. Following the same reasoning, one can see that $M_H(V)$ is also sensitive to the polydispersity of the standard. For standards of narrower MWD, $M_H(V)$ will rotate more toward the horizontal. Of course, a horizontal effective calibration is the extreme case expected of a monodisperse molecular weight standard. Also, the rotation pivot of the $M_H(V)$ line is expected to shift, depending on the MW of the standard used in the calibration.

Knowing how the $M_H(V)$ line will rotate and shift, one can readily predict the trend of molecular weight errors in the Hamielec method in isolated circumstances. If the sample is of higher MW than the standard, the calculated molecular weight values will be too low, and vice versa. If the sample is of broader MWD than the standard, the calculated sample polydispersity will be too small, and vice versa. Therefore, it is apparent that the limitation of the Hamielec method is the lack of proper compensation for column dispersion.

In an improved version of the Hamielec method, GPCV2, compensation is provided for the symmetrical peak broadening caused by column dispersion effects. In GPCV2, the instrumental peak broadening is approximated by a standard deviation (σ), which is assumed to be independent of retention volume (18). The formulations that form the basis of GPCV2 are (see Sect. 4.3 and Refs. 18–21 for derivations):

$$\overline{M}_w = e^{-(1/2)(D_2\sigma)^2} \sum_V [F(V)D_1e^{-D_2V}] \tag{9.12}$$

$$\overline{M}_n = \frac{e^{(1/2)(D_2\sigma)^2}}{\sum_V [F(V)/D_1e^{-D_2V}]} \tag{9.13}$$

These equations relate the average MW values directly to the experimental GPC elution curve $F(V)$ and the true GPC calibration curve constants D_1 and D_2, including an exponential correction factor containing the column dispersion parameter σ. At $\sigma = 0$, Equations 9.12 and 9.13 reduce to Equations 9.6 and 9.7 of the original Hamielec method. To a first approximation, the peak standard deviation σ due to column dispersion is estimated as the minimum experimental value of peak σ for several narrow polystyrene standards (Sect. 3.5). The minimum value of σ is used to avoid overcorrection for column dispersion, since the experimental peak σ may include actual molecular weight separation in addition to column dispersion for some polystyrene standards.

With known $\overline{M}_w$, $\overline{M}_n$, and σ values, Equations 9.12 and 9.13 can be solved in the same manner as Equations 9.6 and 9.7 of the original Hamielec method, except that the computer search now produces values of D_1 and D_2 and thus a more accurate calibration curve for the SEC experiment. Once the calibration curve is obtained. Equations 9.12 and 9.13 can be used again for calculating the MW for unknown samples. Since the GPCV2 method maintains the integrity of the calibration constants D_1 and D_2 and provides correction for column dispersion, it is more accurate than the original Hamielec method for analyzing samples that have very different MW or MWD from that of the standard. Actual GPC analyses have shown that the GPCV2 procedure can provide up to threefold improved MW accuracy over the original Hamielec method (Ref. 18; see also Sect. 9.4).

A more sophisticated version of the Hamielec method, GPCV3 includes a consideration of the skewness of SEC column dispersion (22). The expressions used in GPCV3 are

$$\overline{M}_w = (1 + D_2\tau)e^{-(D_2^2\sigma^2/2)+D_2\tau} \sum_V [F(V)D_1 e^{-D_2 V}] \qquad (9.14)$$

and

$$\overline{M}_n = \left(\frac{1}{1 - D_2\tau}\right) \frac{e^{(D_2^2\sigma^2/2)-D_2\tau}}{\sum_V [F(V)/D_1 e^{-D_2 V}]} \qquad (9.15)$$

where σ and τ are the two peak shape parameters of the assumed exponentially modified Gaussian peak shape model (Sect. 3.5 and Ref. 23). The parameter τ is the time constant of the exponential modifier to the Gaussian component of standard deviation σ and the τ/σ ratio relates to the peak skew. The procedure for extracting σ and τ from experimental SEC peaks is described in Section 3.5. The procedure for using GPCV3 is to insert experimental σ and τ values of column dispersion into Equations 9.14 and 9.15 and then solve for D_1 and D_2 to satisfy the known values of $\overline{M}_w$ and $\overline{M}_n$ for the standard. At $\tau = 0$, Equations 9.14 and 9.15 reduce to the GPCV2 Equations 9.12 and 9.13. Both GPCV2 and GPCV3 can use either one or two calibration standards, as does the original Hamielec method.

In contrast to the integral MWD method, linear calibration methods pose no restrictions on the MWD shape of the standard. In practice, linear calibration methods provide a much better compromise than does the integral MWD method, because the linearity of the SEC calibration curve can be improved by the proper selection of SEC columns, as described in Section 9.6. For SEC columns with nonlinear calibration curves, a modified broad standard method must be used for which the linear calibration approximation is not assumed. However, this approach can only be used when the universal calibration curve is known for the system (24).

9.4 Accuracy of Calibration Methods

The accuracy of the peak position, Hamielec, and GPCV2 calibration methods was tested in a specially designed experiment (18). In the experiment, four polystyrene samples were prepared using blends of commercially available characterized standards. The expected values of MW for the samples are listed as the calculated/reported values in Table 9.1. Four μ-Styragel and four Vit-X columns of different pore sizes were used in the experiment. These column sets were not optimized for range and linearity in MW calibration by the bimodal approach discussed in Section 9.6. The

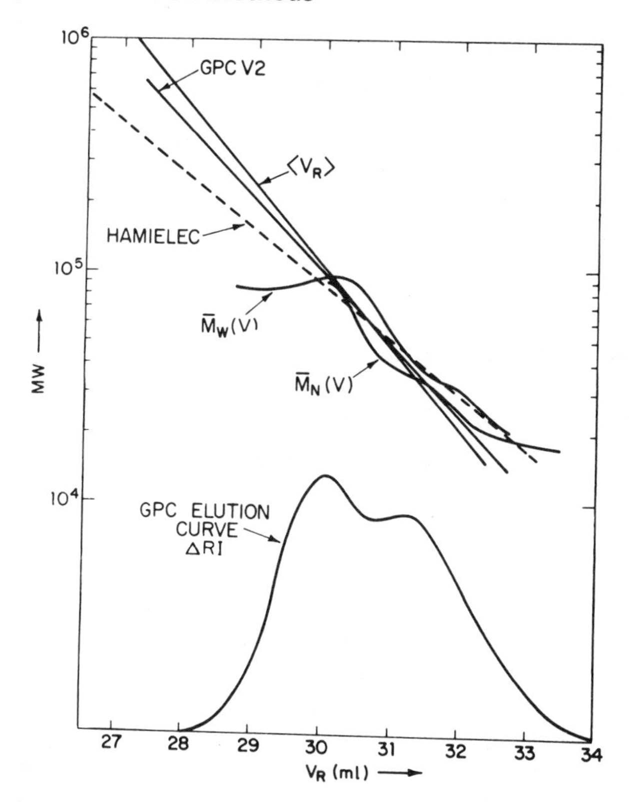

Figure 9.9 Actual MW curves and calibration plots for polystyrene standard. For sample 1 of Table 9.1; see the text for identification of GPCV2, etc. (Reprinted with permission from Ref. 18.)

experimental elution curves obtained on the μ-Styragel column series are shown in Figures 9.9–9.12.

In Figures 9.9–9.12 and in Table 9.1 the peak position calibration was obtained by using the average retention volume $\langle V_R \rangle$ of the individual elution peaks of the narrow polystyrene standards. The Hamielec and the GPCV2 calibration lines (GPCV2 line shown in Fig. 9.9 only) were calculated by using sample 1 (Table 9.1) as the calibration standard. The value for column dispersion σ was determined from the peak broadening of the polystyrene standard 4A (MW 97,200). For the data in Table 9.1, the calculations of $\overline{M}_n$ and $\overline{M}_w$ were made using the appropriate calibration curve, the elution curve $F(V)$, and, where applicable, the values of σ.

The calculated molecular weight results are shown in Table 9.1, in which the percent error in molecular weight for various samples is listed under the

Table 9.1 Effect of GPC Calibration Method on Molecular Weight Accuracy

Method	Sample 1[a]			Sample 2			Sample 3			Sample 4		
	$\overline{M}_w^b$	$\overline{M}_n^b$	Average Error (%)	$\overline{M}_w$	$\overline{M}_n$	Average Error (%)	$\overline{M}_w$	$\overline{M}_n$	Average Error (%)	$\overline{M}_w$	$\overline{M}_n$	Average Error (%)
	A. μ-Styragel Columns[c] (*N* = 13,000, *Toluene*; $\sigma = 0.70$ *ml for 4A*)											
Calculated/reported value	64	44	—	39	28	—	288	137	—	20	20	—
Peak position	74	34	19	47	22	21	454	107	40	23	17	15
Hamielec	64	44	—	46	32	16	210	107	25	30	26	40
GPCV2	64	44	—	42	30	7	314	129	7	23	23	15
	B. Vit-X Columns (*N* = 3500, *Toluene*; $\sigma = 1.05$ *ml for 4A*)											
Calculated/reported value	64	44	—	39	28	—	288	137	—	20	20	—
Peak position	86	42	20	51	26	21	333	136	8	24	18	15
Hamielec	64	44	—	44	31	13	166	105	33	27	23	25
GPCV2	64	44	—	38	27	2	247	143	9	18	19	8

Taken from Reference 18.

[a] Sample 1 was a blend of 3 parts PS 4A, 2 parts 7B, and 1 part 2A. Sample 2 was a blend of 1 part 4A, 2 parts 7B, and 3 parts 2A. Sample 3 was NBS 706 polystyrene (National Bureau of Standards, Washington, D.C.), and sample 4 was the narrow polystyrene standard 2A alone. Narrow standards 4A, 7B, and 2A (Pressure Chemical Co., Pittsburgh, Pa.) have reported polydispersity values of less than 1.06 and reported MW values of 97,200, 37,000 and 19,800, respectively.

[b] Values listed are $\times 10^{-3}$.

[c] Four 30 × 0.76 cm i.d. columns of 10^2, 10^3, 10^5, and 10^6 Å μ-Styragel (Waters Associates, Milford, Mass.) or 84, 171, 660, and 1933 Å Vit-X (Perkin-Elmer Co., Norwalk, Conn.) were used in series.

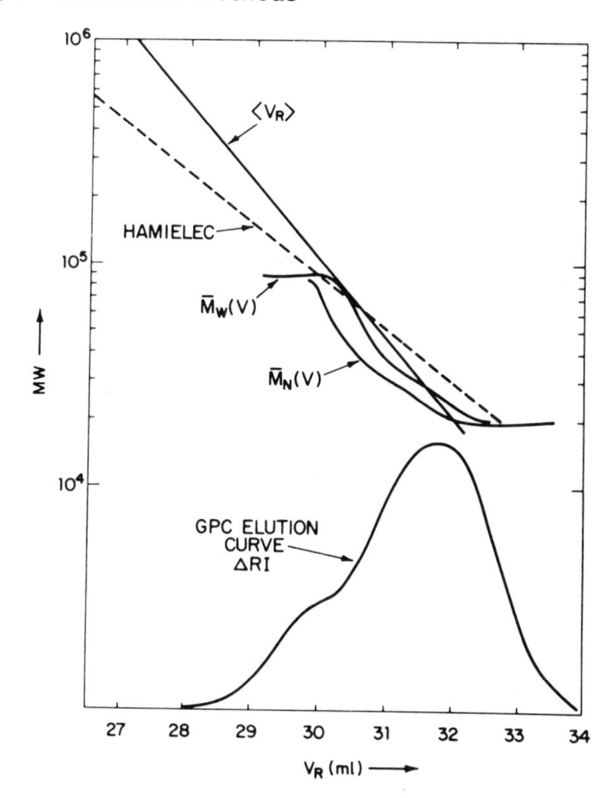

Figure 9.10 Actual MW curves for unknown polystyrene sample and standard calibration plots. For sample 2 of Table 9.1; see the text for identification of $\langle V_R \rangle$, etc. (Reprinted with permission from Ref. 18.)

headings of the three calibration methods. No entry of molecular weight error is made for sample 1 under the Hamielec and GPCV2 methods, because this is the sample chosen as the calibrating standard for these two methods. For the other samples, the molecular weight errors listed under GPCV2 are all more than a factor of two smaller than those under the Hamielec method, clearly showing the superior accuracy of the GPCV2 method. The molecular weight results in Table 9.1 also show the limitation of the Hamielec method for correcting molecular weight errors resulting from column dispersion for samples that do not have a MWD similar to that of the standard.

Although the data are affected by experimental uncertainties, GPCV2 still shows improved molecular weight accuracy over the peak position method.

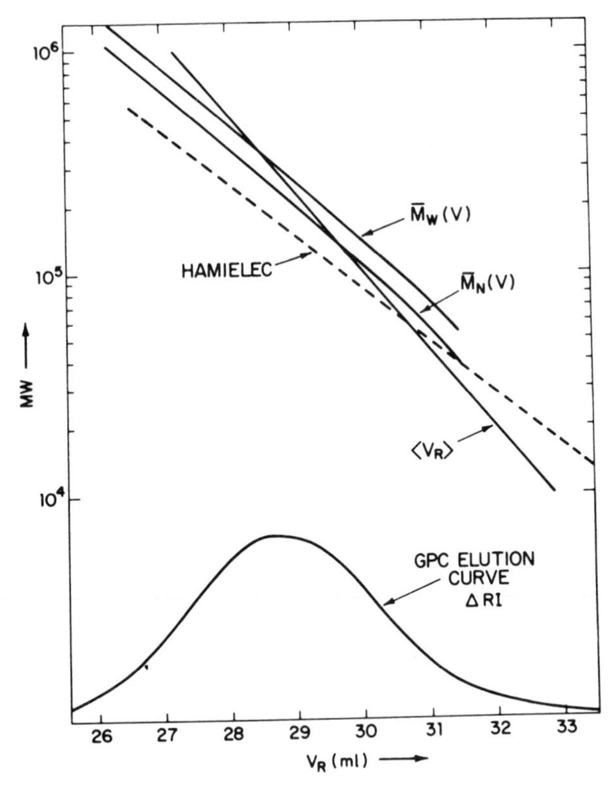

Figure 9.11 Comparison of actual molecular weight curves and the calibration plots. For sample 3 of Table 9.1; see the text for identification of $\langle V_R \rangle$, etc. (Reprinted with permission from Ref. 18.)

Since the calibration curve obtained by the peak position method is not affected by instrumental peak broadening, the molecular weight error of this method comes from the sample molecular weight calculation rather than from the calibration step itself. The molecular weight error in this case is simply caused by the instrumental peak broadening of the elution curves of the samples, not the standards. The residual molecular weight error of the GPCV2 method is rather small considering the possible molecular weight errors due to experimental uncertainties and possible errors in the reported MW values for the polystyrene standards. The utility of the GPCV2 method is also demonstrated by the calibration lines of Figure 9.9, which shows that the true calibration curve $\langle V_R \rangle$ is better approximated by GPCV2 than by the Hamielec calibration line.

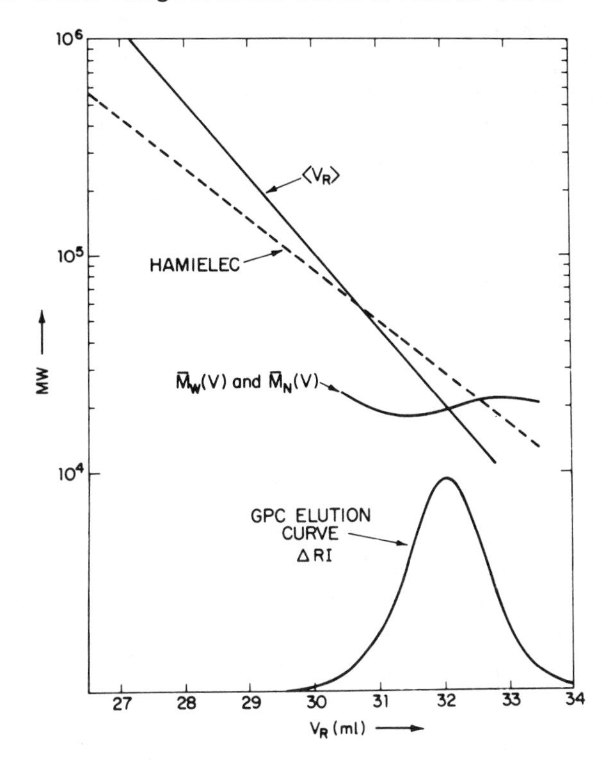

Figure 9.12 Comparison of actual molecular weight curves and the calibration plots. For sample 4 of Table 9.1; see the text for identification of $\langle V_R \rangle$, etc. (Reprinted with permission from Ref. 18.)

9.5 Actual Molecular Weight Across the SEC Elution Curve

The fundamental cause of the molecular weight errors discussed in the last section is the inability of the different calibration curves to accurately describe the actual sample molecular weight eluted at different retention volumes. Large discrepancies between actual and calibration molecular weight values cause large errors in the calculated values of $\overline{M}_n$ and $\overline{M}_w$ for the sample. Therefore, a study of the actual molecular weight across the sample elution curves is important for the full understanding of the SEC calibration problem. In the following we describe how to predict the actual molecular weight and utilize it in studying the SEC calibration methods.

In a GPC experiment for a broad-MWD sample, the species detected at any particular retention volume do not have a specific molecular weight. Neighboring molecules of different molecular weights are moved to the same retention volume by column dispersion. The actual molecular weight at each retention volume is, therefore, not accurately described by the true calibration curve of the GPC columns, except under the hypothetical condition of infinite column resolution. There are mixtures of molecular weights within even infinitesimally small fractions at any retention volume. Expressions were recently developed (18) to examine the actual change of molecular weight with retention volume in the GPC experiment. The weight and the number average molecular weight at any retention volume V have the following functional dependence:

$$\overline{M}_w(V) = \frac{F(V - D_2\sigma^2)}{F(V)} e^{(1/2)(D_2\sigma)^2} M_t(V) \qquad (9.16)$$

and

$$\overline{M}_n(V) = \frac{F(V)}{F(V + D_2\sigma^2)} e^{-(1/2)(D_2\sigma)^2} M_t(V) \qquad (9.17)$$

where $M_t(V)$ is the MW value at V defined by the true or peak position calibration curve; σ is the peak standard deviation of the assumed Gaussian column dispersion function; and $F(V)$ and $F(V \pm D_2\sigma^2)$ are the experimental GPC elution curve heights at the retention volume of V and $V \pm D_2\sigma^2$, respectively. It is interesting to note that the actual molecular weight across the GPC elution curve is not uniquely defined, but varies from sample to sample depending on the elution curve shape $F(V)$ and the values of σ.

With experimental values of $F(V)$, σ, and $M_t(V)$, Equations 9.16 and 9.17 can be used to predict the $\overline{M}_w(V)$ and $\overline{M}_n(V)$ curves. This was done for the four polystyrene samples chromatographed on a set of four μ-Styragel columns used in the experiment described in the last section. The calculated curves for the actual molecular weight variation $\overline{M}_w(V)$ and $\overline{M}_n(V)$ are plotted separately for each sample in Figures 9.9–9.12. Recall that sample 1 shown in Figure 9.9 was the single broad standard used for obtaining the Hamielec and GPCV2 calibration plot, which are shown in the same figure with the peak position line $\langle V_R \rangle$. The same Hamielec and $\langle V_R \rangle$ calibration lines are reproduced in Figures 9.10–9.12. The features of the peak position and the Hamielec calibrations are compared to the actual molecular weight variations $\overline{M}_w(V)$ and $\overline{M}_n(V)$. In Figure 9.9 the Hamielec line encompasses the actual molecular weight curves over most of the molecular weight range. This is understandable since the sample in this case is the Hamielec calibration standard itself. In this case the molecular weight calculation for sample 1 using either GPCV2 or the Hamielec method will by definition give accurate results.

However, the actual molecular weight $\overline{M}_w(V)$ and $\overline{M}_n(V)$ begin to differ from the Hamielec and peak position lines in Figure 9.10. This is the case in which the sample has an MWD different from that of the standard. As the sample differs more and more from the standard, the Hamielec line becomes an increasingly poorer estimate of the actual molecular weight elution behavior. For example, in Figure 9.11 neither the $\overline{M}_w(V)$ nor the $\overline{M}_n(V)$ curve ever intersects with the Hamielec line; and in Figure 9.12 the respective curves are nearly at right angles. Since Figure 9.12 is the chromatogram of a very narrow MWD sample, most of its elution-curve profile is caused by peak broadening, and therefore the calculated $\overline{M}_w(V)$ and $\overline{M}_n(V)$ curves merge and do not vary much with V_R.

The fact that the $\overline{M}_w(V)$ and $\overline{M}_n(V)$ curves vary as a function of column dispersion and the shape of the sample MWD indicates that the SEC elution curve profiles obtained by continuous viscometer or light-scattering detectors (25) would be similarly affected by column dispersion and sample MWD. These effects should be taken into consideration in interpreting results from these molecular-weight-specific GPC detectors to achieve absolute molecular weight calibration.

The preceding results also show that the actual molecular weight variation across an SEC elution curve is a very complex function of the combined effect of column dispersion and elution-curve profile. The $\overline{M}_w(V)$ and $\overline{M}_n(V)$ curves can vary drastically from sample to sample, and can in no way be fitted well by a single linear calibration curve. Generally, tilting and shifting the calibration by way of the Hamielec method cannot properly compensate for column dispersion. The dispersion problem is accounted for, however, in GPCV2 or GPCV3 by including the σ correction in the calibration computation (Eq. 9.14 and 9.15).

GPCV2 or GPCV3 should be helpful in the study of the universal calibration concept (5). With these calibration methods one can get nearly the true calibration curves in various polymer/solvent systems for evaluating the accuracy and the precision of the universal calibration method.

9.6 Linear Calibration Ranges

The discussions in the earlier sections have pointed out that the broad standard linear calibration method is the choice for quantitative GPC-MWD analyses, since this approach provides the best compromise between practical convenience and molecular weight accuracy. For many commercial polymers, linear calibration may be the only workable method that can provide the desired calibration curve for the specific polymer/solvent system. Since this calibration method works best for GPC columns of wide and linear molecular weight separation range, it is important that the GPC columns used in the

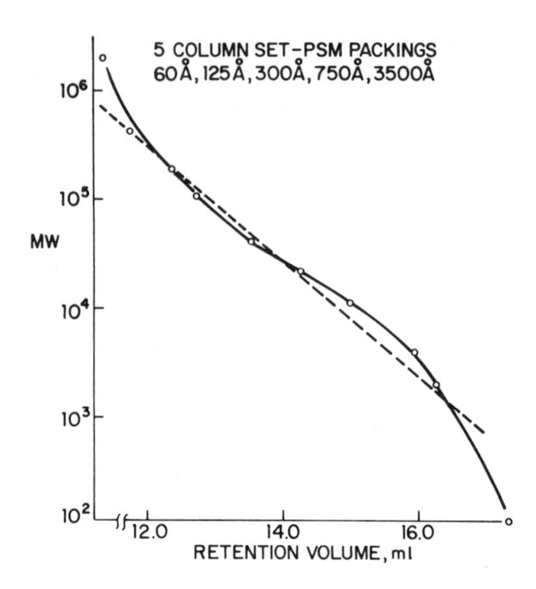

Figure 9.13 Calibration plot from the empirical approach of connecting columns of different pore sizes.

Conditions same as for Figure 9.2. (Reprinted with permission from Ref. 3.)

experiment are optimized for range and linearity. To handle most commercial polymers adequately, a linear molecular weight range of four decades covering molecular weight values of a few hundred to a few million is needed.

A popular empirical guideline for linear calibration recommends the use of columns of each pore size that have finite fractionation capacity in the molecular weight range of interest (26, 27). The calibration in Figure 9.13 illustrates this empirical approach of connecting columns with different pore sizes to obtain fractionation over a wide molecular weight range. Note that this approach produces relatively large deviations from a linear fit with a linearity range of less than three decades. The empirical approach is not as effective as the bimodal-pore-size distribution method (3), the basic theory of which was discussed in Section 4.5. The rules for optimizing bimodal column sets using available GPC columns are described below (see also Sect. 8.5 for experimental practice).

The bimodal concept involves coupling SEC columns containing only two discrete pore sizes and approximately equal pore volumes for the two pore sizes. This column selection rule was developed so that the linear portion of the calibration plot for the individual pore-size columns is substantially parallel but nonoverlapping. The bimodal approach to column selection relies on the proper recognition of two concepts: [1] pores of only one size can fractionate polymers over nearly two decades of molecular weight range; [2] the separation capacities of individual columns are additive in a column set.

The bimodal GPC theory (Sect. 4.5) predicts that wide-range linear calibrations are possible for bimodal column sets with pore sizes differing by about one decade or more. However, there is considerable leeway in selecting the pore-size separation around the optimum bimodal arrangement. Therefore, a detailed knowledge of pore size or pore-size distribution is not necessarily required to assemble a column set with reasonable molecular weight range and calibration linearity. A very wide linear molecular weight calibration range can be obtained at the expense of only slightly poorer linearity by increasing the pore-size separations of the bimodal approach. Figure 9.14

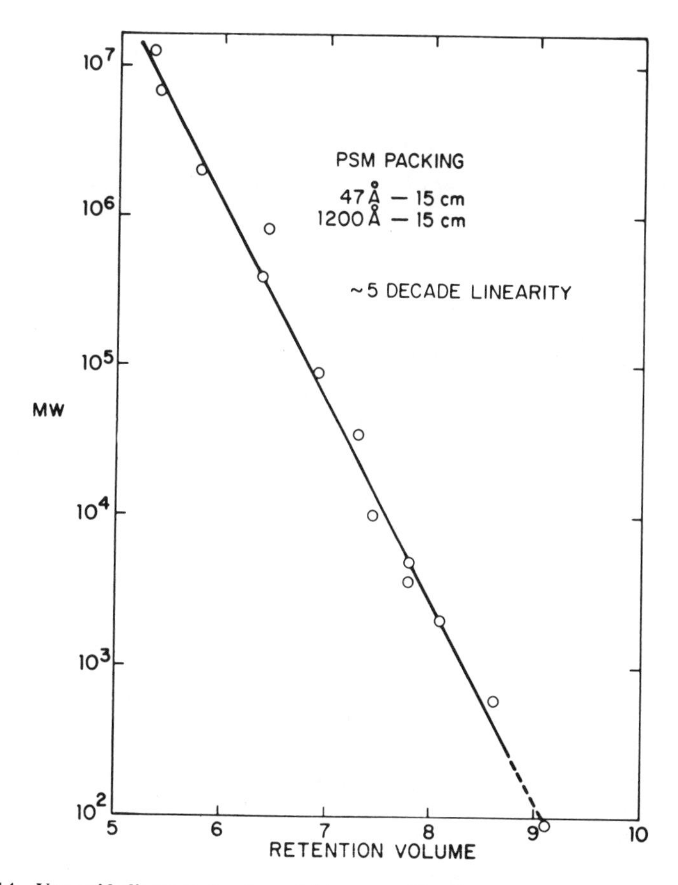

Figure 9.14 Very wide linear molecular weight range calibration curve with bimodal column set. Conditions same as for Figure 9.2. (Reprinted with permission from Ref. 3.)

demonstrates the molecular weight calibration of such a column set which exhibits almost five decades of molecular weight linearity. (It should be noted that at $MW < 10^3$, the random-coil model for describing polymer conformation becomes less definitive. Thus linear extrapolation of the calibration data points to molecular weights of a few hundred should be interpreted with caution.) The increase in molecular weight error caused by the poorer linearity is modest, and in many cases the increased range of linear calibration is more valuable in view of the added versatility and convenience of linear columns of wide range.

A bimodal column set designed for optimum performance for one polymer/solvent system should function equally well in other systems, although the actual calibration curve may shift somewhat.

REFERENCES

1. M. Duval, B. Block, and S. Kohn, *J. Appl. Polym. Sci.*, **16**, 1585 (1972).
2. J. Cazes and D. Gaskill, *Sep. Sci.*, **2**, 421 (1967).
3. W. W. Yau, C. R. Ginnard, and J. J. Kirkland, *J. Chromatogr.*, **149**, 465 (1978).
4. K. K. Unger and N. P. Becker, paper presented at the Pittsburgh Conference on Analytical Chemistry and Applied Spectroscopy, Cleveland, Ohio, 1977, paper 171.
5. Z. Grubistic, R. Rempp, and H. Benoit, *J. Polym. Sci., Part B*, **5**, 753 (1967).
6. J. Brandrup and E. H. Immergut, eds., *Polymer Handbook*, Wiley-Interscience, New York, 2nd ed., 1975.
7. F. W. Billmeyer, Jr., *J. Polym. Sci.*, **4**, 83 (1949).
8. T. D. Varma and M. Sengupta, *J. Appl. Polym. Sci.*, **15**, 1599 (1971).
9. C. Tanford, *Physical Chemistry of Macromolecules*, Wiley, New York, 1961, p. 391.
10. P. J. Flory, *Principles of Polymer Chemistry*, Cornell University Press, Ithaca, N.Y., 1953.
11. O. B. Ptitsyn and Y. E. Eizner, *Zh. Fiz. Khim.*, **32** 2464 (1958).
12. E. F. Casassa, *J. Phys. Chem.*, **75**, 3929 (1971).
13. A. E. Hamielec and A. C. Ouano, *J. Liq. Chromatogr.*, **1**, 111 (1978).
14. A. R. Weiss and E. Cohn-Ginsberg, *J. Polym. Sci., Part A-2*, **8**, 148 (1970).
15. A. H. Abdel-Alim and A. E. Hamielec, *J. Appl. Polym. Sci.*, **18**, 297 (1974).
16. T. D. Swartz, D. D. Bly, and A. S. Edwards, *J. Appl. Polym. Sci.*, **16**, 3353 (1972).
17. S. T. Balke, A. E. Hamielec, B. P. LeClair and S. L. Pearce, *Ind. Eng. Chem., Prod. Res. Dev.*, **8**, 54 (1969).
18. W. W. Yau, H. J. Stoklosa, and D. D. Bly, *J. Appl. Polym. Sci.*, **21** ,1911 (1977).
19. S. T. Balke and A. E. Hamielec, *J. Appl. Polym. Sci.*, **13**, 1381 (1969).
20. A. E. Hamielec, *J. Appl. Polym. Sci.*, **14**, 1519 (1970).
21. T. Provder and E. M. Rosen, *Sep. Sci.*, **5**, 437 (1970).
22. W. W. Yau, H. J. Stoklosa, C. R. Ginnard, and D. D. Bly, 12th Middle Atlantic Regional Meeting, American Chemical Society, April 5–7, 1978, paper PO13.

23. E. Grushka, *Anal. Chem.*, **44**, 1733 (1972).
24. T. Provder, J. C. Woodbrey, J. H. Clark, and E. E. Drott, *Adv. Chem. Ser.*, **125**, 117 (1973).
25. A. C. Ouano, D. L. Horne, and A. R. Gregges, *J. Polym. Sci., Part A*-1, **12**, 307 (1974); A. C. Ouano and W. Kaye, *ibid*, **12**, 1151 (1974).
26. Waters Associates, Technical Bulletin, *Know More about Your Polymers*, N55, 1975.
27. M. R. Ambler, L. J. Fetters and Y. Kesten, *J. Appl. Polym. Sci.*, **21**, 2439 (1977).

Ten

DATA HANDLING

10.1 Introduction

A variety of data-handling procedures is used in HPSEC. Simple inspection of the strip-chart chromatogram often is sufficient for rough evaluation of instrument performance or, for example, to determine whether a preparative separation is adequate in sample purification for toxiological or bioactivity studies. Quick inspection of the chromatogram in Figure 1.1 is all that is needed to determine that component separation has occurred. At other times highly sophisticated computer methods are required for reducing the raw data to molecular weight (MW) and molecular weight distribution (MWD) information. The purpose of this chapter is to provide the reader with the tools to carry out data handling at various levels. Sections 10.3–10.7 specifically relate to the molecular weight analysis of synthetic polymers.

10.2 Simple Curve Inspection

The most elementary data reduction method in HPSEC is simple inspection of the strip-chart record (i.e., the chromatogram). Simple observations reveal the number, size, shape, and retention volume of the peaks, the relative spacing between peaks, the detector response, noise levels, and baseline drift. These observations are used most often to judge the overall quality of the experiment: Are the results consistent with expectations? Are they definitive? Whatever the answers, important data have been obtained, and one of three options then normally will be invoked: [1] repeat the experiment under the "same" or new conditions; [2] the information is sufficient, stop here; or [3] proceed to further interpretive methods with the information at hand.

Several simple parameters are of value in making the inspection technique quantitative. Some of these are defined elsewhere in this book and in the Glossary, but are explained further in the following paragraphs for particular use with this chapter.

The peak retention volume (V_R) is the volume of liquid which has passed through the system from the time of sample injection to the time of the peak maximum. The retention volume, most often used with narrow Gaussian-like peaks, is reproducible and often unique for a given solute for a specified column set and HPSEC conditions. The distance between peaks or peak spacing normally is expressed as the difference in retention volumes (i.e., $V_{R2} - V_{R1}$). Changes in spacing may indicate presence or absence of certain compounds in a series of samples from a similar source (e.g., protein hydrolysates) or an unsuspected change in chromatographic conditions. The peak retention time (t_R) is the time (rather than volume) counterpart of V_R. For unsymmetrical peaks it is useful to represent retention by the center of mass rather than the peak maximum. The center of mass is located at that retention volume which bisects the curve area into equal halves.

The peak height, h_p (line AD in Fig. 10.1), is the linear distance between the baseline (line BC) and the curve maximum. The terms h_i are used to designate heights at points other than the maximum of the peak. The chroma-

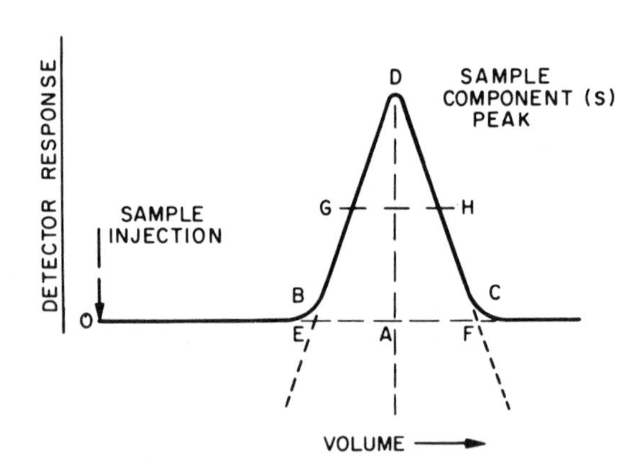

Figure 10.1 Schematic chromatogram of a narrow-MWD polymer sample used for illustrating various curve features.

(Reprinted with permission from Ref. 1.)

tographic heights are proportional to solute concentration and may be quantified by calibration.

The peak width (W_b), illustrated by line EF of Figure 10.1, is the distance in time or volume units between the baseline intercepts of lines drawn tangent to the curve at the points of inflection of the chromatographic peak. W_b frequently is used in column performance expressions. The peak width at half-height ($W_{1/2}$) is illustrated by line GH of Figure 10.1.

The curve overlay technique can be used as an inspection technique in GFC or GPC. By overlaying one chromatogram on another, samples may be compared qualitatively for changes in molecular weight or molecular weight distribution, to note the inclusion or loss of impurities, and to detect changes of component concentrations. The curve overlay comparison is valid if there is no change in the mobile-phase flow rate or variation in detector sensitivity from run to run. Flow rate changes produce artifacts in curve shapes because at constant recorder chart speed the width of the chromatogram is inversely proportional to flow rate. However, such artifacts can be compensated by digitizing and converting the detector response versus time curve to a detector response versus volume curve (Sect. 10.3). Examples illustrating the use of raw-data GPC chromatograms to produce some important information on specific polymer systems are discussed in Chapter 12.

The next level of data handling, useful in both GFC and GPC, involves determination of the sample molecular weight. For monodisperse systems, which include many natural products analyzed by HPGFC, the peak position method can be used (Chapt. 9). The retention volume (V_R) of the peak normally is chosen and the MW is read from an appropriate calibration curve which relates V_R and MW. For example, in Figure 1.2 the molecular weight of an unknown protein can be read directly from the calibration plot after determining the sample V_R. The calibration curve and sample analysis must be developed under identical experimental conditions for this use. Theory shows that the sample must be of the same topography and chemical structure as the standards used in generating the calibration curve, since macromolecules can have different relationships between size and molecular weight in a given solvent. In practice, however, small variations in molecular structures are permitted as attested to by Figure 1.2, where different proteins fall on the same calibration plot.

With monodisperse samples there is only one molecular weight (i.e., $\overline{M}_n = \overline{M}_w = \overline{M}_z$). (MW averages are discussed in Chapt. 1.) With polydisperse systems, where $\overline{M}_n \neq \overline{M}_w \neq \overline{M}_z$, it frequently is desired to calculate one or all of the sample molecular weight averages and the molecular weight distribution. Because molecules differing by a few repeat units are not separated into distinguishable peaks in polydisperse systems of relatively high molecular weight, the chromatogram is a continuous curve, and more sophisticated methods of calculation are required. Here peak summation methods ($\sum h_i$) are used to provide the molecular weight averages. If the chromatographic process and extracolumn effects cause significant band broadening, either the raw chromatographic data or the computed MW values must be corrected as discussed below. The peak summation method for determining MW values employs three steps: [1] calibration, to convert the raw data (detector response versus time) to molecular weight data (concentration versus volume or MW); [2] correction of the molecular weight averages or the chromatogram for curve broadening; and [3] computation of $\overline{M}_n$, $\overline{M}_w$, and MWD. These are essential steps whether the data are reduced manually or by computer. The following section discusses the manual approach to peak summation for calculating sample molecular weights.

10.3 Curve Summation: Manual Method for Computing Molecular Weight Averages and MWD

Chromatographing a sample produces a curve, and the baseline (e.g., BC of Fig. 10.1) must be chosen according to the criteria established in Section 8.7. Digitization of the chromatogram is accomplished manually by drawing vertical lines from the baseline at equally spaced retention volumes, V_i, until they intersect the chromatogram. These chromatogram heights (h_i) should be measured accurately (to ± 0.25 mm, if possible). A minimum of eight heights, determined at equally spaced volume intervals, is required even for very narrow chromatograms, while 20 or more are preferred for broader peaks for good summation accuracy. A plot of a digitized chromatogram (h_i) versus retention volumn (V_i) for a selected, broad-MWD, polystyrene sample is illustrated in Figure 10.2.

After digitization the data should be tabulated as in Table 10.1 for convenient calculation of molecular weights. In Table 10.1, i is simply an integer to keep track of the number of collected data points, V_i is the retention volume of the ith point expressed in milliliters, and h_i is the height from the baseline to the curve expressed in millimeters. The other data in Table 10.1 are obtained by the indicated mathematical operations. For this manual approach, area A, shown in Figure 10.1 as the enclosure BGDHCAB (and in column 5 in Table 10.1), is found by adding one-half of the incremental change in h_i (i.e., one-half of column 3) to the previously summed h_i value (column 4). Thus

$$A = \sum_{i=1}^{N} \left(h_i + \frac{\Delta h_i}{2} \right) \tag{10.1}$$

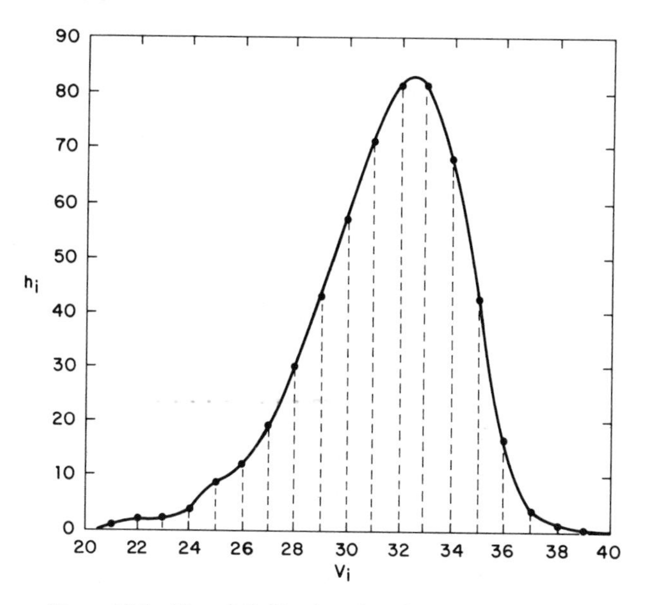

Figure 10.2 Manuel digitization of a selected chromatogram.

The cumulative weight fraction distribution, Cum W_i, is obtained by dividing the partially integrated areas in column 5 by the total area under the curve (in this example, 545.0). Cum W_i equals the weight fraction of polymer having a retention volume greater than V_i and molecular weight less than M_i. The appropriate values for M_i in Table 10.1 were obtained by reading from a molecular weight calibration curve generated separately. (In this case the peak position calibration method was used; however, any valid calibration method may be employed; Chapt. 9.)

Once the table of data is complete, $\overline{M}_n$ and $\overline{M}_w$ are calculated according to Equation 10.2:

$$\overline{M}_n = \frac{\sum_{i=1}^{N} h_i}{\sum_{i=1}^{N} (h_i/M_i)} \quad \text{and} \quad \overline{M}_w = \frac{\sum_{i=1}^{N} (h_i M_i)}{\sum_{i=1}^{N} h_i} \tag{10.2}$$

For the example in Table 10.1, Equation 10.2 gives

$$\overline{M}_n = \frac{545.0}{4.29 \times 10^{-3}} = 127{,}000 \text{ g/mole}$$

and

$$\overline{M}_w = \frac{186.1 \times 10^6}{545.0} = 341{,}000 \text{ g/mole}$$

Table 10.1 Example of Data Used for Calculating Molecular Weight Averages and a Differential Weight Fraction Molecular Weight Distribution[a]

(1)	(2)	(3)	(4)	(5)	(6)	(7)	(8)	(9)	(10)	(11)	(12)	(13)	(14)	(15)
i	V_i (ml)	h_i (mm)	$\sum_{i=1}^{N} h_i$ (mm)	A (mm^2)	Cum W_i	$M_i \times 10^{-6}$ (g/mole)	$h_i/M_i \times 10^6$	$h_i \times M_i \times 10^{-6}$	$\frac{h_i}{\sum h_i}$ (h_i/545.0)	Average $h_i/\sum h_i$ for the Volume Interval	$\Delta M_i \times 10^{-6}$/ml (g/mole)	$\frac{\Delta V}{\Delta M_i} \times 10^6$	$N_i \times 10^6$	X_i ($N_i M_i$)
21	20	0.0	545.0	545.0	1.0	4.709	0.0	0.0	0.0	0.0	1.407	0.7107	0.0	—
20	21	0.0	545.0	545.0	1.0	3.302	0.0	0.0	0.0	0.00073	0.975	1.0256	0.0	0.0
19	22	0.8	545.0	544.6	0.999	2.327	0.34	1.86	0.00147	0.00394	0.687	1.4556	0.005735	0.0133
18	23	3.5	544.2	542.5	0.995	1.640	2.13	5.74	0.00642	0.01681	0.4845	2.0640	0.03469	0.0569
17	24	16.8	540.7	532.3	0.977	1.1555	14.54	19.40	0.0308	0.0543	0.3413	2.9300	0.1590	0.184
16	25	42.4	523.9	502.7	0.922	0.8142	52.08	34.52	0.0778	0.1011	0.2404	4.1597	0.4205	0.342
15	26	67.9	481.5	447.6	0.821	0.5738	118.2	38.90	0.1244	0.13695	0.1735	5.7637	0.7894	0.452
14	27	81.5	413.7	373.0	0.684	0.4003	203.6	32.62	0.1495	0.1495	0.1182	8.4602	1.2647	0.506
13	28	81.4	322.2	291.5	0.535	0.2821	288.6	22.96	0.1494	0.1398	0.0833	12.004	1.6781	0.473
12	29	71.0	250.8	215.3	0.395	0.1988	357.1	14.12	0.1303	0.1174	0.0587	17.035	1.9999	0.398
11	30	57.0	179.8	151.3	0.278	0.1401	406.8	7.98	0.1046	0.09175	0.0413	24.213	2.2215	0.311
10	31	43.0	122.8	101.3	0.186	0.09872	435.6	4.24	0.07890	0.06695	0.02985	33.500	2.2429	0.221
9	32	30.0	79.8	64.8	0.119	0.06887	435.6	2.07	0.0550	0.04493	0.02034	49.164	2.2089	0.152
8	33	19.0	49.8	40.3	0.074	0.04853	391.5	0.92	0.03486	0.02863	0.01433	69.783	1.9979	0.0969
7	34	12.2	30.8	24.7	0.045	0.03420	356.7	0.42	0.02239	0.01945	0.0101	99.009	1.9257	0.0659
6	35	9.0	18.6	14.1	0.026	0.02410	373.4	0.22	0.01651	0.01193	0.00712	140.49	1.6762	0.040
5	36	4.0	9.6	7.6	0.014	0.01698	235.6	0.07	0.00734	0.00605	0.00501	199.60	1.2086	0.020
4	37	2.6	5.6	4.3	0.008	0.01197	217.2	0.03	0.00477	0.00422	0.00353	282.80	1.1934	0.014
3	38	2.0	3.0	2.0	0.004	0.00843	237.1	0.02	0.00367	0.00275	0.00255	392.15	1.0784	9.09×10^{-3}
2	39	1.0	1.0	0.5	0.001	0.00588	170.0	0.01	0.00183	0.00091	0.00173	575.37	0.5276	3.1×10^{-3}
1	40	0.0	0.0	0.0	0.0	0.00414	0.0	0.0	0.0	0.0	—	—	0.0	—

[a] Only the final column of X_i has been rounded off to significant figures. Calculated values from left to right in the table carry excess figures so that the reader can confirm the computations by calculator. Column (15) is from Eq. 10.4. It is computed from col. (11) × col. (7) × col. (13).

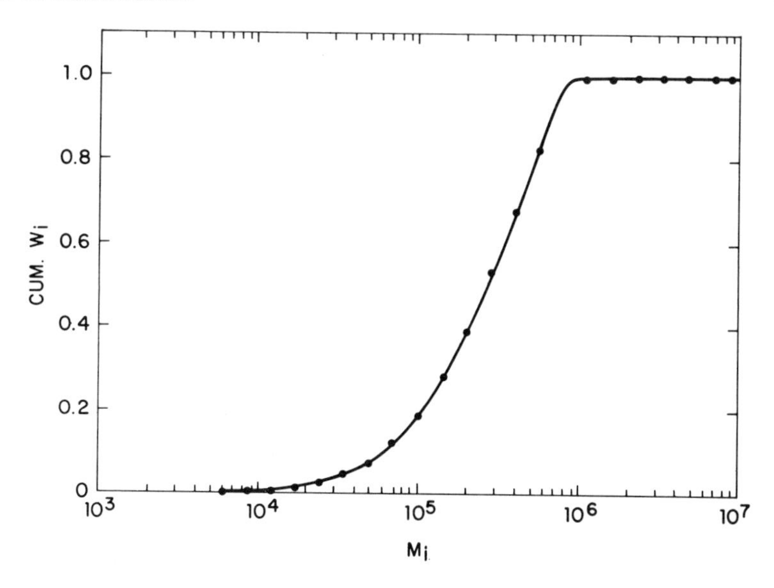

Figure 10.3 Cumulative weight fraction molecular weight distribution for the data of Table 10.1.

The cumulative weight fraction molecular weight distribution is obtained by plotting Cum W_i versus log M_i. The result for the example is shown as Figure 10.3.

In the SEC separation the original recorder trace is expressed as detector response, which is proportional to weight concentration of solute c or $\Delta W_i/\Delta V$, versus retention volume V_R. However, for the differential weight fraction molecular weight distribution, one needs to plot $\Delta W/\Delta \log M$ or X_i versus log M. The calculation of X_i from the original chromatogram can be performed from

$$X_i = \frac{\Delta W_i}{\Delta \log M_i} = \frac{\Delta W_i}{\Delta V} \frac{\Delta V}{\Delta \log M_i} = \frac{h_i}{\sum h_i} \frac{M_i \, \Delta V}{\Delta M_i} \tag{10.3}$$

The derivative $\Delta V/\Delta M_i$ is determined from the molecular weight calibration curve, and the average sum $h_i/\sum h_i$, column 11, for a constant volume interval, (ΔV) is used in Equation 10.3. The data are provided in Table 10.1, columns 10–15. Figure 10.4 shows the weight fraction frequency distribution plot obtained for this example. Note the similarity in curve shape between the raw-data curve of Figure 10.2 and the MW plot of Figure 10.4. (Recall that as V_R increases, MW decreases.) When the sample molecules all elute within the linear calibration range, $\Delta V/\Delta \log M$ is constant and $X_i \propto \Delta V/\Delta \log M$.

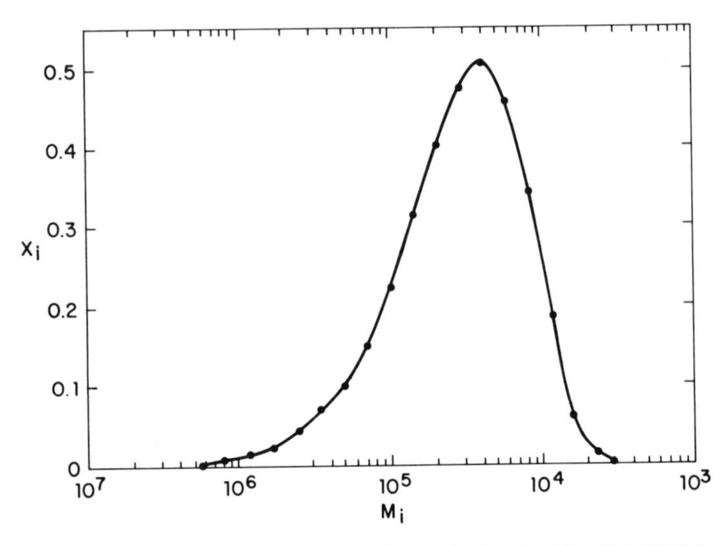

Figure 10.4 Weight fraction frequency distribution for the data of Table 10.1.

The differential MWD curve plotted on the log M scale will then be identical to the raw elution curve plotted in retention volume.

The differential number fraction molecular weight distribution (N_i) is obtained by dividing X_i in Equation 10.3 by M_i:

$$N_i = \frac{X_i}{M_i} = \frac{h_i \ \Delta V}{\sum h_i \ \Delta M_i} \tag{10.4}$$

Equation 10.4 describes the way in which the number of molecules in a sample varies with molecular weight.

10.4 Manual Method for Correcting Molecular Weight Averages

Before discussing the manual method used for correcting MW averages it is necessary to clarify the meanings of certain words and symbols used within this section. When a plug of monomeric sample solution passes through a chromatographic column, it is broadened. The peak shape for such a monomeric species is closely approximated by a Gaussian function. Such broadening occurs mainly because of the solute stationary-phase mass transfer process and because of other mixing processes. The standard deviation of the chromatographic peak, σ, is used as a quantitative measure of symmetrical broadening. Some unsymmetrical band broadening or skewing may also occur as a result of sample dilution effects, and for polymers because of solution viscosity flow effects. The asymmetry of the chromatographic peak is expressed by a quantitative mathematical function, peak skew (γ) (Sect. 3.5). At times in the traditional GPC literature the various processes of peak broadening have been incorrectly given the same (or equivalent) names and symbols.

In GPC some polymer curves are inherently askew and broadened because of the polymer molecular weight distributions they represent. Thus both symmetrical and unsymmetrical peak broadening caused by the chromatographic processes may be superimposed on peaks which, because they were generated by size separation and reflect real molecular weight distributions, already are inherently askew and broad. It has taken the efforts of many workers to unscramble the various curve-broadening effects (3–8).

The magnitude of the symmetrical chromatographic peak broadening depends significantly on the column and other operational parameters, and to a small degree on V_R, but is independent polymer type (5). Peak skewing is less well understood but is known to be affected by the viscosity of both the sample solution and the mobile phase, and is generally considered to be independent of polymer type. Polystyrene standards of very narrow MWD normally are used to generate needed correction terms.

This section does not attempt to clarify further the peak broadening problem, but rather illustrates a simple manual method for obtaining accurate sample MW and MWD values. The approach discussed here uses correction factors obtained by comparing the molecular weights calculated from SEC-broadened curves with the true (or reported) MW values of standards. The correction factors are given the symbol Λ for symmetrically band-broadened peaks and sk for asymmetrically band-broadened peaks, and the method is based on the approach taken in ASTM Standard Method D 3593-77 (6).

Application of the Correction Method for Molecular Weight Averages

Specific molecular weight corrections can be made by determining and using the values of Λ and sk as follows. Chromatograph a series of polystyrene standards whose molecular weights encompass the molecular weight range of the samples to be analyzed. Using the peak position calibration curve, calculate the number- and weight-average molecular weights of these standards (uncorrected for peak spreading), as described in Section 10.3. These uncorrected average molecular weights are given the symbols $\overline{M}_n(u)$ and $\overline{M}_w(u)$. Assume that the reported values for the molecular weights of these standards (as measured by classical methods such as osmometry and light scattering) are the true values and give them the symbols $\overline{M}_n(t)$ and $\overline{M}_w(t)$. Then calculate the quantities Λ and sk as follows:

$$\Lambda = \frac{1}{2}\left[\frac{\overline{M}_n(t)}{\overline{M}_n(u)} + \frac{\overline{M}_w(u)}{\overline{M}_w(t)}\right] \tag{10.5}$$

and

$$\text{sk} = \frac{\Phi - 1}{\Phi + 1} \tag{10.6}$$

where

$$\Phi = \frac{\overline{M}_n(t)}{\overline{M}_n(u)} \frac{\overline{M}_w(t)}{\overline{M}_w(u)} \tag{10.7}$$

Construct the actual plots of Λ and sk versus V_R (such as shown schematically in Fig. 10.5) by using the peak position V_R for the various standards. The top section in Figure 10.5 shows the general form of the relationships between sk and Λ and the retention volume V_R. Typically, different curves are obtained for sk and Λ but with similar V_R dependence. The absolute values of sk and Λ in the curves are dependent on experimental conditions. The Λ and sk values for correcting the sample molecular weight averages should be those which correspond to the unknown sample peak V_R. The correction factors Λ and sk are dimensionless quantities and their values decrease as the instrument

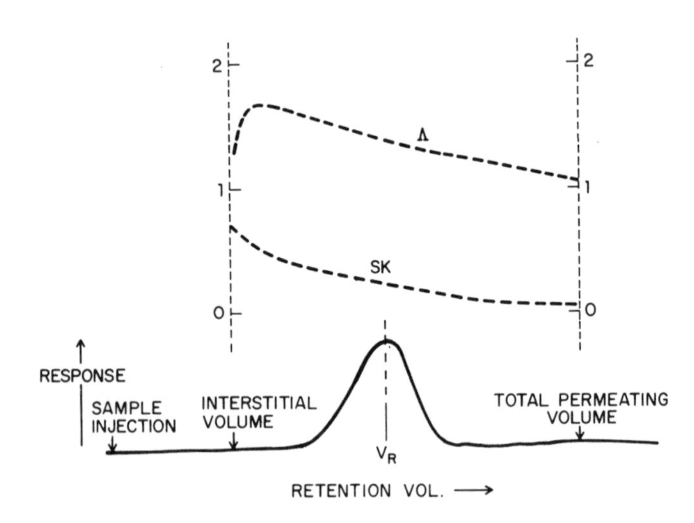

Figure 10.5 Schematic relationship among sk, Λ, and V_R.
The upper curve represents sk and Λ, and the lower is a typical chromatogram.

band-broadening effect decreases near the total permeation volume. The Λ value is always greater than 1.0, whereas the sk value is usually greater than 0. A positive sk value corresponds to a skewed, instrument band-broadened curve shape with extended tailing on the large retention volume side. Under the idealized condition of infinite resolution, $\Lambda = 1$ and sk $= 0$, indicating that no correction is necessary.

If sk ≥ 0.15 at the corresponding sample V_R, generally the molecular weight results cannot be satisfactorily corrected, and an investigation of the SEC experiment should be made in an attempt to improve the experimental conditions. With $\Lambda < 1.05$ and sk < 0.05, corrections to $\overline{M}_n$ and $\overline{M}_w$ are less than 5% and need not be made; uncorrected molecular weight averages may be reported directly.

If there is a significant skewing correction, $0.05 \leq$ sk ≤ 0.15, or significant symmetrical spreading correction, $\Lambda > 1.05$, then the sample $\overline{M}_n$ and $\overline{M}_w$ are corrected by

$$\overline{M}_n(t) = \overline{M}_n(u)(1 + \text{sk})(\Lambda) \tag{10.8}$$

and

$$\overline{M}_w(t) = \frac{M_w(u)}{(1 - \text{sk})\Lambda} \tag{10.9}$$

The correction procedures for molecular weight averages just presented are handled readily by manual manipulation of the data, but these same concepts can be used also in automated data-handling systems with greatly enhanced convenience. With automation, the more sophisticated correction method (e.g., GPCV3) provided in Section 9.3 is preferred. The correction factors Λ and sk are related to the fundamental SEC separation and resolution parameters by

$$\Lambda \simeq e^{(D_2\sigma)^2/2} \tag{10.10}$$

and

$$\text{sk} \simeq D_2\tau \tag{10.11}$$

Here D_2 is the slope of the SEC calibration curve and σ and τ are the symmetrical and the asymmetrical instrument band-broadening parameters, respectively. (Sections 3.5 and 9.3 provide detailed definitions and derivations.)

While corrections to $\overline{M}_w$ and $\overline{M}_n$ can be hand calculated, corrections to each molecular weight in the whole molecular weight distribution are beyond the scope of a manual approach and are discussed in Section 10.6. Some general guidelines can be provided, however. If sk ≥ 0.15, the MWD curves obtained in Section 10.3 should not be used; if $\Lambda \leq 1.05$ and sk ≤ 0.05, the MWD curves are useful without correction. For $\Lambda > 1.05$ and intermediate values of sk, the MWD curves must be corrected for accurate MW data, but the raw-data sample curves for a series of samples may be compared directly.

Other MW and MWD errors may occur which usually are caused by operational variables. Such errors normally are corrected by improving the HPSEC procedure rather than by postcorrecting the data. Problems that should be recognized include:

1. V_R shifts with sample concentration.
2. V_R shifts with flow variation.
3. Insufficient or improper resolution.
4. Changes in refractive index with molecular weight.
5. Operator errors.

10.5 Automated Data Handling for Computing Molecular Weight Averages and MWD

Although the manual procedures described in Sections 10.3 and 10.4 for determining molecular weight are not difficult, they are very time-consuming and involve many operator manipulations. A detailed manual data reduction and calculation generally requires one-half to one full day per sample. In addition, the manual procedure presents many opportunities for operator error, which can lead either to erroneous MW values or additional time spent in checking errors. Automated data-handling systems save considerable time and lead to improve accuracy by eliminating operator errors, allowing the use of more sophisticated routines such as curve smoothing, and permitting higher data rates for improved molecular weight accuracy. High-volume use of HPGPC for determining MW and MWD values (even ≥ 3 samples per day) is not justifiable without automated data processing. There are two major automated data system approaches currently in use for HPSEC: [1] real-time acquisition with off-line data processing and [2] combined real-time data acquisition and processing. (A third type of system, now available with some liquid chromatography systems, includes microprocessor control.) These systems are now discussed to provide the reader with a general set of guidelines for developing a usable system.

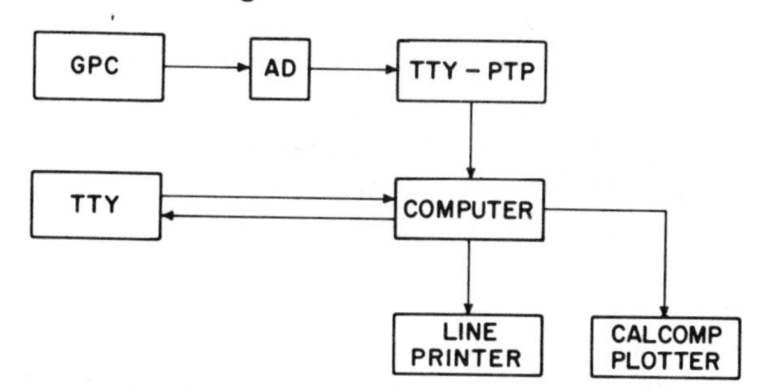

Figure 10.6 Schematic of GPC-data system.
PTP, paper-tape punch; TTY, teletype.

Real-Time Data Acquisition with Off-Line Processing

To accomplish real-time data acquisition with off-line processing, most workers today use software developed in-house with available computers and peripheral hardware. The basic approaches are relatively common, but the maximum output of a given system is limited by the size of the computer and the skills of the worker. For illustrative purposes only, we describe here the system in use in the Central Research and Development Department (CR&D) at the DuPont Experimental Station. This system has the configuration shown in Figure 10.6 and may be used as a guide to develop other systems but we do not make comparisons here. The Digital Equipment Corporation PDP-10 computer system at the Experimental Station has been discussed previously (9) and is not further elaborated here.

To calculate molecular weight averages and the MWD, as well as other SEC parameters such as V_R, h_i, and W, the computer receives the output converted into digital form from the chromatographic detector. For off-line operation the detector is interfaced to a paper tape (other recording devices such as magnetic tape or the computer memory storage itself may be used) for storing the data prior to transfer to the computer. As normally used, the detector puts out a voltage that is proportional to the solute concentration. This voltage is sampled at a known frequency, and the time and magnitude of each voltage signal are recorded. For subsequent calculation, each detector voltage signal is identified with the total volume of mobile phase which has passed through the column between sample injection and the time that the particular signal is taken. Since mobile-phase retention time and not retention volume is measured directly, the accurate conversion of time to volume is required. In addition to the total time-to-volume conversion, short increments of time are also converted to small volume increments to permit compensation for possible flow rate variations (Sect. 7.2). To accomplish both of these conversions, a syphon (Fig. 5.14) and the two clocks described below are used in the DuPont CR&D system.

An in-house constructed analog-to-digital (AD) converter changes the analog voltage signal at the recorder (at a frequency interval preselected by the operator) into voltage pulses whose magnitudes correspond to the incremental heights of the SEC curve. The AD converter is coupled to an amplifier, which filters and then converts 0–100 mV signal pulses into 0–10 V pulses. The final voltage readings are converted to four-digit numbers by an in-house-built teletype coupler. These numbers are printed for display purposes and are punched on paper tape for computer use by a Model 33ASR Teletype (Teletype Corp., Skokie, Ill.).

The two clocks (referred to above) in the teletype coupler determine [1] the total elapsed time from sample injection to the sampling of the detector voltage, and [2] the time interval from the previous syphon dump. The first clock is ganged to the sample injection valve, while the second is coupled to an optical detector triggered by the emptying of the syphon. The data from these clocks are transmitted to the teletype display and paper tape on a regular basis (about once every eight voltage readings). Computer program software makes the necessary time-to-volume changes. A somewhat similar system with different voltage sampling hardware and a Hewlett-Packard 2116C computer has been described (10).

The DuPont CR&D computer software program permits the operator to choose the beginning and ending of data acquisition at arbitrary points to eliminate taking extraneous baseline data. For flexibility both positive and negative peaks are accommodated easily and the operator can choose whether every data point or some fraction of them will be used. The raw data can be averaged (smoothed) and a baseline can be generated automatically. These options all are incorporated into one data-conversion program (11).

A second part of the computer program accomplishes molecular weight calibration and permits any of the calibration procedures described in Chapter 9 to be used. A third part uses an in-house-modified version of the Chevron Program (12) for MW and MWD calculations. Figures 10.7–10.9 illustrate output of exemplary MW and MWD information from the Chevron Program. All the sample identification, calibration information, and raw- and transformed-sample data are also printed in the output, which, in the DuPont CR&D case, permits the operator to use a Calcomp plotter and CRT in interacting with the data. Batching of results, especially plotting the MWD on one Calcomp plot, is also permitted and is very useful for sample comparisons.

The system described above was used extensively in our laboratory for several years prior to the development of modern high-performance GPC systems. With HPGPC, where the throughput is high (e.g., one sample per

	Average Molecular Weight					
	Number	Viscosity	Weight	Z	Z + 1	Reduced Area
	MN	MV	MW	MZ	M(Z+1)	RED. AREA ...
	1.455E 05	1.643E 05	1.671E 05	1.890E 05	2.166E 05	9.000E 01
	UN		UW	UZ		
Inhomogeneities	1.482E-01		1.308E-01	1.461E-01		
	SIGMA N		SIGMA W	SIGMA Z		
Standard Deviations	5.603E 04		6.044E 04	7.222E 04		

MOL WT	CUM AMT	D AMT/DM
1.862E 04	-C.	5.673E-16
2.007E 04	3.909E-05	5.214E-08
2.162E 04	1.563E-04	9.675E-08
2.330E 04	3.522E-04	1.349E-07
2.510E 04	6.247E-04	1.651E-07
2.705E 04	9.718E-04	1.907E-07
2.914E 04	1.403E-03	2.205E-07
3.140E 04	1.946E-03	2.613E-07
3.383E 04	2.637E-03	3.063E-07
3.645E 04	3.496E-03	3.475E-07
3.928E 04	4.516E-03	3.719E-07
4.232E 04	5.668E-03	3.874E-07
4.560E 04	7.010E-03	4.407E-07
4.913E 04	8.791E-03	5.791E-07
5.294E 04	1.137E-02	7.826E-07
5.704E 04	1.501E-02	9.870E-07
6.146E 04	1.974E-02	1.153E-06
6.623E 04	2.552E-02	1.274E-06
7.136E 04	3.233E-02	1.388E-06
7.688E 04	4.045E-02	1.569E-06
8.284E 04	5.065E-02	1.874E-06
8.926E 04	6.388E-02	2.262E-06
9.618E 04	8.102E-02	2.729E-06
1.036E 05	1.038E-01	3.452E-06
1.117E 05	1.358E-01	4.560E-06
1.203E 05	1.809E-01	5.874E-06

MOL WT	CUM AMT	D AMT/DM
1.296E 05	2.418E-01	7.144E-06
1.397E 05	3.190E-01	8.155E-06
1.505E 05	4.109E-01	8.705E-06
1.883E 05	7.097E-01	8.546E-06
2.028E 05	7.903E-01	7.701E-06
2.186E 05	8.535E-01	6.318E-06
2.355E 05	9.004E-01	4.749E-06
2.946E 05	9.710E-01	3.332E-06
3.174E 05	9.791E-01	2.271E-06
3.420E 05	9.844E-01	1.480E-06
3.685E 05	9.888E-01	8.775E-07
4.609E 05	9.924E-01	4.718E-07
4.966E 05	9.951E-01	2.647E-07
5.351E 05	9.970E-01	1.851E-07
5.766E 05	9.980E-01	1.452E-07
6.213E 05	9.986E-01	1.092E-07
6.694E 05	9.991E-01	7.099E-08
7.213E 05	9.995E-01	4.055E-08
7.771E 05	9.998E-01	2.107E-08
	9.999E-01	1.290E-08
	1.000E 00	9.828E-09
		7.349E-09
		4.486E-09
		2.081E-09
		.434E-14

Labels: Differential Molecular Weight Distribution; Cumulative Molecular Weight Distribution; Molecular Weight

Figure 10.7 **Example of output of calculations from the Chevron Program.** (Reprinted with permission from Ref. 12.)

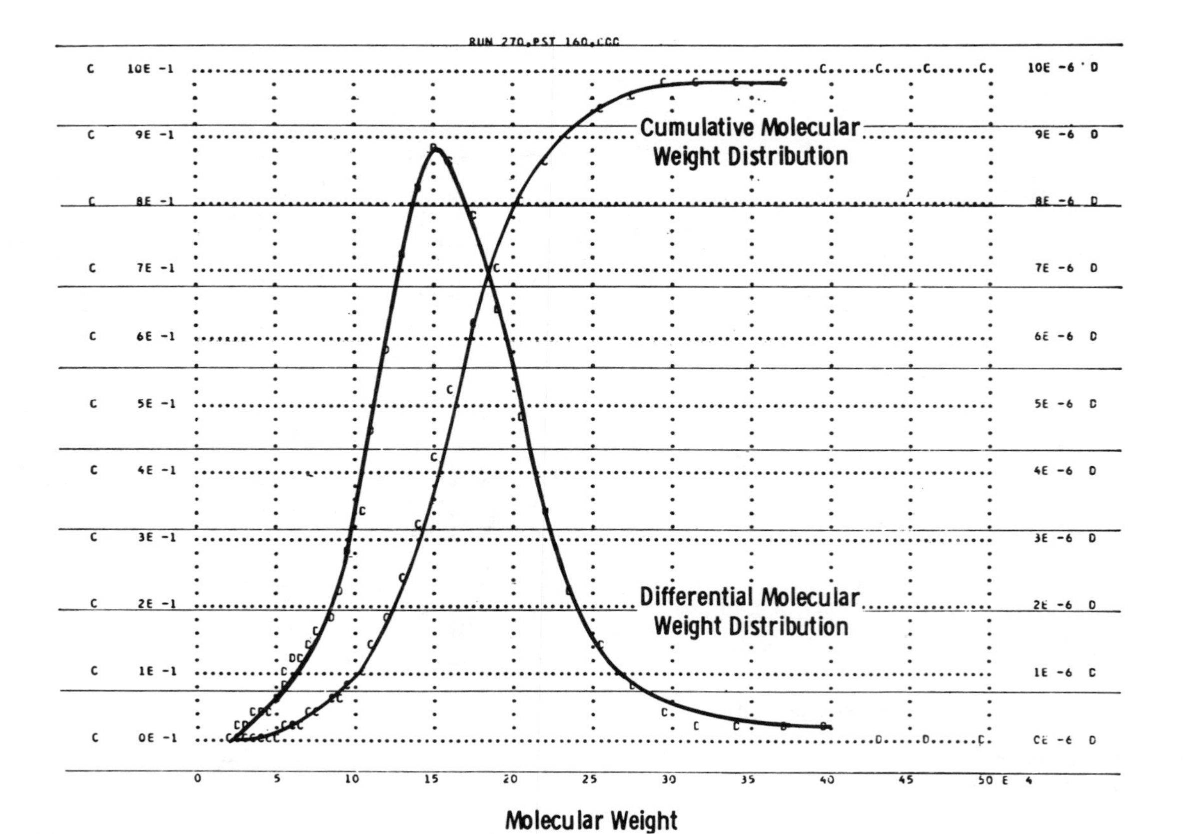

Figure 10.8 Example of output of graphs from the Chevron Program.
(Reprinted with permission from Ref. 12.)

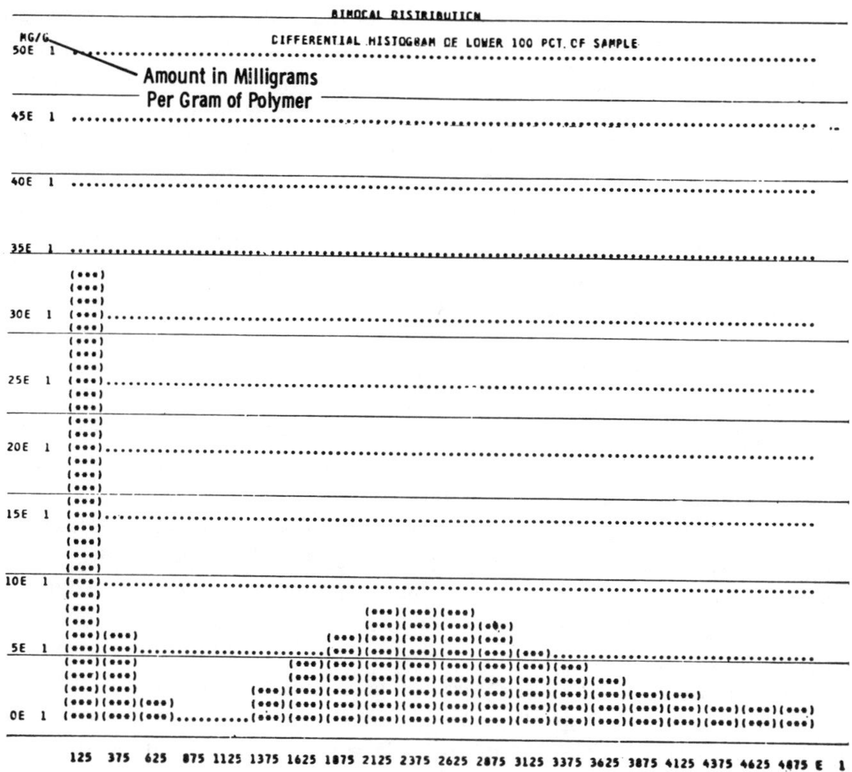

Figure 10.9 Example of sample output for a bimodal distribution; differential histogram from the Chevron Program.

(Reprinted with permission from Ref. 12.)

15–20 min), the off-line approach is both slow and cumbersome, and this led to the development of a real-time system.

Real-Time System

The real-time system uses the same Experimental Station PDP-10 computer for the analog-to-digital conversion of the chromatograms, as does the off-line system. ("Real-time" commonly refers to systems in which data are calculated on the fly. While the DuPont CR&D system acquires and processes data automatically, it is interactive in nature, does postprocessing of the data only after acquisition is complete, and permits the operator to override computer-selected parameters if desired.) An acquisition data rate of 1 point/sec is commonly used for HPGPC runs of 15–20 min. All information entered by the operator about the sample (sample name and number, etc.) and the digitized chromatogram are stored in a data file.

The analysis of the data is divided into two programs, GPCN3 and GPCV3. GPCN3 examines the data, sets default values for operating and calibration parameters, selects peak start and end values, and evaluates the flow rate from cumulative syphon dumps. The default procedure permits the computer programs to continue with data evaluation if overriding values are not provided. Peak start and end values are selected by the computer using an algorithm of matching slopes. Thus this method is limited to single peaks or those with only slight shoulders. The computer-generated peak start and end parameters, used later for the MW calculations, are printed out in time-based units, and these parameters are compared visually with the strip-chart recording of the chromatographic curve (obtained as detector response versus time) to ascertain their acceptability.

Program GPCV3 performs the actual molecular weight and molecular weight distribution analysis. This program uses data from the first program, GPCN3, and incorporates the elements of the modified Chevron program mentioned above. GPCV3 also incorporates the corrections for column dispersion, σ (Sect. 10.4), and permits use of the linear calibration procedures described in Chapter 9. The program also generates a report on a line printer (optionally on a Calcomp plotter) and sends a summary page back to the operator's teletype for immediate use. While the chromatographic analysis of a polymer sample takes about 15 min, approximately 2 min of computational time is required to complete the data analysis.

10.6 Computer Method for Correcting Band-Broadened SEC Curves

Section 10.4 discusses a method for correcting HPSEC-derived molecular weight averages using manual calculations. However, correcting the SEC data for accurate MWD is a complex task and normally requires a computer. Also, the complex band-spreading theory does not lead to simple or universal correction parameters for all MW because it is derived from simplified models (e.g., Gaussian band spreading). We present here one of the more accepted and useful literature methods for correcting MWD data for peak spreading (5, 13). A review of other approaches up to 1968 may be of interest (14). Other more recent approaches are cited in the ASTM Bibliographies discussed in Section 1.5, but none appear to have general superiority.

The problem of instrument band spreading in SEC can be represented mathematically by expressing the experimental chromatogram, $F(V)$, as a convolution between the true chromatogram $W(y)$ and the function $G(V)$, which describes the instrument band spreading. This convolution is described by Equation 4.11 and is repeated in Equation 10.12. It is often called *Tung's integral equation*, in honor of the author's pioneering work (5, 13):

$$F(V) = \int_{-\infty}^{+\infty} G(V - y)W(y)\,dy \tag{10.12}$$

where V and y both represent elution volume. This equation explains, as is also illustrated in Figure 10.10, that the experimental $F(V)$ curve is necessarily broader than $W(y)$, owing to $G(V - y)$ broadening. The deconvolution problem is to solve for $W(y)$ from Equation 10.12.

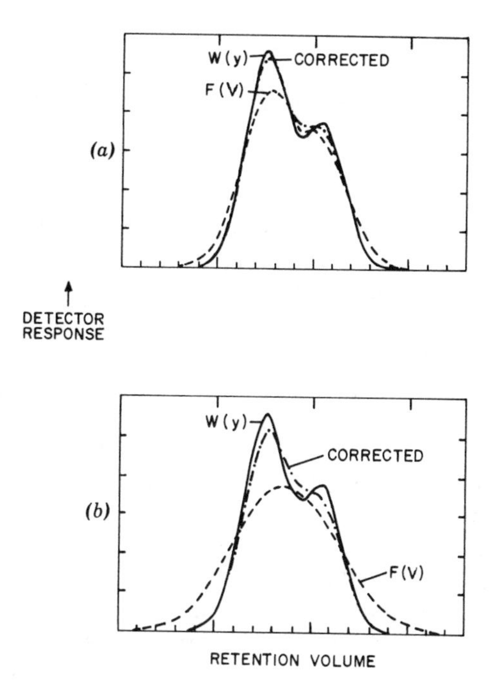

Figure 10.10 Comparison of true, broadened, and corrected chromatograms using the polynomial correction method.

$W(y)$, true; $F(V)$, broadened; corrected, indicated. (*a*) Moderate band broadening; (*b*) extensive band broadening. (Redrawn in part and reprinted with permission from Ref. 13.)

In Tung's computer method for solving for $W(y)$, a fourth-degree polynomial of the type given in Equation 10.13 is used to describe the experimental chromatogram, $F(V)$:

$$F(V) = e^{-q^2(V-V_0)^2} \sum_{i=0}^{4} U_i(V_i - V_o)^i \tag{10.13}$$

where q, V_o, and U_i are adjustable coefficients. A similar polynomial is assumed for $W(y)$:

$$W(y) = e^{-p^2(y-V_0)^2} \sum_{i=0}^{4} R_i(y_i - V_o)^i \tag{10.14}$$

where p and R_i are unknown coefficients to be determined. Values of p and R_i are directly calculable from q, V_o, U_i, and σ by using the simple equations derived in Reference 13. The instrument band spreading is assumed to be Gaussian:

$$G(V) = \frac{1}{\sigma\sqrt{2\pi}} e^{-V^2/2\sigma^2} \tag{10.15}$$

where σ is the standard deviation of this Gaussian spreading function. A variable σ value (for different SEC retention volumes) can be used in Tung's polynomial method.

The algorithm of this computer program is as follows: [1] determine the q, V_o, and U_i values that cause $F(V)$ from Equation 10.11 to fit the experimental SEC chromatogram—the computer curve fitting is done by the least-squares residue method; [2] calculate p and R_i values from the q, V_o, and U_i values from step 1; and [3] calculate $W(y)$ and generate the spreading-corrected SEC curve point by point at different retention volumes.

The results of testing the method on computer-simulated SEC chromatograms are shown in Figure 10.10. The solid and dotted curves in the figure are, respectively, the synthesized $W(y)$ and $F(V)$ chromatograms obtained by summing two Gaussian distributions. For the case of moderate instrument band spreading (Fig. 10.10*a*), the method completely deconvolutes the SEC chromatogram and the corrected chromatogram is in excellent agreement with the original $W(y)$. For the case of extensive instrument band spreading (Fig. 10.10*b*), the method does not completely resolve the two peaks; however, it does remove much of the band broadening. Furthermore, it more than doubles the SEC resolution, making it equivalent to that gained by increasing experimental column length by a factor of four. Alternatively, resolution enhancement permits more rapid flow rates and shorter analysis times. This polynomial resolution enhancement method produces more accurate molecular weight averages (calculated from the corrected chromatograms) and can be used to generate broad MWD calibration curves directly without the need of σD_2 corrections (Chapts. 4 and 9).

The polynomial method has been used on experimental SEC curves with good success. Unfortunately, not all chromatograms (e.g., multimodal curves) can be fitted by a polynomial of reasonable size. Also, for cases where the elution peaks are relatively sharp, or the instrument band spreading is relatively large, or the detected signal is noisy, oscillating peaks in the corrected chromatogram may occur. If oscillation occurs, it can be damped out by using a polynomial of one degree lower than that which gives the best fit to the experimental $F(V)$.

10.7 Special Methods and Information

Specific techniques can sometimes be used with SEC data to obtain additional sample structural information. The methods described below relate to polymer analyses, and readers with other interests may wish to omit this section. The universal calibration method can be used in certain instances for molecular weight calculations where needed calibration standards are not available and also in polymer branching calculations. Simple chromatogram comparison techniques can sometimes be employed for calculation of polydispersities, and the intrinsic viscosity $[\eta]$ can be computed from SEC data. The principal concepts of these approaches are discussed below, and Chapters 12 and 13 can be consulted for specific examples.

Use of Universal Calibration

The concepts behind the universal calibration (U.C.) methods are discussed in more detail in Chapter 9. In this chapter we point out how the approach is used and what information is required.

$M[\eta]$, which is proportional to the hydrodynamic volume, can be used as a universal parameter in SEC calibration (15–16), because it is a basic parameter governing size separation. A plot of $M[\eta]$ versus V_R for all polymers should be identical; at any retention volume, the hydrodynamic volumes of two polymer samples will be equal according to

$$M_1[\eta]_1 = M_2[\eta]_2 \qquad (10.16)$$

The viscosity $[\eta]$ can be related to the viscosity-average molecular weight, $\overline{M}_v$, through the familiar Mark-Houwink relationship (17):

$$[\eta] = \mathbf{K}\overline{M}_v^{\mathbf{a}} \qquad \text{or} \qquad \overline{M}_v[\eta] = \mathbf{K}\overline{M}_v^{\mathbf{a}+1} \qquad (10.17)$$

where **K** and **a** are empirical constants.

Table 10.2 Mark-Houwink Parameters for Polymers in Tetrahydrofuran

Polymer	Temperature (°C)	K × 10^4 (dl/g)	a	Range of MW × 10^{-3}
Polystyrene	25	1.60	0.706	>3
	25	1.41	0.700	50–1000
	25	1.60	0.700	—
Polystyrene (comb)	23	2.2	0.56	150–11,200
Polystyrene (star)	23	0.35	0.74	150–600
Poly(vinyl chloride)	25	1.63	0.766	20–170
	25	1.50	0.77	10–120
	25	1.60	0.77	10–1000
	25	4.98	0.69	—
Poly(vinyl bromide)	25	1.59	0.64	20–100
Poly(vinyl acetate)	25	3.5	0.63	10–1000
Poly(vinyl ferrocene)	30	0.72	0.72	—
Poly(methylmethacrylate)	25	21.1	0.406	<31
	25	1.04	0.697	≥31–1700
	—	1.28	0.69	150–1200
	23	0.93	0.72	170–1300
Amylose propionate	25	248.0	0.61	20–500
Poly(octadecyl methacrylate)	30	0.25	0.75	230–1670
Poly(octadecyl vinyl ether)	30	22.4	0.35	9–110
Polycarbonate	25	4.9	0.67	7–77
		3.89	0.70	
Polydioxalane	25	0.937	0.874	—
Polyisoprene	25	1.77	0.735	40–500
Polychloroprene	30	0.418	0.83	69–511
Poly(*cis*-isoprene) (natural rubber)	25	1.09	0.79	10–1000
Butyl rubber	25	0.85	0.75	4–4000
Poly(1,4-butadiene)	25	76.0	0.44	270–550
	40	5.78	0.67	10–100
Polybutadiene (20% *cis*, 20% vinyl)	25	2.36	0.75	3–6
SBR(25% S)	40	3.18	0.70	70–1000
	25	4.1	0.693	24–40
SBR 1507	30	3.0	0.70	10–1000
SBR 1808	30	5.4	0.66	10–1000
Cellulose nitrate	25	25.0	1.00	95–2300
Cellulose trinitrate	25	3.21	0.83	(Dp)60–6000
Amylose acetate	25	108.0	0.70	20–500
Amylose butyrate	25	111.0	0.70	20–500
Poly(vinyl carbazol)	25	1.44	0.65	10–45

Table 10.2 *Continued*

Polymer	Temperature (°C)	K × 10⁴ (dl/g)	a	Range of MW × 10^{-3}
Poly(acrylonitrile costyrene), (38.3/61.7 mole)	25	2.15	0.68	100–780
Poly(styrene-comonoethyl-maleate)	25	0.75	0.695	200–1800
Polyoxypropylene	20	5.5	0.62	0.5–3.3
Poly(oxycarbonyloxy-1,4-phenylene)	25	3.89	0.70	10–70
Isopropylidene-1,4-phenylene	25	3.99	0.70	10–270
Poly(oxymaleoyl-oxyhexamethylene)	20	4.37	0.66	13–66
Poly(oxysuccinyl-oxyhexamethylene)	20	4.43	0.69	15–50
Poly(butyl isocyanate)	20	0.0457	1.18	18–210

Taken in part from Reference 6.

Equation 10.18, obtained by rearranging Equations 10.16 and 10.17, is used to obtain M_2, the molecular weight for one or more specific samples of unknown polymer. The method uses a series of narrow-MWD polymer standards with molecular weights (M_1) and appropriate values of **K** and **a**:

$$\log M_2 = \frac{1}{1 + \mathbf{a}_2} \log \frac{\mathbf{K}_1}{\mathbf{K}_2} + \frac{1 + \mathbf{a}_1}{1 + \mathbf{a}_2} \log M_1 \tag{10.18}$$

Frequently, critically evaluated M, **K** and **a** values from the literature for polystyrenes in organic solvents and dextrans or sulfonated polystyrenes in aqueous solvents are used to solve for M_2. After the M_2 values are obtained, they are used for the needed calibration using the peak position method or GPCV2 (Chapt. 9). Table 10.2 provides a list of **K** and **a** values from the literature for the frequently used solvent, THF, while Table 8.2 lists other **K** and **a** values which may be useful. Note that the Mark-Houwink constants (**K** and **a**) for both polymers must be known for this treatment, or they must be determined for the polymer, solvent, and temperature of interest by using standard viscometric techniques.

Some authors have noted exceptions to the universal calibration concept, but these are not common (18–19). There also are several possible sources of error which can affect the results, especially mechanism bias (e.g., adsorption and inaccurate or inappropriate **K** and **a** values from the literature). For example, in some systems **a** is very dependent on molecular weight and the Mark-Houwink expression is valid only over a narrow molecular weight range. The source of and limitations on accuracy of **K** and **a** should always be noted when they are used in the calculations.

Calculation of Intrinsic Viscosity [η] from SEC Data

The viscosity average molecular weight ($\overline{M}_v$) can be calculated from HPSEC data using

$$\overline{M}_v = \left[\frac{\sum_{i=1}^{N} h_i (M_i)^{\mathbf{a}}}{\sum h_i}\right]^{1/\mathbf{a}} \tag{10.19}$$

After $\overline{M}_v$ is computed, the intrinsic viscosity [η] is determined from Equation 10.17, where the **K** and **a** values used in the computations must have been determined from narrow MWD polymers. However, equations 10.17 and 10.19 are valid for calculating the viscosity of even a broad-MWD unknown sample because the sample is fractionated into a series of narrow-MWD samples by the SEC process. Table 10.3 shows that very reliable values of [η] can be obtained via SEC (20). The advantage of the approach is its convenience. Once the polymer solution is injected onto the SEC columns, no further experimental work is necessary.

It has been suggested that under certain conditions the viscosity of the sample in one solvent can be estimated by using the sample chromatogram generated in another solvent and the values of **K** and **a** for the new solvent (21). This approach should be valid if the **a** values are nearly the same for both solvents; however, few test data have been published.

Table 10.3 Comparison of Intrinsic Viscosities Obtained by SEC and by Direct Measurements

Sample (Polystyrene in Tetrahydrofuran, 25°C)	Polydispersity	[η] Measured	[η] SEC
Fraction 1	1.30	3.92	4.16
Fraction 2	1.10	1.28	1.23
Fraction 3	1.06	0.14	0.14
NBS-705	1.07	0.74	0.73
NBS-706	2.10	0.93	1.00
Polyethylene in trichlorobenzene, 130°C	2.90	1.01	1.01[a]

Taken from Reference 20.

[a] Individual values for five replicate runs were 0.970, 1.010, 1.012, 1.016, and 1.014.

Values of **K** and **a** for the universal calibration method can be obtained from Equations 10.17 and 10.19. For such determinations only narrow-MWD standards ($\overline{M}_w/\overline{M}_n \leq 1.2$) should be used, where $\overline{M}_n \simeq \overline{M}_v \simeq \overline{M}_w$. Substituting $\overline{M}_w$ (or $\overline{M}_n$) in Equation 10.19 for $\overline{M}_v$ along with the calibration data (M_i) permits the calculation of **a**. After **a** is determined, the intrinsic viscosity of the standard sample is measured in the same SEC solvent for which M_i was obtained, using conventional viscometric techniques; **K** is calculated from Equation 10.17. Only one narrow MWD standard and a set of calibration data are required for this approach, but additional standard runs can provide data to refine the values of **K** and **a** obtained. This method usually is used for special sample solvents, where for polystyrene values of **K** and **a** have not been published.

Comparative Technique for MWD

A rapid and simple method for comparing sample MWDs without the necessity of calibration would be desirable. The empirical method of Equation 10.20,

$$\frac{W_1}{d_1} = \frac{W_2}{d_2} \tag{10.20}$$

where W is the elution curve width as defined in Section 10.2 and d is the sample polydispersity defined as $\overline{M}_w/\overline{M}_n$, has been suggested to calculate values of d of unknowns from the sample curve width and the respective values for W and d for the standard. Equation 10.20 is suggested for normal and Gaussian-like MWD of linear-chain polymers where the SEC curves fall within the linear region of the molecular weight calibration curve (22).

Equation 10.21 is a similar expression from theory which has been suggested for relating curve width and polydispersity (23).

$$\frac{W_1^2}{\log d_1} = \frac{W_2^2}{\log d_2} \tag{10.21}$$

Very little experimental data have been published to evaluate the applicability of Equations 10.20 and 10.21 for polymer characterization. Calculations show that markedly different results are obtained for very small and very large sample polydispersities, but such polydispersities are difficult to confirm experimentally. Similar results ($\pm 20\%$ relative between the methods) are predicted by calculation for the two equations in the low- to midpolydispersity range common to many commercial polymers, (e.g., $d = 1.2$–3). These equations are simple and are suggested for use with single experiments when calibration is not possible or not justified because of the time required.

The Q-Factor Method

To define the molecular parameter governing size separation, the *Q factor* was used by early workers in gel permeation chromatography prior to development of the hydrodynamic volume theory. *Q* is defined as the molecular weight per extended chain length in angstroms or the "molecular weight per angstrom chain length." The angstrom length represents the size of the molecule in a fully extended conformation and is calculated for each polymer from appropriate bond angles and bond lengths. It is further assumed that a given retention volume corresponds to a given angstrom size, so that if the molecular weight per angstrom length is known, the molecular weight for any polymer can be determined by relating it to a standard calibration curve.

Unfortunately, the assumptions of this approach are unrealistic and the *Q*-factor method should not be used. The sizes of different kinds of polymer

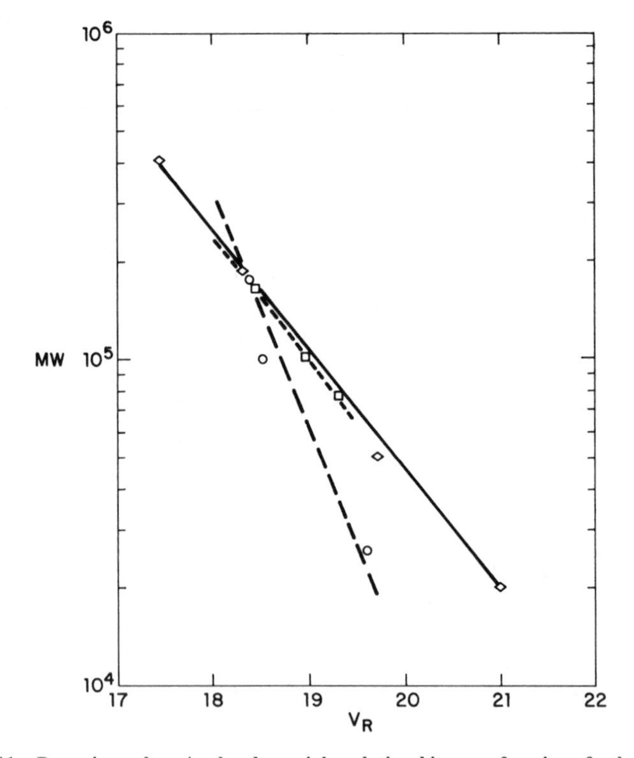

Figure 10.11 Retention volume/molecular weight relationships as a function of polymer type. Columns, 120 × 0.8 cm, Styragel, $10^3 + 10^5 + 10^6$ Å; mobile phase, *m*-cresol; flow rate, 1 ml/min; temperature, 100°C; detector, differential refractometer; sample, 2 ml of 0.25%: ×, polystyrene; □, poly(methyl methacrylate); ○, polycaprolactam. (Drawn from Table I of Ref. 24.)

molecules in solution depend on experimental conditions and do not bear constant relationships to one another. As shown by the data of Figure 10.11, the relationship between the sizes in solution of polycaprolactam and polystyrene or poly(methyl methacrylate) is not a constant. The slopes of the calibration curves are not the same and no constant Q-value relationship exists between the molecular weights of these polymers. A more basic and measurable parameter such as $M[\eta]$, discussed above, is required for molecular weight calculations.

REFERENCES

1. *ASTM Method D* 3016-79, "Standard Recommended Practice for Use of GPC Definitions and Relationships," ASTM, 1916 Race St., Philadelphia, Pa. 19103.
2. P. Andrews, *Br. Med. Bull.*, **22**, 109 (1966).
3. W. W. Yau, J. J. Kirkland, D. D. Bly, and H. J. Stoklosa, *J. Chromatogr.*, **125**, 219 (1976).
4. A. E. Hamielec and W. H. Ray, *J. Appl. Polym. Sci.*, **13**, 1319 (1969).
5. L. H. Tung and J. R. Runyan, *J. Appl. Polym. Sci.*, **13**, 2397 (1969).
6. *ASTM Method D* 3593-77, "Standard Method of Test for the Determination of Molecular Weight Averages and the Molecular Weight Distribution of Certain Polymers by Liquid Exclusion Chromatography Using Universal Calibration," ASTM, 1916 Race St., Philadelphia, Pa. 19103.
7. S. T. Balke and A. E. Hamielec, *J. Appl. Polym. Sci.*, **13**, 1381 (1969).
8. T. Provder and E. M. Rosen, *Sep. Sci.*, **5**, 437 (1970).
9. J. S. Fok and E. A. Abrahamson, *Chromatographia*, **7**, 206 (1974).
10. G. Braun, *J. Appl. Polym. Sci.*, **15**, 2321 (1971).
11. H. J. Stoklosa, E. I. Du Pont de Nemours and Company, private communication, 1977.
12. H. E. Pickett, M. J. R. Cantow, and J. F. Johnson, *J. Appl. Polym. Sci.*, **10**, 917 (1966).
13. L. H. Tung, *J. Appl. Polym. Sci.*, **13**, 775 (1969).
14. J. H. Duerksen and A. E. Hamielec, *J. Polym. Sci., Part C*, **21**, 83–103 (1968).
15. H. Benoit, F. Grubisic, P. Rempp, D. Decker, and J. G. Zilliox, *J. Chim. Phys.*, **63**, 1507 (1966).
16. K. A. Boni, F. A. Sliemers, and P. B. Stickney, *J. Polym. Sci., Part A*-2, **6**, 1567 (1968).
17. A. Tanford, *Physical Chemistry of Macromolecules*, Wiley, New York, 1961, p. 407.
18. J. V. Dawkins, *J. Macromol. Sci.*, **B2**, 623 (1968).
19. J. Pannell, *Polymer*, **13**, 277 (1972).
20. M. Y. Hellman, in *Liquid Chromatography of Polymers and Related Materials*, J. Cazes, ed., Dekker, New York, 1977, p. 29.
21. J. Cazes, *Polym. Lett.*, **8**, 785 (1970).
22. D. D. Bly, *J. Polym. Sci., Part C*, **21**, 13 (1968).
23. T. Williams, Y. Udagawa, A. Keller, and I. M. Ward, *J. Polym. Sci., Part A*-2, **8**, 35 (1970).
24. D. D. Bly, *J. Polym. Sci., Part A-1*, **6**, 2085 (1968).

Eleven

SPECIAL TECHNIQUES

11.1 HPSEC Separation of Small Molecules

Well-packed columns with $<10\ \mu m$ particles often permit rapid separation of small molecules (<5000 MW) with good resolution. While other high-performance LC methods generally afford higher resolution, HPSEC can often be applied successfully with good results to a surprisingly large variety of analytical problems as long as there is sufficient difference in solute size. Whereas isomers usually cannot be resolved, close homologs and oligomers can often be separated adequately by HPSEC.

HPSEC is often a logical first approach for the liquid chromatographic analysis of many unknown samples. A specific advantage of HPSEC is that many analytical problems can be solved in about 20 min with essentially no separations development work. The sample needs only to be dissolved, injected onto the column, and chromatographed under isocratic (constant solvent composition) conditions. All components elute within a predetermined interval, amounting to the time required for a single column of mobile phase to elute. Using this simple approach, relatively extensive information can be gained in a short time, for example, estimations of molecular weight and sample complexity. Extensive use is now being made of HPSEC for the separation of small molecules, not only for the rapid survey of unknown samples, but also for the rapid quality-control or routine assay of a wide variety of materials. The general advantages and disadvantages of using SEC for small molecule separations are summarized in Table 11.1.

Table 11.1 Use of SEC for Small-Molecule Separations

Advantages
Run time predetermined
Total elution isocratic
Low dilution for better detection
No development time
Elution correlation to molecular size
Mild interactions
Disadvantages
Size differences required
Low resolution
Limited peak capacity
Other retention mechanisms possible

An inherent limitation to small molecule separation by HPSEC is the limited peak capacity, n, which may be determined by Equation 4.4 and estimated by (1):

$$n \simeq 1 + 0.2N^{1/2} \tag{11.1}$$

where N is the column plate count. For example, a column set that has a total plate count of 20,000 would have a theoretical peak capacity of about 30 for solutes of the same diffusion coefficient. Theoretically, this means 30 peaks of equal width can be placed within the total permeation and total exclusion volumes at peak resolutions of unity. Peak capacity for the other LC methods can be much larger because the mechanism of retention is different (see discussion for Fig. 4.4).

Experimental

Typically, HPSEC systems that have 10,000–30,000 total theoretical plates are used. To obtain this plate count, several HPSEC columns must be connected in series, using the technique described in Section 8.6. When separations are exclusively in the low-molecular-weight range, the column set will be a series of like columns with the smallest available pore size, generally <100 Å. The main purpose of separating samples of small molecules by HPSEC is to determine the number of components of different molecular weight or to determine the amount of one or more specific components in the sample quantitatively.

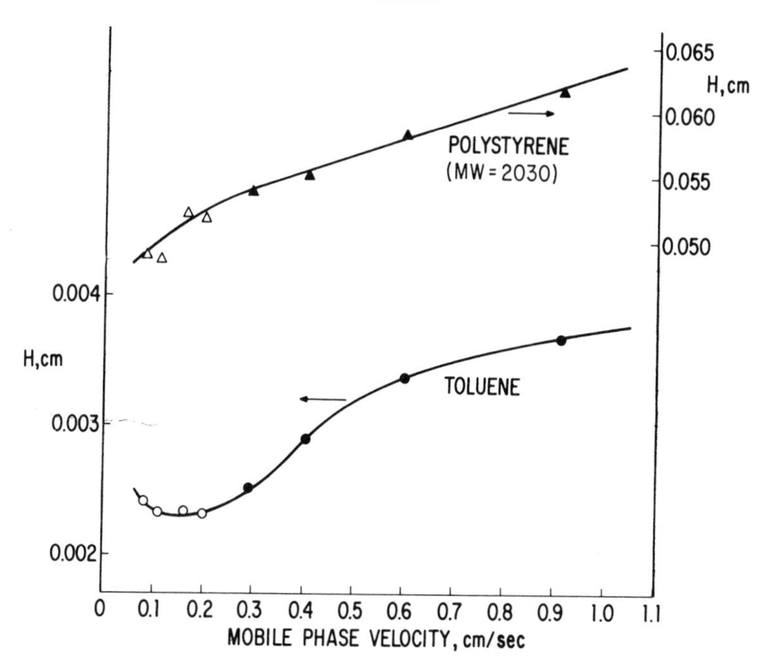

Figure 11.1 Plate height versus velocity plots for small molecules.
Column, 25 × 0.62 cm porous silica microspheres, 45 Å (silanized); mobile phase, tetrahydrofuran; detector, UV, 0.1 AUFS, 254 nm; sample 6.7 μl, 4 mg/ml PS 2030, 1 mg/ml toluene; temperature, 23°C. (Reprinted with permission from Ref. 2.)

HPSEC separations of small molecules can be made quickly without significant sacrifice in resolution because of the rapid solute equilibrium with the < 10 μm particles. This conclusion is documented by the data of Figure 11.1, which illustrate the effect of mobile-phase velocity on plate height for a single column of 8 μm silica particles with 45 Å pores. The relatively flat plots found for both toluene and a low-molecular-weight polystyrene standard make it feasible to operate such a column at relatively high mobile-phase velocities.

While adsorption of solutes to the surface of silica substrates is recognized to occur with improper substrate/mobile phase combinations (see below and Sect. 13.3), it is not a widely known fact that the retention of a variety of solutes beyond total permeation may also occur with columns of organic gels. For example, in one material balance study, 60 wt.% of a mixture of unknown substituted aromatic compounds was not eluted from 100 Å μ-Styragel columns using tetrahydrofuran as a mobile phase (3). To make certain that all compounds elute from HPSEC columns, it is desirable to establish a material balance (Sect. 8.8).

Experience has shown that a single mobile phase provides the most satisfactory use of microparticulate, small-pore gel columns. Repeated changing of the mobile phase greatly increases the chance of collapse of the gel particles and failure of the column set. Thus it is recommended that small-pore, microparticulate gel columns (e.g., μ-Styragel 100 Å) be used with only a *single* mobile phase out of the recommended list (Table 8.5) for small-molecule separations. Tetrahydrofuran is often a preferred solvent for many cases. If a sample is not totally soluble in tetrahydrofuran, a different column set, permanently equilibrated with a different solvent (e.g., dichloromethane), should be used for best results.

Experience with columns of rigid inorganic particles (e.g., silica) for separating small molecules is limited, because of their recent availability. However, rigid silica packings have several significant advantages. As discussed in Chapter 6, small silica particles are relatively easily packed into stable, high-efficiency columns. More importantly, compared to columns of organic gels, a much wider range of mobile phases may be used, resulting in significant advantages in versatility and convenience. Solvent changeover with columns of rigid particles is accomplished quickly and easily, since equilibration is rapid. Columns of silica particles can also be used at high temperatures if required (Sect. 12.4). Finally, approximate MW calibration curves can be predicted from pore-size distribution data, which is not possible with organic gels. At times, columns of rigid siliceous particles exhibit undesirable retention of solutes by adsorption, but in situations that exhibit this effect, a different mobile phase or additive can be used to eliminate the problem. Alternatively, the silica particles can be modified with appropriate organic functional groups to eliminate adsorptive interactions that interfere with the desired size-exclusion process (see Sects. 6.2 and 13.3).

Molecular Weight Estimations

Several approaches have been used to correlate the retention of small molecules to a basic size parameter. One approach involves the use of molecular volume. Effective molar volumes of unknowns are obtained by comparing their retention volumes with the retention volumes of normal hydrocarbon standards whose molar volumes can readily be calculated (4). The factors to be considered for predicting the retention volume of small molecules in tetrahydrofuran have been generalized (5, 6).

Correlations between retention volume and molecule chain length have also been examined (7, 8). With this approach molecular sizes are calculated from atomic radii and bond angles for various functional groups, again using hydrocarbons as standards. However, other compound types often do not follow these calculated "carbon-number" values, and therefore corrections are needed to establish "effective carbon numbers." Also, factors must be introduced to allow for solute/solvent associations.

Thus the effective molar volume, effective carbon number, or the effective chain length of simple molecules all can be derived from a calibration curve obtained from normal alkanes, and it is simple to convert from the calibration curve based on one parameter to that based on another. The choice of solute-size parameter and the type of calibration to be used for characterizing small molecules is arbitrary.

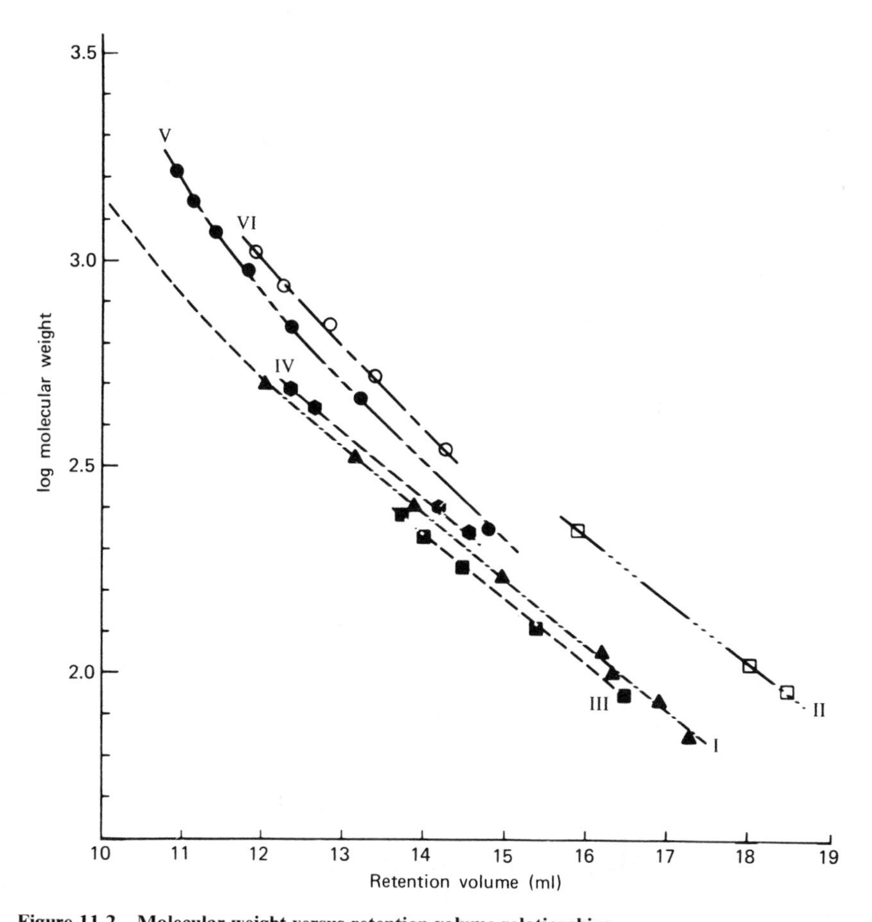

Figure 11.2 Molecular weight versus retention volume relationships.
I, *n*-alkanes, C_5H_{12} to $C_{36}H_{74}$; II, toluene, *p*-xylene, diethylphthalate; III, *n*-alcohols, C_4H_9OH to $C_{16}H_{33}OH$; IV, 2,6-di-t-butyl-*p*-cresol, dibutyladipate, *n*-didecylphthlate, *n*-didodecylphthlate; V, nonylphenol-formaldehyde adducts; VI, 2,2,4-trimethyl-1,2-dihydroquinoline oligomers, dimer to hexamer. (Reprinted with permission from Ref. 9.)

Reasonably linear plots also frequently occur when the retention volumes of small molecules are plotted versus log molecular weight, as illustrated in Figure 11.2 for series of similar compounds. To predict the identity or characteristics of an unknown from an easily obtainable HPSEC chromatogram, it is often much simpler and just as effective to use molecular weight values rather than effective carbon numbers or molar volumes, as illustrated by the example in Figure 11.3 for a series of drugs. Although the structures of these compounds are grossly dissimilar, a reasonable correlation of molecular weight with retention volume is obtained.

A molecular weight calibration is very useful for predicting the identity of an unknown component in a particular series. However, it is important in the separation of small molecules to recognize that HPSEC retention is only a rough function of solute molecular weight. That small molecules are not retained in HPSEC as a strict function of molecular weight is illustrated in

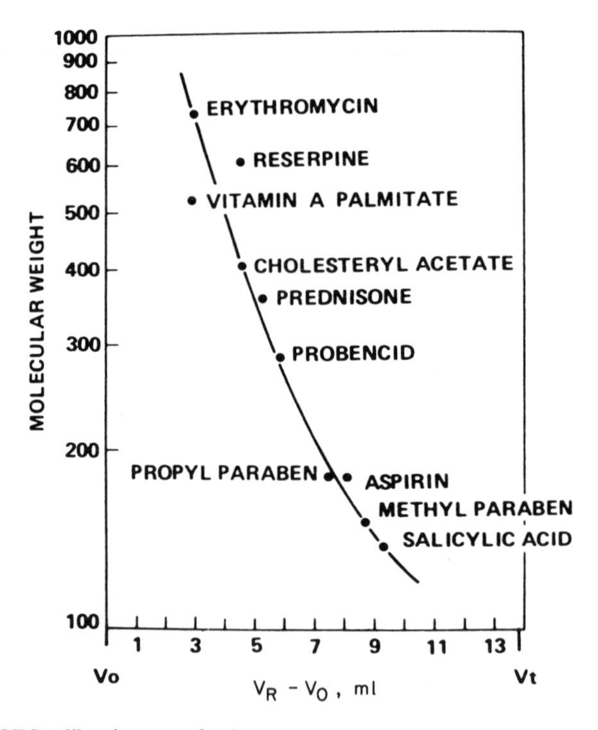

Figure 11.3 SEC calibration curve for drugs.
Column, 90 × 0.8 cm μ-Styragel 100 Å; mobile phase, tetrahydrofuran; flow rate, 2 ml/min; detector, UV, 0.32 AUFS, 254 nm. (Reprinted with permission of Waters Associates.)

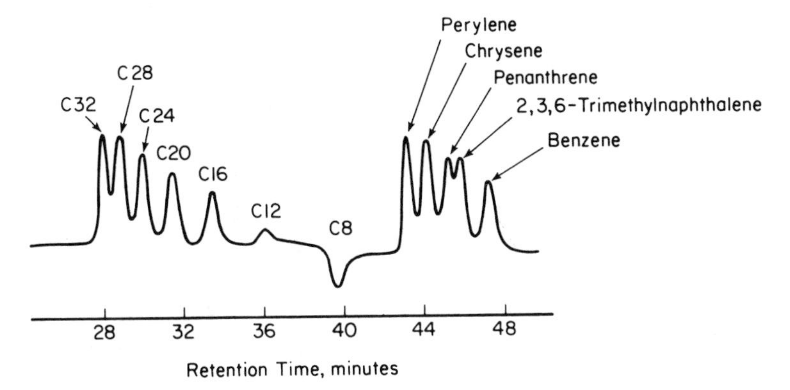

Figure 11.4 HPSEC separation of hydrocarbons.
Column, 150 × 0.8 cm μ-Styragel 100 Å; mobile phase, tetrahydrofuran; flow rate, 1 ml/min; detector, RI. (Reprinted with permission from Ref. 10.)

Figure 11.4, where a series of alkanes of low-molecular-weight elute prior to aromatics of higher molecular weight. This retention pattern is determined by the fact that aliphatic compounds have a significantly larger size per unit molecular weight than do the ring-structured aromatic compounds.

Applications

Recently there has been a rapid increase in the use of SEC for separating small molecules, and more than 50 reports dealing with the characterization of fossil fuels, refined products, by-products, fats and oils, additives in plastics, and various other kinds of materials have been tabulated (11). SEC of small molecules is used to make qualitative comparisons between samples, to isolate fractions for subsequent characterization by other techniques, and to perform quantitative analyses. The uncertainty in quantitative component analysis by HPSEC was determined in one study to be 1%, while the repeatability of retention volumes was 0.3% (9).

Whether to use columns of microparticulate organic gels (e.g., μ-Styragel) or rigid silica microparticles (e.g., Zorbax PSM-60) depends on the individual problem. While such column packings are complementary, with some overlap in separating capabilities, there are strong reasons for having columns of both materials available. Inspection of the calibration curves for small-pore HPSEC packings (Fig. 6.3 and 6.4) shows that gel columns (e.g., μ-Styragel 100 Å and TSK-G2000 H8 or MicroPak BKG-1000H) fractionate the lowest-molecular-weight range. As illustrated in Figure 11.5, size separations

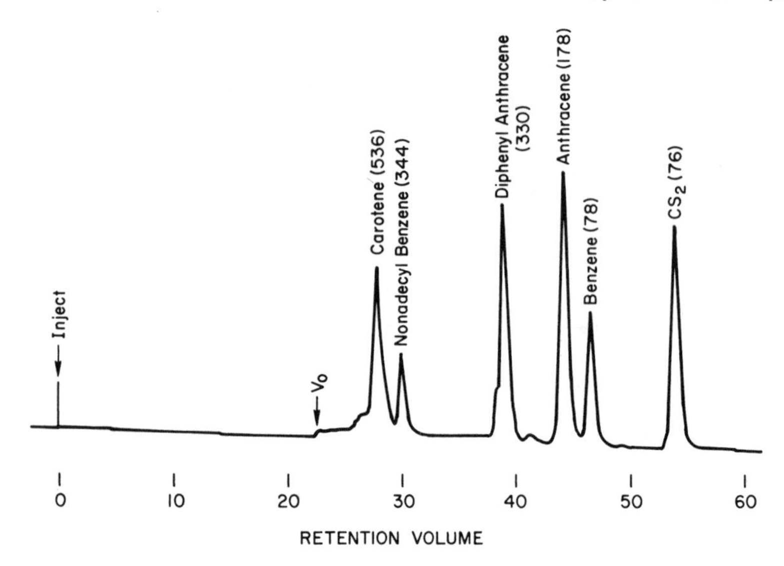

Figure 11.5 High-resolution of small molecules with microparticulate organic-gel column. Column, 150 × 0.8 cm μ-Styragel 100 Å; mobile phase, tetrahydrofuran; flow rate, 1 ml/min; detector, RI. (Reprinted with permission from Ref. 10.)

with μ-Styragel 100 Å are possible from MW < 100 up to about MW 1000. Small-pore silica columns are better suited for samples with a wide molecular weight range (including low-molecular-weight polymers) and for water-soluble solutes.

For separating organic-soluble samples with MW < 1000, microparticulate organic gel columns with the smallest pore sizes (e.g., μ-Styragel 100 Å) are often preferred, since these columns have a lower useful fractionation range than do current porous-silica columns. Porous-gel columns are often useful for the rapid qualitative comparison of various organic-soluble materials (e.g., in Figure 11.6, in defining "good" and "bad" samples of an antioxidant). Microparticulate gel columns are also useful for certain quality-control analyses. For example, in Figure 11.7 a separation is shown that could be used for the quality-control analysis of vitamin A palmitate formulations, by direct injection of a solution without the pretreatment that is often required for samples of this type.

Microparticulate-gel columns are also useful for separating extremely labile substances, particularly those that are sensitive to water, alcohols, Si—OH groups, and so on. Figure 11.8 is a calibration plot for a series of

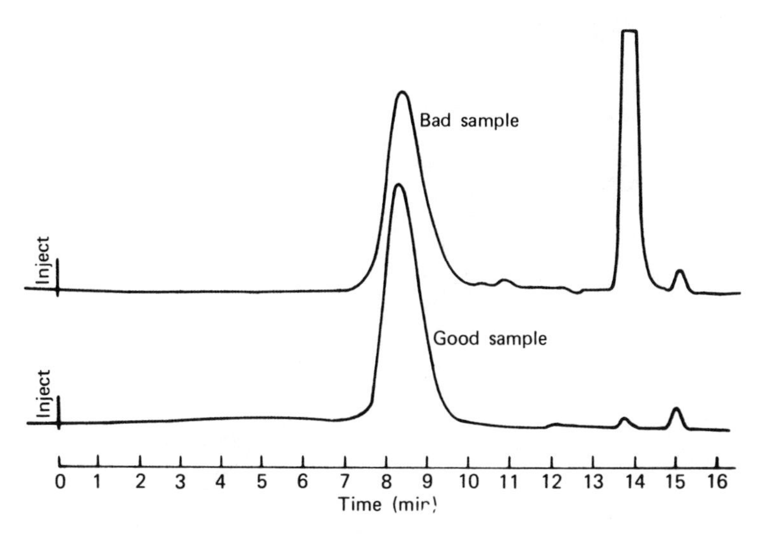

Figure 11.6 HPSEC comparison of antioxidant samples.

Conditions same as in Figure 11.4 except sample, 10 mg/ml nonoxynol; detector, UV, 0.32 AUFS, 254 nm. (Reprinted with permission of Waters Associates.)

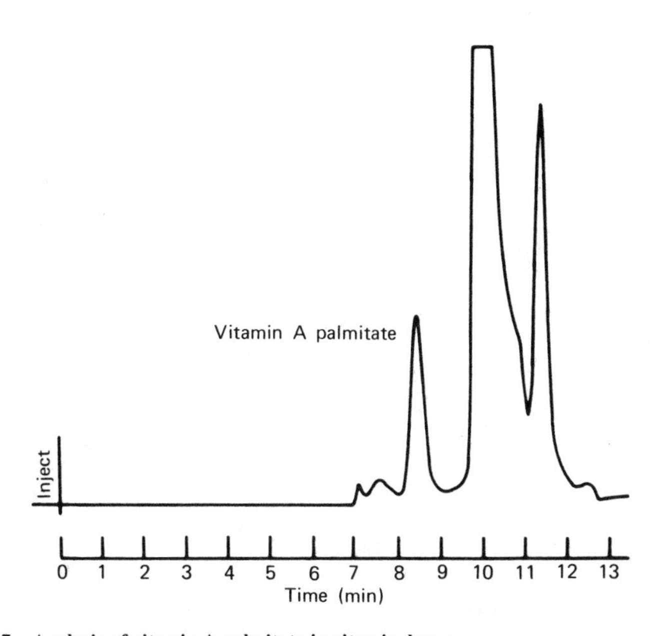

Figure 11.7 Analysis of vitamin A palmitate in vitamin drops.

Same conditions as in Figure 11.4. (Reprinted with permission of Waters Associates.)

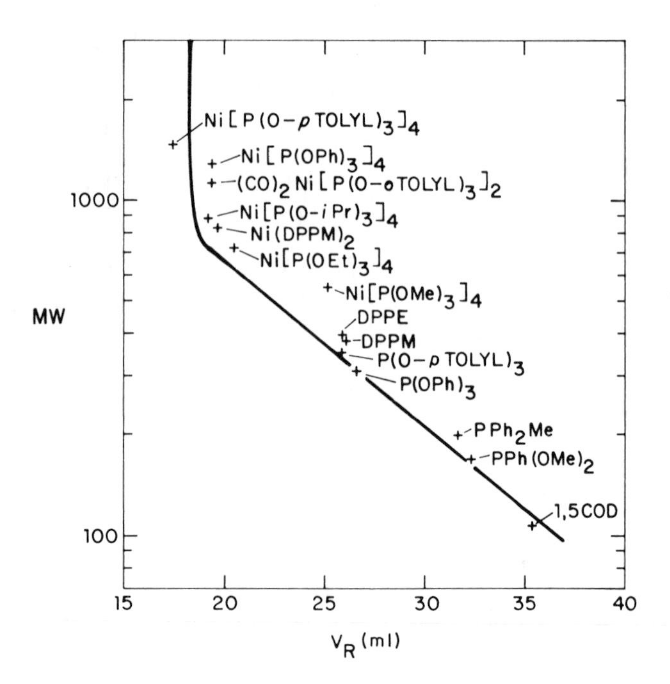

Figure 11.8 Molar volume calibration for nickel complexes.

Column, 120 × 0.8 cm μ-Styragel 100 Å; mobile phase, tetrahydrofuran (dry); flow rate, 3.4 ml/min; detector, UV and RI; sample, 25 μl, ~2% solution in THF. (Reprinted with permission from Ref. 12.)

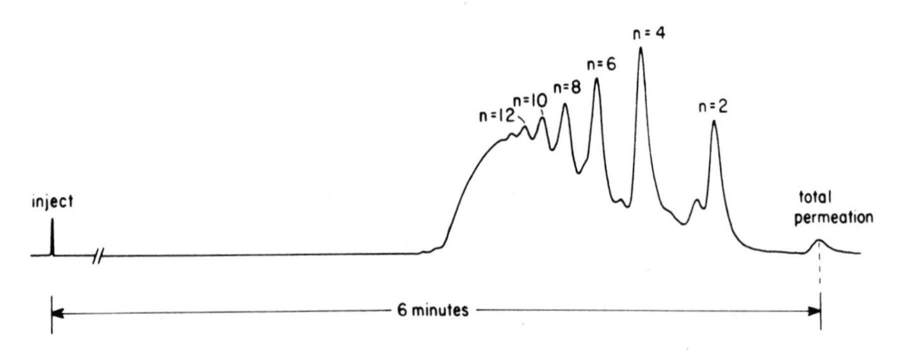

Figure 11.9 Separation of epoxy resin oligomers with porous silica.

Column, 25 × 0.62 cm DuPont-SE 60; mobile phase, tetrahydrofuran; flow rate, 0.9 ml/min; temperature, ambient; detector, UV, 254 nm. (Reprinted with permission from Ref. 13.)

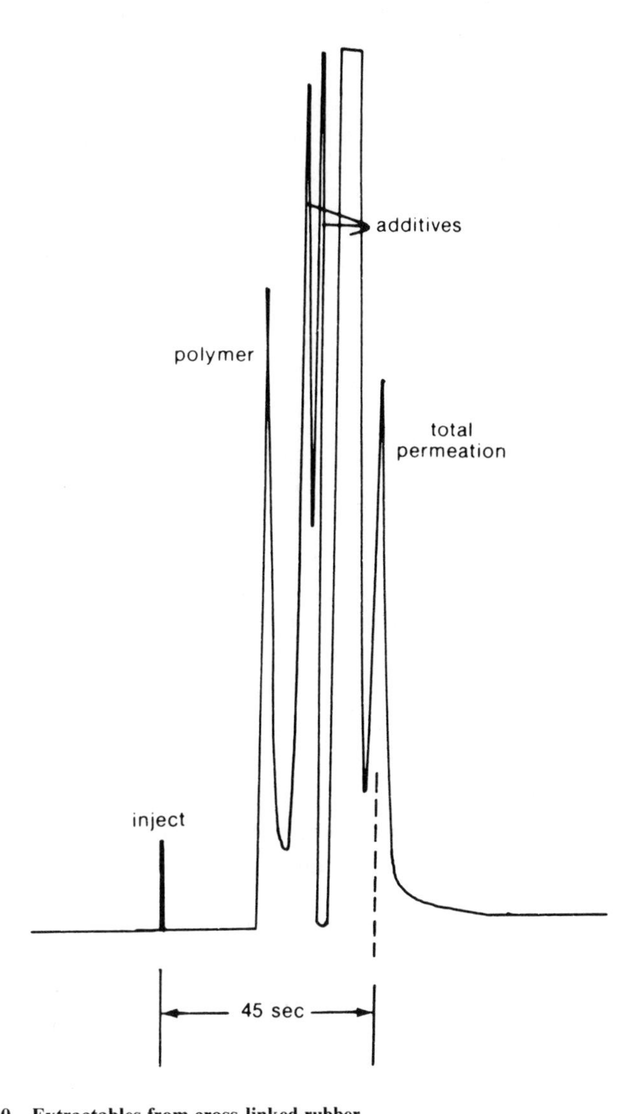

Figure 11.10 Extractables from cross-linked rubber.
Conditions same as for Figure 11.9 except flow rate, 7.2 ml/min. (Reprinted with permission from Ref. 13.)

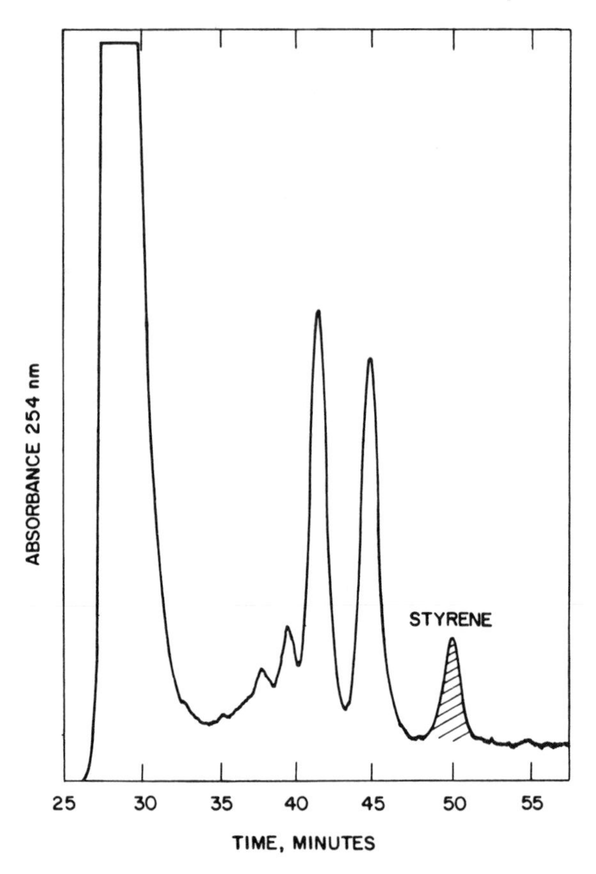

Figure 11.11 Determination of styrene monomer in copolymer.
Column, 2 × 500 Å and 3 × 100 Å μ-Styragel (150 cm total); mobile phase, chloroform; flow rate, 0.8 ml/min; detector, UV, 254 nm; sample, 20 μl, 0.5% copolymer in chloroform. (Reprinted with permission from Ref. 14.)

labile nickel complexes which are unstable and easily hydrolyzed. This separation allowed a series of these and other labile compounds to be characterized according to their SEC retention (12) and was the only separation approach found applicable for this class of compounds.

HPSEC with rigid porous-silica packings is useful for the high-speed separation of small molecules, because of the high flow rates that can be used without collapsing the packing. Figure 11.9 shows a 6-min analysis of an epoxy resin, where the various peaks correspond to each even-numbered

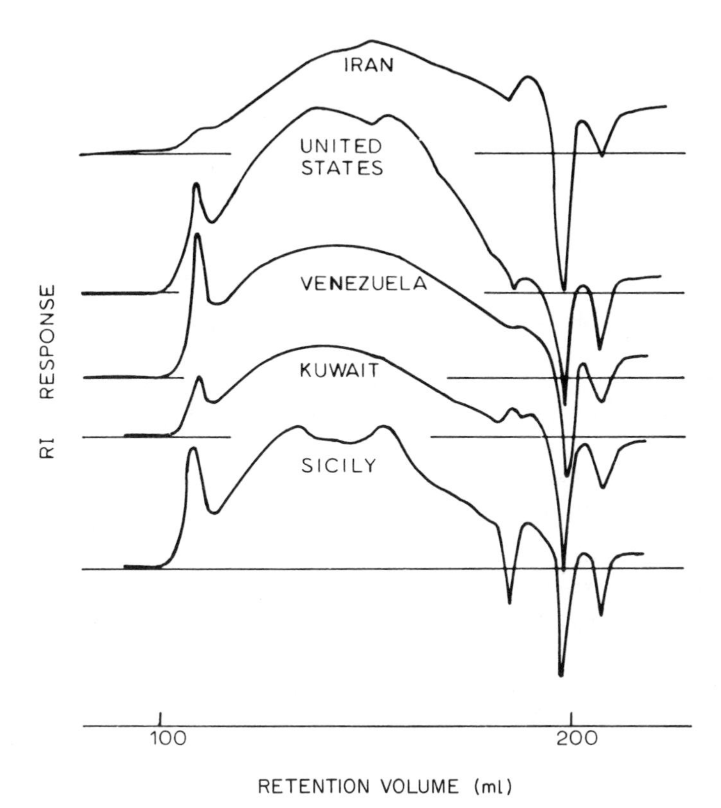

Figure 11.12 Comparison of crude oils by SEC.
Columns, five 120 × 0.8 cm Styragel 500 Å; mobile phase, tetrahydrofuran; flow rate, 1.0 ml/min; temperature, 30°C; detector, RI; sample, 8 mg. (Reprinted with permission from Ref. 15.)

oligomer. In this case, by measuring the magnitude of each peak to determine component concentration and by accounting for the number of reactive sites per molecule, a reactivity index may be calculated to provide the proper curing parameters for thermosetting resins used in hard-molded parts. Additives in polymer systems can also be monitored very rapidly for quality control. Figure 11.10 shows that a separation of additives extracted from

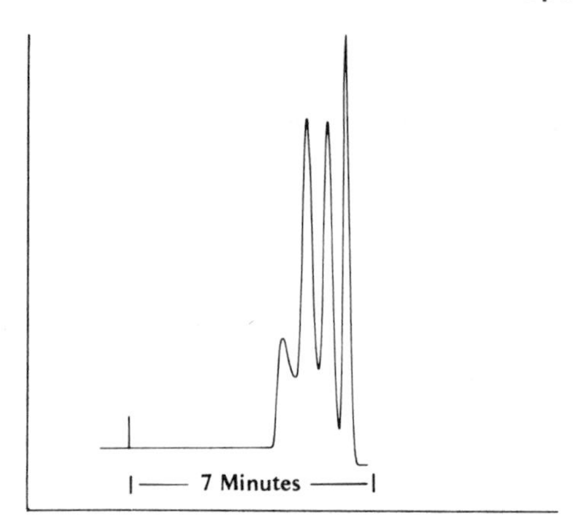

Figure 11.13 Separation of low-molecular-weight poly(propylene glycol) standards. Column, 30 × 0.39 cm, μ-Porasil GPC 60 Å; mobile phase, water/methanol (10/90); flow rate, 0.5 ml/min; detector, RI, 16 X; sample, poly(propylene glycol) standards, 3900 MW, 2000 MW, 800 MW, and 2,3-butanediol. (Reprinted with permission from Ref. 16.)

cross-linked rubber was completed in about 1 min by using the very high flow rates permitted by porous silica columns.

HPSEC is particularly useful for determining minor concentrations of small molecules in synthetic polymeric systems. Figure 11.11 shows the analysis of residual monomeric styrene in a copolymer sample containing mostly 70,000–200,000 MW material and some oligomeric species. Using SEC, the molecular weight ranges of crude oils can also be estimated and different crude oils qualitatively differentiated, as illustrated in Figure 11.12.

HPSEC separations of small molecules in very polar mobile phases can be made with columns of rigid particles (see Fig. 5.19). For example, Figure 11.13 shows the separation of a series of low-molecular-weight polypropylene glycols using a methanol/water mobile phase with an ether-modified silica packing. In such cases, methanol is sometimes required to eliminate reverse-phase retention by the modifying organic phase (17). Figure 11.14 shows the separation of basic compounds on porous silica with an aqueous mobile phase. The major component of this chromatogram was collected for characterization. With this approach only a very minor component of this basic sample was adsorbed to the (acidic) silica packing surface. However, as discussed in Section 7.3, solute retention by processes other than size-exclusion is generally more of a problem when using unmodified porous silica columns than it is with surface-deactivated silica.

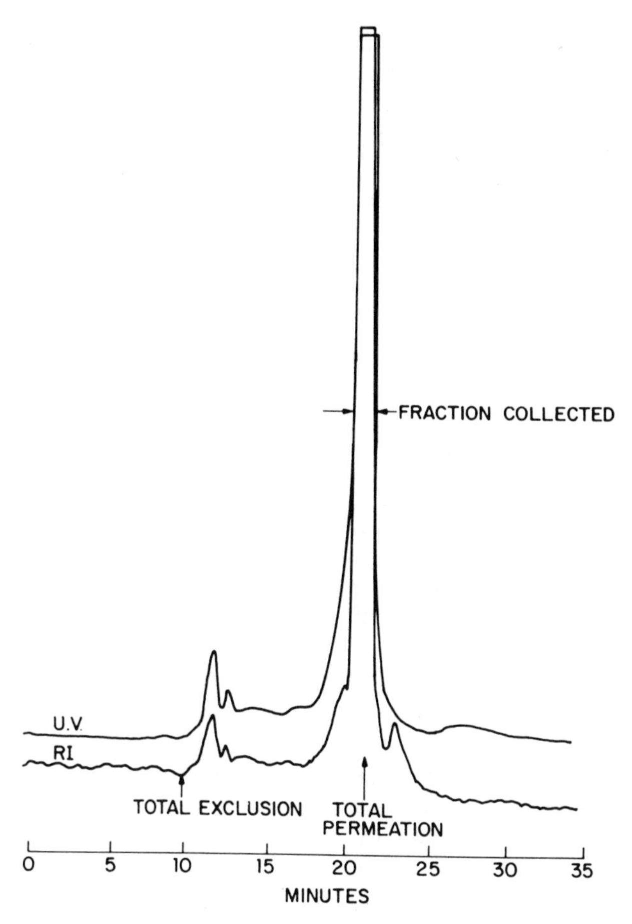

Figure 11.14 Separation of basic compounds with unmodified porous silica columns. Column, 100 × 0.62 cm porous silica microspheres, 50 Å; mobile phase, 3% water in *N,N'*-dimethylformamide; flow rate, 1.0 ml/min; detectors: UV, 0.32 AUFS, 254 nm; RI, 2×10^{-5} RIUFS; sample, 250 μl, 6.8 mg/ml in mobile phase; temperature, 37°C. (Reprinted with permission from Ref. 2.)

11.2 Preparative SEC

Other sections in this book emphasize the data-acquisition aspects of SEC, in which either qualitative or quantitative information is obtained on a sample. This section discusses another aspect of SEC, the preparative technique, which is effective and convenient for isolating relatively large amounts of purified components for molecular weight standards, testing, materials characterization, and so on. While sufficient quantities of materials for identification may often be obtained with analytical systems, the larger quantities of purified samples needed for other studies must normally be prepared with large-diameter columns, and with conditions that are different from those used for analytical SEC. Sample capacity is the goal of preparative HPSEC and variations of technique and equipment from those for SEC analysis must be used.

Experimental

Commonly, large-diameter, low-pressure columns and lower-cost column packings are employed for preparative SEC studies. This section describes some of the specialized aspects of preparative SEC, while more extensive general treatments of preparative LC are given in References 18 and 19.

Columns. In preparative SEC, sample capacity is increased by using columns of larger internal diameter (i.d.). Increasing the diameter of SEC columns does not necessarily reduce chromatographic resolution (Sect. 6.4). In fact, separation efficiencies with large-diameter columns are frequently superior to those obtained with narrow-bore columns of the same column length, provided that the same ratio of sample weight to cross-sectional area is maintained (for nonoverloaded systems). The practical upper limit of column i.d. for SEC has not yet been established. Laboratory columns of 5.7 cm i.d. have been reported (20), and production apparatus of up to 2500 liter capacity have long been commercially available for carrying out preparative separations of biologically important substances by gel filtration chromatography (21). An illustration of the superior performance of large-diameter columns is shown in Figure 11.15, where resolution units per minute based on a separation of ethylene glycol (MW 62, total permeation) versus pentaethylene glycol (MW 238) is shown for columns of 20–44 μm organic gels. In this study, resolution steadily increased with increasing column diameter, as has been noted with columns of other LC packings (23). As with analytical columns (Sect. 6.4), preparative SEC columns prepared in straight sections are connected when higher efficiency units are needed.

Both organic gel and porous silica column packings for preparative SEC are listed in Table 6.3. The choice of packing material is based on the same considerations as for analytical SEC separations. In GFC, preparative columns of semirigid gels are also used, with the specific pressure limitations noted in Section 6.2.

Sample capacity increases proportionally to cross-sectional area, regardless of particle size. In cases that require the highest resolution, preparative SEC separations are carried out with columns of fine (e.g., ≤ 10 μm)

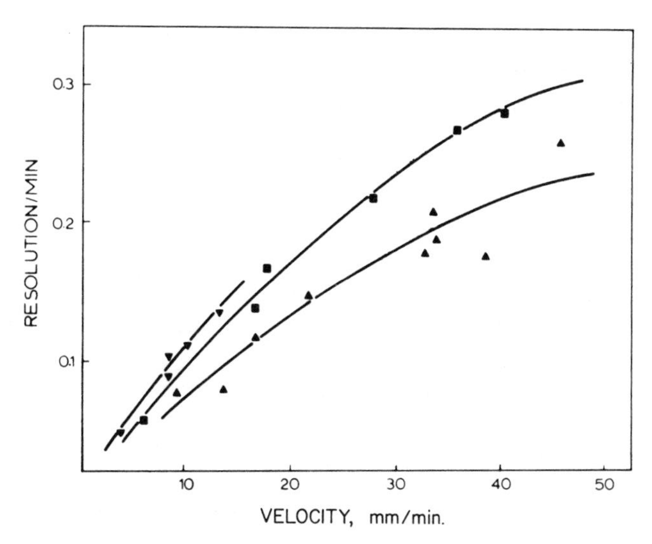

Figure 11.15 Effect of column internal diameter and velocity on SEC resolution. Columns, 16.5 cm long, 20–44 μm Sephadex G-25; mobile phase, water; temperature, ambient; sample, 10 μl ethylene glycol/pentaethylene glycol (1 : 1 : 5) by volume in water; columns, 6 (▲), 10 (■), and 21 (▼) mm. i.d. (Reprinted with permission from Ref. 22.)

column packings. However, because of high cost, large-diameter columns of small-particle column packings have not been widely used, and columns of small particles have been used only to prepare relatively small amounts (e.g., 100–300 mg per run) of purified materials. Most preparative SEC separations are made with longer columns of larger particles (e.g., 30–60 μm). Studies on other LC methods (24) suggest that at high sample loading, long, large-diameter columns filled with coarse, uniform particles should be employed. Higher sample capacity is primarily a result of the larger amount of packing available in larger-diameter or longer columns. It should be noted that columns of larger particles require relatively low mobile-phase velocities for efficient operation; therefore, for the same column volume, significantly longer separation times are required for long columns of large particles relative to shorter columns of smaller particles.

To summarize, SEC columns of small particles should be used for rapidly isolating relatively small, highly purified samples at high resolution; columns of large particles should be used to prepare large amounts of the purified material. The resolution afforded by each approach can be made equivalent (at comparable sample size) by adjusting column length, but separation time is longer for larger particles, for which longer columns are needed for equivalent resolution.

Equipment. Equipment requirements for satisfactory preparative SEC are not as critical as for analysis; lower-cost, less sophisticated systems generally are used. However, to optimize preparative SEC separations it is necessary to use different pumps, sampling systems, and detectors than those that are normally required for analytical work.

Pumping systems should deliver solvent up to 100 ml/min for large-bore (e.g., ≥ 2 cm i.d.) columns. Very high pressure capability is not required in preparative studies, and pressure limits of about 2000 psi (140 bar) are usually adequate. Since analytical information from the preparative chromatogram is of less interest, the precision and accuracy specifications of the pump are not as critical as discussed in Section 7.2. Pneumatic-amplifier and reciprocating pumps provide satisfactory pumping rates and a continuous solvent output. Since the pumping systems of commercial analytical LC instruments often deliver no more than about 10 ml/min, separations with this equipment use only the narrower-bore preparative columns (e.g., ≤ 0.8 cm i.d.). Larger-i.d. columns used with analytical pumping systems require very long separation times because of the low volume flow rates. Pulsations from certain pumps (e.g., reciprocating) which affect detector baselines are usually not a serious disadvantage in preparative applications. To supply the large volume of mobile phase used in preparative SEC, relatively large solvent reservoirs (e.g., >2 liters) are required.

In preparative SEC, solute concentrations are generally high and sensitive detectors used for analysis are not required. High sample concentrations can cause problems, since it may be difficult to determine whether overlapping peaks are due to column overload or to a nonlinear detector response. The RI detector is generally suitable for preparative SEC; UV detectors with a short-path-length cell (e.g., 1 mm) are also useful. Using both the UV and RI detectors in series helps to ensure that all the components of interest are monitored. UV spectrophotometric detectors are often "detuned" from the wavelength of solute absorption maximum to decrease detection sensitivity and reduce the potential for a nonlinear detector response.

Use of low dead-volume tubings and fittings is not as critical in preparative SEC as in analytical applications, because of the relatively large internal volumes of large-diameter preparative columns. High-volume flow rates are needed for wide-diameter columns, and detectors for handling this flow should not be constructed from narrow-bore tubing, which can severely limit the flow of mobile phase and cause excessive back pressure. (Both RI and UV detectors with larger bore tubing are commercially available.) Alternatively, a stream splitter on the exit of the large-diameter column can be used with an analytical detector.

Sample volumes up to about 10 ml are conveniently introduced into preparative SEC columns with a sampling valve and without interrupting the mobile-phase flow. If a syringe is used, the sample is introduced at atmospheric pressure by the stop-flow method. Very large volumes (e.g., 100–200 ml) can be delivered with a syringe, but experimental difficulties may occur (e.g., piece of elastomeric septum material obstructing the column inlet). For very large sample volumes (e.g., >100 ml), the sample can be loaded into the column by means of a low-volume sample-metering pump, using a sampling valve in the stop-flow mode. The pump is attached to the sample loop with the sampling metering pump turned off and the valve in the "inject" position (Fig. 5.12). After the required sample volume has been pumped through the loop into the column inlet, the valve is then rotated to the bypass position and the mobile-phase pump is restarted.

When only a few components are to be isolated, manual collection of the fraction is adequate, particularly when fast, small-particle columns are employed. However, when long, repetitive runs are needed, it is more convenient to use automatic fraction collectors.

Operating Variables. Volumetric flow rates for preparative columns must be increased linearly with cross-sectional area, to maintain the same linear flow velocity as in analytical SEC columns. Table 11.2 shows typical column diameters in both analytical and preparative SEC and the corresponding volumetric flow rates at equivalent linear flow velocities.

Dispersive effects of sample injection in preparative SEC are not well understood. Loading the sample across the entire column cross section is preferred, since this permits more effective use of the total column packing with reduced column overload (Sect. 7.4). When possible, samples should be injected as relatively large volumes of a lower concentration rather than as

Table 11.2 Typical Column Diameters in SEC and Corresponding Flow Rates (ml/min)[a]

Analytical Column, $\frac{3}{8}$ in. O.D.	Preparative Column, 1 in. O.D.	Preparative Column, 2.5 in. O.D.
0.1	0.8	5.5
1.0	7.7	55
5.0	38.2	275
10.0	76.4	549

Taken from Reference 25.

[a] At equivalent mobile phase velocities.

smaller volumes of more concentrated solutions. Improved sample loadability and column performance results from this approach, since the effect of overloading the packing at the column inlet is minimized.

The volume of sample that can be introduced into a preparative column will depend on column internal diameter and length, the solute and solute solubility, the mobile phase/stationary phase combination, and the resolution required. As in analytical SEC (Sect. 7.4), for highest resolution, sample volume should not exceed about one-third the volume of a totally permeating monomer. However, much larger sample volumes are often used in preparative SEC if resolution permits (e.g., >20 ml for a 3.7 cm i.d. column).

The sample loading limit in SEC is dependent on solute MW. The loading capacity of columns can be increased by increasing column length or diameter. However, increasing column length also increases the solute resolution and retention volume; this requires additional separation time and mobile phase.

Figure 7.16 and the discussion in Section 7.4 indicate that plate heights increase as sample volumes are increased to the point where band dispersion occurs largely because of the sample volume alone. Column performance can also be affected by sample weight if the column is overloaded. Experimental results suggest that 0.1–1 mg of a solute per gram of packing can be injected into a column without significant change in either retention volume or plate height (2). However, for preparative studies, much larger sample weights are often employed to obtain the desired weight of isolated component at the required purity. As noted in Section 7.4, the column overload phenomenon in SEC is not well understood. However, in preparative SEC, as well as for the other LC methods (18, 19), sample loads should be increased to the point where there is just adequate resolution, even though such sample loads affect both retention volume and column efficiency. Since under overload conditions solute retention is a function of sample size, it is important to maintain a constant concentration when performing preparative SEC separation of polymer fractions. Only in this manner can the expected molecular weight (as determined by analytical SEC) be obtained for the isolated polymer fractions.

The effect of flow rate on preparative LC column efficiency is the same as for analytical SEC (Sect. 7.2) when columns are operated at small sample loads. However, at high sample loadings (e.g., with the column in overload) the effect of flow rate is less important (23). In sample-overloaded preparative columns, mobile-phase velocity can be substantially increased to reduce separation time, without significant sacrifice in resolution. Thus, preparative SEC separations should be carried out at the highest practical mobile-phase velocity which still allows adequate resolution.

As in analytical SEC, mobile phases of relatively low viscosity are favored in preparative SEC to maintain high column efficiency. The solvent must be compatible with the detector and should be volatile for convenient removal from isolated fractions. Of special importance is that the mobile phase be highly purified, since nonvolatile impurities are concentrated when the solvent is removed from a fraction and significant contamination results. This problem is minimized by using a freshly distilled or "distilled-in-glass" solvent. Higher column temperatures usually enhance solubility if needed, but high-temperature operation is less convenient.

Isolated fractions can be concentrated by evaporation of solvent under a stream of pure, dry nitrogen while warming (e.g., with an infrared lamp). Operations that tend to condense water in the isolate (e.g., heating on a steam bath) should be avoided. Large volumes of mobile phase can often be conveniently removed with a rotary vacuum evaporator. Freeze drying is effective for some solvents, such as water, dioxane, and benzene.

The cut points that are used to collect the fraction largely determine purity and yield. For two overlapping components of equal amounts, Figure 11.16*a* shows that at the equal purity cut point (i.e., valley between peaks, $R_s = 0.6$ in this example), fractions obtained are 88% A and 12% B, and 88% B and 12% A, respectively. If it is desired to improve the purity of both components, the overlapping center portion of the bands can be rejected (crosshatched in 11.16*b*). For the fraction collection in Figure 11.16*b*, composition of the peak on the left would be 98% A, 2% B, while the composition of the peak on the right would be 98% B, 2% A (assuming equivalent detector response). Total yield of purified material obtained by this technique is ~61% of that injected, compared to 100% yield in the previous example.

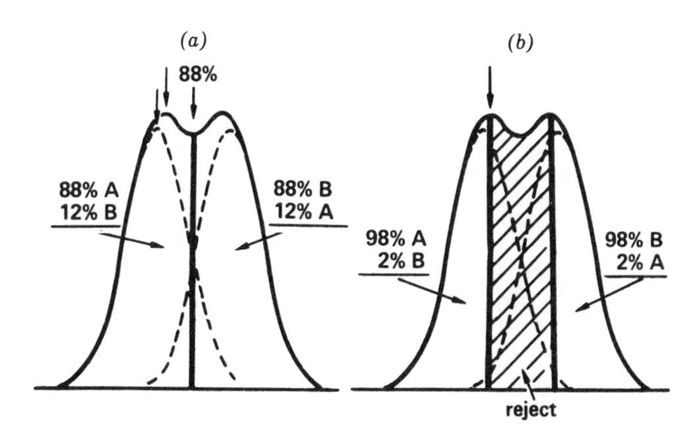

Figure 11.16 Effect on fraction purity by rejecting overlapped peaks.
(Reprinted with permission from Ref. 26.)

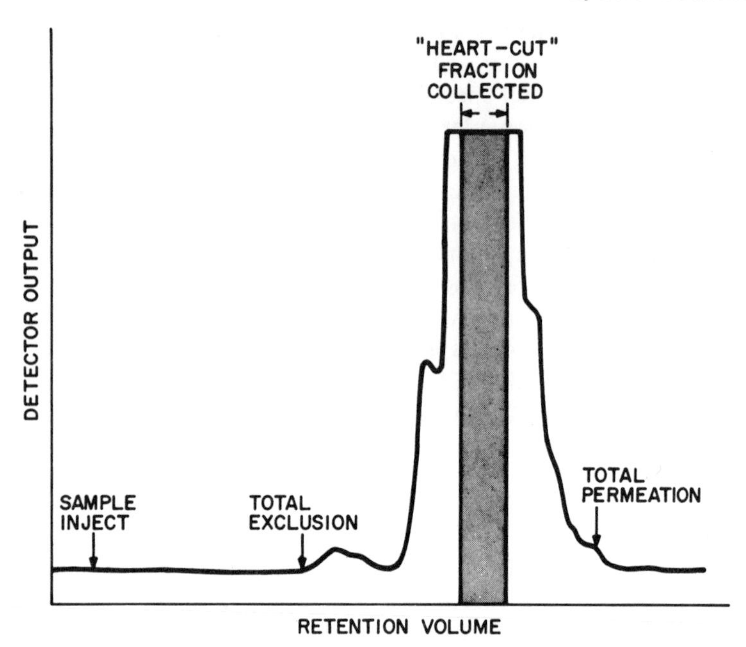

Figure 11.17 Heart-cut technique in preparative SEC.

If required, the rejected overlapping fraction may be rechromatographed to obtain components of the same purity.

For major components it is often desired to use a "heart-cut" technique, which produces a highly purified component with a modest yield loss. This approach is illustrated by the hypothetical separation in Figure 11.17, where a heart cut of a major component overlapped by unwanted contaminants is selected. Here, by rejecting the impure "wings" of the main peak, overall product yield is decreased, but with a significant improvement in product purity. If insufficient amounts of purified material are obtained from any collection approach, it may be necessary to make replicate runs and accumulate the desired fractions. Following the final collection, purity should be analyzed by HPSEC or another appropriate analytical technique. If the isolated component is not of the desired quality, it can be fractionated again for higher purity.

Applications

Occasionally, narrow-MWD polymer standards are prepared by fractionation of a broad-MWD polymer. Preliminary runs on the broad MWD polymer

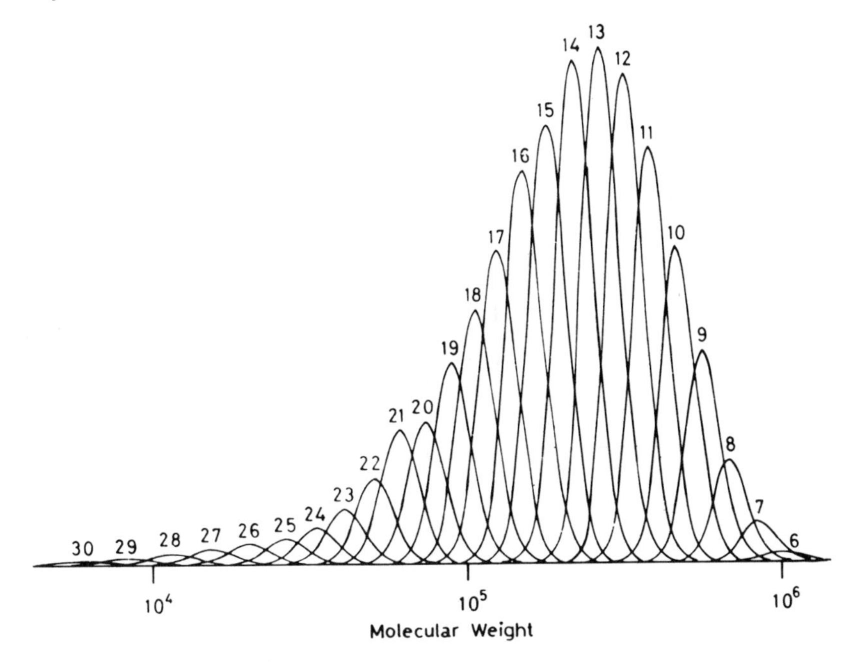

Figure 11.18 MWD curves of fractions from standard polystyrene NBS 706.
Fraction numbers are indicated. Column, 240 × 0.8 cm total TSK-GEL, Type G, 10 μm, 10^6, 10^4, 10^3 Å porosity; mobile phase, methyl ethyl ketone/methyl alcohol (88.7:11.3-theta solvent at 25°C); flow rate, 8.5 ml/min; temperature, 25°C; detector, RI; sample, 20 ml, 6.5 mg/ml of polymer in mobile phase. (Reprinted with permission from Ref. 27.)

determine the cut points that give fractions with the desired MW and MWD. Figure 11.18 shows MWD curves of fractions from NBS 706 polystyrene. The polymer fractions obtained in this preparation showed polydispersities of 1.017–1.035, which represents material suitable for molecular weight calibration by the peak position calibration method (Sect. 9.2; see also Sect. 11.3 for a method used to determine the polydispersity of very narrow MWD fractions).

In addition to preparing narrow MWD standards of polymers, high-resolution preparative SEC is useful to prepare purified low-molecular-weight compounds (e.g., monomers). As illustrated in Figure 11.19, 150 mg of three components (molecular weight indicated on peaks) was injected into a small-particle analytical HPSEC column. Such a sample load is easily fractionated at high resolution with this column. Larger amounts of purified samples can be obtained by increasing sample size and using the heart cut technique.

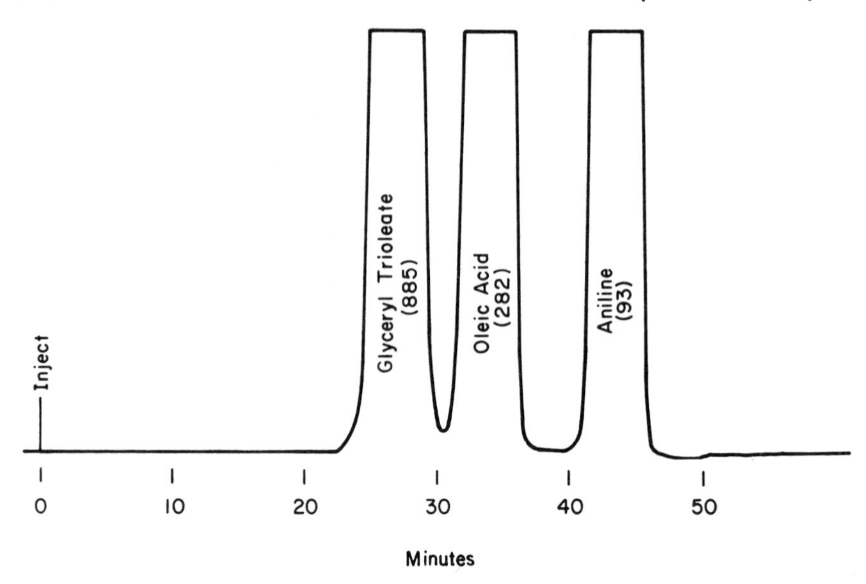

Figure 11.19 Preparative SEC of small molecules.
Column, 150 × 0.8 cm μ-Styragel 100 Å; mobile phase, tetrahydrofuran; flow rate, 1.0 ml/min; detector, RI; sample, 150 mg each compound. (Reprinted with permission from Ref. 10.)

Preparative SEC is often valuable for isolating and identifying trace concentrations of high-molecular-weight additives in a low-molecular-weight matrix, or for measuring a low-molecular-weight additive (e.g., plasticizer) in a polymer. For example, Figure 11.20 shows a chromatogram in which a polymeric additive in lubricating oil is well separated and readily available for collection and subsequent characterization by a suitable auxiliary technique. In favorable cases, parts per million of such additives can be isolated and identified in a single run.

Preparative SEC has been carried out on a wide variety of water-soluble macromolecules (as illustrated by one example in Figure 13.17). Until now, most of these separations have been made with columns of soft gels at low pressures and long separation times. The recent availability of more rigid organic gels and rigid inorganic packings should greatly promote interest in the more rapid, high-efficiency GFC preparative separation of water-soluble solutes.

Finally, SEC can also be carried out on a process scale to prepare commercial quantities of purified materials. For example, certain water-soluble biological compounds (e.g., enzymes) have long been purified by process-

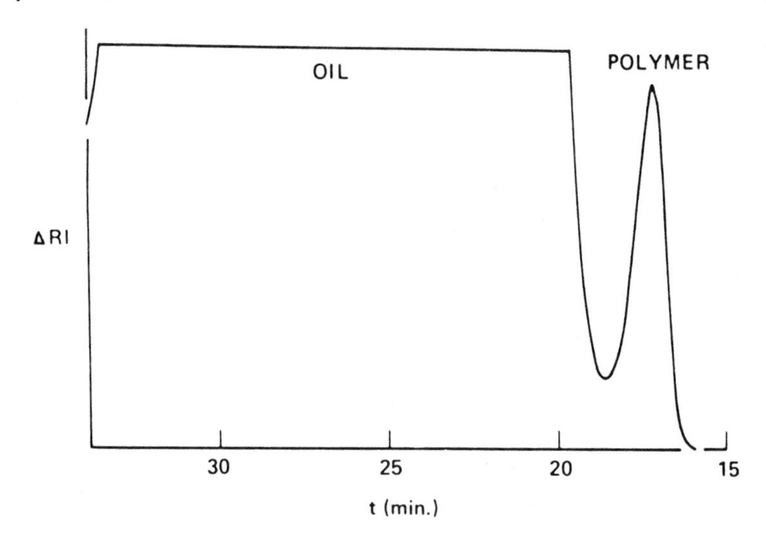

Figure 11.20 Polymer additive in lubricating oil.
Column, 120 × 0.8 cm Styragel 2–10^2 Å, 1–10^3 Å, 1–10^4 Å; mobile phase, tetrahydrofuran; temperature, room; detector, RI. (Reprinted with permission from Ref. 29.)

scale GFC using relatively soft hydrophilic gels (21). The theory and apparatus for continuous preparative chromatography of a binary mixture of polymers by SEC also has been developed, but no commercial application has been reported (28, 30).

11.3 Recycle SEC

Very high resolution is needed occasionally for certain SEC separations, and the required increase in resolution can be obtained by adding many extra columns to the system. Resolution increases linearly with the square root of column length, but for well-packed columns the back pressure increases linearly with length. Therefore, in practice there is a finite restriction on maximum column length as a result of pressure limitations. In addition, well-packed columns are relatively expensive, and the inventory of columns needed for extending the column length may not be immediately available. One solution to these problems is to recycle the sample through the same column set one or more times to increase the effective column length. Increased resolution by recycle is obtained just as if extra column lengths were added, but the attendant increase in pressure is not experienced. There are several advantages and pitfalls of recycle:

Advantages:

1. Additional resolution is obtained without the need for additional columns.
2. With some arrangements the sample profile is recorded by the detector after each pass through the column.
3. The method can be made semiautomatic.

Disadvantages or pitfalls:

1. Some commercial equipment will permit recycle only after substantial modifications.
2. Extracolumn effects must be more carefully minimized (particularly if recycling is carried out through the pump).
3. Complex or broad-MWD samples permit very few cycles before the front edge of the retention curve in one cycle overtakes the trailing edge of the retention curve from the previous cycle.

The recycle method can be used for many applications to: [1] Increase the accuracy of molecular weights calculated from the chromatogram; [2] determine true $\overline{M}_w/\overline{M}_n$ values of very narrow MWD standards; [3] purify fractions of materials for other studies; [4] increase column resolution to bring out the fine features of a sample (e.g., to distinguish individual oligomers).

Theory

Chromatographic peak separation [i.e., the distance between peak retention volumes, $(V_{R2} - V_{R1})$ or ΔV_R] in SEC is linearly proportional to the column length L. In addition, $D_2 = \Delta \ln M/\Delta V_R$, which is proportional to the slope of the molecular weight calibration curve. For a change in column length, $D_2 \propto 1/L$. In recycle, for n passes through the column,

$$D_{2,n} = \frac{D_2}{n} \tag{11.2}$$

However, peak or band spreading as measured by the peak width σ varies with $\sqrt{L}$ so that for n passes through the column, σ varies as $\sqrt{n}$.

SEC column resolution, R_s, makes use of the parameters for peak separation (ΔV_R) and broadening (σ) and is expressed as (Sect. 4.2)

$$R_s = \frac{\Delta V_R}{4\sigma} = \frac{\Delta \ln M}{4 D_2 \sigma} \tag{11.3}$$

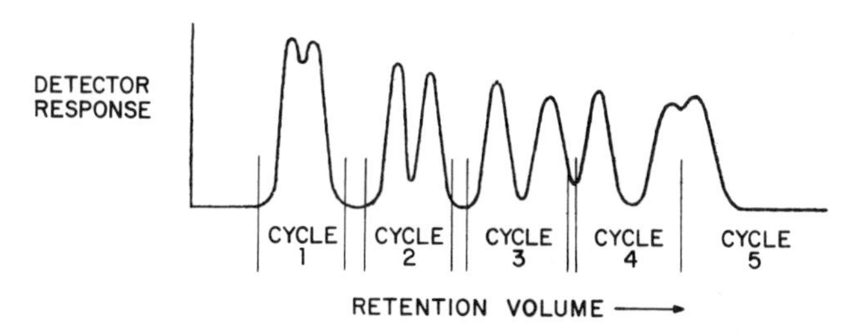

Figure 11.21 Separation of tyranine and dopamine using recycle SEC.
Biogel P-2 columns; UV detector, 254 nm. (Reprinted with permission from Ref. 31.)

For n passes through the column (recycle), the column resolution becomes

$$R_{s,n} = \frac{n\,\Delta V_R}{\sqrt{n}\,4\sigma} = \sqrt{n}\,R_s \tag{11.4}$$

If a more rigorous expression of resolution is needed, the σ terms must be broken down into the components. The values of σ due to injection are not repeated on additional passes through the column and values of σ due to pump mixing do not exist in the first pass but are introduced in subsequent cycles.

One main limitation to the recycle method is that the fastest-moving peak eventually overtakes the slowest and remixing occurs. An example of this is shown in Figure 11.21, where optimum separation occurs in the third cycle and remixing is evident in the fourth cycle. For separating two components the optimum number of cycles, n_{opt}, is given by (31)

$$n_{opt} = \frac{V_{R1}}{2(V_{R2} - V_{R1})} \tag{11.5}$$

where V_{R2} is the slower-moving peak and n_{opt} is a roundoff integer. For example, if the relative peak distance is 0.25, then $n_{opt} = 2$, while for a relative peak distance of 0.05 (5% separation), $n_{opt} = 10$. Optimum resolution for n passes through the column is calculated by (31)

$$R_{s,opt} = \frac{1}{5.7} N^{1/2} \left(\frac{V_{R2} - V_{R1}}{V_{R1}} \right)^{1/2} \tag{11.6}$$

where in this case the values of V_R and N are those obtained at the n_{opt} pass through the column. As described below, for multipeaked samples a "draw-off" procedure can be used to eliminate unwanted materials, permitting increased separation of other components by additional cycles.

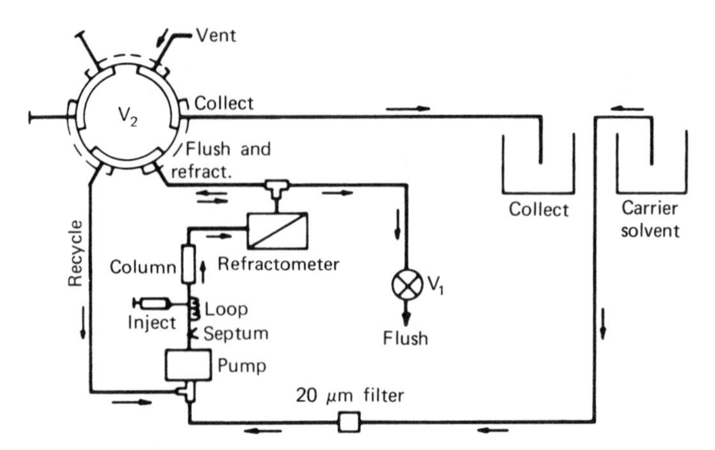

Figure 11.22 Schematic diagram of the closed-loop method of recycle operation. (Reprinted with permission from Ref. 32.)

Equipment

There are two approaches for carrying out recycle SEC. The first and simplest is the closed-loop method shown in Figure 11.22, for which the solvent-flow options are indicated by the arrows. The sample is passed through the column and detector and back through the pump in a closed loop for the required number of times. Each pass is monitored by the detector, and a switching valve (V_2) then permits the operator to collect or discard peaks as they emerge. Although very simple in concept, this approach is sometimes difficult in practice. The position of several valves must be carefully coordinated, and the detector must be capable of withstanding the high operating pressure of the system without leaking (many commercial detectors have not been so constructed). Additionally, peak dispersion due to the connector tubing and pump chamber must be carefully minimized or the advantages of recycle will not be obtained.

The other recycle method employs alternate pumping of dual columns, which has the advantage that the sample does not pass repeatedly through the pump chamber, thus minimizing peak broadening. An understanding of the details of this method is aided by reference to Figure 11.23. The sample containing the peaks to be resolved is introduced to the system via the injection valve (six-port). As the sample passes through column 1, it is monitored by cell 1 of a dual-cell UV photometer before it passes into column 2. In the first valve position, cell 1 is at high pressure while cell 2 is at ambient pressure. While the sample is in column 2, the valve is switched to divert the flow back

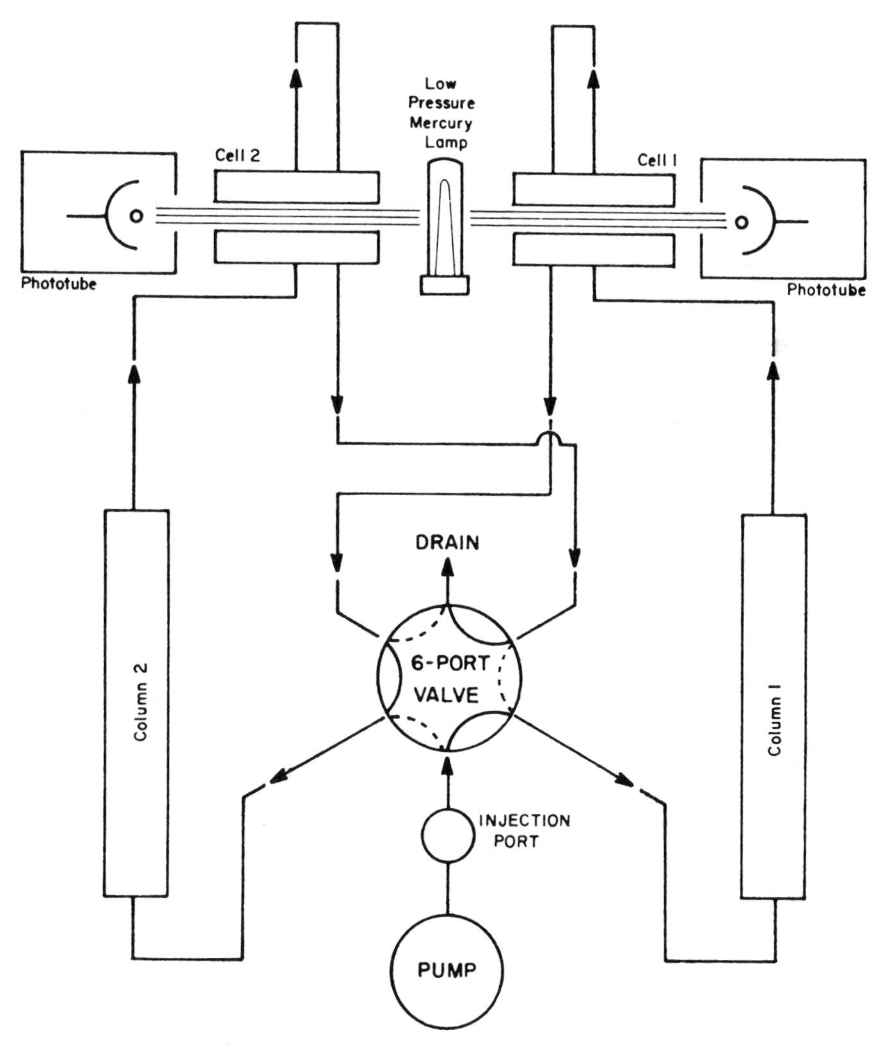

Figure 11.23 Schematic diagram for dual column, alternate pumping recycle method. (Reprinted with permission from Ref. 33.)

into column 1. In this valve position, cell 2 becomes the high-pressure cell. As the sample emerges from column 2, the output from cell 2 provides a record to indicate whether adequate resolution has been obtained. A sample can be cycled through such a column system until the peaks broaden and completely occupy one column volume, or until adequate resolution has been obtained. Actually, it is not essential to monitor the peaks as they emerge from each column, if matched-performance columns of essentially the same elution volume are used. In this case the switching cycle of one column can be calculated based on data obtained with the other matched column.

Uses of the Recycle Method

Improvement of Molecular Weight Accuracy. Recycle can be used to decrease the errors in calculated molecular weight averages. The errors in $\overline{M}_n$ and $\overline{M}_w$ are given by Equations 4.18 and 4.19. Since for n passes through the column, the resolution increases by $\sqrt{n}$ (Eq. 11.4), the molecular weight *error* becomes

$$\overline{M}^*_{w,n} \text{ or } \overline{M}^*_{n,n} = e^{\pm(1/2n)(\sigma D_2)^2} - 1 \tag{11.7}$$

This relationship shows that for low values of σD_2 (typical values for HPSEC range from 0.2 to 0.5), the molecular weight error decreases nearly linearly with n. For example, $\overline{M}^*_w$ and $\overline{M}^*_n$ for $\sigma D_2 = 0.4$ are 8%, 4%, 3%, and 2% for $n = 1, 2, 3$, and 4, respectively. Thus a molecular weight error based on a single pass is halved by a second pass through the column. However, the use of recycle to improve molecular weight accuracy assumes that no additional extracolumn peak broadening occurs.

When using the recycle method to improve accuracy, either the peak position method or GPCV2 or GPCV3 described in Section 9.3 must be used for molecular weight calibration. Also, the values of σ and D_2 determined for the standards must be for the same number of passes n as for the unknown samples. Increasing column length can also improve molecular weight accuracy, but this approach requires higher pressures and additional column inventory. On balance, it is usually more convenient to add column length than to use recycle if well-packed columns are available and the pressure is not excessive.

Determination of the MWD of a Very Narrow MWD Material. Recycle chromatography is probably the only method available for accurately determining the molecular weight distribution of a very narrow MWD material. The approach takes advantage of the increased resolution of the sample peak in each pass through the column and permits extrapolation of the band broadening to zero so that the final peak width is due to MWD alone. As discussed in Section 3.1, the variance of the total chromatographic curve width is given by the sum of variances of each of the contributors, i.e.,

$$\sigma_T^2 = \sigma_{\text{inj}}^2 + \sigma_{\text{disp}}^2 + \sigma_{\text{MWD}}^2 + \sigma_{\text{ex. col.}}^2 \tag{11.8}$$

where σ_T^2 is the total peak dispersion, σ_{inj}^2 is that dispersion due to sample injection, σ_{disp}^2 is the chromatographic band dispersion, σ_{MWD}^2 is the spreading due to the natural MWD of the sample, and $\sigma_{\text{ex. col.}}^2$ consists of the spreading caused by all extracolumn sources.

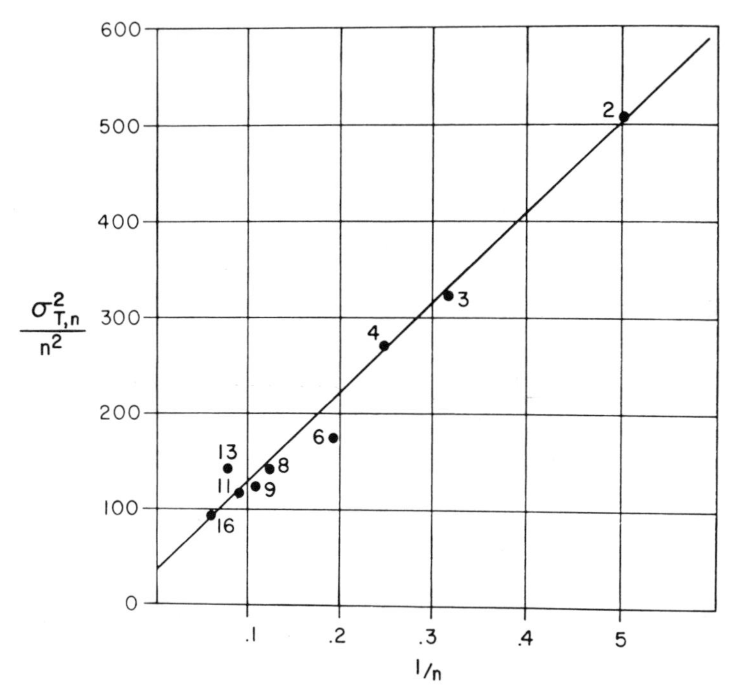

Figure 11.24 Relationship between number of cycles and total curve broadening for a narrow-MWD polystyrene sample in the recycle SEC mode.
(Reprinted with permission from Ref. 34.)

In optimum recycle SEC experiments, σ^2_{inj} becomes insignificant and $\sigma^2_{\text{ex. col.}}$ is also minimized, so that only the chromatographic and molecular weight dispersions are important. Since σ_{disp} is proportional to $\sqrt{n}$ and σ_{MWD} is proportional to n,

$$\sigma^2_{T,n} \simeq n\sigma^2_{\text{disp}} + n^2\sigma^2_{\text{MWD}} \tag{11.9}$$

or

$$\frac{\sigma^2_{T,n}}{n^2} \simeq \frac{\sigma^2_{\text{disp}}}{n} + \sigma^2_{\text{MWD}} \tag{11.10}$$

A plot of $\sigma^2_{T,n}/n^2$ versus $1/n$ yields a straight line with the intercept σ^2_{MWD}. (This corresponds to an extrapolation to an infinite number of cycles, i.e., infinite resolution.) To obtain values of MW and MWD from the σ^2_{MWD} data, the MW calibration curve for the columns is required and the shape of the molecular weight distribution must be assumed. Figure 11.24 presents the data for a very narrow MWD in-house-fractionated polystyrene. Using a peak position calibration curve and assuming a Gaussian distribution

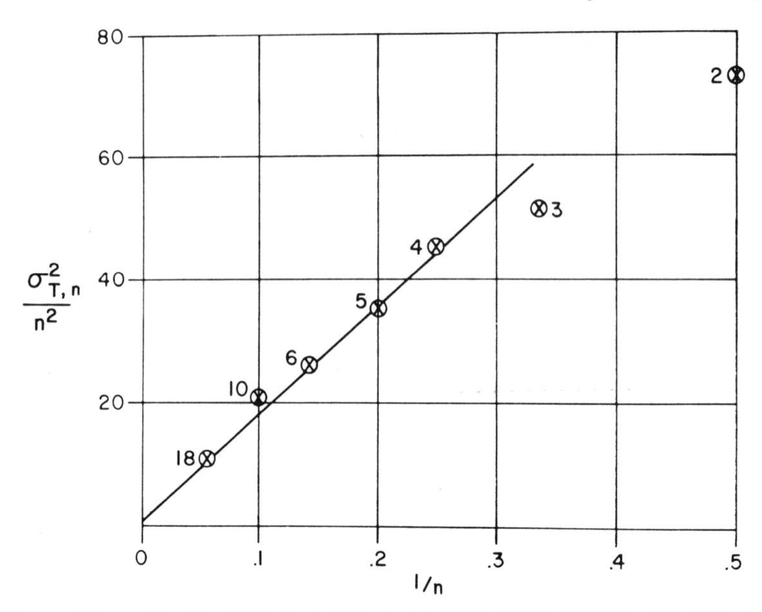

Figure 11.25 Relationship between number of cycles and total curve broadening for hexane in the recycle SEC mode.

(Reprinted with permission from Ref. 34.)

of molecular weights for the sample, a polydispersity value of 1.00248 was obtained (34). Further verification of the method is provided by Figure 11.25, which shows the data obtained for hexane. Here the sample is monodispersed and the extrapolated $\sigma^2_{T,n}/n^2$ value approaches the expected value of zero.

Preparative SEC Separations by Recycle. Recycle SEC can be used to obtain the additional resolution needed to separate materials of nearly the same size in preparative SEC (Sect. 11.2). Using the equipment shown in Figure 11.22, Figure 11.26 illustrates the recycle and draw-off method for isolating various components in Triton X-45, a complex surfactant based on alkylaryl polyether alcohols, sulfonates, and sulfates. In this case there were six cycles with the low molecular weight end resolved first. The increase in resolution with cycle number is apparent, and the characterization of the components probably could be accomplished after cycle three or four since adequate resolution was obtained. To collect purified fractions, peak 1 could be drawn off in cycle three; peaks 2 and 3 must be drawn off at cycle five since they overtake the highest-molecular-weight peak in the next cycle. Collection and further recycling of peak 3 for five times (Fig. 11.26) indicates that 92% of peak 3 and 8% of peak 4 were obtained in the original peak 3 cut.

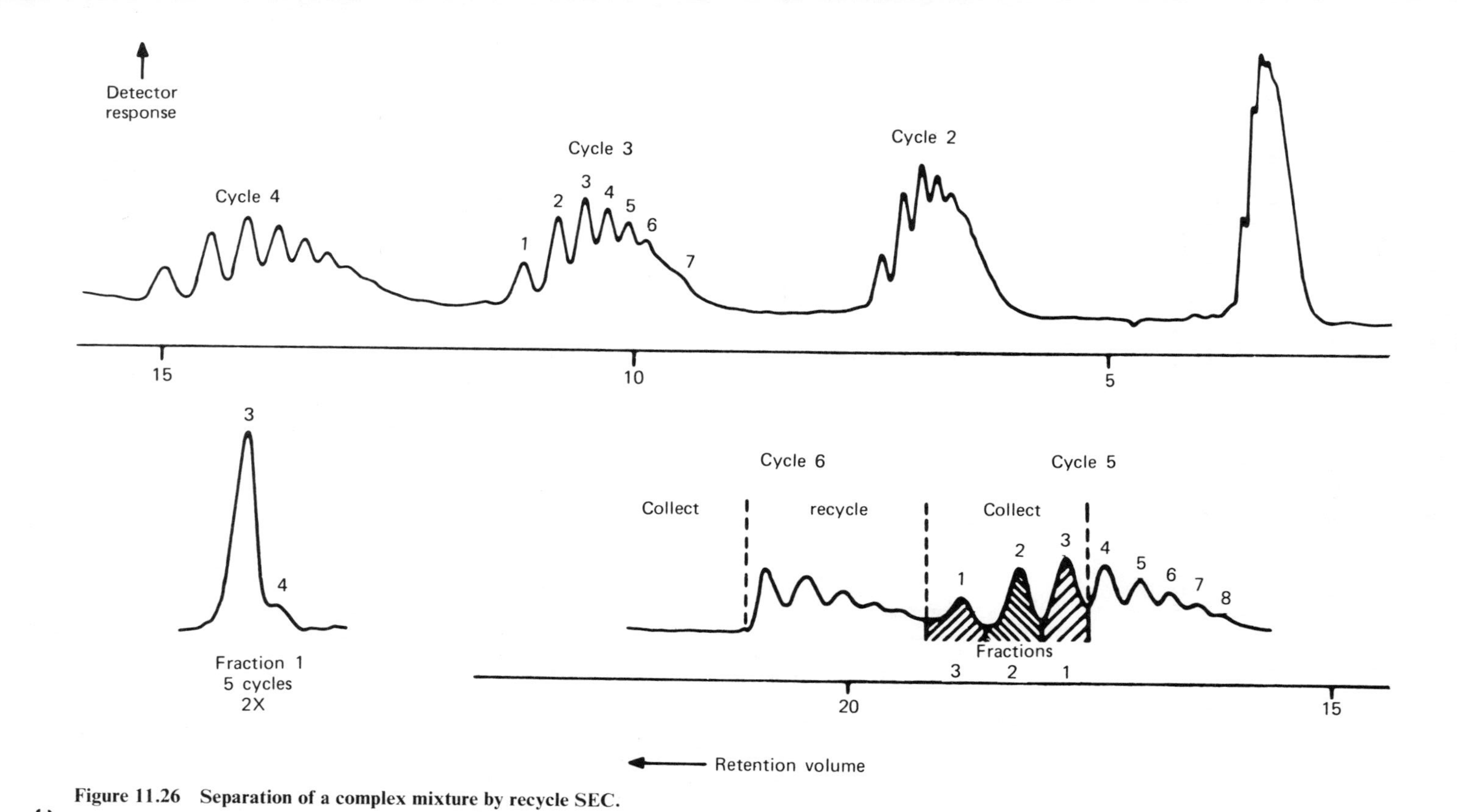

Figure 11.26 Separation of a complex mixture by recycle SEC.
Columns, 450 × 0.8 cm Styragel 60 Å; mobile phase, tetrahydrofuran, 25°C; flow rate, 0.48 ml/min; sample, 50% Triton X-45, 30 μl; detector, differential refractometer. (Reprinted with permission from Ref. 32.)

11.4 Vacancy and Differential SEC

In vacancy size-exclusion chromatography, a sample of pure mobile phase is injected into columns that have been equilibrated with a dilute mobile-phase solution of the solute to be analyzed. Under ideal conditions a chromatogram is obtained that is nearly the exact mirror image of a conventional SEC chromatogram. Figure 11.27 illustrates that the conventional and

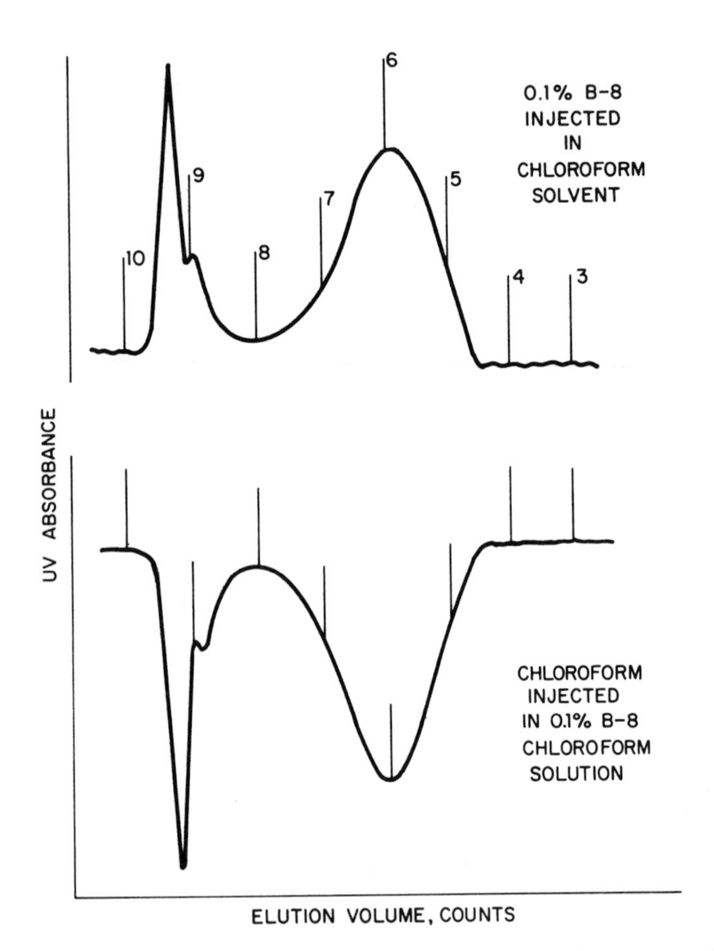

Figure 11.27 Comparison of conventional and vacancy size-exclusion chromatograms for polystyrene.

Upper curve, conventional chromatogram: column, 120 × 0.8 cm Styragel 10^4 Å; mobile phase, chloroform, 25°C; flow rate, 1 ml/min; sample, polystyrene Dow B-8, 0.1% (amount injected unspecified); detector, UV. Lower curve, vacancy chromatogram: same as upper curve except mobile phase, 0.1% Dow B-8 polystyrene in chloroform; sample, pure chloroform. (Reprinted with permission from Ref. 35.)

vacancy chromatograms for a polystyrene polymer are very similar but not identical mirror images, as are the calibration curves obtained by the two methods (35). The small difference between vacancy and conventional SEC curves can be attributed to kinetic effects, as illustrated in the calibration curves of Figure 11.28 for a swellable gel packing. In conventional SEC the diffusion rates for the larger molecules are too slow for equilibrium to be reached at higher flow rate velocities. This effect is minimized in the vacancy mode, where the actual diffusing species are the rapidly diffusing

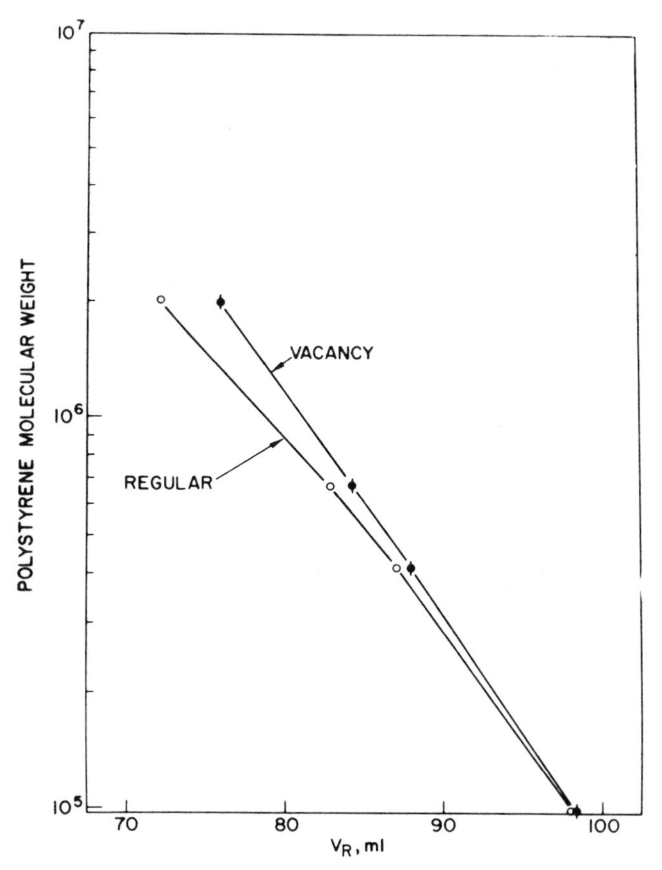

Figure 11.28 Comparison of calibration data for conventional and vacancy SEC on swellable column packing.

Conventional: columns, set of four 30 × 0.8 cm each of 10^6, 10^5, 10^4, and 10^3 Å Styragel, respectively; mobile phase, chloroform, 25°C; flow rate, 1.5 ml/min; samples, narrow-MWD polystyrenes (Pressure Chemical Co.), total solids 0.1%, 2 ml; detector, differential refractive index. Vacancy: same as conventional except mobile phase, total solids 0.01% in chloroform; sample, pure chloroform. (Reprinted with permission from Ref. 36.)

solvent molecules. The curve differences in Figure 11.28 could also be due to sample concentration effects. The much smaller particle sizes used in modern HPSEC packings would be expected to minimize the solute diffusion effect.

In the differential SEC technique, a sample solution is injected into columns equilibrated with a dilute mobile-phase reference solution. Any small difference between the sample solution and reference mobile-phase solution is detected in the differential chromatogram. Advantages of using the vacancy or differential methods occur [1] in process control, where the control or reference material sample is used in the mobile phase and the test sample is then injected, and [2] in problem systems, where the column packing surface can be deactivated by the solute-containing mobile phase. Unfortunately, very little information on the application of either of these methods has been reported (35, 36).

11.5 High-Speed Process SEC

Other discussions in this book show that it is now possible to obtain the MWD of many polymers in 10–15 min, often faster than many polymers can be dissolved. The key to these very fast separations is the small size of the column packing particles, which allow very fast permeation, and the appropriate instrumentation, which maintains narrow band dispersion and accurately monitors narrow peaks.

Under favorable circumstances very rapid separations such as those shown in Figure 11.29 are possible. To obtain accurate molecular weight measurements in such separations, the components of interest must not be degraded by the shear force of the relatively high mobile-phase velocities used (Sect. 7.2). While this can be a practical problem with very high molecular weight materials, it is not a problem with small molecules. To utilize the high mobile-phase velocities shown in Figure 11.29, it is generally necessary to use the mechanically strong, rigid inorganic supports, since organic gel packings often compress at high mobile-phase velocities.

Even though in many cases the speed of the analysis of a high-molecular-weight polymer is limited by the time required for the solute to dissolve, there are other instances in which solubilized polymer streams or process streams of small molecules could be analyzed on-line by HPSEC. It is expected that molecular weight determinations or other analytical measurements could be made in 1–2 min (e.g., as in Figs. 11.29 and 6.16*c*) automatically with good analytical accuracy. This type of on-line SEC analysis has yet to be exploited, but the basic aspects of this approach now are well recognized. It can be anticipated that on-line process measurements by HPSEC will soon become an important analytical tool.

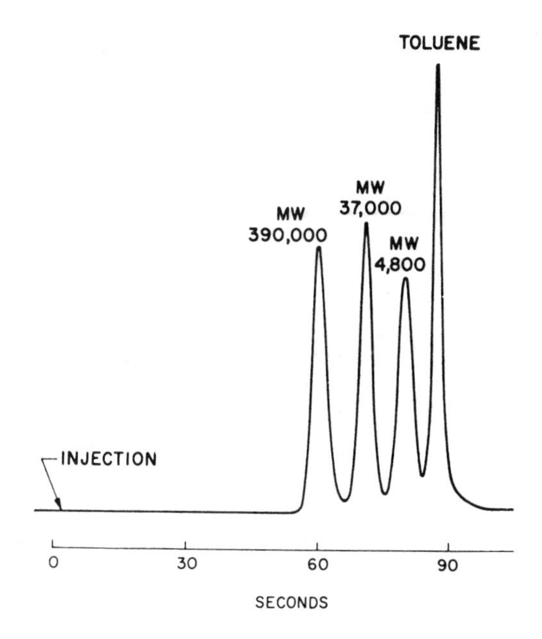

Figure 11.29 Rapid GPC: a 90 sec separation of polystyrene on PSM columns.
PSM columns, 40 × 0.79 cm, 10 cm each of 50S, 800S, 1500S, and 4000S [see Ref. 38]; mobile phase, tetrahydrofuran; flow rate, 6.7 ml/min; temperature, 22°C; sample, 25 μl, 2.5 mg/ml polymer, 1.2 mg/ml toluene. (Reprinted with permission from Ref. 37.)

REFERENCES

1. J. C. Giddings, *Anal. Chem.*, **39**, 1027 (1967).
2. J. J. Kirkland and P. E. Antle, *J. Chromatogr. Sci.*, **15**, 137 (1977).
3. P. E. Antle, E. I. du Pont de Nemours & Co., private communication, 1976.
4. W. B. Smith and A. Kollmansberger, *J. Phys. Chem.*, **69**, 4157 (1965).
5. A. Lambert, *J. Appl. Chem.*, **20**, 305 (1970).
6. A. Lambert, *Anal. Chim. Acta*, **53**, 63 (1971).
7. J. G. Hendrickson and J. C. Moore, *J. Polym. Sci., Part A-1*, **4**, 167 (1966).
8. J. G. Hendrickson, *Anal. Chem.*, **40**, 49 (1968).
9. A. Krishen and R. G. Tucker, *Anal. Chem.*, **49**, 898 (1977).
10. A. P. Graffeo, Association of Official Analytical Chemists' Meeting, Washington, D.C., Oct. 19, 1977.
11. V. F. Gaylor and H. L. James, *Anal. Chem.*, **48**, 44R (1976).
12. C. A. Tolman and P. E. Antle, *J. Organomet. Chem.*, **159**, C5 (1978).
13. DuPont Instrument Products Bulletin E-14063, 1977.

14. J. M. Pacco, A. K. Mukherji, and D. L. Evans, *Sep. Sci. Technol.*, **13**, 277 (1978).
15. E. W. Albaugh and P. C. Talarico, *J. Chromatogr.*, **74**, 233 (1972).
16. Waters Associates, Bulletin 833199, D76, Sept. 1977.
17. F. A. Buytenhuys and F. P. B. van der Maeden, *J. Chromatogr.*, **149**, 489 (1978).
18. L. R. Snyder and J. J. Kirkland, *Introduction to Modern Liquid Chromatography*, 2nd ed., Wiley-Interscience, New York, 1979, Chapt. 15.
19. J. J. DeStefano and J. J. Kirkland, *Anal. Chem.*, **47**, 1103A (1975); **47**, 1193A (1975).
20. A. R. Cooper, A. J. Hughes, and J. F. Johnson, *J. Appl. Polym. Sci.*, **19**, 435 (1975).
21. J. Curling in *Chromatography of Synthetic and Biological Polymers, Vol. 2*, R. Epton, ed., Ellis Horwood Ltd., Chichester, England, 1978, Chapt. 6.
22. T. A. Maldacker and L. B. Rogers, *Sep. Sci.*, **6**, 747 (1971).
23. J. J. DeStefano and H. C. Beachell, *J. Chromatogr. Sci.*, **10**, 654 (1972).
24. A. Wehrli, *Z. Anal. Chim.*, **277**, 289 (1975).
25. A. R. Cooper, A. J. Hughes, and J. F. Johnson, *Chromatographia*, **8**, 136 (1975).
26. L. R. Snyder and J. J. Kirkland, *Introduction to Modern Liquid Chromatography*, Wiley-Interscience, New York, 1974, Chapt. 3.
27. Y. Kato, T. Kametani, K. Furukawa, and T. Hashimoto, *J. Polym. Sci.*, *Part A*-2, **13**, 1695 (1975).
28. P. E. Barker, F. J. Ellison, and B. W. Hatt, in *Chromatography of Synthetic and Biological Polymers, Vol. 1*, R. Epton, ed., Ellis Horwood Ltd., Chichester, England, 1978, Chapt. 13.
29. L. R. Snyder and J. J. Kirkland, *Introduction to Modern Liquid Chromatography*, Wiley-Interscience, New York, 1974, Chapt. 10.
30. P. E. Barker, B. W. Hatt, and A. N. Williams, *Chromatographia*, **10**, 377 (1977).
31. H. Kalasz, J. Nagy, and J. Knoll, *J. Chromatogr.*, **107**, 35 (1975).
32. K. J. Bombaugh and R. F. Levangie, *Sep. Sci.*, **5**, 751, (1970).
33. R. A. Henry, S. H. Byrne, and D. R. Hudson, *J. Chromatogr. Sci.*, **12**, 197 (1974).
34. J. L. Waters, *J. Polymer Sci.*, *Part A-2*, **8**, 411 (1970).
35. C. P. Malone, H. L. Suchan, and W. W. Yau, *J. Polym. Sci.*, Part B, **7**, 781 (1969).
36. E. P. Otocka and M. Y. Hellman, *J. Polym. Sci.*, *Part B*, **12**, 439 (1974).
37. W. W. Yau, J. J. Kirkland, D. D. Bly, and H. J. Stoklosa, *J. Chromatogr.*, **125**, 219 (1976).
38. J. J. Kirkland, *J. Chromatogr.*, **125**, 231 (1976).

Twelve

GEL PERMEATION CHROMATOGRAPHY APPLICATIONS

12.1 Introduction

While most of this book is devoted to the common features of GPC and GFC in liquid size-exclusion chromatography, this chapter is devoted principally to GPC applications for synthetic polymers. It is problem-orientated and illustrates the value of and types of information available from GPC. No attempt has been made to provide detailed bulk-property or structural information for the materials used in the illustrations. Laboratory techniques for high-performance GPC are covered in Chapter 8, special GPC methods such as those for small molecules and preparatory work in Chapter 11, and applications of GFC in Chapter 13.

Traditionally, GPC has been used for the analysis of synthetic polymers. A sample solution is passed through the column and size sorting takes place in the packing material by an exclusion process (Chapt. 2), the largest molecules exiting from the column first, followed by those of decreasing size. For a polymer sample, the detector response versus retention volume is a measure of the molecular size distribution. Conversion of molecular size to molecular weight by calibration (Chapt. 9) permits calculation of sample molecular weight (MW) and molecular weight distribution (MWD) (Chapt. 10). Even without further computation, however, the raw-data chromatograms are very informative, and often provide significant insights into polymer products and processes. In fact, the data from the raw chromatogram accounted for much of the early success of GPC in the polymer analysis field. Section 12.2 describes some significant examples of this type of GPC analysis.

Traditional or conventional GPC and modern high-performance GPC (HPGPC) yield the same information about samples, although with HPGPC the data are more accurate and available in much less time. The examples used in this chapter have been chosen to illustrate the solutions of real sample problems rather than to illustrate differences between conventional and high-performance GPC. While many classic problem solutions by conventional GPC have not yet been repeated with HPGPC, it should be apparent that the HPGPC method would normally be preferred because of its increased accuracy and time savings.

12.2 Value of the Strip-Chart Chromatogram

Inherently, GPC provides a useful description of the size distribution of the sample molecules. The strip-chart recording of detector response versus retention time (or volume) is a visual display of that size distribution. This section illustrates how these displays (chromatograms) have been used to characterize samples.

In one case it was needed to know whether a dispersed polymer additive remained inert or became chemically bound during polymer extrusion. To answer this question three test samples were analyzed by GPC, as shown in Figure 12.1. Curve A is a homopolymer with no additive, curve B the test blend (after extrusion), and curve C an unheated mixture of polymer and

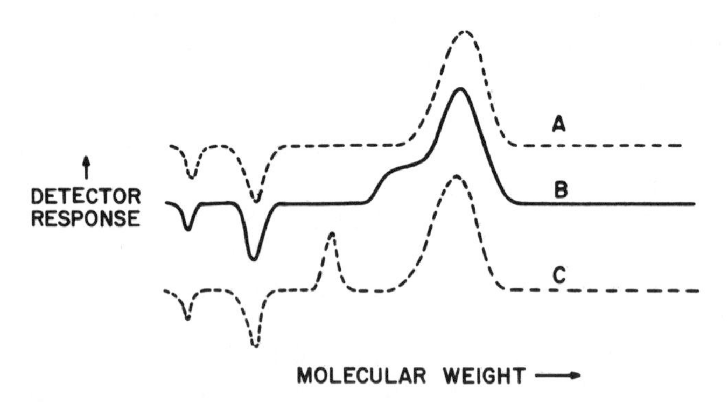

Figure 12.1 Gel permeation chromatograms illustrating fate of additives. (Reprinted with permission from Ref. 1.)

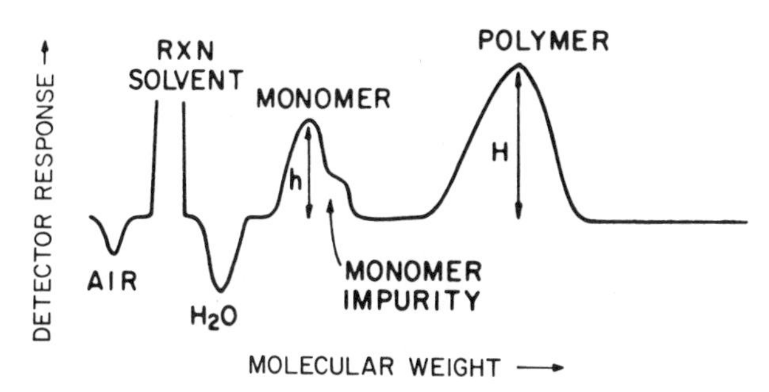

Figure 12.2 Gel permeation chromatogram illustrating a study of reaction rate. (Reprinted with permission from Ref. 1.)

additive. Curve inspection quickly leads to the conclusion that the additive is not inert to the extrusion process. In curve B, the additive has either homopolymerized or added to the sample polymer, since the peak attributable to the additive has moved to a higher molecular weight region of the chromatogram.

Figure 12.2 illustrates how different kinds of information can be obtained by inspection of GPC curves. In a study of the products from a photolytic reaction, several samples were chromatographed and the following conclusions drawn: [1] a polymeric product formed via the photolysis reaction; [2] the ratio of the amount of polymer (H) to monomer (h) increased with photolysis time and also correlated with light intensity; [3] an unsuspected and photoactive impurity was found in the monomer; and [4] a qualitative estimate of the product polymer MW and MWD could be made. In this chromatogram the water peak appears at an "anomalous" retention volume. Water sometimes appears at a lower retention volume than a monomer because it readily forms "clusters" with additives or with the chromatographic mobile phase.

The value of obtaining a complete MWD chromatogram by GPC is shown in a study of the degradation of natural rubber during milling. In Figure 12.3, curve A represents 8 min milling time, while the other curves represent increasingly longer milling times, up to 76 min (curve F). This study indicates that as the rubber is milled, the MWD is narrowed, probably as a result of the selective breaking down of the very large molecular weight molecules by mechanical shear. The peak of the distribution curve shifts to lower and lower molecular weight with increased milling time and appears to approach a limiting value. In some of the curves there is a "hump" in the high-molecular-weight region which becomes larger and moves to lower molecular weight with increased milling time; then it disappears completely (curve F).

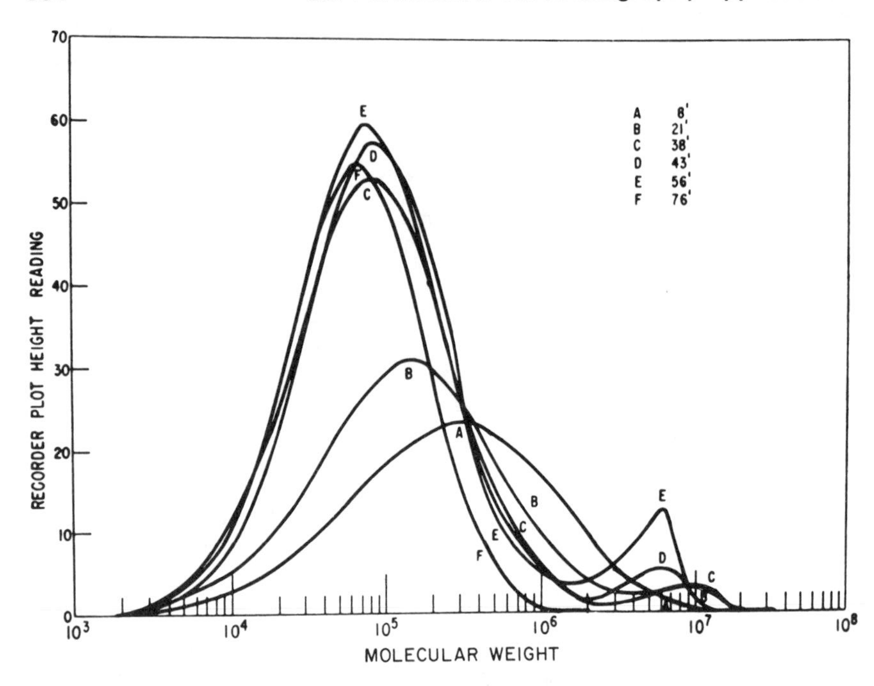

Figure 12.3 Gel permeation chromatograms showing the effect of natural rubber mastication. Mastication times for samples A, B, C, D, E, and F; 8, 21, 38, 43, 56, and 76 min, respectively. (Reprinted with permission from Ref. 2.)

One possible explanation for the appearance of this high-molecular-weight material is that, as the large molecules are broken down by the shearing action, they form free radicals, some of which are large and recombine to form even larger molecules than were originally present. Evantually, these large molecules are broken down by the shear processes and finally disappear.

Figures 12.4 and 12.5 provide additional examples of the value of knowing the whole MWD as determined by GPC. Viscosity-index improvers are polymers that have suitable solubility characteristics for reducing variation in the viscosity of automotive engine petroleum oils with temperature. For such uses a broad-MWD polymer has two weaknesses. The low-molecular-weight end of the MWD contributes very little to viscosity and is therefore an expensive filler, while the high-molecular-weight end is degraded by engine shear forces and rapidly loses its effectiveness under conditions of normal use. By using narrow-MWD oil-soluble polymers in the correct molecular-weight range, as illustrated in Figure 12.5, these deficiencies are overcome and smaller amounts of the additive are required (4).

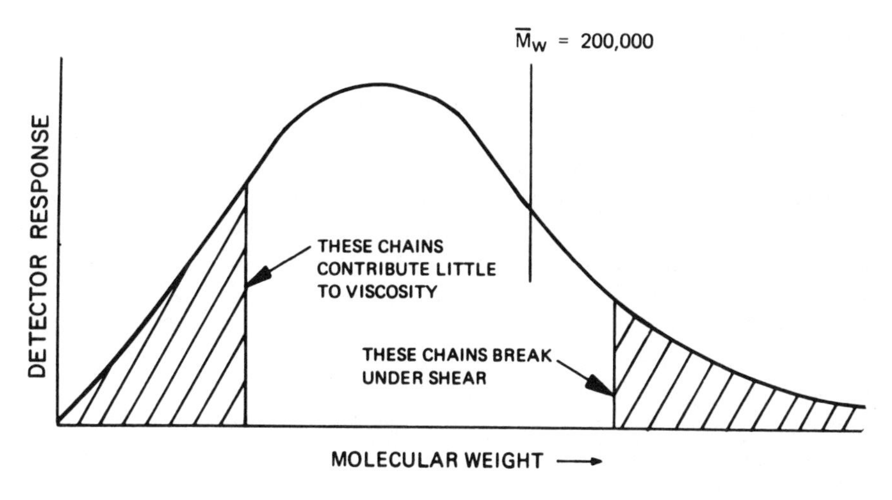

Figure 12.4 Useless components in a broad-MWD viscosity-index improver.
(Reprinted with permission from Ref. 3.)

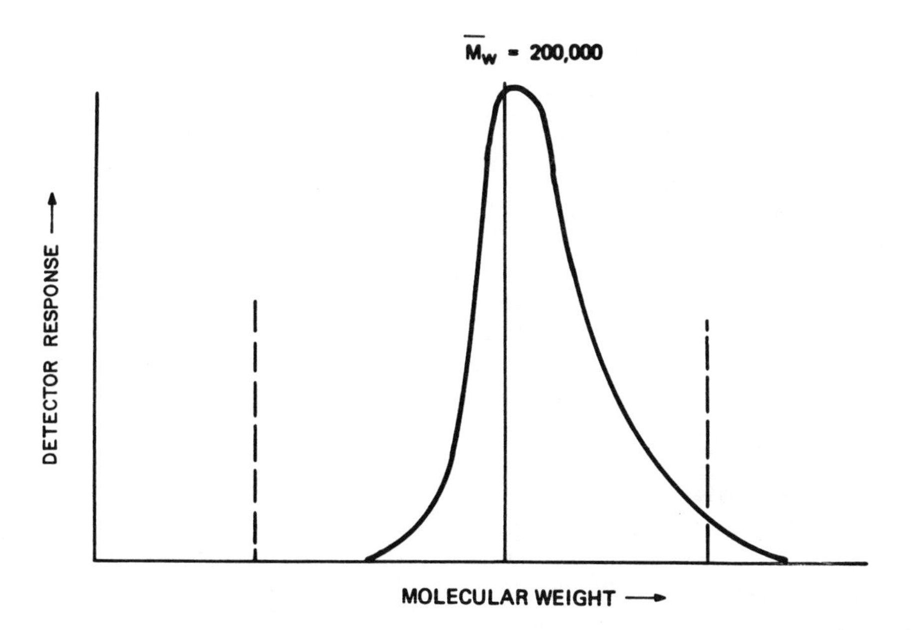

Figure 12.5 Narrow-MWD viscosity-index improver.
(Reprinted with permission from Ref. 3.)

12.3 Tetrahydrofuran-Soluble Polymers

A large number of commercial polymers are soluble in tetrahydrofuran (THF), which constitutes a good "first choice" mobile phase for GPC. It is therefore convenient to mention THF systems in a separate section.

The low viscosity of THF permits high GPC resolution and relatively low operating back pressures. THF has a low refractive index, 1.4072 (20°C), so most polymer solutes are readily detected by R.I., and it swells organic gel packings, facilitating its use in SEC separations. THF is a fairly polar solvent and neutralizes most of the active sites in both organic and inorganic column packings. However, THF does form peroxides, and is highly flammable, so it must be used with caution. Section 8.2 and Table 8.3 list other safety considerations, and Tables 8.2 and 10.2 list many of the polymer types known to be soluble in THF.

Figure 12.6 illustrates a typical chromatogram which was obtained on an organic gel column set in THF. The bimodal form of the MWD in this illustration is clearly revealed. Chromatograms of this kind can be used without further processing for direct comparison with other sample chromatograms in a series. Probably the most desirable features of THF are its overall high-mobile-phase performance and its broad utility as a polymer solvent.

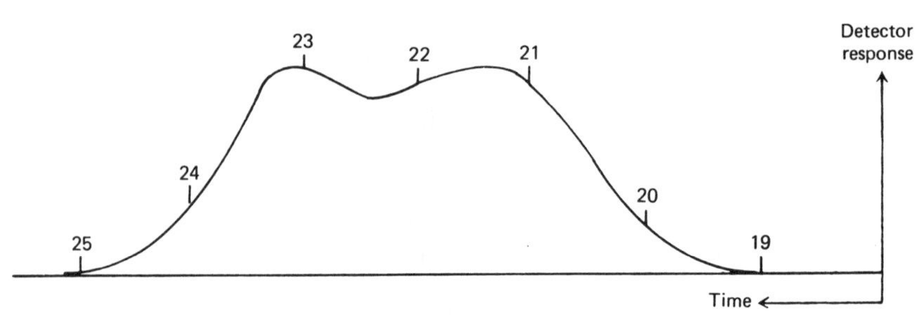

Figure 12.6 Bimodal polybutadiene sample in THF.
(Reprinted with permission from Ref. 19.)

12.4 Polyolefins

High-molecular-weight polyolefins are relatively difficult to handle in GPC, primarily because of solubility problems. With high-molecular-weight solid polyethylenes (PE), polypropylenes (PP), poly(ethylene/vinylacetates) (EVA), and similar polymers, large percentages of the polymer molecules are in the crystalline state. The degree of crystallinity decreases with increasing branch and/or copolymer contents, but there are no known solvents for many of these polymers at room temperature. It usually becomes necessary to heat the polymer almost to its melting point (in the solvent) to break up the crystalline bond forces before dissolution occurs. Typically PE, PP, and EVA polymers are analyzed by GPC in 1,2,4-trichlorobenzene (TCB) or *o*-dichlorobenzene (ODCB) at 130–150°C.

There are problems with handling hot solutions in GPC analyses, since they can cause serious thermal burns. Thus polymer dissolution should be accomplished in stirred bottles in a metal (not glass) oil bath or in round-bottom flasks heated with a mantle and stirred with a magnetic bar. The operator should wear rubber gloves, protective clothing, and a face shield or safety glasses when handling such hot solutions. While an antioxidant (e.g., Santonox-R, Monsanto) is usually added to prevent oxidative degradation of the polymer during the several-hour dissolution process, the temperature should be kept as low as possible to minimize polymer degradation. The sample solution must be injected hot, or the polymer will precipitate and plug the injector or the column-inlet frits. Oven-preheated syringes should be used to fill the heated valve injectors. Since high-molecular-weight polyolefin solutions can be very viscous even at high temperatures, sample solutions should be dilute (typically 0.1 % w/v) to facilitate injection and passage through GPC columns.

While analyses by GPC or HPGPC at 130–150°C are usually feasible, certain technical limitations are imposed. Organic gels can be used since they swell readily in these hot solvents. However, if the column is cooled, the gel may shrink, with attendant consolidation of the packing. This shrinking may lead to channels in the packed bed and reduced resolution. Loss of resolution is especially noticeable with high-performance columns, even for small changes in the packing structure. Since the original packed structure is not recovered in the next heat-up, organic gel columns are best kept hot even when not in use.

Rigid column packings (e.g., silica) do not suffer from a swelling problem, but adsorption of the polyolefins sometimes occurs, which can bias the SEC separation mechanism. Such difficulties can be overcome by using silanized packings (Sect. 6.2) or by modifying the mobile phase to neutralize the active sites on the column packing. Figure 12.7 shows the analysis of EVA copolymer on a series of porous silica columns using 1 % Carbowax-200 as a modifier in ODCB. Without the solvent modifier, the EVA did not elute from these columns. However, various MW and MWD PEs can be chromatographed on the same columns in unmodified ODCB (Figure 12.8). It is more convenient to operate without the modifier if possible, since its presence may contribute to baseline noise and interference in subsequent characterization steps (e.g., IR); however, at times it is essential for successful analyses. Figure 12.7 also illustrates the advantage of using two detectors for composition analysis. In this example one IR detector monitored the acetate

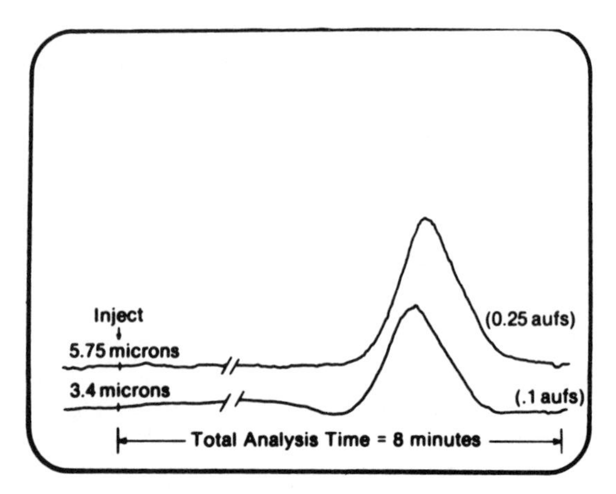

Figure 12.7 Poly(ethylene/co-vinyl acetate) chromatogram obtained at high temperature. Three DuPont SE columns in series: 1000 Å, 500 Å, 100 Å; solvent, *o*-dichlorobenzene containing 1% Carbowax at 135°C; flow rate, 2 ml/min; IR detection, 0.25 AUFS at 5.75 μm (upper curve) and 0.1 AUFS at 3.4 μm (lower curve). (Reprinted with permission from Ref. 5.)

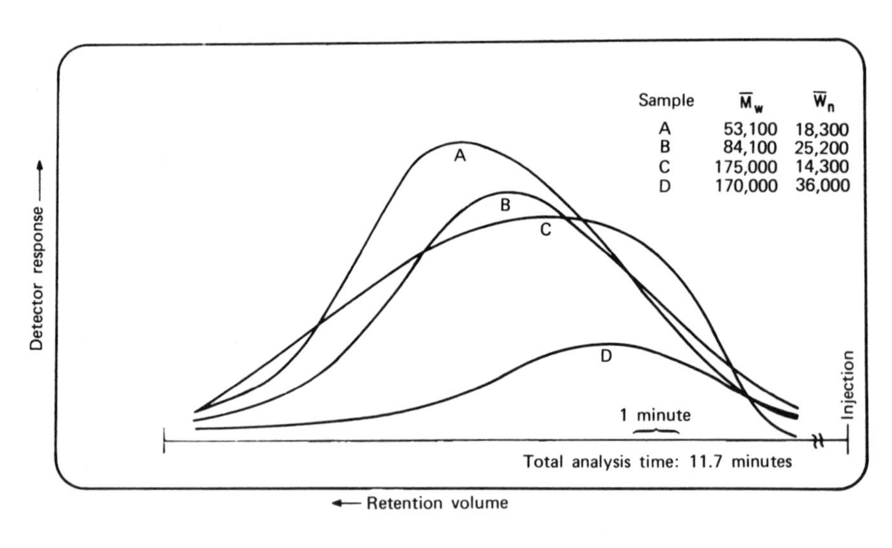

Figure 12.8 High performance/high temperature chromatograms of linear polyethylenes. Three DuPont SE columns in series: 1000 Å, 500 Å, 100 Å; solvent, *o*-dichlorobenzene at 135°C; flow rate, 1.3 ml/min; IR detection, 0.025 AUFS at 3.40 μm. (Reprinted with permission from Ref. 5.)

carbonyl absorbance at 5.75 μm while another monitored the total alkyl absorbance at 3.4 μm. From these two outputs the copolymer composition as a function of molecular-weight could be calculated (Sect. 12.9).

The type and extent of branching greatly affect the properties of polyethylene (PE). Conventional GPC curves for a series of five high-pressure synthesized branched PE resins, obtained on a four-column set of Styragel, are shown in Figure 12.9, while properties for these resins are shown in Table

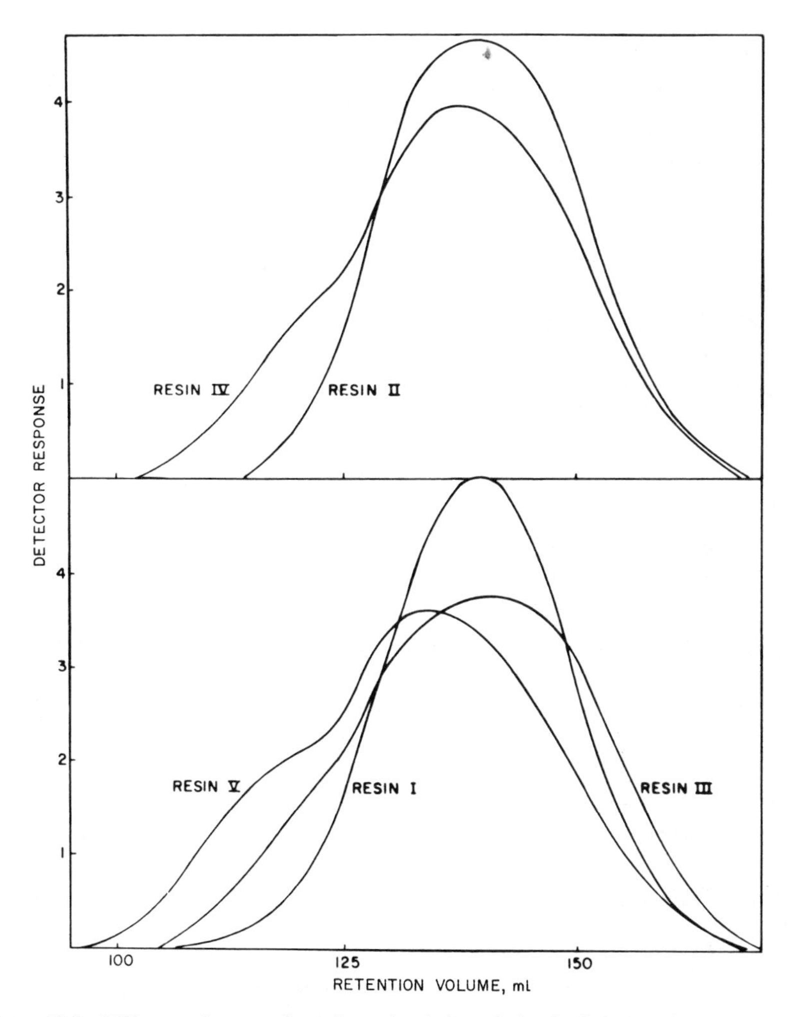

Figure 12.9 Differences in curve shape for variously branched polyethylene resins. Reprinted with permission from Ref. 6.)

Table 12.1 Properties of Polyethylene Resins

Resin	MI[a] (g/10 min)	Density[b] (g/cm^3)	Mean Synthesis Temperature (°F)	Expected Long-Chain Branching Level
I	2.44	0.934	350	Low
II	2.74	0.929	400	Moderate to low
III	1.88	0.924	420	Moderate
IV	2.82	0.924	440	Moderate to high
V	2.96	0.918	490	High

Taken from Reference 6.
[a] Melt index; ASTM D 1238-57T.
[b] Annealed density, ASTM D 1505-57T.

12.1. Even though the melt index values (MI) indicate that all the resins have similar molecular weights, the GPC curves are quite different. This anomaly can be explained in terms of structural differences in the resins caused by long-chain branching. A more quantitative description of polymer branching is given in Section 12.8.

12.5 Polyamides and Polyesters

Many polyamides and polyesters are produced commercially for use as fibers, films, and molding resins. Chief among these are 66 nylon [poly-(hexamethylene adipamide)], 6 nylon [polycaprolactam], and PET [poly-(ethylene terephthalate)]. These highly crystalline, polar polymers are soluble in only a few solvents which break up the polymer crystallinity. For analytical methods (e.g., GPC, osmometry, light scattering, end-group titration and viscosity), *m*-cresol, phenol/water, phenol/methanol, α-chlorophenol, benzyl alcohol (boiling), and formic acid have been used traditionally. Of these, only *m*-cresol has been used extensively for GPC; however, more recently trifluoroethanol, hexafluoroisopropanol (HFIP), and trichloroethane/nitrobenzene (95.5/0.5) have proven useful.

The two now most commonly used solvents for polyester and polyamide GPC are *m*-cresol and HFIP. Care must be taken to use these solvents safely. Neither solvent is highly toxic (Table 8.3), but both should be used only in well-ventilated areas. *m*-Cresol is normally used at 100–125°C, and as such can cause both chemical and thermal burns. The severity of a chemical burn from *m*-cresol is a function of contact time. Rubber gloves, protective clothing, and chemical goggles should be used. If contact with the skin does occur, immediate scrubbing with soap and water usually prevents serious injury. HFIP generally is less toxic than *m*-cresol (Table 8.3), and while HFIP should be handled with rubber gloves, it is not very corrosive. *However, HFIP is very dangerous to the eyes, and chemical goggles should always be worn when handling this solvent.*

Most polyamides and polyesters can be dissolved in *m*-cresol by stirring for 1 hr or so at 100°C or by shaking at room temperature overnight. Provided that freshly distilled *m*-cresol is used (eliminating dissolved H_2O and acid contaminants), polyester and polyamide solutions are stable to hydrolysis, transesterification or transamidation for several hours at $\leq$100°C (7, 8). Styragel, μ-Styragel, or porous silica column packings may remain stable for years in this solvent. However, column efficiencies are poor even at elevated temperatures because of high solvent viscosity. Figure 12.10 illustrates that GPC easily displays the significant differences in MWD between an as-synthesized and melt-transesterified broad-MWD polyester sample, while Figure 12.11 shows that 66 nylon in *m*-cresol yields a GPC curve with the expected symmetrical curve shape.

Hexafluoroisopropanol (HFIP) (E. I. du Pont de Nemours & Co.) is a very polar good solvent for most polyesters and polyamides at room

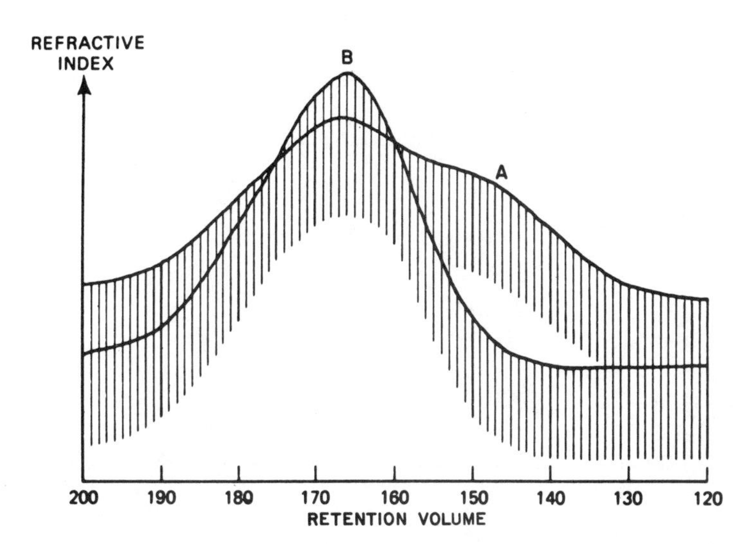

Figure 12.10 Comparison of (A) as-synthesized and (B) melt equilibrated poly(ethylene terephthalate) polymers.

Solvent, *m*-cresol stabilized with 0.5% dilauryl thiopropionate; temperature 100°C. (Reprinted with permission from Ref. 7.)

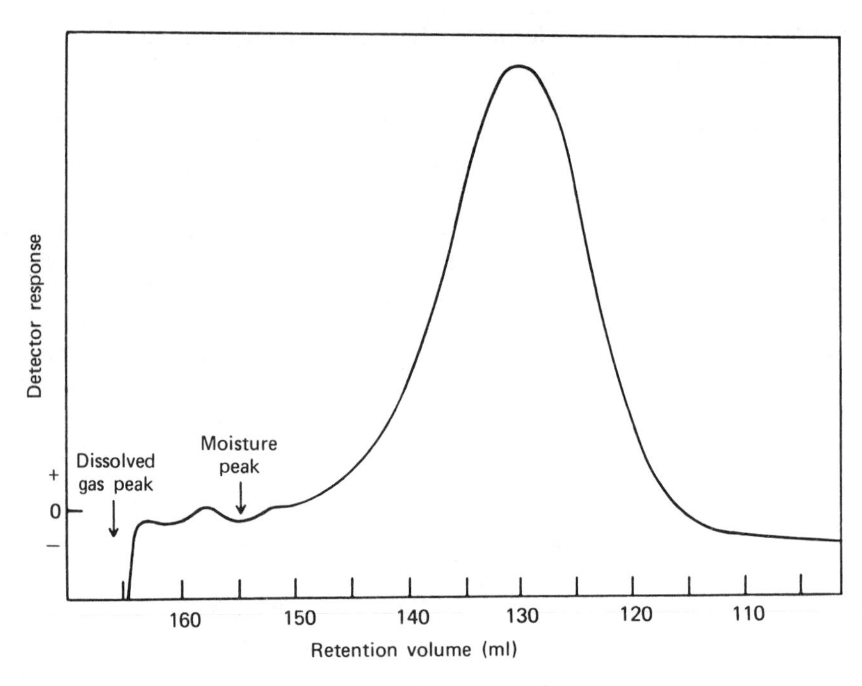

Figure 12.11 Typical chromatogram of 66 nylon in m-cresol.
(Reprinted with permission from Ref. 8.)

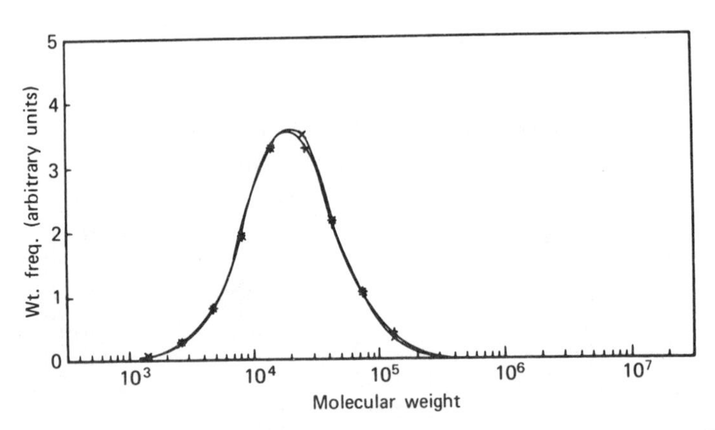

Figure 12.12 Illustration of the reproducibility of 66 nylon data obtained in HFIP on μ-Styragel columns.
+, Run on 12/10/75; ×, run on 7/16/76. Columns, four 30 × 0.78 cm μ-Styragel, 10^3 + 10^4 + 10^5 + 10^6 Å; mobile phase, 0.01 *M* NATFAT in HFIP; flow rate, 2 ml/min; temperature, 25°C; detector, RI. (Reprinted with permission from Ref. 9.)

temperature, even for those samples of high molecular weight and crystallinity. The low viscosity and refractive index of HFIP make it a desirable GPC mobile phase (9). HFIP should be distilled prior to use to eliminate water and hydrofluoric acid, both of which are reagents for sample hydrolysis (see earlier in this section regarding handling safety). With purified HFIP, column packings and samples are stable for long periods, as documented by Figure 12.12, which shows the GPC curves of 66 nylon run on the same μ-Styragel column in HFIP at 7 month intervals. Since HFIP does not swell the organic gel packings (in fact, it shrinks them), rigid, silica-based column packings are generally preferred for use with this solvent. Shrinking of the organic gel in the column leads to channeling and chromatographic band broadening. Because of the high cost of HFIP, it is normally recovered and redistilled for reuse.

It has been found that sodium trifluoroacetate (NATFAT) effectively suppresses some polyelectrolyte effects in HFIP. Figure 12.13 shows that gross polyelectrolyte effects cause a 66 nylon GPC curve to appear bimodal in pure HFIP, while the addition of NATFAT completely eliminates these effects, yielding the expected monomodal curve. Polyelectrolytes are discussed in more detail in Section 12.7.

MW calibration of polyesters and polyamides can be made in HFIP by using whole polymer standards and GPCV2, GPCV3, or the Hamielec method (Sect. 9.3). The fact that polystyrene is not soluble in HFIP causes some

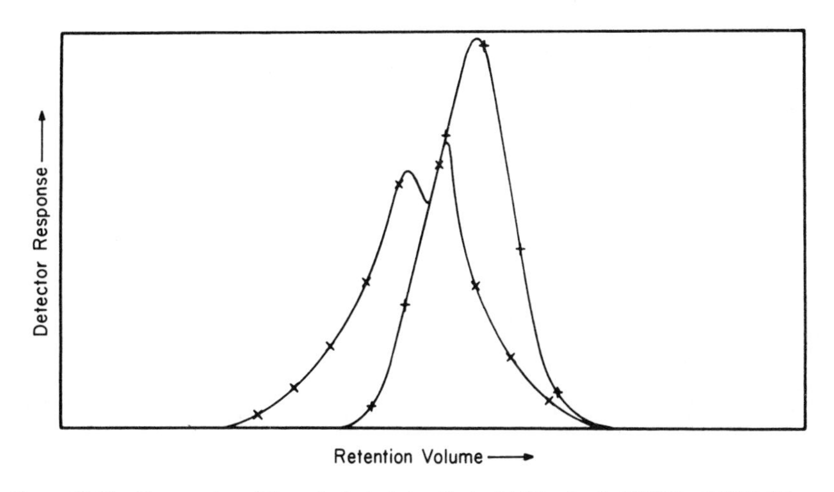

Figure 12.13 Suppression of the polyelectrolyte effect with 66 nylon in HFIP by NATFAT. ×, HFIP alone; +, HFIP plus NATFAT. Other conditions same as for Figure 12.12. (Reprinted with permission from Ref. 9.)

difficulties in using the universal calibration method. With rigid packings polystyrene calibration can be made in another solvent prior to use of the column with HFIP (10).

12.6 Synthetic Water-Soluble Polymers

For certain synthetic polymers, water is used as the GPC mobile phase, and water is the only mobile phase used in GFC. Chapter 13 is devoted to a discussion and examples of GFC, while this section covers briefly the GPC analysis of some synthetic polymers, for which water is the preferred solvent. Examples of water-soluble synthetic polymers include poly(vinyl alcohol), poly(vinyl-2-methoxyethylether), poly(vinylsulfonic acid) and its salts, other polyvinyl acids, poly(acrylic acid), poly(methacrylic acid), certain polyacrylamides, low MW melamine-formaldehyde resins, certain cellulose ethers (cold water only), and hydroxy celluloses. Some inorganic polymers, such as polyphosphates and those formed from the hydrolysis of iron and ruthenium nitrates, have also been studied (11, 12). Figure 12.14, for example, shows

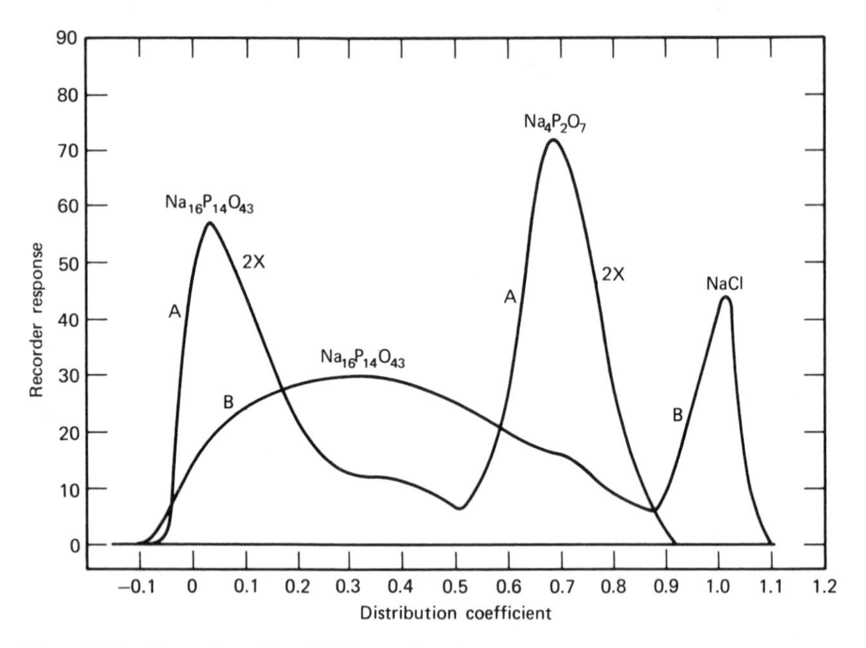

Figure 12.14 Separation of low MW inorganic polymers.
A, Sephadex G-25 column packing; mobile phase, 0.0165 *M* $Na_4P_2O_7$. B, Sephadex G-50 column packing; mobile phase, 0.010 *M* NaCl. (Reprinted with permission from Ref. 11.)

that it is possible to observe directly certain polyphosphate ions via size-exclusion chromatography with two different column packings, permitting a more quantitative understanding of the formation reaction.

Until recently, very little HPSEC in water was reported. There are no theoretical limitations to using water in HPSEC, but in practice there have been experimental difficulties with traditional high-pressure SEC column packings so that alternative solvents have usually been sought. Water has not been used with Styragel packings in traditional GPC because it neither wets nor swells this packing material. At pH greater than 8, water also dissolves glass and silica packings. Some water-soluble macromolecules adsorb onto inorganic packings so that deactivating agents must be added to the mobile phase, or the surface of the packing must be chemically modified (Sect. 13.3). Because of these limitations, aqueous GPC traditionally has most often been carried out on macroparticle organic polydextrans (Sephadex G-10 and G-25, Pharmacia Fine Chemicals) and cross-linked polyacrylamides such as Bio-Gel P-2 (Bio-Rad Laboratories). These soft-gel column packings do not permit the use of pressures exceeding about 150–200 psi, and the resolution of these macroparticle columns is poor compared to HPSEC columns with $\leq 10\ \mu m$ particles.

The fractionation of poly(vinyl alcohol) in water at 65°C is shown in Figure 12.15. The dependence of MWD on polymerization conditions is readily apparent from this chromatogram. The column set used was calibrated with a series of dextran polymers (Pharmacia Fine Chemicals). The

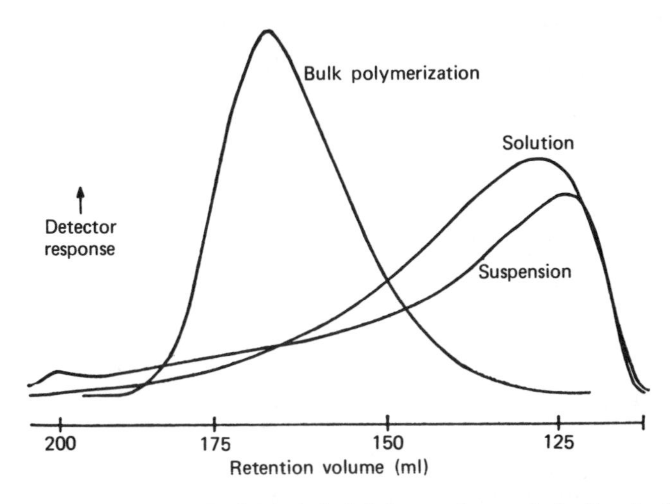

Figure 12.15 Chromatograms of poly(vinyl alcohol) from various methods of production. (Reprinted with permission from Ref. 18.)

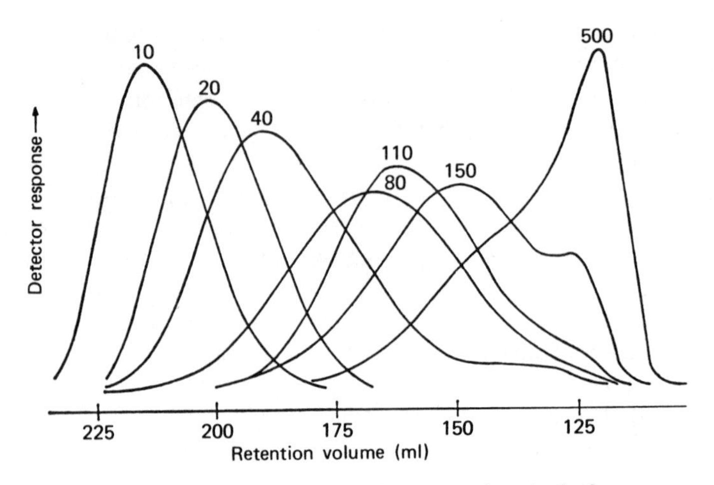

Figure 12.16 Gel permeation chromatograms of dextrans used as standards.

Numbers on the curves are nominal $\overline{M}_w \times 10^{-3}$. Columns, 120 × 0.8 cm each of 1000, 400, 250, 60 Å Porasil; solvent, water; temperature, 65°C; flow rate, 1 ml/min; detector, RI. (Reprinted with permission from Ref. 18.)

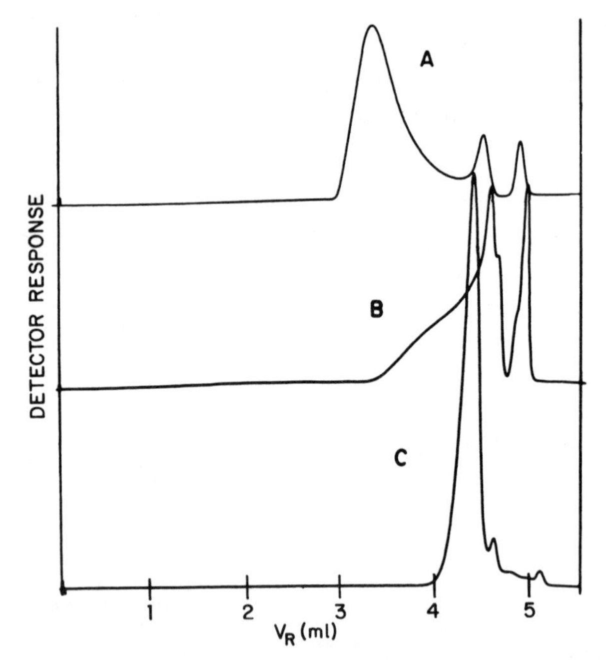

Figure 12.17 Gel permeation chromatograms of sulfonated polystyrenes in water.

A, B, C samples from different suppliers; four 30 × 0.40 cm columns of μ-Bondagel E linear; mobile phase, H_2O; flow rate 1.0 ml/min. (Reprinted with permission from Ref. 13.)

Table 12.2 Average Molecular Weights of Dextran Standards

Dextran Standard No.	$\overline{M}_w$	$\overline{M}_n$	$\overline{M}_w/\overline{M}_n$
500	510,000	185,000	2.75
250	236,000	109,000	2.17
150	150,000	—	—
110	100,500	62,000	1.69
80	85,800	43,700	1.95
40	39,800	25,600	1.55
20	21,800	14,500	1.50
10	11,200	5,700	1.94

Taken from Reference 18.

chromatograms of these useful dextran standards are shown in Figure 12.16 and the average molecular weights in Table 12.2. Note that these dextrans are broad-MWD standards. They may be used with the GPCV2, GPCV3, or Hamielec methods of molecular weight calibration (or if necessary with the peak position method—see Section 13.5).

Recently, porous silica microparticles have been prepared which can be used readily with water as the mobile phase in HPGPC. Commercial availability, preparation, and deactivation are discussed in Chapter 6. These microparticles provide high resolution and permit much more rapid analyses at high pressures. Figure 12.17 illustrates the advantages of speed gained through using the microparticles. Here an HPGPC separation of three sulfonated polystyrene samples in water was made in <6 min.

Even with modified silica packings it is relatively common to add an active site inhibitor (e.g., 1% sodium dodecyl sulfate) to the water mobile phase to prevent solute adsorption effects. Salt must also be added sometimes to overcome polyelectrolyte effects, as discussed in the next section.

12.7 Polyelectrolytes

Polyelectrolytes can normally be fractionated by HPGPC, but care must be taken to assure that the process is reproducible. Polyelectrolytes are polymers with anionic or cationic groups located along the polymer chain or on pendant groups attached to the polymer chain backbone. The ionic groups inherently may be part of the chemical structure, or their presence may be induced by the solvent. (For example, when 66 nylon is dissolved in formic acid, the amide group is protonated by the solvent and the polymer backbone becomes a polyelectrolyte in this medium, with formate ion as the counterion).

The problem of reproducibly fractionating a polyelectrolyte by SEC is that the polymer size in solution is governed not only by molecular-weight but also by the number of attached ionic groups, the type of counterion (its charge and mobility), and the polarity and electrical screening properties of the solvent. Abnormal chain expansion occurs when an ionic polymer is dissolved, because the number of electrical charges within the polymer coil generally exceeds that in the bulk solvent. Osmotic forces drive solvent into the coil to expand it and cause the counterions to diffuse out away from the backbone chain into the bulk portions of the solvent. This process leaves a net residue of charged groups (cationic or anionic) on the polymer chain. The charged groups remaining on the polymer chain are responsible for large intramolecular repulsive forces and lead to further chain expansion. The addition of a strong electrolyte (such as LiBr) to the solvent suppresses the loss of counterions from the charged sites on the polymer and permits a return of the polymer to normal physical and thermodynamic solution properties, in which state the polymer can be fractionated reproducibly.

Changes in salt concentration made to suppress the polyelectrolyte effect may also change the column packing affinity for the solute. While a successful size-exclusion separation may be made for one polyelectrolyte in a given solvent mixture, unwanted partition may dominate for another polymer in the same solvent mixture. A study of GPC substrate/polymer interactions

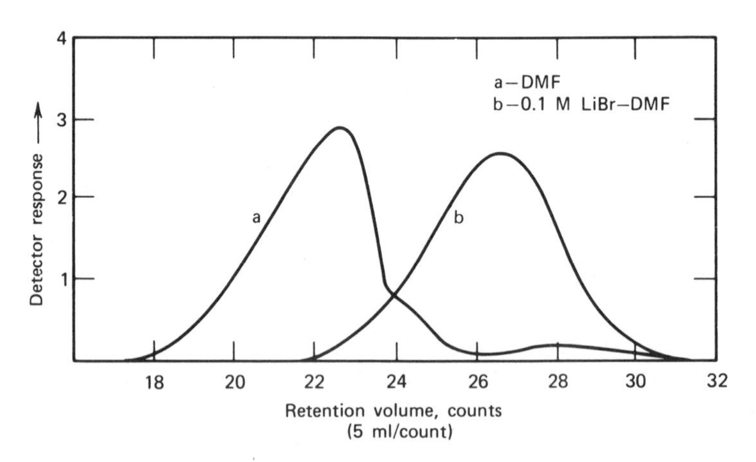

Figure 12.18 Effect of LiBr on the chromatogram of a sulfonated polystyrene in DMF. a, DMF alone; b, DMF plus 0.1 *M* LiBr. (Reprinted with permission from Ref. 15.)

in *N,N*-dimethylformamide (DMF) provides further details for those concerned with this subject (14).

An example of the suppression of the polyelectrolyte effect is illustrated in a study of polyacrylonitrile with and without pendant sulfonate groups (PAN-S, PAN). The fractionation of PAN by GPC in DMF leads to normal unimodal Gaussian-like chromatograms (15). Chromatographing PAN-S (containing only 0.2 % sulfur as sulfonate) leads to the multimodal chromatogram (curve a) of Figure 12.18. The multimodal nature of the curve is due to differing charge distributions or charge numbers on the polymer chains. Upon addition of 0.1 *M* LiBr to the mobile phase, the polyelectrolyte effect for PAN-S is eliminated, and the expected monomodal curve (b) of Figure 12.18 is obtained.

Additional information on polyelectrolyte effects in GFC is presented in Chapter 13, and Figure 12.13 illustrates a polyelectrolyte effect with 66 nylon in HFIP and its suppression with a salt (NATFAT).

12.8 Branching and Chain Folding

This section has a different emphasis from the preceding ones in that attention is directed to the determination of special polymer properties: chain branching and folding. The concepts of chain branching and folding are simple, but quantitative expression is difficult. Figure 12.19 illustrates schematically some of the types of branching and folding that occur in synthetic polymers. While some polymers contain chain branches attached randomly to the polymer backbone, others have branches that are attached in a repeating fashion (stereoregular). In most cases the type and extent of branching is significant in determining the physical properties of the bulk polymer.

A short overview of the methodology for determining branching by GPC followed by an example is provided below. Because of the complexity and unique aspects of each polymer branching problem, the original literature should be consulted for details. A bibliography of the GPC determination of long-chain branching has been published (16).

Branching

The objectives of branching analysis by GPC are to calculate from experimental data a set of branching parameters that will correlate with bulk-polymer physical properties. To be useful, the GPC method should determine the functionality of the branch points and one or more of the following parameters: [1] the lowest molecular weight polymer molecule that can be measured which contains at least one branch point, [2] the average

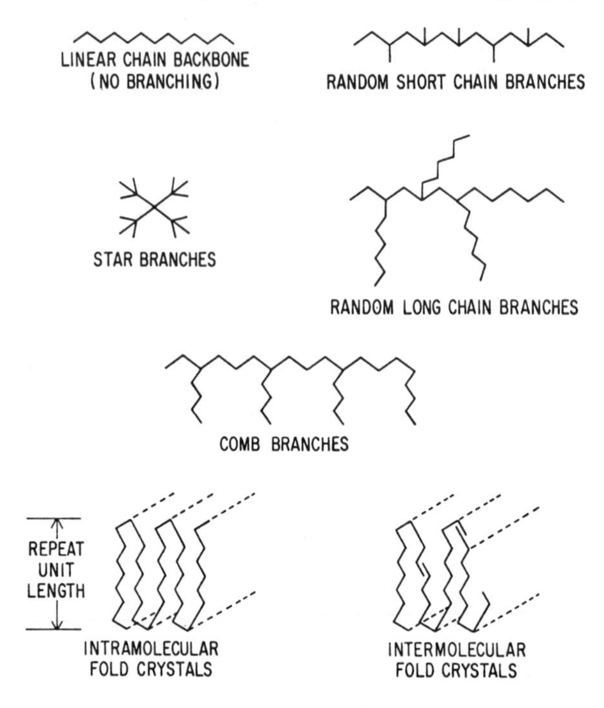

Figure 12.19 Schematic illustration of branches and folds in carbon-backbone polymers.

molecular weight between branch points, and [3] the weight percentage of polymer that is branched.

The basic concept behind the GPC approach is that branches on a polymer chain in solution reduce the hydrodynamic volume of the polymer molecule relative to that of a linear molecule of the same molecular weight. Thus the GPC calibration curve of log M versus V_R for a branched polymer in GPC is displaced from the calibration for the same polymer of a linear-chain type. For the same reasons the intrinsic viscosity is smaller for the branched polymer than for the linear-chain polymer of the same average molecular weight. The basis for determining branching by SEC, however, is that both branched and linear polymers have the same universal calibration curve, log $M[\eta]$ versus V_R.

The polychloroprene system serves as a good model for illustrating the principles of the method for determining branching by GPC. The following experimental design has been employed for this purpose (17):

1. A linear, broad-MWD polychloroprene standard was synthesized and fractionated according to molecular weight.

2. The weight-average molecular weight (by light scattering) and intrinsic viscosity in THF were determined for each fraction.

3. The Mark-Houwink parameters **K** and **a** were determined in THF using the results obtained in item 2 and equation $[\eta] = \mathbf{K}\overline{M}_w^{\mathbf{a}}$ (Sect. 10.7).

4. Each fraction was chromatographed (SEC) to obtain the retention volume.

5. It was confirmed that polychloroprene adheres to the universal calibration curve concept by showing that corresponding values of $[\eta]M$ for polystyrene and the polychloroprene had the same retention volumes.

6. The intrinsic viscosities for various samples of whole polychloroprene polymers (some of which were branched) were than calculated from GPC data using an approach suggested by Equation 10.19. The full calculation involves parameters derived from the linear standard, the universal calibration curve, and a series of essential theoretical branching functions which include the branching index λ, defined as the number of branch points per unit molecular weight. The value of λ is varied by computer iteration until $[\eta]_{calc} = [\eta]_{obs}$, yielding a single numerical value of λ. The theoretical branching functions vary with expected branching types (e.g., long-chain, short-chain, star), and the original literature must be consulted to select the functions for various polymer systems (16).

7. The various molecular weight averages of the samples were then calculated using the selected value of λ.

The molecular weight and branching parameter results for a series of laboratory synthesized polychloroprenes are shown in Table 12.3 and Figure 12.20. The samples were prepared by emulsion polymerization at 40°C using diethyl xanthogen disulfide (EXD) as a chain transfer agent, and the data obtained are consistent with the expected increase in branch content with conversion. Note that the polymer chains are essentially linear ($\lambda \simeq 0$) up to 30 % conversion, after which λ rises exponentially. $\overline{M}_w$ parallels this trend while $\overline{M}_n$ decreases, giving rise to a rapid increase in polydispersity. The behavior of $[\eta]$ (inset, Fig. 12.20) is also as expected, the minimum being caused by two competing forces: [1] as long-chain branching increases, $[\eta]$ decreases; but [2] as $\overline{M}_w$ increases, $[\eta]$ increases. Eventually, the increasing $\overline{M}_w$ dominates as the molecules become very large with conversion and $[\eta]$ increases accordingly.

Chain Folding

It is known that the molecular chains in some polymer solids fold back and forth in rows to form crystalline regions (lamellae) which are bounded by amorphous regions at some of the crystal surfaces. (Fig. 12.19). Digestion

Table 12.3 EXD-Modified Polychloroprenes

Percent Conversion[a]	11.7	33.6	55.9	62.8	71.9	82.3
$[\eta]_{THF}^{30°C}$	1.50	1.49	1.47	1.49	1.52	1.54
$[\eta]_{\lambda=0}$	1.49	1.48	1.60	1.67	1.82	1.92
$\lambda \times 10^5$	—	—	0.15	0.23	0.36	0.52
$\overline{M}_w \times 10^{-5}$	3.25	3.25	4.05	4.44	5.49	6.15
$\overline{M}_n \times 10^{-5}$	1.44	1.44	1.19	1.07	1.05	1.26
$\overline{M}_w/\overline{M}_n$	2.3	2.3	3.4	4.2	5.2	4.9
γ	0	0	0.38	0.50	0.66	0.76

Taken from Reference 17.

[a] $[\eta]_{THF}^{30°C}$ = experimentally determined intrinsic viscosity.
$[\eta]_{\lambda=0}$ = calculated $[\eta]$, assuming linear polymer.
λ = number of branch points per unit molecular weight.
$\overline{M}_w$ = calculated weight-average molecular weight of the branched polymer.
$\overline{M}_n$ = calculated number-average molecular weight of the branched polymer.
γ = branching parameter from $\lambda\overline{M}_w = \gamma/(1 - \gamma)$.

with nitric acid and subsequent GPC analysis can be used in some cases to obtain quantitative information about lamellar size and fold periodicity (20). With polyethylene, for example, nitric acid preferentially destroys the amorphous phase and the chain folds on the lamellae surfaces. The percent residue provides a rough estimate of crystallinity and crystalline perfection. The MW and MWD of the residue obtained after exhaustive etching of the sample permit estimates of the distribution of lamellar thicknesses. Figure 12.21 shows the chromatograms of etched polyethylene as a function of etching time. The distributions of the incompletely degraded products are bimodal or multimodal. The major peak at MW = 2×10^4 is attributable to the extended-chain lamellae and the smaller peaks (2–5×10^3 MW) to the folded-chain lamellae. The folded-chain crystals are eventually nearly destroyed by digestion, as exhibited by the disappearance of the 10^3 MW peaks with time. If it is assumed that the chain folds are degraded first by the nitric acid, leaving the linear portions intact, the lamellar thickness (in angstroms) can be approximated by: $c\text{MW}/\text{MW}_e$, where $c = 2.55°\text{A}$, the c-axis unit cell length, MW is the molecular weight of the residue, and MW_e is 28, the molecular weight of ethylene. Results for a series of samples of varying initial crystallinity are shown in Table 12.4. The percent

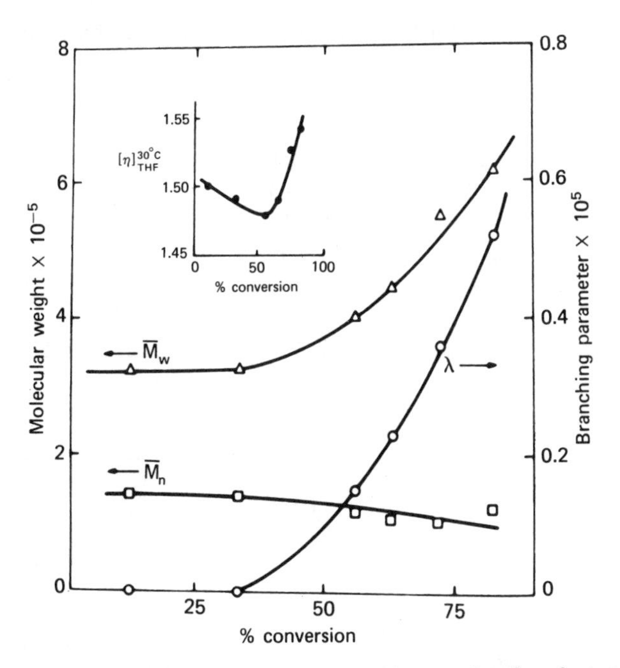

Figure 12.20 Viscosity, molecular weight, and branching as a function of conversion in polychloroprene.

$[\eta]$ relationship is in the inset, and λ is the branching parameter. (Reprinted with permission from Ref. 17.)

Table 12.4 Polyethylene Crystallinity and HNO_3 Etching Results

Sample Number	Original $\overline{M}_w$	Crystallinity (%)	HNO_3 Etched Results			
			$\overline{M}_n$[a]	$\overline{M}_w$[a]	M_{peak}[a]	% EC[b]
1	4.9×10^4	100	1.2	3.6	2.8	98
2	8.5×10^4	98	1.7	4.1	3.0	98
3	1.7×10^5	98	2.2	5.1	3.3	97
4	2.7×10^5	93	1.8	5.3	5.0	90
5	4.2×10^5	93	2.4	5.7	5.0	95
6	1.2×10^6	79	1.2	2.1	2.0	86
7	1.6×10^6	87	1.9	6.0	5.5	88
8	2.2×10^6	90	2.3	5.9	5.0	91
9	2.3×10^6	94	2.4	5.3	4.5	90
10	4.6×10^6	74	1.3	2.4	2.5	87

Taken from Reference 21.

[a] Molecular weights ($\times 10^{-4}$) after exhaustive etching (72 hr).

[b] Percent extended chain lamellae in residue after 8 hr of etching.

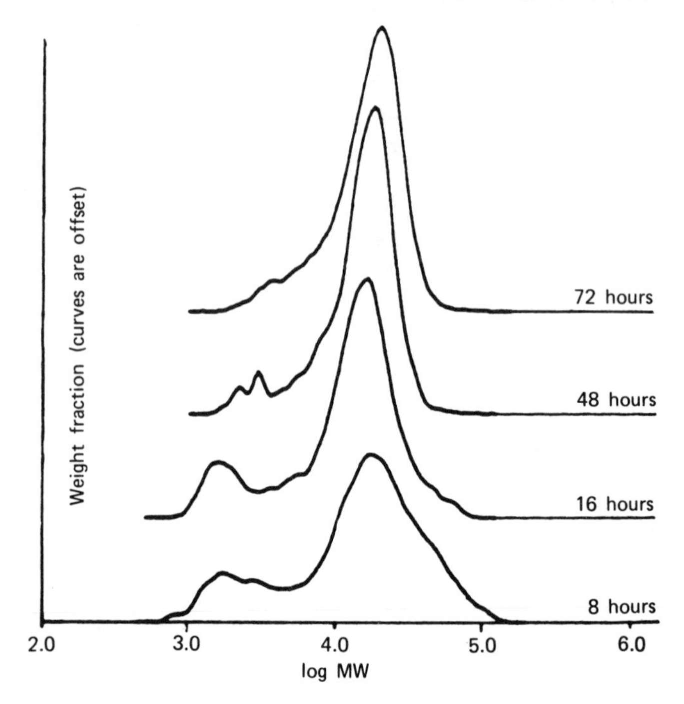

Figure 12.21 Gel permeation chromatograms of etched polyethylene.
PE ≤0.5 mm thick chips (0.3 g), digested at 80°C in fuming nitric acid for time indicated. Chromatographed on Styragel columns in 1,2,4-trichlorobenzene at 135°C. (Reprinted with permission from Ref. 21.)

of the crystalline phase in the extended chain form was estimated from the relative areas of the resolved peaks in the MWD of the residues after 4–8 hr of etching.

12.9 Copolymers

Bulk properties of polymeric materials vary with composition; for example, the physical and chemical properties of nylon are very different from those of polyethylene. With multicompositional polymers (e.g., copolymers, terpolymers) a change in properties also occurs with changes in concentration of the various components (monomers) comprising the polymer chains. Even when the bulk composition of copolymers or terpolymers is fixed, the relative location in the polymer of the comprising components can markedly affect the overall physical or chemical properties. Some different relative locations of polymer segments in one backbone are found in: [1] random and block copolymers, [2] block distributions in block copolymers (e.g., A-B, A-B-A, and B-A-B copolymers, where the A's and B's represent segment lengths of homopolymers joined together), and [3] polymers where the composition is neither random nor block but varies in some statistical manner with molecular weight. In example [3] the composition distribution is often a function of monomer conversion and is usually dependent on reactant concentrations and process conditions, such as time, temperature, mixing rates, catalysts, and the reaction mechanism.

To properly characterize a co- or terpolymer sample, the composition distribution should be specified along with the MW and MWD. GPC has greatly increased the speed and accuracy with which this can be done. Historically, most workers have used bulk fractionation techniques based on solubility to obtain fractions which were then characterized for composition by spectroscopic means (e.g., NMR, UV, IR, and x-ray) and thermal methods (such as DTA, DSC, and TGA). Fractionation from solution most often has been accomplished through precipitation by gradual addition of nonsolvent to rapidly stirred solutions or by gradual cooling. Since solubility

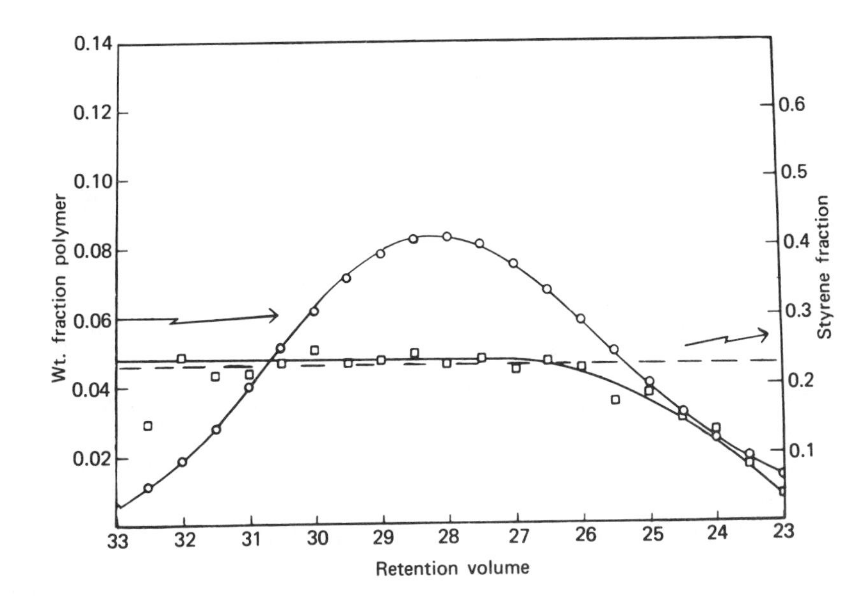

Figure 12.22 Comparison of molecular composition and size distributions in a free-radical-polymerized styrene/butadiene copolymer.

○, RI-detected MW curve; □, UV-detected styrene concentration curve. Dashed line is theoretical composition distribution. (Reprinted with permission from Ref. 22.)

is a function of both molecular weight and composition, this approach at times has yielded useless or misleading information.

GPC separates molecules while they remain in solution, and because of this offers the advantage of rapid analysis and *in situ* characterization of the "fractions" by UV, IR, and LS detection techniques; isolation of fractions is unnecessary. Since both molecular weight and composition affect the size of a copolymer molecule, it is necessary, however, to use at least two detectors to extract the needed information from the chromatograms.

The styrene/butadiene system is illustrative of a copolymer study by GPC. UV detection can be used to monitor the styrene content of the polymer chains as a function of molecular size (polybutadiene is transparent at 260 nm), while refractive index measures the total amount of copolymer present at each size. The accuracy of determining the copolymer composition at each molecular size depends on the extent and sophistication of the detector calibration for the various components present. Calibration of detector response versus concentration is normally made with standard reference homopolymers and sometimes with blends to determine synergistic effects (e.g., polystyrene and 1,4-polybutadiene, Pressure Chemical Co.).

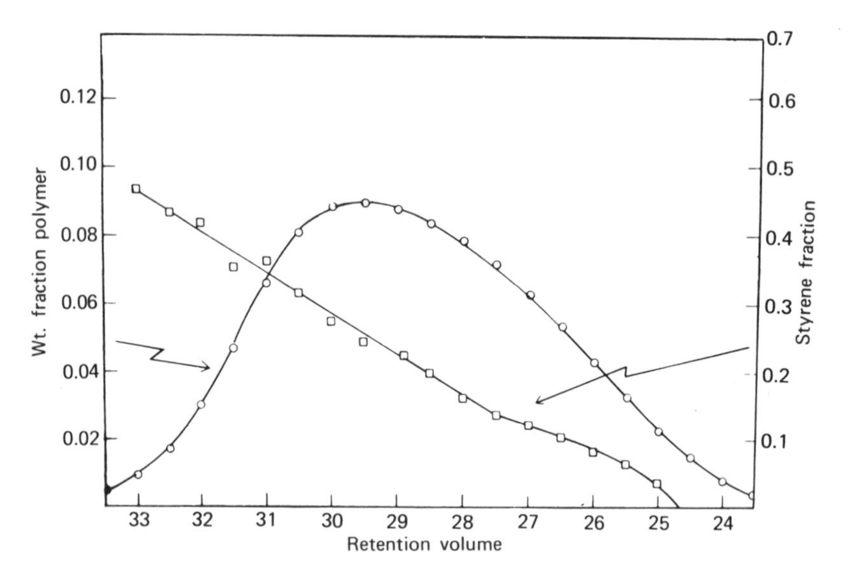

Figure 12.23 Comparison of molecular composition and size distributions in anionically polymerized styrene/butadiene copolymer.

○, RI-detected MW curve; □, UV-detected styrene concentration curve. (Reprinted with permission from Ref. 22.)

From the accepted mechanism of free-radical polymerization, styrene/butadiene rubber should have a fairly broad MWD, and a very uniform composition as a function of molecular weight. This hypothesis is supported by the data in Figure 12.22. If this type of polymer is prepared by anionic initiation, however, both composition and molecular weight are expected to vary with conversion. Figure 12.23 illustrates the variation in composition that occurs with the same type of copolymer initiated anionically. The copolymer sample in Figure 12.23 was prepared under conditions designed [1] to prevent the formation of polystyrene blocks and [2] with poor monomer control conditions to accentuate overall composition heterogeneity. Figure 12.24 illustrates the composition/molecular weight distribution of a block styrene/butadiene polymer of the A-B type as defined above. Notice that the polymer is relatively narrow in MWD and that the composition is uniform with MW for most molecular weights. This polymer was prepared by "living end" anionic initiation and the uniform composition proves that nearly all the initial blocks were still active when the second monomer was added. Styrene/butadiene/styrene polymers of the A-B-A type also can be prepared by the anionic living end method. The example presented in Figure

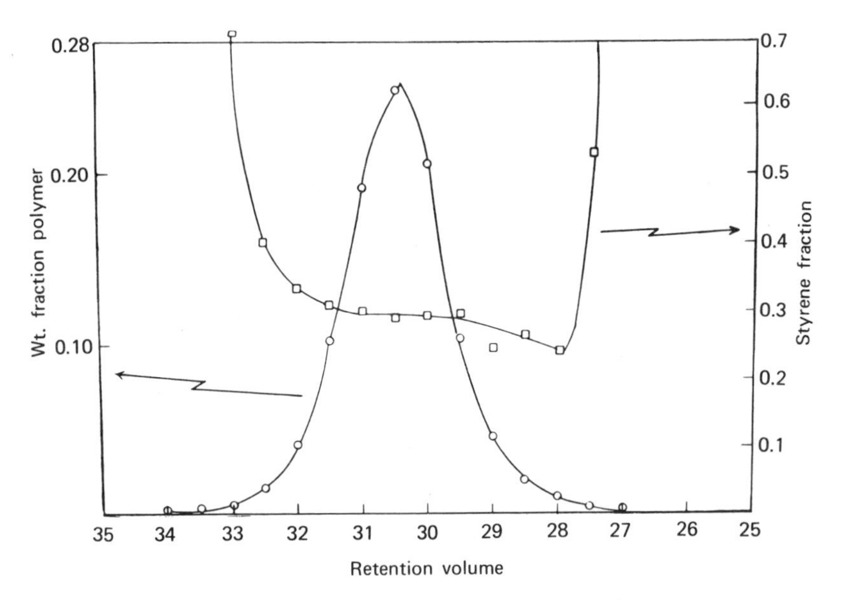

Figure 12.24 Comparison of molecular composition and size distributions in a styrene/butadiene block copolymer.

○, RI-detected MW curve; □, UV-detected styrene concentration curve. (Reprinted with permission from Ref. 22.)

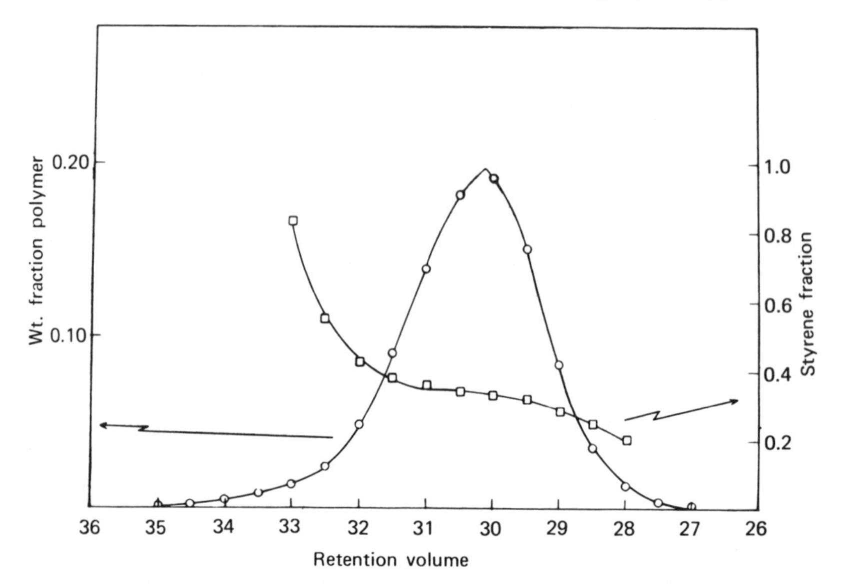

Figure 12.25 Comparison of molecular composition and size distributions in a styrene/butadiene/styrene block copolymer.

○, RI-detected MW curve; □, UV-detected styrene concentration curve. (Reprinted with permission from Ref. 22.)

12.25 shows that there is a uniform composition in all but the low-molecular-weight regions, where the relative percent of styrene in the copolymer chains increases.

A two-peaked chromatogram for an A-B-A styrene/butadiene/styrene block copolymer, prepared with an organolithium catalyst, is shown in Figure 12.26. Until this anomalous chromatogram was obtained the reaction was believed to have proceeded properly and smoothly. Isolation of the two peaks (Fig. 12.26) and subsequent infrared analysis showed the first peak to be the expected A-B-A copolymer while the second unexpected peak was polystyrene homopolymer. It was later determined that a small impurity in the butadiene terminated some of the polystyryl lithium catalyst, leaving polystyrene in the reaction mixture.

Figure 12.27 represents a copolymer prepared by grafting styrene onto a polybutadiene backbone via an intermediate metalation reaction. The shape of the elution curves suggest that metalation of the backbone and subsequent grafting by the polystyrene were nonuniform or incomplete. Some of the backbone apparently escaped metalation with the formation of polystyrene homopolymer (count 27).

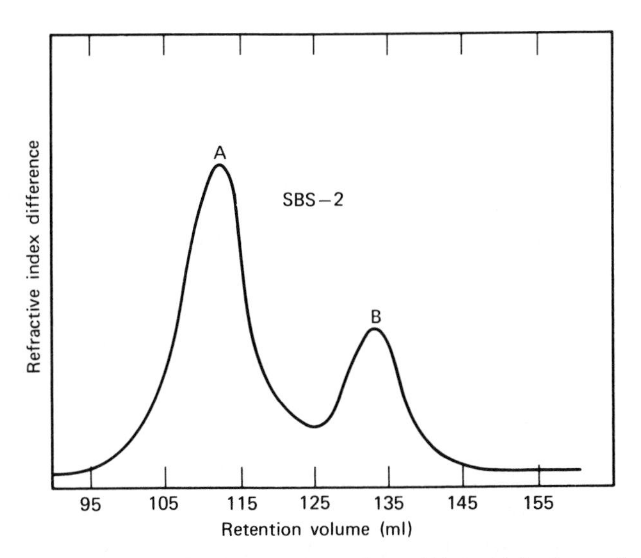

Figure 12.26 Styrene/butadiene/styrene block copolymer (A) containing low-molecular-weight styrene homopolymer contaminant (B).

(Reprinted with permission from Ref. 23.)

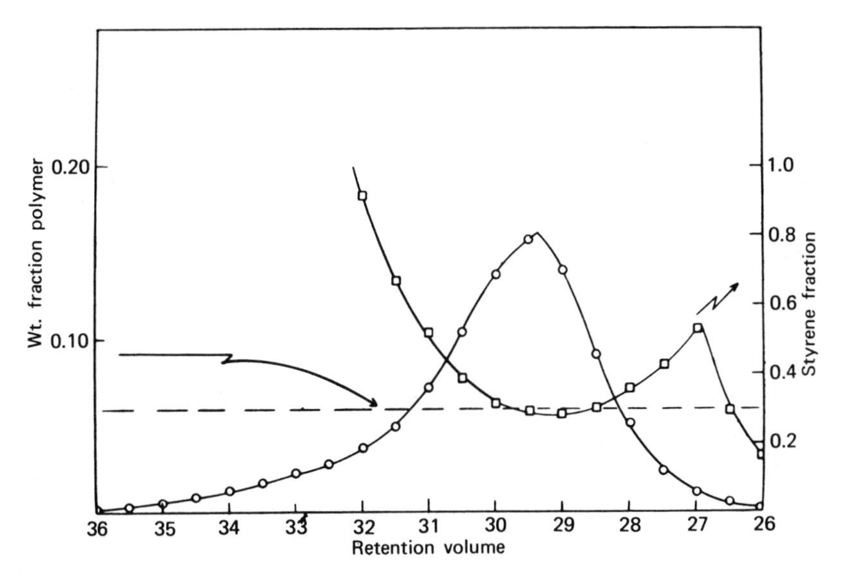

Figure 12.27 Comparison of molecular composition and size distributions in a grafted styrene/butadiene copolymer.

○, RI-detected MW curve; □, UV-detected styrene concentration curve. Dashed line is theoretical composition distribution. (Reprinted with permission from Ref. 22.)

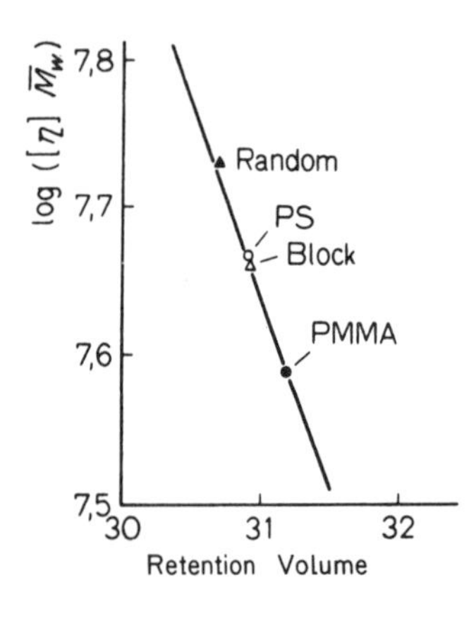

Figure 12.28 Universal calibration curve for homopolymers and random and block copolymers.

Solid line, narrow-MWD polystyrene standards; ○, polystyrene homopolymer; ●, PMMA homopolymer; △, MMA/S/MMA triblock copolymer; ▲, random copolymer of S/MMA, where S is styrene and MMA is methyl methacrylate. (Reprinted with permission from Ref. 24.)

Because the molecular size of dissolved molecules varies with composition and structure (block, random), there has been a problem in establishing accurate molecular weight calibrations for polymer systems of varying composition and for systems where composition is a function of molecular weight. Recently, a limited study comparing random and block copolymers of styrene and methyl methacrylate and the corresponding homopolymers showed that all these polymer molecular weights fit the same universal calibration plot (24). Figure 12.28 shows the universal calibration ($[\eta]M$) obtained with narrow MWD polystyrene standards and the very good fit of the other polymers to it, while Table 12.5 lists the properties of the polymers studied. Additional information on other copolymer systems, especially those containing heteroatoms in the polymer chain backbone, is needed to

Table 12.5 Homo- and Co-polymers Used in a Universal Calibration Study

Polymer	Weight % Styrene	$\overline{M}_w$ (g/mole)	$[\eta]$ (cm^3/g)	$\log([\eta]\overline{M}_w)$ (cm^3/mole)	$0.2V_R$ (cm^3)
Polystyrene	100	400,000	116.5	7.67	30.90
Poly(methyl methacrylate)	0	425,000	92.0	7.59	31.20
Block copolymers[a]	53	425,000	109.0	7.66	30.90
Random copolymers[b]	52	389,000	139.0	7.73	30.70

Taken from Reference 24.

[a] Methyl methacrylate/styrene/methyl methacrylate triblock copolymer.

[b] Random copolymer of styrene and methyl methacrylate.

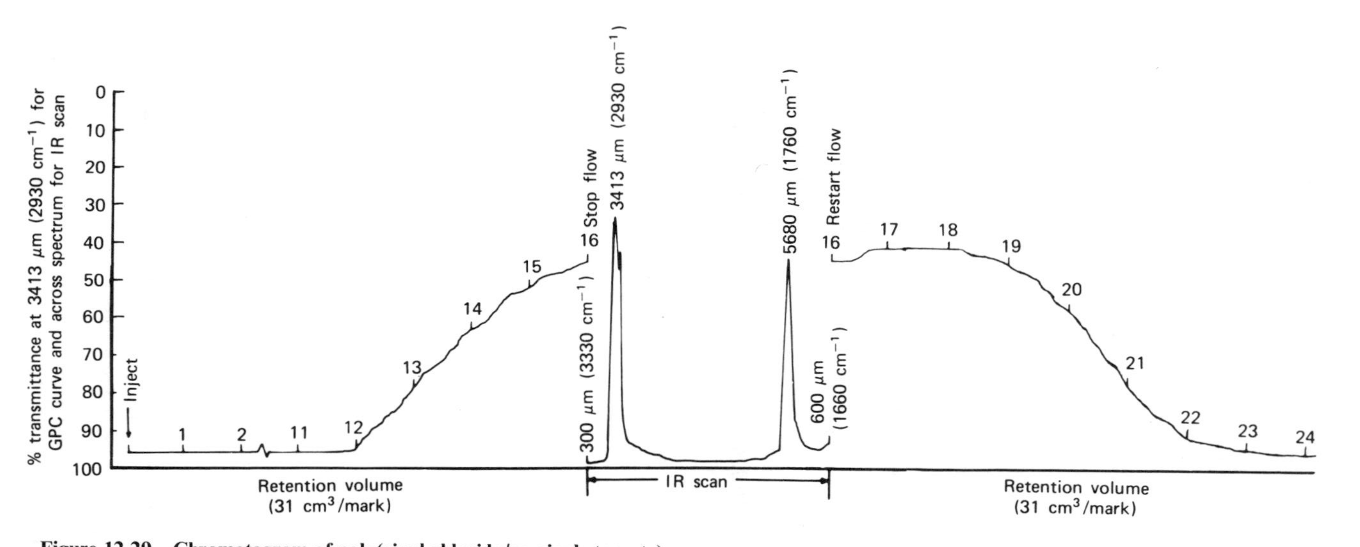

Figure 12.29 Chromatogram of poly(vinyl chloride/co-vinyl stearate).

Monitored with stop-flow infrared detector at C—H stretch (3.413 μm) and scanned at the peak from 3.00 to 6.00 μm. (Reprinted with permission from Ref. 26.)

specify the limitations on this universal calibration method. If proven general, the approach could provide a useful solution to the molecular weight calibration problem with copolymers, because a single broad standard (Sect. 9.3) can be employed for any representative composition for a copolymer of interest.

In addition to the on-line single-wavelength photometric methods discussed above, stop-flow GPC-IR can be used to analyze polymer compositions. The method involves simply attaching an IR flow-through cell assembly to the GPC column outlet and monitoring the effluent at a selected wavelength. When a peak of interest is within the cell, the mobile-phase flow is stopped and a complete IR scan is taken. After the IR scan, the flow of mobile phase is resumed. With proper matching of columns, sample size, and cell volume the stop-flow process can be repeated several times during one chromatographic run. Figure 12.29 illustrates the monitoring of the C-H absorption of poly(vinyl chloride/co-vinyl stearate) at 3.413 μm in tetrachloroethylene during a GPC run. The flow was stopped at the peak maximum and an IR scan made through the carbonyl stretch frequency of the stearate. Finally, the monitoring at 3.413 μm was completed. Using this approach the distribution of vinyl stearate in the copolymer was determined as shown in Figure 12.30.

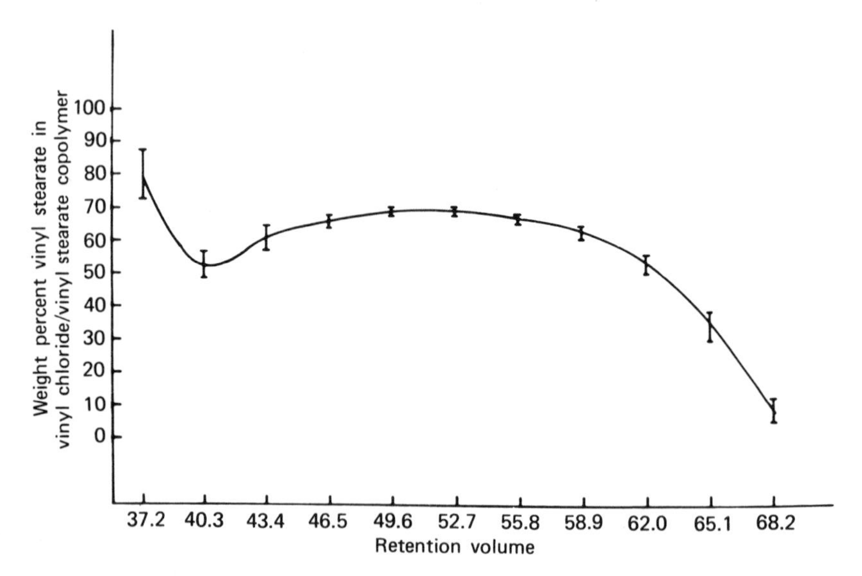

Figure 12.30 Weight percent vinyl stearate in poly(vinyl chloride/co-vinyl stearate) as function of retention volume (MW).

(Reprinted with permission from Ref. 25.)

12.10 Miscellaneous

Polymer and Additive Blending

Size exclusion chromatography can be used just as well to aid in the creation of desired composites or blends as it is in the (traditional) analysis of complex mixtures which already exist (26–28). Figure 12.31 illustrates the size distribution of the components of a selected piece of chewing gum. In this formulated product many of the components of interest have different molecular sizes so that each component can be identified and its concentration estimated by SEC. Furthermore, the MW and MWD of the polymer component can be determined readily.

Formulations or blends are composites of two or more diverse materials. Such composites exhibit some properties which are different from those of any of the individual components, and they usually are constructed to meet some specific need. When SEC is used to develop a formulation, it usually is in the context of helping to replicate some existing composite material whose chromatogram contains a number n of peaks. By properly blending several materials, each of which contain some or all the n components

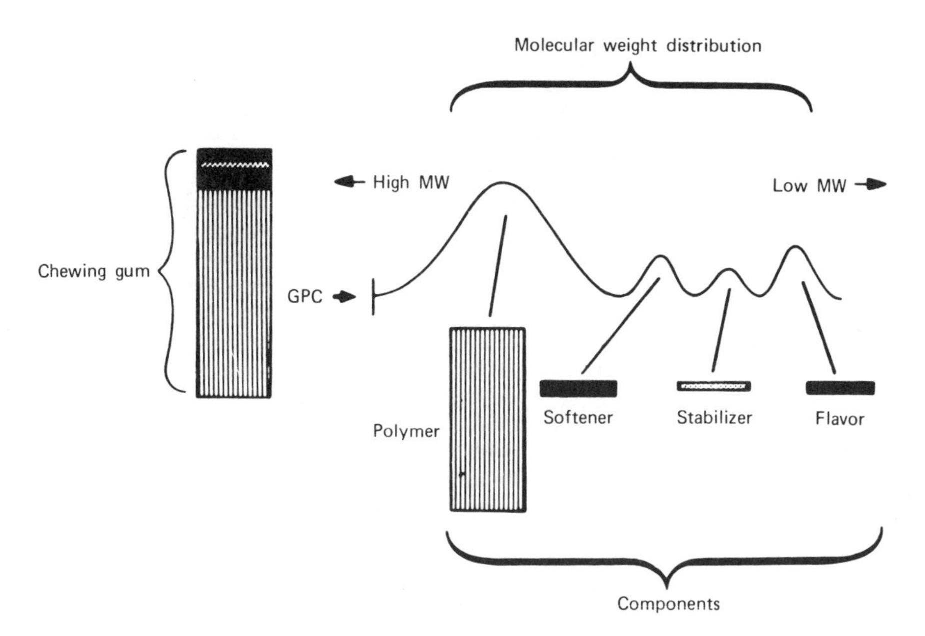

Figure 12.31 Size-distribution chromatogram of a selected piece of chewing gum. (Reprinted with permission from Ref. 26.)

of the composite but at different concentrations, the desired composite can be obtained as in the chewing gum example. A general and quantitative procedure for composite formulations via SEC is available for those workers concerned with the details (28).

Quality Control

That SEC can be used in some cases for on-line product quality control is discussed in Section 11.5. Historically, simpler and more rapid quality control measures often have been used, but there are occasions when a more detailed analysis such as that provided by SEC is required. An example is with the production and use of epoxy-fiberglass printed circuit boards. Variations in board cure times and resultant warpage are caused by batch-to-batch variations in the epoxy resins (26). Traditional methods for estimating the resin reactivity (e.g., epoxide equivalent or hydroxyl number) are often not sufficiently sensitive and do not yield values that correlate well with the observed variations. It was found that the extent of MW variation in the medium- and high-molecular-weight regions was

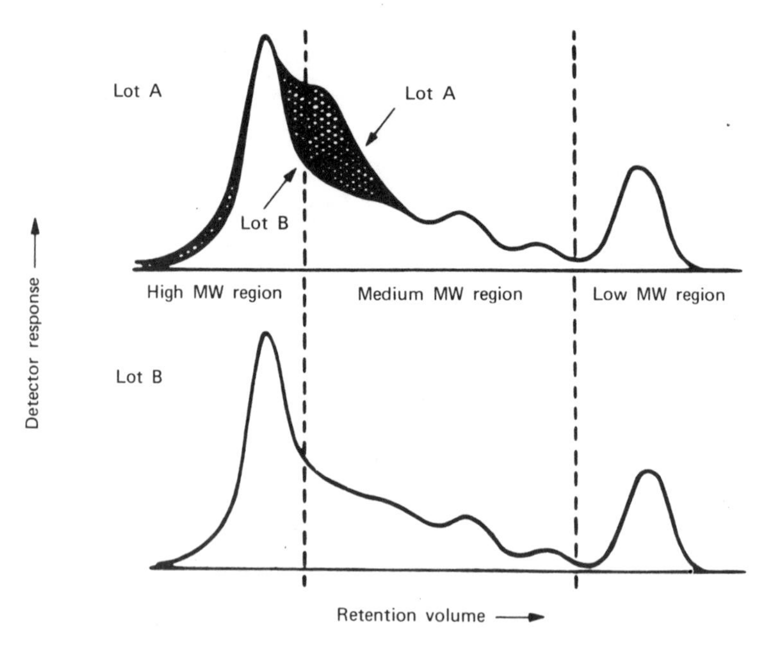

Figure 12.32 Chromatograms of two lots of epoxy resin.
(Reprinted with permission from Ref. 26.)

significant for resin formulation. Two lots of typical resins are shown in Figure 12.32, where resin A contains much more material in the medium- and high-molecular-weight region than does lot B. Using these kinds of chromatograms, acceptable limits on the variation in resin MWD can be established. The resins that lie within the acceptable limits are then cured with a hardener (e.g., dicyandiamide). The amount of hardener used, however, depends on the level of the medium and high molecular weights present in the resin. SEC can be used again to monitor the presence and level of added hardener as shown in Figure 12.33.

Cleanup Procedures

Because SEC is inherently simple to perform and the process is noncontaminating (i.e., requires little method development, simple solvents are used, and there is no column bleed), it sometimes makes a very valuable sample cleanup procedure. Used in tandem with other procedures, SEC provides a very rapid, efficient first-round separation. Figure 12.34 shows

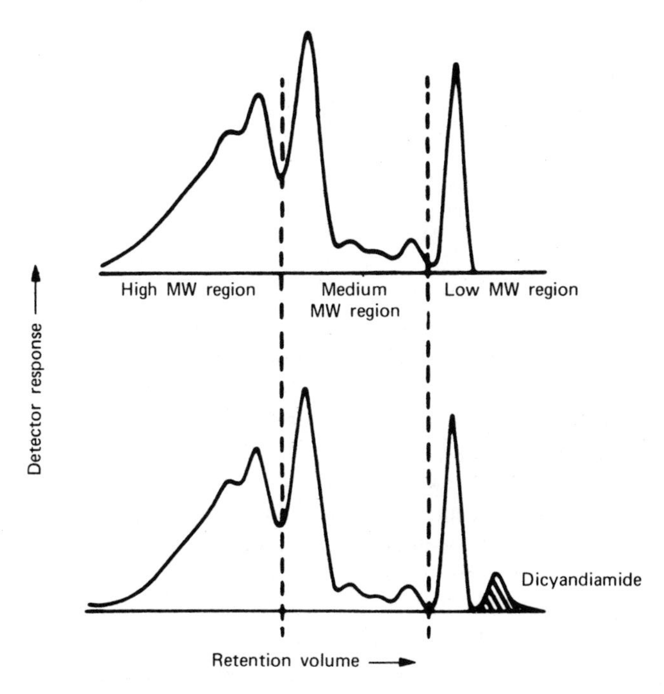

Figure 12.33 Chromatogram of epoxy formulation showing presence of hardener. (Reprinted with permission from Ref. 26.)

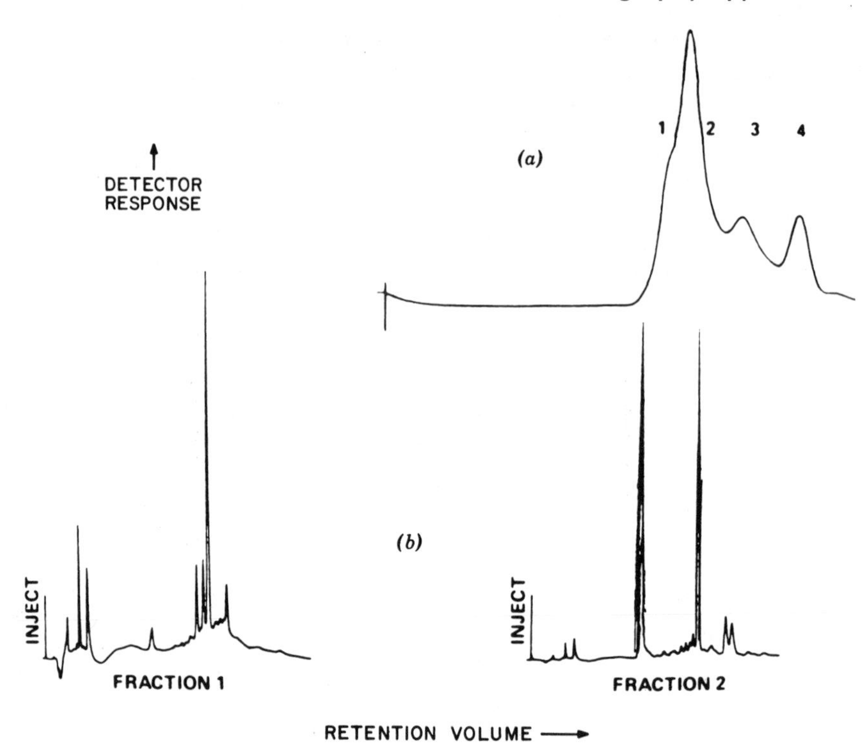

Figure 12.34 Size and partition separations of aloe plant extract.
(a) Size: three columns, 30 × 0.8 cm μ-Styragel 100 Å; solvent, tetrahydrofuran; flow rate, 1 ml/min; temperature, 25°C; detector, RI. (b) Partition: column, 30 × 0.8 cm μ-Bondapak C18; solvent, water/methanol; detector, UV, 254 nm. (Reprinted with permission from Ref. 29.)

the size separation of an extract of the aloe plant, which is used in drug formulations for sunburn treatment. The first run separates the components of the extract into size categories. These fractions were collected and then further classified by partition chromatography. Without the size separation the partition chromatograms would have been very complex and more difficult to interpret or to correlate with activity. Cleanup approaches are closely related to preparative SEC, which is discussed in Section 11.2.

Particle Dispersions

A new application of SEC is the characterization of colloidal particles. Figure 12.35 shows a particle diameter calibration for a series of polystyrene latex standards (and a silica sol). This technique has been used to investigate

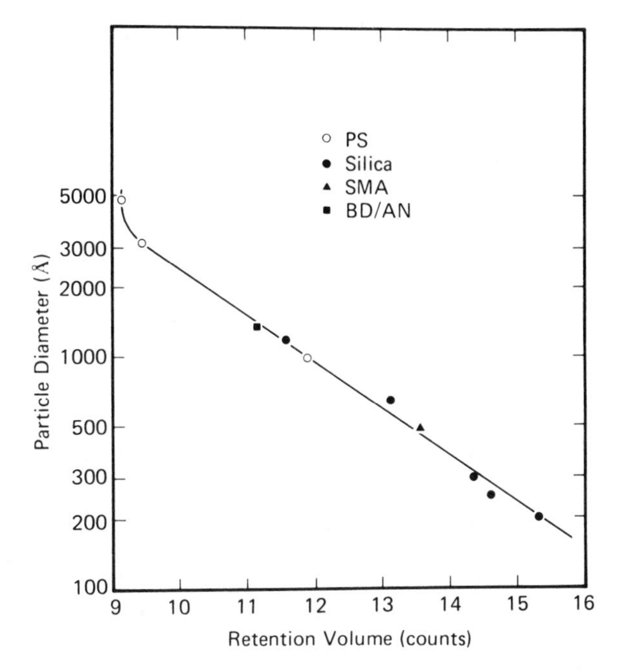

Figure 12.35 Particle diameter calibration for polystyrene lattices and silica sols. Column, 120 × 0.8 cm, one each: Fractosil 25,000, 10,000, 5000; one CPG-10-1250; one Bioglass 2500, 1500, 1000 (equal portions); one Porasil A, C, D (equal portions); mobile phase, 1 g/liter KNO_3 and 1 g/liter Aerosol OT in water; flow rate, 7.6 ml/min; detector, UV, 254 nm. (Reprinted with permission from Ref. 30.)

the growth of polymer particles in the emulsion polymerization of styrene and of vinyl acetate, in which particles in the size range 200–6500 Å were characterized (30). Certain particulates of ≤ 1 μm can be fractionated by SEC, and particle-size distribution can be determined with packings of the appropriate pore size. Inorganic colloids of <100 nm also can be size-characterized with high-efficiency porous silica columns (31).

REFERENCES

1. D. D. Bly in *Physical Methods in Molecular Chemistry, Vol. 2*, B. Carroll, ed., Dekker, New York, 1972.
2. D. J. Harmon and H. L. Jacobs, *J. Appl. Polym. Sci.*, **10**, 253 (1976).
3. R. F. Boyer, in *Seminar Proceedings, Sixth International Seminar on Gel Permeation Chromatography*, Miami Beach, Fl., 1968.
4. H. W. McCormick and W. R. Nummy, U.S. Patent 3,318,813, May 9, 1967.

5. *Liquid Chromatography Review: Size-Exclusion Chromatography*, DuPont Instrument Products Division, 1977.
6. L. Wild and R. T. Guliana, in *Seminar Proceedings, Sixth International Seminar on Gel Permeation Chromatography*, Miami Beach, Fl., 1968.
7. H. L. Browning, Jr., and J. R. Overton, *Polym. Prepr.*, **18**, 237 (1977).
8. M. A. Dudley, *J. Appl. Polym. Sci.*, **16**, 493 (1972).
9. E. E. Drott, in *Chromatographic Science Series, Vol. 8, Liquid Chromatography of Polymers and Related Materials*, J. Cazes, ed., Dekker, New York, 1977, p. 41.
10. T. Provder, J. C. Woodbrey, J. H. Clark, and E. E. Drott, *Adv. Chem. Ser.*, **125**, 117 (1973).
11. P. A. Neddermeyer and L. B. Rogers, *Anal. Chem.*, **41**, 94 (1969).
12. I. Kitayevitch, M. Rona, and G. Schmuckler, *Anal. Chim. Acta*, **61**, 277 (1972).
13. R. Vivilecchia, B. Lightbody, N. Thimot, and H. Quinn, in *Chromatographic Science Series, Vol. 8, Liquid Chromatography of Polymers and Related Materials*, J. Cazes, ed., Dekker, New York, 1977, p. 11.
14. P. L. Dubin, S. Koontz, and K. L. Wright, III, *J. Polym. Sci.*, **15**, 2047 (1977).
15. C. Y. Cha, *Polym. Lett.*, **7**, 343 (1969).
16. E. E. Drott, in *Chromatographic Science Series, Vol. 8, Liquid Chromatography of Polymers and Related Materials*, J. Cazes, ed., Dekker, New York, 1977, p. 161.
17. M. M. Coleman and R. E. Fuller, *J. Marcromol. Sci.—Phys, B*, **11**, 419 (1975).
18. K. J. Bombaugh, W. A. Dark, and J. N. Little, *Anal. Chem.*, **41**, 1337 (1969).
19. N. C. Billingham, *Molar Mass Measurements in Polymer Science*, Wiley, New York, 1977.
20. T. Williams, Y. Udagawa, A. Keller, and I. M. Ward, *Polym. Sci., Part A-2*, **8**, 35 (1970).
21. R. C. Ferguson, H. H. Hoehn, and R. R. Hebert, *Polymer Prepr.*, **18** (2), 309, 1977.
22. H. E. Adams, in *Gel Permeation Chromatography*, K. H. Altgelt and L. Segal, eds., Dekker, New York, 1971, p. 391.
23. R. D. Mate and M. R. Ambler, in *Gel Permeation Chromatography*, K. H. Altgelt and L. Segal, eds., Dekker, New York, 1971, p. 377.
24. A. Dondos, P. Rempp, and H. Benoit, *Macromol. Chem.*, **175**, 1659 (1974).
25. F. M. Mirabella, Jr., and E. M. Barrall, II, *J. Appl. Polym. Sci.*, **19**, 2131 (1975).
26. J. Cazes and G. Fallick, *Polym. News*, **3**, 295 (1977).
27. J-P Menin and R. Roux, *J. Chromatogr.*, **64**, 49 (1972).
28. M. Schrager, *J. Appl. Polym. Sci.*, **17**, 3357 (1973).
29. R. V. Vivilecchia, B. G. Lightbody, N. Z. Thimot, and H. M. Quinn, *J. Chromatogr. Sci.*, **15**, 424 (1977).
30. S. Singh and A. E. Hamielec, *J. Liq. Chromatogr.*, **1**, 187 (1978).
31. J. J. Kirkland, *J. Chromatogr.*, in print.

Thirteen

TECHNIQUES OF MODERN GEL FILTRATION CHROMATOGRAPHY

13.1 Introduction

Although gel filtration chromatography is a form of SEC carried out in aqueous mobile phases, primarily by those interested in the biological sciences, separation principles and equipment are essentially the same as for all SEC methods. Some applications and laboratory techniques are different for GFC, and this chapter is especially concerned with the unique features of GFC as a separating tool.

As with GPC (Chapt. 12), there are substantial differences between the traditional and modern GFC approaches, mainly due to the use of columns of small ($< 10\ \mu m$) particles for much improved resolution. Another distinctive aspect of modern GFC is the use of columns with rigid particles (e.g., silica), as compared to the relatively soft organic gels which have been widely used in traditional GFC. While columns of these stable, highly efficient, inorganic particles have led to substantially improved GFC performance, retention by adsorption and ionic effects do sometimes occur which are not as significant with the traditional soft organic gels. These undesirable characteristics necessitate a somewhat different set of separating conditions for the columns of inorganic particles, and this aspect of modern GFC is detailed in Section 13.3. Novel theories for the GFC process have often been advanced, but we emphasize that all size-exclusion separations have a common basis (Chapt. 2). Also, in GFC the influence of the mobile phase is particularly important because of its effect on the conformation (and size) of the solute. Therefore, in this chapter special attention is given to this effect.

The utility of modern GFC is focused in several areas: [1] for size classification of high-molecular-weight, water-soluble mixtures for analysis; [2] for determination of molecular weight and molecular weight distribution (e.g., for proteins); [3] as a preparative tool; and [4] as a clean-up or prefractionation step, or for classifying complex mixtures into separate fractions for testing or subsequent high-resolution separation.

13.2 Column Packings

Packings for modern GFC columns have been discussed broadly in Section 6.2, and Tables 6.1–6.3 specifically list some materials which are now supplied for this method. Inorganic particles, mainly unmodified or organic-modified silicas, are often utilized for modern GFC, since most organic gels are not suitable at the higher pressures required.

Preparation of Surface-Modified Substrates

Unmodified silica (or alumina) of the proper pore size can be used satisfactorily for many GFC separations, provided that the proper mobile phase/substrate combination is used (Sect. 13.3). An adequate variety of unmodified silica particles is available, but some of the more important packings for GFC are the porous silicas that have been chemically modified with neutral, hydrophilic functional groups to reduce or eliminate unwanted solute adsorption. Surface modification may be conveniently carried out by the reaction

$$\text{Si—OH} + (CH_3O)_3\text{Si—}(CH_2)_3\text{—X} \longrightarrow \text{Si—O—}\overset{\overset{\diagdown O}{|}}{\underset{\underset{\diagup O}{|}}{\text{Si}}}\text{—}(CH_2)_3\text{—X} \qquad (13.1)$$

where X is a highly polar group such as $—NH_2$, $—CHOHCH_2OH$, $-\overset{|}{N}-\overset{\overset{O}{\|}}{C}-CH_3$, and so on. With this general approach, stationary phases modified with several different functional groups have been prepared. For the materials in Table 13.1, silica SI-100 with average pore diameter of 100 Å,

Table 13.1 Characteristics of Substrates Chemically Modified with $Si—CH_2—CH_2—CH_2—X$

		Elemental Analysis of Bonded Phases			
Modifying Group (X)[a]	Nomenclature	C (%, *w/w*)	N (%, *w/w*)	Surface Concentration (μmoles/m^2)	Average Spatial Requirement ($Å^2$)
$—NH_2$	Amine	6.28	2.25	5.0	33
$—NH—CO—CH_3$	Amide	8.44	1.66	4.0	41
$—NH—CO—CF_3$	Trifluoroamide	8.11	1.70	3.9	43
$—NH—SO_2—CH_3$	Sulfonamide	8.82	1.98	5.2	32
$—NH—CO—CH_2—NH—CO—CH_3$	Glycinamide	8.84	2.57	3.0	55
$—O—CH_2—CHOH—CH_2OH$	Glycol	5.48	—	2.2	76
$—(CH_2)_{14}—CH_3$	$RP—C_{18}$	21.03	—	2.8	60

Taken from Reference 1.
[a] Substrate, SI-100 silica gel.

pore volume of 1 ml/g, and specific surface area of about 350 m^2/g, was reacted with various alkoxysilanes to produce the desired dense monomolecular coating. Because of the lower reactivity of alkoxysilanes, the conditions used for the more reactive chlorosilanes (e.g., chlorotrimethylsilane, Sect. 6.2) must be modified to obtain complete surface coverage. However, the lower reactivity of the alkoxysilanes does permit more polar functional groups to be utilized.

To react alkoxysilanes to completeness (full monomeric surface coverage), trace amounts of water must be excluded, and fresh alkoxyorganosilane (free of hydrolysis products) should be used, as with the chlorosilane reactions (Sect. 6.2). Distillation of the alcohol formed during the reaction forces the surface reaction to completion (1). Reaction rate can also be accelerated by increasing temperature or by the addition of an acid catalyst (e.g., *p*-toluene sulfonic acid).

A useful procedure for surface modification with amine, amide, glycol, glycinamide, and sulfonamide groups follows:

> Treat about 10 g of silica support of the desired porosity with 100 ml of 6 *N* HCl ~90°C for 24 hr. Filter off the support and wash repeatedly with distilled water to neutrality. Dry for several hours at 200°C in a vacuum oven. Suspend this fully hydrolyzed silica support in 100 ml of *dry* toluene (dried over Type 4A molecular sieves), and add at least a 2 *M* excess of the alkoxysilane to the suspension. Heat the stirred mixture at a temperature above the boiling point of the alcohol reaction product to distill it from the mixture. Reactions of this type generally are complete within 12 hr, the actual time depending on the particular organosilane used and the boiling point of the suspending solvent. (Xylene may be used instead of toluene to increase the boiling point of the mixture if desired.)

A useful hydrophilic silica support for GFC is that which has been modified with glyceryl (or diol) groups. The reaction proceeds (2, 3) according to

$$\text{Si—OH} + (\text{MeO})_3\text{Si—(CH}_2)_3\text{OCH}_2\overset{\text{O}}{\text{CH—CH}_2} \xrightarrow{\text{pH } 3.5}$$

(glycidoxypropyltrimethoxysilane)

$$\text{Si—O—}\underset{\text{O}}{\overset{\text{O}}{\text{Si}}}\text{—(CH}_2)_3\text{OCH}_2\text{—}\overset{\text{OH}}{\text{CH}}\text{—CH}_2\text{OH} \qquad (13.2)$$

A useful procedure for carrying out this reaction is as follows:

> Heat 10 g of hydrolyzed silica (see above) with 100 ml of a 10% aqueous solution of glycidoxypropyltrimethoxysilane (adjusted to pH 3.5 with dilute hydrochloric acid) for several hours at 90°C with occasional stirring. Filter off the bonded phase support and wash with distilled water and acetone. Reflux for 15 min in 150 ml of acetone and dry. The pH 3.5 condition used is sufficient to convert the oxirane ring in the starting silane to a glycol during the coupling reaction. The functionality on surface prepared by this or similar procedures is referred to as Glycophase-G or diol by various suppliers.

Properties of Surface-Modified Substrates

The surface-modification reaction can result in significant alteration of the mean pore diameter for the starting porous silica. For example, when a support with 507 Å pores was coated with 3-propyl-glyceryl groups by the procedure just given, the pore diameter was found to have decreased to 470 Å. This 37 Å decrease in pore diameter suggests that the organic layer is 18–19 Å thick, while the theoretical layer for a 3-propyl-glyceryl monolayer is 14 Å (2). The level of surface deactivation of porous silica or glass can also be determined by measuring the extent of solute adsorption. The recovery of enzymatic activity from test solutions has also been used (2). Both the change in pore dimensions and the fact that labile enzymes are not denatured (i.e., their activity is unchanged) during separation suggest that acidic SiOH groups on the surface are effectively eliminated by a cohesive layer of neutral glyceryl groups with Reaction 13.2.

13.3 Substrate/Mobile Phase Combinations

GFC with silica-based small particles can be carried out by one of three approaches. First, unmodified porous silica particles can be used in aqueous systems for separating less adsorptive solutes. If required to eliminate adsorption, a modifier that is more strongly adsorbed to the packing than the solute can be added to the mobile phase. Second, the adsorption of some solutes can be eliminated by forming certain complexes prior to the separation. Finally, GFC separations can be made with a support surface that is permanently modified with neutral, hydrophilic groups to eliminate solute adsorption. In all cases the mobile phases must meet the usual criteria of purity, detector compatibility, viscosity, and so on (Sect. 8.2).

Unmodified Porous Silicas

Unmodified porous silica can be used with solutes that do not strongly adsorb onto the packing in aqueous mobile phases. Typically, it is convenient to use a buffer to reduce adsorption or to protect biological substances from denaturation (i.e., to prevent a loss in biological activity). For example, to prevent adsorption of bovine serum albumin on controlled porosity glass, a mobile phase of amino acid buffers containing glycine, *dl*-alanine, and β-alanine at pH 8.0 was used to effect solute recoveries of 68–76% (4).

With unmodified porous silicas, an increase in mobile phase ionic strength generally decreases solute adsorption. Polyvalent ions, such as phosphate and sulfate, seem to be more effective than monovalent anions in reducing adsorption. Effective ionic strengths are generally ≥ 0.05 *M*, depending on the solute and the anion, with 0.1 – 1.0 *M* being satisfactory. However, with unmodified porous silicas and glasses the effect of mobile-phase ionic strength on retention can be anomalous and unpredictable with some solutes. For example, certain proteins elute at very low ionic strengths (0.001 *M*), but adsorb when the ionic strength is greater than about 0.05 *M* (5). Such effects suggest that the elution of electrically charged solutes from silica pores that also can assume a charge is very much influenced by the nature and concentration of the mobile-phase ions. The relationships between retention volume and mobile-phase ionic strength for proteins on controlled-pore glasses appear to follow the same general phenomena as found for agarose gels (6). In summary, it appears that the ionic strength of the aqueous mobile phase must be optimized experimentally for each system.

If the sample permits, adjustment of pH is sometimes effective in eliminating adsorption. For example, poly(vinyl alcohol) has been successfully characterized on unmodified porous silica using a mobile phase of 1:1 0.025 *M* tetramethylammonium nitrate (pH 3.0)/methanol (7). Adjusting the pH to <3.5 inhibits SiOH ionization, making these groups less available for unwanted ion exchange. However, at low pH, hydrogen bonding to the unionized SiOH groups may increase.

Unwanted interactions of solutes with organic gels are also widely recognized. For example, Sephadex and Bio-Gel P contain a small number of carboxylic acid groups which can ionize and exhibit cationic exchange properties in eluents of low ionic strength at pH > 4. Basic proteins such as lysozyme may be retained by adsorption on these gels, and elution can only be effected with a dilute salt solution (e.g., 0.05–0.1 *M* sodium chloride). In some cases ionic interaction between the solute and the carboxylic groups in the matrix is so strong that suppression of the ionization of the carboxylic groups is necessary (e.g., by working with 0.02 *N* HCl). Adsorptive losses of certain solutes (e.g., proteins) on freshly packed organic gel columns can sometimes be eliminated by first saturating the binding sites on the gel with the solute.

Other mechanisms of interaction between organic gels and aromatic and heterocyclic solvents suggest involvement of the π-electron system of the solute and an electron-deficient or electronegative portion of the gel matrix. Hydrogen bonding between substituents on aromatic solutes, or the heteroatom of heterocyclic solutes, and certain functionalities on the organic gel is also possible. For example, at low pH aromatic acids hydrogen-bond to Sephadex G-10, while at high pH the acids are ionized and excluded from the gel.

Adsorption of some proteins to porous silica can be eliminated by coating the surface with highly polar macromolecules such as Carbowax-6000 or Carbowax-20M. In this approach (8, 9), the untreated porous silica is washed with 6–8 bed volumes of methanol at 60°C. The temperature is raised to 90°C and the column washed with three bed volumes of water. Three hundred milliliters of 1% Carbowax-20M (for a 120 × 0.3 cm column) in water is next passed through the column at 90°C, and the temperature is then lowered to 9°C. The 1% Carbowax-20M solution is passed through the column during the cooling period. When the temperature of the column reaches 9°C, the column is washed with five bed volumes of water and before use is equilibrated with three bed volumes of the appropriate buffered mobile phase. Used columns may be regenerated by washing the column with 3–5 bed volumes of methanol and repeating the procedure described above. Carbowax-20M typically is used to deactivate large-pore (e.g., $\geq$250 Å) substrates and Carbowax-6000 for smaller-pore packings.

EM-gel type SI-1000 treated with Carbowax-20M is satisfactory for separating proteins without apparent change in solute properties (12). Figure 13.1 shows the SEC separation of a fatty acid synthetase from *Corynebacterium diphtheriae* using a Carbowax-20M modified substrate. This separation shows two protein peaks, the first peak at the shorter retention time (MW 2.5–3 × 10^6) exhibiting enzyme activity of fatty acid synthetase which coincides with UV absorbance. The second eluting protein peak (MW 1.25–1.5 × 10^6) contains no enzyme activity and is presumed to represent a subunit of the enzyme. Recovery was 70–80% for these labile enzymes in various runs.

Since Carbowax-treated porous silicas and glasses must be renewed periodically to ensure lack of adsorption by solutes, a more convenient approach is to maintain a continuous concentration of the Carbowax modifier (e.g., 0.01–0.1%) in the mobile phase to ensure a reproducible level of substrate deactivation. The ionic strength level of the mobile phase buffer is important even with Carbowax deactivation of porous glasses. Certain

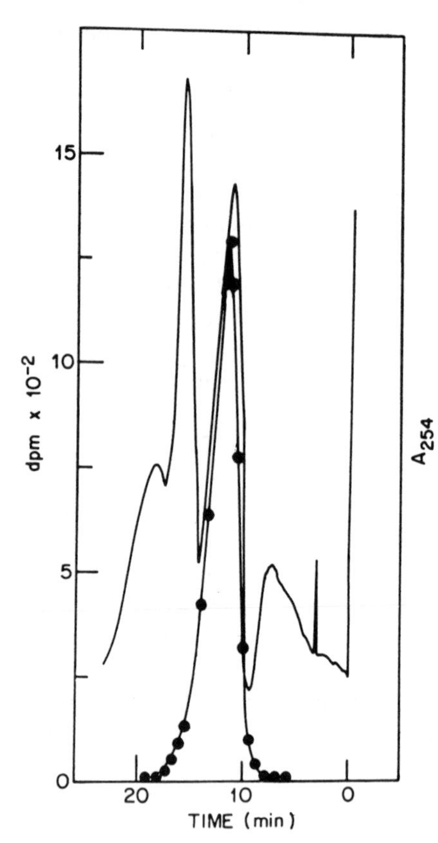

Figure 13.1 HPGFC of a fatty acid synthetase using Carbowax-20M modified porous silica.

Column, 120 × 0.3 cm, EM-silica gel Type SI-1000 (exclusion limit, 10^6 MW), deactivated with Carbowax-20M; mobile phase, 0.4 *M* potassium phosphate buffer, pH, 7.3; temperature, 9°C; pressure, 300 psi; flow rate, 3 ml/min; sample, 100 μl containing 200 μg of protein. ●-●, Enzyme activity (disintegrations per minute × 10^{-2}); solid line, UV detector at 254 nm. (Reprinted with permission from Ref. 10.)

protein separations are successful only when the concentration of the eluting buffer is greater than about 0.1–0.3 *M* (8).

Solute Complexation

Adsorption of certain solutes to unmodified porous silicas can be eliminated by forming stable solute complexes and operating with appropriate mobile phases. For example, sodium dodecylsulfate (SDS) complexes of denatured low-molecular-weight proteins have been separated on unmodified porous silica packings specifically for molecular weight determination. With this approach the proteins are placed in the same conformation so that a linear MW versus V_R relationship is obtained for accurate MW determination [see. Ref. 10]. Initial attempts to separate the free, low-molecular-weight proteins with 8 *M* urea mobile phase resulted in either partial or complete adsorption to untreated controlled porosity glass. This problem is not completely overcome by increasing pH or ionic strength, or by adding alcohols, Carbowax-6000, or Carbowax-20M. However, by preparing the protein/SDS complexes, and operating in a mobile phase of 6 *M* urea, 0.5% SDS and 0.5 *M* sodium phosphate buffer (pH 7), all proteins are eluted without adsorption. Protein-SDS complexes are typically prepared by incubating approximately 10 mg of protein with 20 mg of SDS in 1 ml of the mobile phase at 90°C for 10 min. For those proteins which are not in the *S*-carboxymethyl form, 2% β-mercaptoethanol is included in the incubation mixture. With this approach a unique plot of MW versus distribution coefficient K_d was obtained for proteins (12,000–140,000 MW) chromatographed on controlled-porosity glass of 225 Å (22.5 nm) pore diameter (10). The fact that all the proteins lie on the same well-defined calibration plot suggests that in the presence of SDS, each polypeptide chain adopts the same conformation. A reliable calibration curve was also established for polypeptides in the molecular weight range 3500–12,000 with 12.3 nm pore diameter columns, supporting the contention that protein-SDS complexes exhibit a common conformation, approximating a sphere, or more likely an oblong football-like shape (14). Thus, complexation of proteins with SDS places these solutes in a common shape, and in a high concentration of urea, it is possible to obtain accurate molecular weight estimations by GFC. On the other hand, apparent correlation of the molecular weight of "free" proteins with retention volume is largely fortuitous, since the shape (and therefore the size) of uncomplexed proteins varies significantly (Table 2.4).

Surface-Modified Packings

The most recent activity in modern GFC has been with porous silica substrates modified with a dense monolayer of a neutral, hydrophilic organic group (e.g., μ-Bondagel E) to eliminate adsorption. Work in this area is still limited and surface-modified packings suitable for modern GFC have just been introduced, so critical evaluation is not yet available. Nevertheless, surface-modified materials appear to be the most generally useful approach for HPGFC, and it is anticipated that developments in this area will become increasingly important.

To be effective in GFC, the surface-modified porous packing must be readily wetted by water. This can easily be tested by placing a little packing on the surface of water and noting whether the material quickly wets and sinks to the bottom of the container. Modified surfaces with poor wetting characteristics can engage in hydrophobic interactions that can result in the reverse-phase retention of certain molecules in addition to the desired size exclusion.

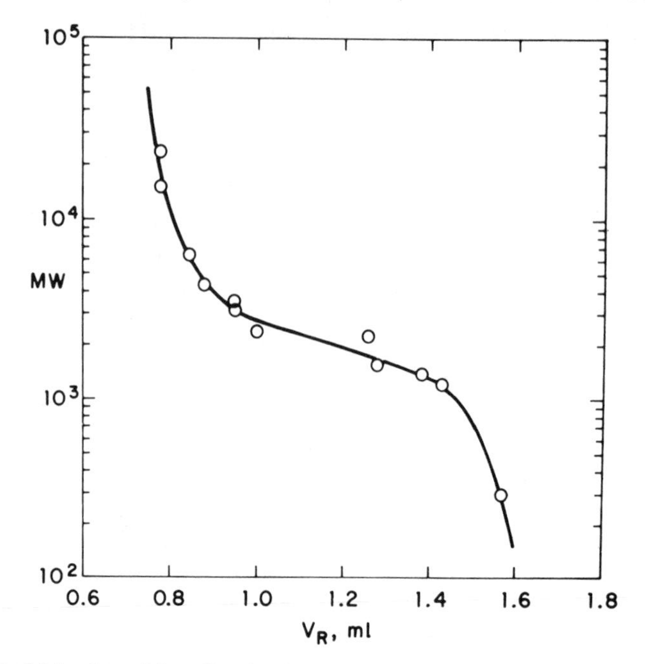

Figure 13.2 Molecular weight calibration for proteins with "diol"-modified porous silica. Column, 30 × 0.4 cm LiChrospher SI-100 "diol," 10 μm; mobile phase, 0.05 *M* sodium phosphate buffer, 0.1 *M* sodium chloride, pH 7.5; linear velocity, 0.95 cm/s; temperature, ambient; sample, 10^{-4} *M* or lower in 0.07 *M* phosphate buffer, pH 6.8; detector, UV at 210 nm. (Data from Ref. 13.)

Glyceryl-modified supports have been developed primarily for the size separation of biological substances. When properly prepared, such packings appear to be especially suited for labile solutes such as proteins. Figure 13.2 shows that retention of various enzymes chromatographed on a diol-modified silica substrate decreases with log MW. Recoveries of enzymes on diol packings were made under static and dynamic conditions, and the results indicate that a neglibly small irreversible adsorption or denaturization of the enzymes does occur (Table 13.2).

Unfortunately, coverage of silica surfaces with glyceryl or diol groups appears to be imperfect in some cases. For example, controlled-pore glass substituted with glyceryl groups (Glycophase G/CPG, Electro-Nucleonics) showed a slight anionic character which affected the size-exclusion elution of certain enzymes at ionic strengths below 0.1 (14). However, at ionic strengths >0.1, this residual ionic character had no effect on the elution of proteins and the packing is judged to be a good medium for some GFC

Table 13.2 Recovery of Enzymes from LiChrospher-100/Diol

			Recovery (%)			
				Dynamic[b] (Column)		
Enzymes	MW	pI	Static[a] (Incubation)	I	II	Retention Time (s)
Glutathione (red.)	307	—	—	—	98.1	376
Cytochrome c (horse heart)	12,500	9.2	98.5	87.3	99.3	342
Lysozyme (human)	14,300	10.5	98.7	94.7	99.1	333
Haemoglobin (human)	16,000	7.0	94.6	97.9	99.3	282
β-Lactoglobulin (cow)	36,000	5.2	84.4	96.3	96.6	228
Trypsin	23,300	8.5	97.1	98.5	98.9	303
Chymotrypsinogen a (beef)	25,000	9.2	89.6	93.7	99.4	240
Pepsin	33,000	2.9	98.9	95.2	97.6	228
Albumin (hen)	45,000	4.6	—	—	100	210
Albumin (ox)	67,000	5.1	96.7	93.3	100	204
Aldolase (rabbit)	158,000	9.5	—	—	98.1	186
Catalase (beef)	240,000	8.0	—	—	98.6	186

Taken from Reference 15.

[a] *Test conditions.* Support: 100 mg; incubation time 1 hr; sample: protein (enzyme) 10^{-6} *M* in 0.07 *M* phosphate buffer + 0.1 *M* NaCl pH 6.8.

[b] *Test conditions.* I—column: 30 by 0.4 cm i.d.; room temperature; detector: UV 210 nm; sample: 10^{-4} *M* or lower in 0.07 *M* phosphate buffer, pH 6.8; linear velocity: 0.29 cm/s. II—same as I but linear velocity 0.95 cm/sec and 0.05 phosphate buffer +0.1 *M* NaCl pH 7.5.

separations. Glyceryl-modified porous silicas also are suitable for determining the molecular weight of protein polypeptide chains in a variety of denaturing solvents, for example, 8 *M* urea and 6 M guanidine hydrochloride with 0.1% sodium dodecyl sulfate (15).

Silica surfaces modified by treatments with λ-aminopropyltriethyoxysilane and 3-(2-aminoethylamino)propyltrimethoxysilane also have been successfully utilized for eliminating irreversible protein adsorption (16). Figure 13.3 shows the isotherms for albumin and lysozyme adsorption from 0.1 *M* acetic acid on a hydroxylated silica and that same substrate modified by λ-aminopropyltriethyloxysilane. For the modified silica, adsorption of these proteins

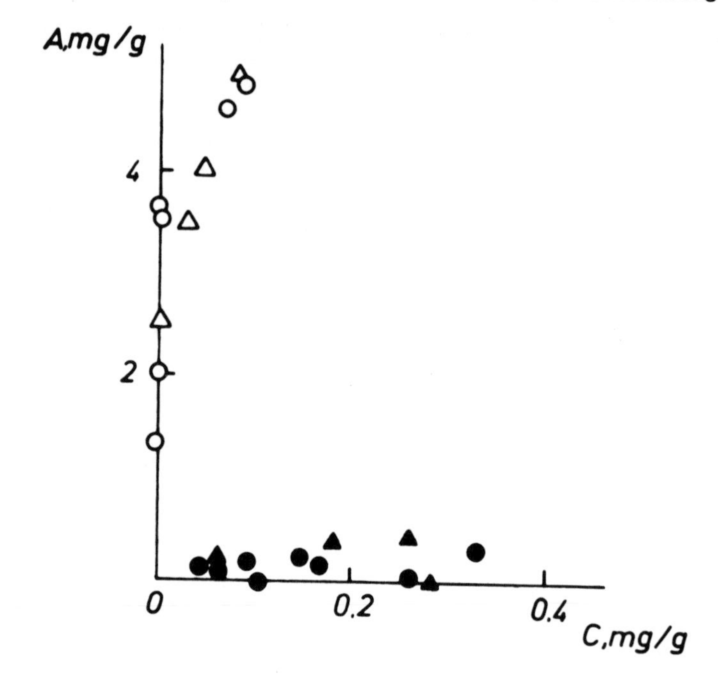

Figure 13.3 Adsorption isotherms on λ-aminopropyltriethoxysilane-treated porous silica. Lysozyme (circles) and serum albumin (triangles); open points, hydroxylated Silochrom C-80, 80 m^2/g, 500 Å; filled points, amino-Silochrom; 0.1 *M* acetic acid solutions. A, adsorbed concentration; C, initial concentration. (Reprinted with permission from Ref. 16.)

is negligible; the surface silanol groups are essentially deactivated by the organosilyl group. Because of the high concentration of amino groups, the charge on the particle surface is strongly positive in acid solution. Since net charges on albumin and lysozyme in 0.1 *M* acetic acid are also positive, the decrease in protein adsorption on the modified silica may not only be connected with the screening of the residual surface silanol groups by an organic film, but may also be due to electrostatic repulsion of the positively charged macromolecules, preventing penetration of the pores by the protein molemolecules—called the ion-exclusion effect. If salt is added (e.g., 0.3 *M* sodium chloride), the electrical double layer around the protein particles and on the adsorbent surface is neutralized, with attendent reduction of electrostatic repulsion. The pores then become at least partially available to the protein molecules.

Unfortunately, free "amine" stationary phases are relatively unstable in some aqueous systems. This is not surprising, since aqueous solutions of amine packings are alkaline (pH 9 or so), which is sufficient to attack the silica substrate. However, if the amine stationary layer is neutralized with acid (pH ≤ 7), the modifying layer is relatively stable. Amine-modified silica supports need additional study to determine general applicability for the HPGFC separation of proteins and other biologically important macromolecules.

Other neutral hydrophilic functional groups (Table 13.1) have also been used for separating biologically important macromolecules. Supports modified with "amide" groups are particularly promising, as amide has been used with only water as the eluent. In this case, proteins are eluted prior to totally permeating species such as D_2O (1). Such findings hold considerable promise for the usefulness of polar-modified supports; however, more experience with the stability of these materials is required.

Hydrophobic interactions can become a problem with some polar-bonded phases. For example, some proteins are retained on ether-modified μ-Bondagel columns beyond the total permeation volume (17). Apparently the bonded ether layer on μ-Bondagel, although water-wettable, is weakly hydrophobic. Increasing the concentration of salt in the mobile phase leads to increased retention, and changing pH has little effect on retention, both characteristics being properties of hydrophobic (reverse-phase) chromatography. In this case, addition of 0.2–1% sodium dodecylsulfate to the mobile phase often gives quantitative recovery of proteins, as illustrated in Figure 13.4 for a human plasma profile. Adding ethylene glycol to buffers can also result in quantitative elution of proteins for μ-Bondagel columns.

To summarize, hydrophilic-modified supports have significant applicability in HPGFC, providing the proper mobile phase/substrate combination is employed. Where possible, it is desirable to use a buffer with the modified packings, to fix the pH of the system, and to maintain ionic strength in a range such that adsorption and ion exchange are minimized. As with the unmodified silicas, ionic strengths of >0.1 are preferred, and polyvalent anions may be more effective in eliminating adsorption. Tetraalkylammonium salts may also be useful in reducing adsorption, because of the steric shielding effect of the large tetraalkylammonium ion in the adsorbed ionic double layer. In some systems, adjustment of pH helps to reduce adsorption. Whether or not a low or high pH is required depends on the nature of the undesired adsorption process and must be determined experimentally. The recognition that unwanted retention can occur with solutes on all types of packings, both inorganic and organic, indicates that each system should be investigated to ensure that the mobile phase/substrate combination allows the separation to be carried out by the desired size-exclusion process. Nonelution of a solute, or elution after the total permeation volume, indicates that the separating system must be appropriately modified.

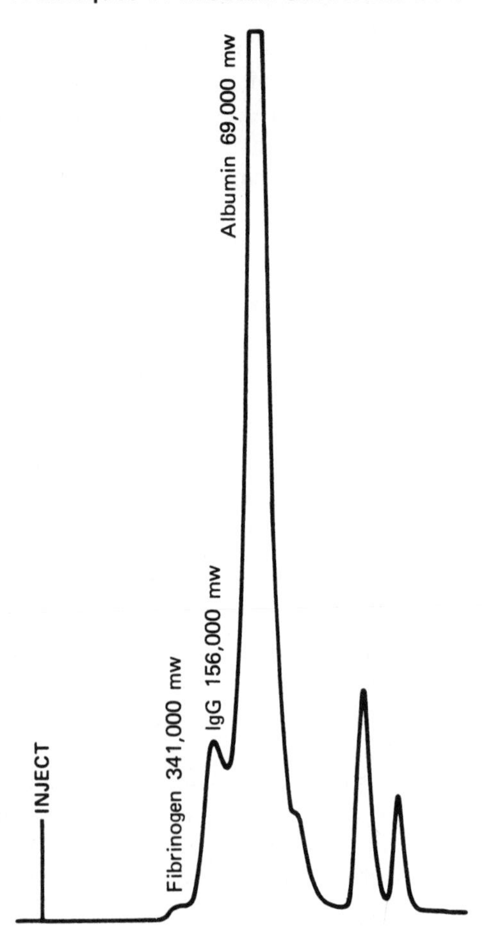

Figure 13.4 Separation of a diluted human blood plasma sample.
Columns: four 30 × 0.4 cm μ-Bondagel (two 1000 A, one 500 A, one 125 A); mobile phase, 0.5 *M* Trizma acetate, pH 7.4 with 1% sodium dodecylsulfate; flow rate, 0.3 ml/min; detector UV, 280 nm. (Reprinted with permission from Ref. 17.)

13.4 Operating Variables and Technique

The effect of particle size d_p, and mobile phase velocity, v, on the efficiency of GFC separations are the same in all of SEC (Sect. 3.4). Figure 13.5 shows plate height versus velocity plots for ovalbumin (MW 43,000) on various size Glycophase-G controlled-porosity glass packings. As expected, small-diameter supports are more efficient, and this effect is most apparent at

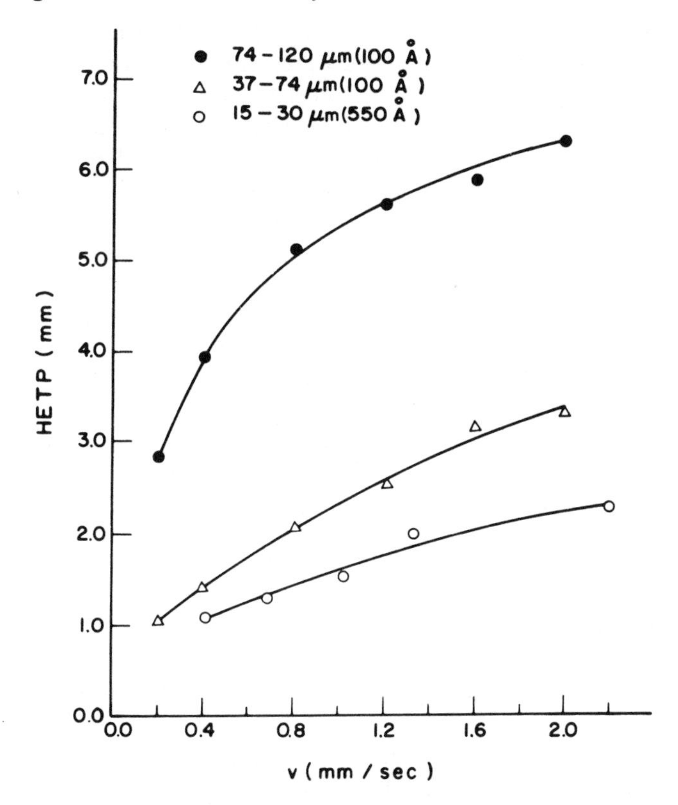

Figure 13.5 Plate height versus mobile-phase velocity plots for ovalbumin.
Column, 100 × 0.46 cm Glycophase-CPG; mobile phase, 0.1 *M* sodium phosphate buffer, pH 7; detector, UV, 254 nm. (Reprinted with permission from Ref. 2.)

high mobile-phase velocity. Clearly, the decrease in diffusion distance (pore depth) within the particle improves high-speed GFC analyses.

When allowed by the sample, elevated temperatures should be used in GFC separations to decrease the viscosity of the mobile phase and enhance column performance. However, in many instances, biologically important macromolecules are unstable at higher temperatures, and therefore room temperature operation is required. It is rare that below-ambient temperatures are necessary in modern GFC, since separations are carried out rapidly (e.g., <20 min) before significant solute decomposition occurs. As with all SEC separations, minor changes in plots of molecular weight versus retention volume can occur with changes in temperature, owing to alterations in solute conformation.

The stability of high-performance GFC packings depends to a large extent on the conditions in which they are utilized. Organic gels are little affected by pH change, but the more mechanically stable silica-based substrates must be used at pH < 8 (Fig. 8.1). There are indications that hydrophilic-bonded phases are more susceptible to degradation than reverse-phase packings in aqueous mobile phases (18). This may be a function of the greater ability of water to attack the bond connecting the organic phase to the silica because of superior wetting of the hydrophilic surface. It is also likely that, just as in the case of reverse-phase packings, partially modified silica surfaces are less stable than are fully reacted surfaces. Inserting a precolumn of the same packing prior to the sample injector is at times effective in prolonging column life (18a). More data on the stability of hydrophilic-modified supports for GFC are needed.

The effect of sample size on column performance in GFC is discussed in Section 7.4. Total sample loads should be maintained within a range where constant retention volumes and constant plate count (peak widths) are maintained.

13.5 Utility

Modern gel filtration chromatography is used for several purposes: analytical classification of complex mixtures by size, molecular weight distribution determination, preparative fractionations, and prefractionations or "cleanup" separations. Each of these application areas requires different experimental techniques, and sometimes different equipment. HPGFC is the newest of the LC methods, and therefore the number of published applications is relatively limited. However, most if not all of the traditional GFC separations can undoubtedly be performed in the modern HPGFC mode with greatly improved separation speed and resolution. For information on traditional GFC with organic gel columns, appropriate references in the Bibliography for this chapter should be consulted.

Size Classification for Analysis

Classification of samples into discrete peaks or fractions is perhaps the area of largest utility for GFC. A wide variety of sample types can be used, from very high molecular weight macromolecules in naturally occurring systems to monomolecular species in drugs and foods. While rigid silica-based particles are generally preferred because of their superior resolution and mechanical stability, organic gels have been utilized historically for most separations because of their earlier availability. For example, in Figure 13.6 the separation of nucleotides from bases and nucleosides on a column of

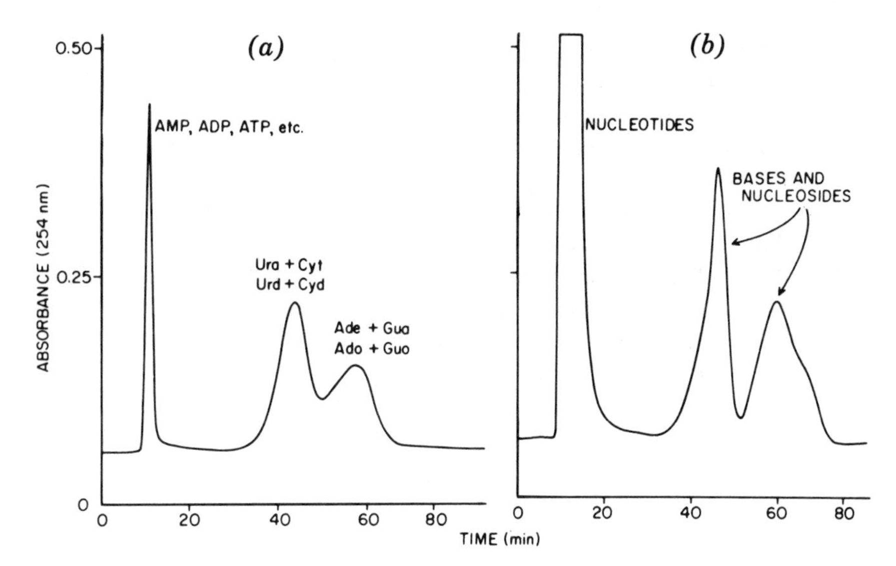

Figure 13.6 GFC separation of nucleotides from bases and nucleosides on Bio-Gel P-2.
Column, 75 × 0.63 cm, 44–73 μm Bio-Gel P-2; mobile phase, 0.02 M NH_4HCO_3, pH 8.2; flow rate, 0.5 ml/min at room temperature; pressure, 50 psi; detector, UV; sample; (a), ~5 nmol each of 5′-mono-, 5′-di-, and 5′-triphosphates of adenosine, guanosine, uridine, and cytidine, and ~10 nmol each of indicated bases and nucleosides dissolved in 100 μl of mobile phase; (b), 200 μl of 0.025 M citrate buffer (pH 8.2) containing acid-soluble constituents derived from mouse neuroblastoma. (Reprinted with permission from Ref. 19.)

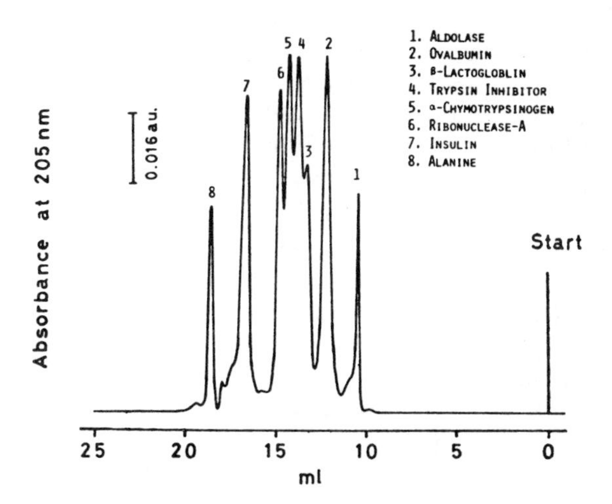

Figure 13.7 Separation of proteins on small-particle gel column.
Column, 60 × 0.75 cm TSK-Gel 2000 SW; mobile phase, 0.01 phosphate buffer (pH 6.5) with 0.2 M sodium sulfate; flow rate, 0.3 ml/min; detector, UV, 205 nm; sample, about 1 μg each. (Reprinted with permission from Ref. 16a.)

Bio-Gel P-2 is illustrated. Figure 13.6*a* is a chromatogram of known mixtures, and Figure 13.6*b* is a chromatogram of acid-soluble constituents derived from mouse neuroblastoma. Figure 13.7 shows the separation of a complex mixture of protein standards using a small-particle (8–12 μm) column of hydrophilic gel packing. The high performance of columns of such packings permits the detection of nanogram amounts of proteins by monitoring in the 200–200 nm wavelength range.

Modified substrates are usually quite versatile for separating a wide variety of natural macromolecules. For example, proteins and enzymes can be separated by high-speed GFC, as illustrated in Figure 13.8. Here a mixture is separated on a single 30 cm column of glyceryl-modified silica packing

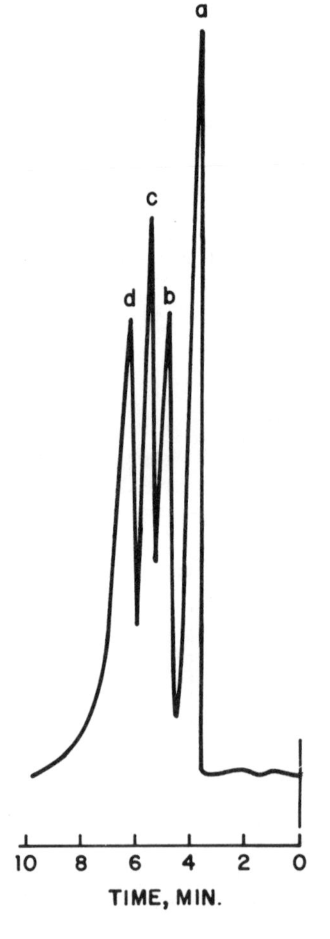

Figure 13.8 Separation of enzymes on "diol"-modified porous silica.

Conditions same as for Figure 13.2 except flow rate, 2.85 ml/min; pressure, 35 bars; sample: a, aldolase (rabbit); b, chymotripsinogen a (beef); c, lysozyme (human); d, reduced glutathione. (Reprinted with permission from Ref. 18.)

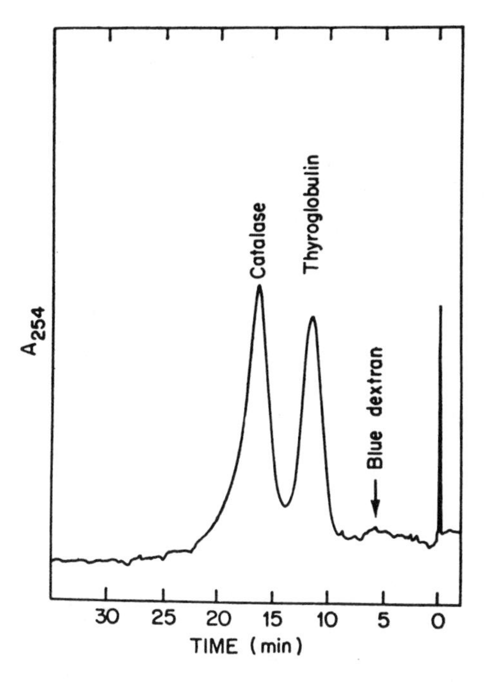

Figure 13.9 Separation of high-molecular-weight proteins on Carbowax-20M deactivated porous silica.

Conditions same as for Figure 13.1 except mobile phase, 0.3 Tris-HCl buffer, pH 7.6; flow rate, 3.5 ml/min; detector, UV, 254 nm. (Reprinted with permission from Ref. 10,)

in less than 10 min, making feasible the rapid assay of such compounds in a variety of matrices. Certain high-molecular-weight proteins can also be separated on unmodified porous silica packings which are deactivated with Carbowax-20M prior to use (Sect. 13.3). In Figure 13.9 we see the HPGFC of two high-molecular-weight proteins, thyroglobulin and catalase, the column void volume being determined with blue dextran in an earlier run. Compared to conventional chromatography on a soft SEC gel column, this separation is much faster. For this run it was necessary to use relatively high buffer concentrations (0.3 *M* Tris), since at lower concentrations peak tailing was apparent.

Nucleic acids have been characterized on columns of Glycophase G/CPG as illustrated in Figure 13.10. This separation was carried out with packing particles somewhat larger than normally used for modern GFC. Consequently, much improved performance can be expected with a similar column of 10 μm particles. The first peak in Figure 13.10 to elute from the column was

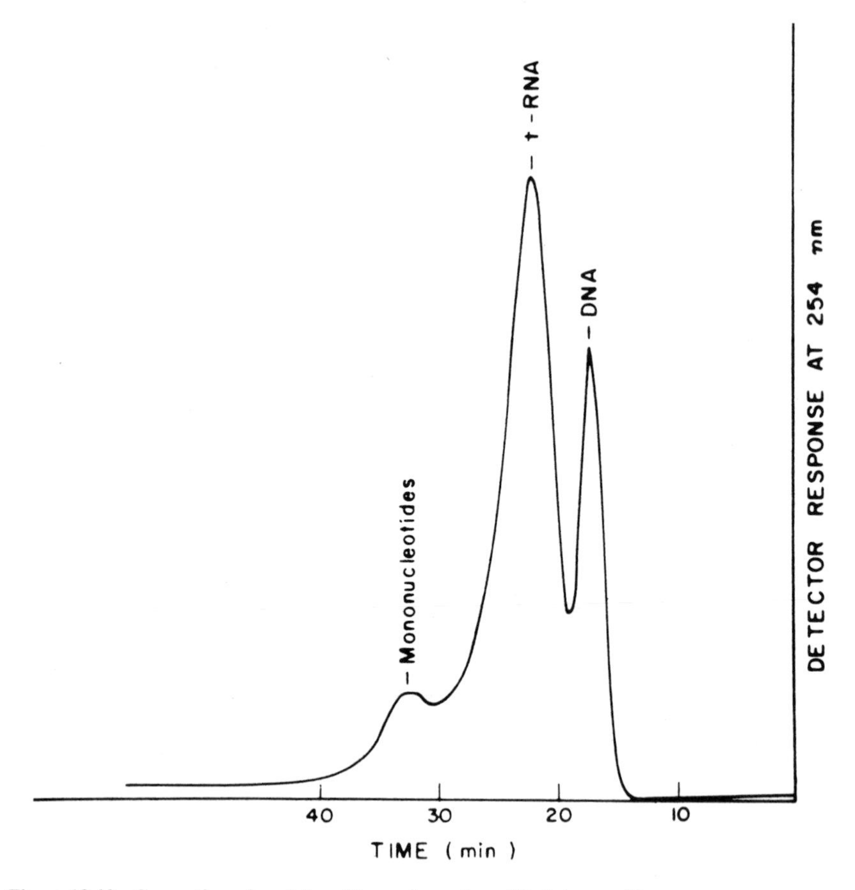

Figure 13.10 Separation of nucleic acids on glyceryl-modified porous silica.
Column, 100 × 0.48 cm Glycophase G/CPG 100 Å, 37–74 μm; mobile phase, 0.2 *M* phosphate buffer, pH 7.0; mobile-phase velocity, 0.1 cm/s; pressure, <100 psi; detector, UV, 254 nm, 0.2 AUFS. (Reprinted with permission from Ref. 2.)

salmon sperm DNA followed by a mixture of *t*-RNA species and mononucleotides. Note that elution was carried out with 0.1 *M* sodium phosphate buffer at a pH of 7, a system that is compatible with many samples of biological consequence. A variety of materials (e.g., drugs, flavors) in naturally occurring substrates can be readily measured by HPGFC. In Figure 13.11 we see a fractionation of beer using a deactivated porous silica packing. Constituents were size-classified into flavor components, sugars, and low-molecular-weight compounds. In Figure 13.12 a separation of filtered coffee on an "amide"-modified substrate is illustrated. In this case 1 μl of filtered

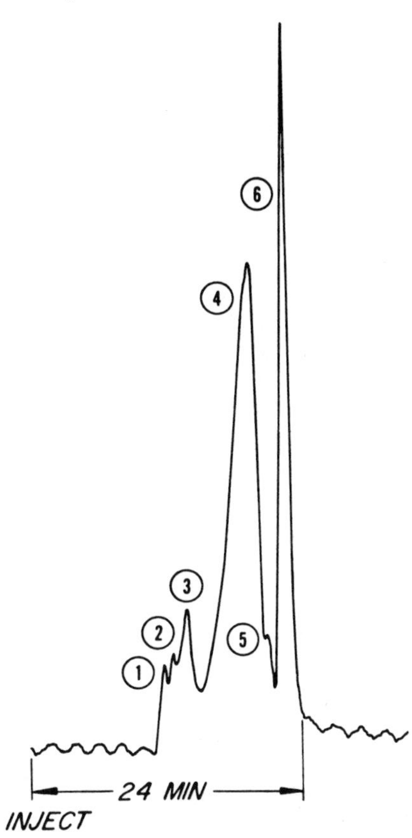

Figure 13.11 Fractionation of beer with deactivated porous silica.
Column, 200 × 0.26 cm Vit-X-328, 84 Å; mobile phase, distilled water; flow rate, 0.48 ml/min; detector, RI, sensitivity, 2 X; peaks; 1–3, flavor components; 4, sugars; 5, unknown; 6, ethanol. (Reprinted with permission from Ref. 20.)

coffee produced peaks eluting after the total permeating volume V_e when water was used as the mobile phase. These adsorbed peaks, identified as chlorogenic acid and caffeine, respectively, probably could have been made to elute only by size exclusion with an appropriately buffered mobile phase.

Molecular Weight and Molecular Weight Distribution

As with the GPC analysis of synthetic polymers, GFC can be used to determine the molecular weight (MW) or molecular weight distribution (MWD) of neutral, water-soluble macromolecules. The same possibilities and limitations exist in GFC as for GPC. When macromolecules with a single molecule weight or a narrow distribution of molecular weights are involved, (e.g., proteins), the simple peak position calibration method can be utilized (Sect. 9.2). If good standards are available, accurate molecular weight

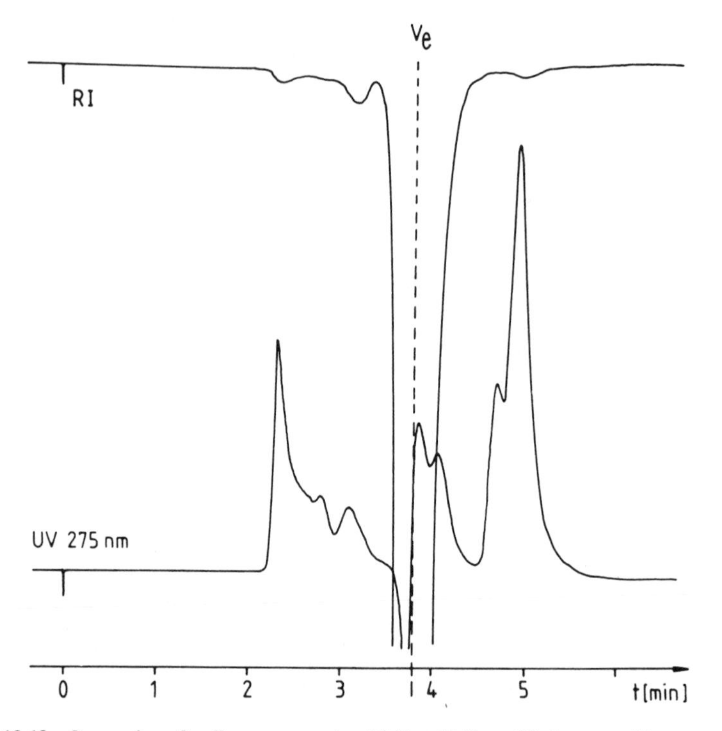

Figure 13.12 Separation of coffee components with "amide"-modified porous silica. Column, 30 × 0.43 cm amide-modified LiChrosorb SI-100; mobile phase, water; flow rate, 0.93 ml/min; pressure, 35 bar; detectors, UV, and RI; sample, 1 μl of filtered coffee. (Reprinted with permission from Ref. 1.)

information can be obtained. With this approach, a series of protein standards of known molecular weight is chromatographed using column packings whose pore structures are optimized for the molecular weight range of interest (Sect. 8.5).

In Figure 13.13, calibration curves are shown for protein polypeptides using denaturing mobile phases to prevent adsorption (see also Table 13.3). In Figure 13.13*a* the mobile phase was 6 *M* guanidine hydrochloride/0.15 *M* phosphate buffer, pH 7, used with hydrophilic-modified porous silica columns of three mean pore diameters. Recently, this mobile phase has replaced 8 *M* urea as the solvent of choice for protein denaturation, and only a few exceptionally stable proteins fail to adopt a conformation close to a random coil under these conditions. The judgment is based on the sharp, symmetrical peaks eluting in order of known molecular weight with values of the distribution coefficient *K* less than unity. Figure 13.13*b* shows calibration curves for two of

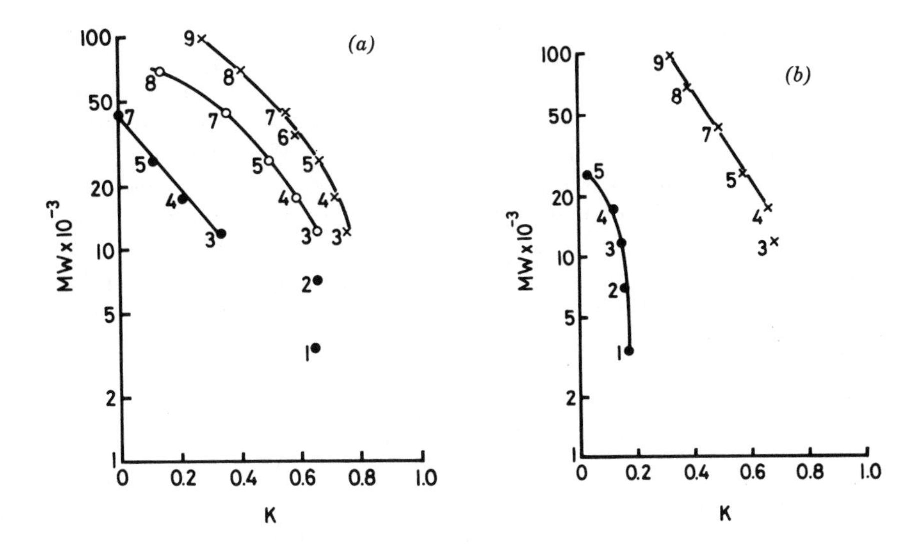

Figure 13.13 MW calibration plots for proteins with glyceryl-modified controlled-porosity glass. Columns, 165 × 0.9 cm, 74–125 μm. ○, Glycophase-CPG, 22.7 nm pores; ●, Glycophase CPG, 11.8 nm pores; ×, Glycophase CPG, 36.8 nm pores. (*a*) Mobile phase, 6 *M* guanadine hydrochloride/0.15 *M* phosphate buffer, pH 7. (*b*) Mobile phase, 0.1% sodium dilauryl sulfate (SDS)/0.05 *M* phosphate buffer, pH 7. (Reprinted with permission from Ref. 15.)

Table 13.3 Molecular Weights of Proteins Used for Figure 13.13

Number	Protein	Subunit Molecular Weight
1	Insulin B chain	3,400
2	High-tyrosine component	7,000
3	Cytochrome c	11,700
4	Apomyoglobin	17,200
5	*a*-Chymotrypsinogen a	25,700
6	Rabbit tropomyosin	33,500
7	Ovalbumin	43,000
8	Bovine serum albumin	68,000
9	Oyster paramyosin	97,000

Taken from Reference 1.

the pore diameters (36.8 and 11.8 nm) using a different mobile phase—0.1% SDS/0.05 *M* phosphate buffer, pH 7, as eluent. Note the change in the calibration plots with the two different mobile phases, due to the change in protein conformations between the two different mobile phases. Unknown proteins can be chromatographed with either system, and the elution volumes of sample peaks compared to the calibration curve for estimation of molecular weight. Using the column of 36.8 nm mean pore diameter, molecular weight results have been obtained for several purified low-sulfur protein components from wool which agree well with those obtained by analytical ultracentrifugation (15). As stated in Section 2.4, the accuracy of this approach is dependent on constant conformation and structures between the calibration standards and the unknown solutes. Under these conditions, errors of <5–10% in molecular weight are obtainable if single species are involved.

HPGFC can be used to obtain MW and MWD for bulk sodium heparins using the peak position procedure with narrow-MWD sodium heparin standards, as shown in Figure 13.14. Figure 13.15 is a chromatogram of mucosal sodium heparin with both refractive index and ultraviolet detection. Using the approach of Equation 10.2, $\overline{M}_n$ and $\overline{M}_w$ were calculated for a series of unknown sodium heparin samples in the MW range of 10^4. Reproducibility of data showed standard deviations of 0.8% for $\overline{M}_w$ and 1.2% for $\overline{M}_n$, and analyses on duplicate sets of columns gave results within 1%.

When samples involve a wide molecular weight distribution, the simple peak position calibration method must be modified. If no narrow-MWD

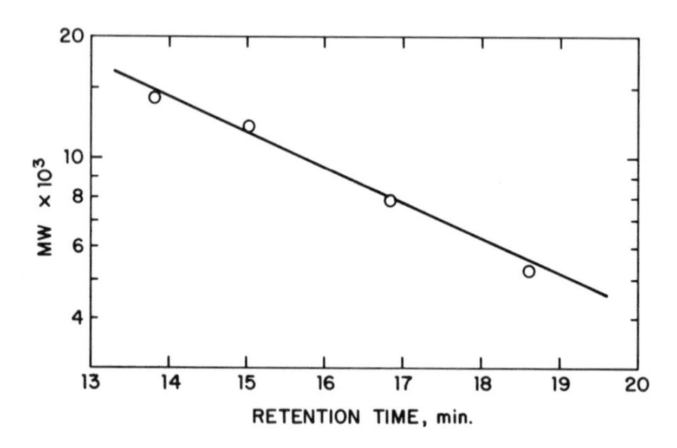

Figure 13.14 Molecular weight calibration for sodium heparin.

Columns, 0.23 cm i.d., 10 cm of 40 Å, 25 cm of 100 Å, 60 cm of 250 Å, Glycophase G-CPG; mobile phase, 0.1 *M* sodium acetate with 0.01% NaN_3; flow rate, 0.1 ml/min; pressure, 2500 psi; detectors, UV at 254 nm and RI; sample, 9 μl of 15 mg/ml solutions in mobile phase. (Data from Ref. 21.)

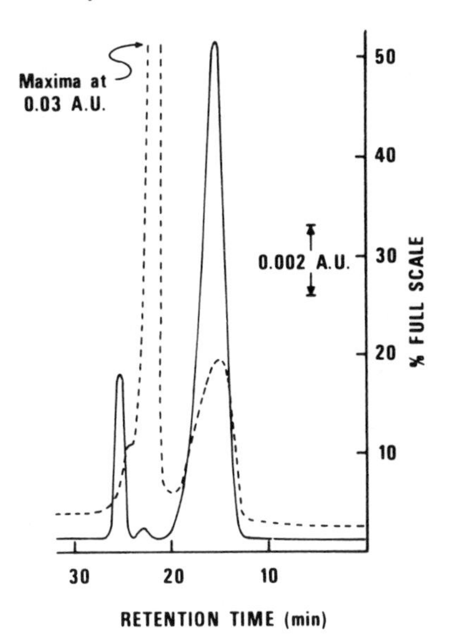

Figure 13.15 GFC chromatogram of mucosal sodium heparin.

Conditions same as in Figure 13.14. Solid line, RI; dashed line, UV. (Reprinted with permission from Ref. 21.)

standards are available, the single broad standard approach discussed in Section 9.3 can be used, for it is with this approach that the most accurate molecular weight distribution data can be obtained. In this case the bimodal pore-size column arrangement to establish a wide, linear calibration range (Sect. 8.5) should be used. When a computer is not available, it is still possible to establish approximate calibrations that can provide useful molecular weight estimations. Let us now examine how this method might be used on real samples to estimate molecular weight. Figure 13.16 shows an elution curve for an unknown chitosan, a carbohydrate polymer manufactured from chitin in shrimp and crab wastes. The composite elution curves for a series of broad molecular-weight-distribution dextrans are also shown. With this approach, two (or more) of these polymers are selected as "standards," using samples that are very different in molecular weight. If the values of weight average molecular weight, $\overline{M}_w$, for these standards are unknown, they can be determined by light scattering or ultracentrification (Sect. 1.4). A calibration curve is constructed by plotting $\overline{M}_w$ versus the SEC peak position for these two (or more) broad standards ($\overline{M}_w$ being more representative of the peak position molecular weight than the values of other molecular weight averages). Values of $\overline{M}_w$ for the unknown sample can then be estimated by comparing the observed V_R for the unknown with the calibration curve. Estimations of values of $\overline{M}_n$ for the unknown biopolymer can then be made according to Equation 10.2. The results for the chitosan

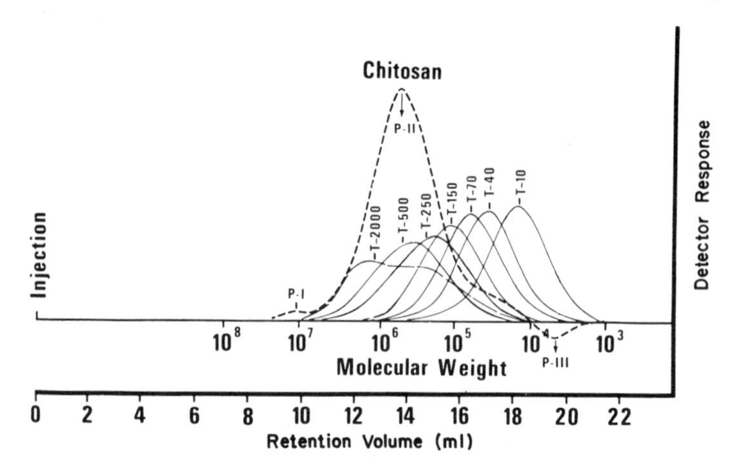

Figure 13.16 Elution patterns of dextran standards and chitosan.
Column, 30 × 0.23 cm lengths in the ratio 2 : 4 : 6 : 2 : 2 : 2 of 2500, 1500, 550, 250, 100, and 40 Å, respectively (total, 540 cm), 37–74 μm, Glycophase G-CPG; mobile phase, 2% acetic acid; flow rate, 1 ml/min; temperature, ambient; detector, RI; sample, 50 μl at 5 mg/ml in mobile phase. (Reprinted with permission from Ref. 22.)

sample in Figure 13.16 are $\overline{M}_w = 2.1 \times 10^6$, $\overline{M}_n = 9.4 \times 10^5$, and a polydispersity $(\overline{M}_w/\overline{M}_n) = 2.16$.

It should be stressed that this method produces only *approximate* molecular weight data and should only be used when more accurate methods are not feasible. The molecular weight accuracy of the method just discussed is dependent on the unknown polymer having the same structure and molecular weight distribution as the standards. In the case shown in Figure 13.16 chitosan standards were not available; therefore, readily procured dextran standards of similar structure were substituted. In this case errors in MW due to structural differences in the standards and unknown could be predicted. Differences in the MW distribution profiles for the standards and unknown in Figure 13.16 would also cause errors in estimated molecular weights; better molecular weight accuracy would have been obtained if standards with a MWD similar to that of unknown could have been used.

Isolation by GFC

As discussed in Section 11.2, preparative SEC is used to isolate significant amounts of purified solutes for characterization or further testing. HPSEC has not yet been widely used for the preparative isolation of biologically important compounds, but it is likely that this will be an area of considerable

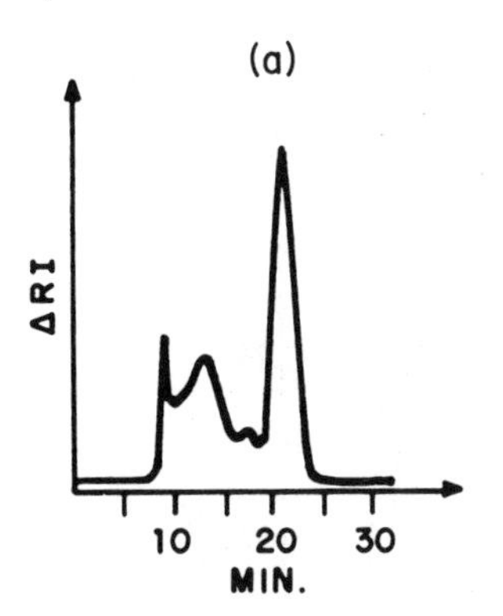

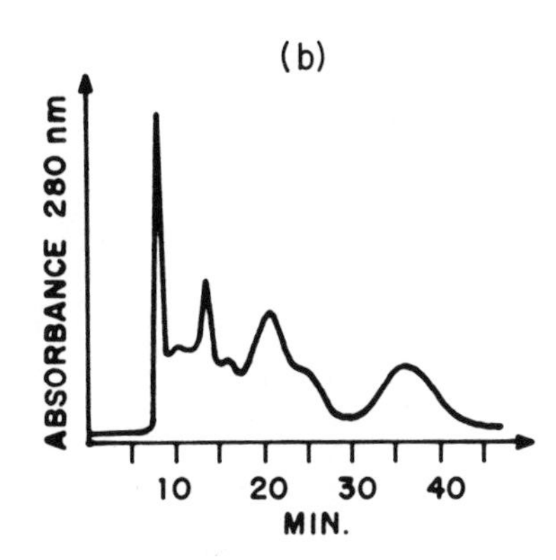

Figure 13.17 Fingerprinting of a diet beer with organic gel column.
Column, 100 × 0.4 cm EM Gel PGM-2000, 100–125 μm; mobile phase, water; pressure, 150 psi; sample, 5 μl beer; detector: (*a*) refractive index; (*b*) UV at 280 nm. (Reproduced with permission from Ref. 26.)

interest once appropriate column packings become more widely available. Traditionally, preparative GFC has mostly been carried out with wide, gravity-fed columns of rather large organic gel particles. As discussed in Section 6.2, certain of these organic gels can be used at higher pressures (~150 psi) as a result of their greater rigidity. In Figure 13.17 we see the separations of beer with a column of poly(ethylene glycol dimethacrylate) gel operated at 150 psi (~10 atm). Use of both the RI and UV detectors permitted the rapid characteristic "fingerprinting" of a variety of beers by this approach (compare this profile of beer to that shown in Fig. 13.11).

Columns of porous glass have also been used for preparative HPGFC isolations. Shown in Figure 13.18 is the separation of RNA-directed DNA-polymerase (from avian myeloblastosis virus) from a sample of *Escherichia coli* alkaline phosphatase that had been added to virus in plasma. Despite the labile nature of such enzymes, good recovery of activity is obtained, probably due to the rapid separation, in this case only 5 min.

Prefractionation and Sample Cleanup

GFC columns have been widely used as a sample clean-up step prior to subsequent studies. For example, with gels of the proper pore size, salts and lower-molecular-weight materials of no interest may be separated from desired higher-molecular-weight components in a few minutes. Alternatively, contaminating higher-molecular-weight components may be removed from desired lower-molecular-weight constituent fractions, which can then be

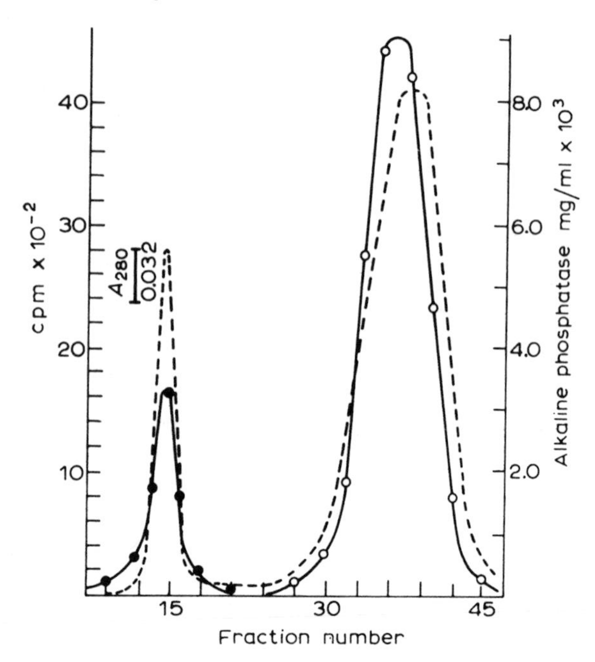

Figure 13.18 Isolation of avian myeloblastosis virus (AMV).
Column, 200 × 0.48 cm controlled-porosity glass, GPG-10-1250; mobile phase, 0.01 *M* Tris buffer, pH 8.3; flow rate, 30 ml/hr; pressure, 100 psi. ●, Detergent-requiring polymerase activity; ○, alkaline phosphatase activity; cpm, counts per minute for polymerase activity; dashed line, UV, 280 nm. (Reprinted with permission from Ref. 24.)

subjected to further high-resolution separation. While gravity-fed columns of large, soft gels have mostly been used for this purpose, it is desirable, when possible, to use smaller gel particles and operate at higher pressures for faster separations. Since Bio-Gel P2 and Sephadex G-10 and G-25 packings can be utilized at pressures up to about 150 psi, rapid prefractionation or sample clean-up may be accomplished with smaller molecules (e.g., pesticide metabolites) using these materials (25).

With small silica-based packings, prefractionations or clean-ups of both high- and low-molecular-weight components can be performed rapidly with a variety of mobile phases. For example, Figure 13.19 shows the isolation of a fraction from a plant tissue extract. This fraction was subsequently resolved further by high-performance liquid chromatography and tested for biological activity. For this separation, the isopropanol plant extract was injected directly into the column. The component of interest was monitored with a refractive index detector, and significant separation from

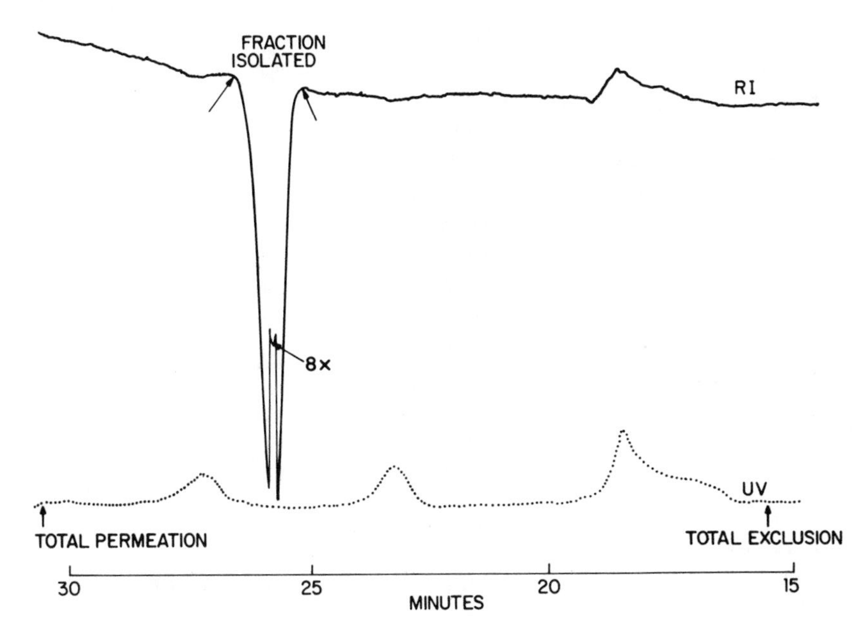

Figure 13.19 Plant extract cleanup with porous silica

Column, 100 × 0.62 cm porous silica microspheres (silanized), 50 Å pores, 1·1 μm; mobile phase, 1 : 1 isopropanol/water; flow rate, 0.78 ml/min; detector, UV, 0.04 AUFS, 254 nm; RI, 4 X; sample, 50 μl of 1.38 mg/ml in mobile phase; temperature, 22°C. (Reprinted with permission from Ref. 27.)

UV-contaminating materials was obtained. Hydrophilic-modified (deactivated) porous silica packings should be used for clean-up separations with samples having desired components which may be strongly adsorbed by unmodified silica. With unmodified porous silica columns, modifying the mobile phase (e.g., with salts, or Carbowax-20M) generally is undesirable in clean-up studies, since these additives will be present in the isolated fraction to complicate subsequent studies.

REFERENCES

1. H. Engelhardt and D. Mathes, *J. Chromatogr.*, **142**, 311 (1977).
2. F. E. Regnier and R. Noel, *J. Chromatogr. Sci.*, **14**, 316 (1976).
3. F. E. Regnier, U.S. Patent 3,983,299, Sept. 28, 1976.
4. T. Mizutani and A. Mizutani, *J. Chromatogr.*, **111**, 214 (1975).
5. H. D. Crone, R. M. Dawson, and E. M. Smith, *J. Chromatogr.*, **103**, 71 (1975).

6. H. D. Crone, *J. Chromatogr.*, **92**, 127 (1974).
7. F. A. Buytenhuys and F. P. B. Van der Maeden, *J. Chromatogr.*, **149**, 489 (1978).
8. K. W. Williams, *Lab. Pract.*, **21**, 667 (1972).
9. A. N. Glazer and D. Wellner, *Nature*, **194**, 862 (1962).
10. J. Schechter, *Anal. Biochem.*, **58**, 30 (1974).
11. G. L. Hawk, J. A. Cameron, and L. B. Dufault, *Prep. Biochem.*, **2**, 193 (1972).
12. M. J. Frenkel and R. J. Blagrove, *J. Chromatogr.*, **111**, 397 (1975).
13. K. K. Unger and N. P. Becker, Pittsburgh Conference on Ananytical Chemistry and Applied Spectroscopy, Cleveland, Ohio, March, 1977, abstract of paper 171.
14. H. D. Crone and R. M. Dawson, *J. Chromatogr.*, **129**, 91 (1976).
15. R. J. Blagrove and M. J. Fenkel, *J. Chromatogr.*, **132**, 399 (1977).
16. Y. A. Eltekov, A. V. Kiselev, T. D. Khokhlova, and Y. S. Nikitin, *Chromatographia*, **6**, 187 (1973).
16a. S. Rokushika, T. Ohkawa, and H. Hatano, Joint U.S.-Japan Seminar on Modern Techniques of Liquid Chromatography, Boulder, Colo., June 28–July 1, 1978.
17. R. Vivilecchia, B. Lightbody, N. Thimot, and H. Quinn, in *Chromatographic Science Series*, Vol. 8, *Liquid Chromatography of Polymers and Related Materials*, J. Cazes, ed., Dekker, New York, 1976, p. 11.
18. K. K. Unger, University of Mainz, German Federal Republic, private communication, 1977.
18a. F. E. Regnier, Purdue University, private communication, 1979.
19. J. X. Khym, J. W. Bynum, and E. Volkin, *Anal. Biochem.*, **77**, 448 (1977).
20. M. J. Telepchak, *J. Chromatogr.*, **83**, 125 (1973).
21. H. J. Rodriguez, *Anal. Lett.*, **9**, 497 (1976).
22. A. C. M. Wu, W. A. Bough, E. C. Conrad, and K. E. Alder, Jr., *J. Chromatogr.*, **128**, 87 (1976).
23. D. Randau, H. Bayer, and W. Schnell, *J. Chromatogr.*, **57**, 77 (1971).
24. T. Darling, J. Albert, P. Russell, D. M. Albert, and T. W. Reid, *J. Chromatogr.*, **131**, 383 (1977).
25. J. Harvey, Jr., J. C-Y. Han, and R. W. Reiser, *J. Agric. Food Chem.*, **26**, 529 (1978).
26. Product News, L-8/74, Electro-Nucleonics, Inc., Fairfield, N.J., 1974.
27. J. J. Kirkland and P. E. Antle, *J. Chromatogr. Sci.*, **15**, 137 (1977).

BIBLIOGRAPHY

Ackers, G. K., *Adv. Protein Chem.*, **24**, 343 (1970). (Analytical aspects of traditional GFC.)

Andrews, P., *J. Biochem.*, **96**, 595 (1965). (Review of classical gel chromatography procedure for determining protein molecular weights).

Bio-Rad Laboratories, *Gel Chromatography*, Bio-Rad Laboratories, Richmond, Calif., 1977. (Practical aspects of conventional gel chromatography, particularly GFC).

Cameron, B. F., "Gel Filtration Chromatography," in *Gel Permeation Chromatography*, K. H. Altgelt and L. Segal, eds., Dekker, New York, 1971. (Review of traditional GFC.)

Determan, H., *Gel Chromatography*, 2nd ed., Springer-Verlag, New York, 1969. (A laboratory handbook of conventional gel chromatography, with special emphasis on GFC).

Determan, H., and J. E. Brewer, *Gel Chromatography*, in *Chromatography*, 3rd ed., E. Heftmann, ed., Van Nostrand Reinhold, New York, 1975. (Review of conventional gel chromatography).

Hjerten, S., "Molecular Sieve Chromatography of Proteins," in *New Techniques in Amino Acid, Peptide and Protein Analysis*, A. Niederweiser and G. Pataki, eds., Ann Arbor–Humphrey Science Publishers, Inc., Ann Arbor, Mich., 1973. (Review with conventional GFC of proteins).

Pharmacia Fine Chemicals, *Sephadex—Gel Filtration in Theory and Practice*, Pharmacia AB, Uppsala, Sweden, 1970. (Practical aspects of GFC with Sephadex gels).

SYMBOLS

General symbols, used throughout this book, are indicated by •. For those symbols used mainly in one or two sections, the section numbers are given at the end of the definition. Minor symbols (e.g., those used only once or used as lesser constants) are not shown.

Symbols		Definitions	SI or cgs Units (Common Units)
A	•	Eddy-diffusion contribution to plate height	cm
A	•	Peak area	—
$\bar{a}$		Effective pore radius, $\bar{a} = 2 \times$ pore volume/pore surface (2.4)	nm (Å)
$\mathbf{a}$	•	Exponent constant for the Mark-Houwink relation, $[\eta] = \mathbf{k}M^{\mathbf{a}}$	—
a_c		Inside radius of a cylindrical pore (2.4)	nm (Å)
B	•	Coefficient for longitudinal molecular diffusion contribution to plate height	cm^2/s
C	•	Coefficient for mobile-phase mass transfer, lateral diffusion contribution to plate height	s^{-1}
C_M		Interparticle C-term coefficient (3.2)	s^{-1}
C_S		LC stationary-phase C-term coefficient (3.2)	s^{-1}
C_{SM}	•	Stagnant mobile phase, "SEC stationary phase," C-term coefficient	s^{-1}
C_1, C_2	•	Intercept and slope of linear SEC calibration, $V_R = C_1 - C_2 \log \mathrm{MW}$	cm^3 (ml)
C'_1, C'_2		Effective linear calibration constants (9.3)	cm^3 (ml)
c	•	Concentration of sample solution	g/cm^3 (g/ml)

c_m		Solute concentration in mobile phase (3.3)	g/cm^3 (g/ml)
c_s		Solute concentration in stationary phase (3.3)	g/cm^3 (g/ml)
D_E		Eddy-diffusion coefficient (3.3)	cm^2/s
D_M	•	Solute-diffusion coefficient in interparticle mobile phase	cm^2/s
D_S		Solute-diffusion coefficient in LC stationary phase (3.2)	cm^2/s
D_{SM}	•	Solute-diffusion coefficient in stagnant mobile phase (i.e., "SEC stationary phase")	cm^2/s
D_1	•	Intercept of true linear SEC calibration, $M = D_1 \exp(-D_2 V_R)$	g/mol (g/mole)
D_2	•	Slope of true linear SEC calibration, $M = D_1 \exp(-D_2 V_R)$	$1/cm^3$ (1/ml)
D'_1, D'_2		*Effective* linear calibration constants (9.3)	g/mol (g/mole), $1/cm^3$ (1/ml)
d, d_1, d_2		Polydispersity ($\overline{M}_w/\overline{M}_n$) of polymer samples (10.6)	—
d_p		Particle diameter	μm (cm)
d_f		Film thickness of LC stationary phase (3.2)	μm (cm)
F	•	Eluent volume flow rate	cm^3/s (ml/min)
$F, F(V_R), F(V)$	•	Experimental SEC elution curve height as a function of retention volume	—
$G(V_R - y)$		Instrument-column-dispersion function which describes the weight fraction of a solute that should have been at the retention volume y but is actually dispersed and detected at retention volume V_R (4.3)	—
ΔG°		Standard free energy difference (2.3)	cal
ΔG		Free energy of mixing (7.2)	cal
H, HETP	•	Height equivalent to a theoretical plate	cm (mm)

H_F		Eddy-diffusion plate height (3.2)	cm
H_L		Longitudinal-diffusion plate height (3.2)	cm
H_M		Plate height due to interparticle mobile-phase effects (3.2)	cm
H_S		Plate height due to LC stationary-phase effect (3.2)	cm
H_{SM}	•	Plate height due to stagnant-mobile-phase (SEC stationary-phase) effects	cm
H_{min}		Minimum value of H in the plate height vs. velocity plot (3.2)	cm
ΔH°		Standard enthalpy difference (2.3)	cal
ΔH		Enthalpy of mixing (7.2)	cal
h, h_i	•	Chromatogram height	—
$\mathbf{h}$	•	Reduced plate height, $\mathbf{h} = H/d_p$	—
h_p		Peak height at apex	—
I_L		Linearity index; goodness of the linear fit to SEC calibration (4.5)	—
I_R		Separation range index; MW separation range of the SEC calibration curve	—
K, K_e		Equilibrium solute distribution coefficient between two chromatographic phases (2.4)	—
K_{LC}		LC solute distribution coefficient (2.2)	—
K_{SEC}, K_{GPC}, K_{GFC}	•	Solute distribution coefficient in SEC, including GPC, GFC	—
K°	•	Column permeability (6.4)	cm^2
$\mathbf{K}$	•	A proportional constant for the Mark-Houwink relation, $[\eta] = \mathbf{K}M$	$m^3\ kg^{-1}$ (dl/g)

k		Boltzmann's constant (2.4)	erg deg^{-1} $molecule^{-1}$
k'	•	LC solute capacity factor, $k' = K_{LC} V_S/V_M$	—
L	•	Column length	cm (ft)
M, MW	•	Molecular weight	g/mol (g/mole)
$\overline{M}_n$	•	Number-average molecular weight	g/mol (g/mole)
$\overline{M}_w$	•	Weight-average molecular weight	g/mol (g/mole)
$\overline{M}_z$		Z-average molecular weight (1.3)	g/mol (g/mole)
$\overline{M}_v$		Viscosity-average molecular weight	g/mol (g/mole)
$(\overline{M}_n)_{true}$, $(\overline{M}_w)_{true}$ or $\overline{M}_n(t)$, $\overline{M}_w(t)$		True values of number- and weight-average molecular weight (4.3, 10.4)	g/mol (g/mole)
$(\overline{M}_n)_{exp}$, $(\overline{M}_w)_{exp}$ or $\overline{M}_n(u)$, $\overline{M}_w(u)$		Experimental values of number and weight-average molecular weight uncorrected for instrument spreading (4.3, 10.4)	g/mol (g/mole)
$\overline{M}_n^*$, $\overline{M}_w^*$		Percent error in $(\overline{M}_n)_{exp}$ and $(\overline{M}_w)_{exp}$ due to instrument spreading (4.3)	—
M^*		Average of absolute $\overline{M}_n^*$ and $\overline{M}_w^*$ (4.3)	—
$\overline{M}_n(V)$, $\overline{M}_w(V)$		Actual $\overline{M}_n$ and $\overline{M}_w$ (as in a detector cell) as a function of retention volume (9.4)	g/mol (g/mole)
N	•	Column plate count or plate number, number of theoretical plates	—
N_i, N_t		Plate count of ith column and column set, respectively (8.6)	—
N_0		Avogadro's number (3.4)	molecules $mole^{-1}$

N_X		Number fraction of chains with degree polymerization X (1.3)	—
n	•	General sequential indexing integer	—
n		Peak capacity (4.1)	—
n_{opt}		Optimum number of recycles (11.3)	—
dn/dc		Specific refractive index increment (7.4)	cm^3/g (ml/g)
P	•	Pressure	Pa (psi, bar)
PS		Pore size (4.5)	nm (Å)
PSD		Pore-size distribution; standard deviation of the log-normal PSD curve (4.5)	—
p		Extent of reaction (9.3)	—
Q		Q-factor; molecular weight/extended chain length (10.7)	g/mol-Å (g/mole-Å)
R	•	Gas constant	cal deg^{-1} $mole^{-1}$
R_g	•	Radius of gyration of solute molecules, $R_g = kM^{\alpha}$	nm (Å)
R_s	•	Resolution of two peaks	—
$R_{s, opt}$		Optimum resolution in recycle SEC (11.3)	—
R_{sp}	•	Specific resolution, $R_{sp} = R_s/\Delta \log M$	—
R_{sp}^*	•	Packing resolution factor, $R_{sp}^* = R_{sp}/\sqrt{L}$	$cm^{-1/2}$
$r, \bar{r}$		Radius and average radius of hard-sphere solutes (2.4)	nm (Å)
r_e		Equivalent radius of a polymer solute (2.5)	nm (Å)
ΔS°		Standard entropy difference (2.3)	cal/K
ΔS		Entropy of mixing (7.2)	cal/K
s		Surface area per unit pore volume (2.4)	1/cm (cm^2/ml)
sk		Molecular weight correction factor for band-broadening asymmetry (10.4)	—
T	•	Temperature	K (°C)
T_c		Consolute temperature (7.2)	K (°C)
t	•	Time	s (min)

t_0		Retention time of unretained peak (2.1)	s (min)
t_R	•	Retention time	s (min)
V	•	Volume	cm^3 (ml)
Var		Variance, Var $= \sigma_x^2$ (3.1)	cm^6 (ml^2)
V_e		Total permeation volume (13.5)	cm^3 (ml)
V_h		Hydrodynamic volume of an equivalent sphere (9.2)	$(nm)^3$ ($Å^3$)
V_i	•	Total accessible liquid volume contained within the pores of the SEC packing	cm^3 (ml)
V_i		Variable retention volume used in the integral-MWD calibration method (9.3)	cm^3 (ml)
V_{inj}		Injected sample volume (7.4)	cm^3 (ml)
V_M	•	Total liquid volume, $V_M = V_0 + V_i$	cm^3 (ml)
V_0	•	Volume of mobile phase in the interstices between the SEC packing particles	cm^3 (ml)
V_R	•	Retention volume	cm^3 (ml)
$\langle V_R \rangle$		Average retention volume (9.4)	cm^3 (ml)
V_s		Equivalent liquid volume of a LC stationary phase (2.1)	cm^3 (ml)
v	•	Mobile-phase velocity	cm/s
v_{opt}		Optimum velocity at $H = H_{min}$ (3.2)	cm/s
$\mathbf{v}$	•	Reduced velocity, vd_p/D_M (3.2)	—
W, $W(V_R)$	•	True SEC elution curve height at ideal infinite resolution as a function of V_R	—
W_b	•	Peak width at the base, the distance between the baseline intercepts of lines drawn tangent to the points of inflection of the elution peak trace	cm^3 (ml)
W_1, W_2		W_b of polymer peaks (10.6)	cm^3 (ml)
$W^{1/2}$	•	Peak width measured parallel to baseline at one-half of the peak height	cm^3 (ml)
W_X		Weight fraction of chains with degree of polymerization X (1.3)	—
X		Degree of polymerization, number of repeating monomer units in a polymer chain (1.3)	—
X_i		$\Delta W/\Delta \log M$, differential weight fraction (10.3)	—

$\overline{X}_n$, $\overline{X}_w$	Number- and weight-average degree of polymerization	—
α	Separation factor, $\alpha = k'_2/k'_1$ (4.1)	—
α	Exponent constant in the MW dependence of R_g, $R_g = kM^\alpha$ (2.4)	—
α	Expansion factor of polymer solute R_g (7.2)	—
γ	Peak skew (3.5)	—
γ'	Peak skew calculated from values of σ and τ (6.4)	—
γ_{SM}	Peak skew due to stationary mass transfer effect (3.4)	—
δ	Solubility parameter (7.2)	cal/cm^3 (J/cm^3)
δ_d	Solubility parameter due to dispersion force (7.2)	cal/cm^3 (J/cm^3)
δ_h	Solubility parameter due to hydrogen bonding (7.2)	cal/cm^3 (J/cm^3)
δ_p	Solubility parameter due to polar force (7.2)	cal/cm^3 (J/cm^3)
δ_v	$\sqrt{\delta_d^2 + \delta_p^2}$ (7.2)	cal/cm^3 (J/cm^3)
δ_0	Total solubility parameter, $\delta_0 = \sqrt{\delta_d^2 + \delta_h^2 + \delta_p^2}$	cal/cm^3 (J/cm^3)
δ_s, δ_m	Solubility parameters of solvent and macromolecules, respectively (7.2)	cal/cm^3 (J/cm^3)
ε	Polymer solution parameter; $\varepsilon = \dfrac{2\mathbf{a} - 1}{3}$ with $\mathbf{a}$ = Mark-Houwink viscosity constant (2.4)	—
η	• Viscosity	dyne sec/cm^2 (cP)
η_{rel}	η_{rel} = relative viscosity = $\eta_{solution}/\eta_{solvent}$ (9.2)	—
η_{sp}	η_{sp} = specific viscosity = $\eta_{rel} - 1$ (9.2)	—
η_0, η_1	Solvent and solution viscosity (3.5) (5.12)	dyne sec/cm^2 (cP)
$[\eta]$	• Intrinsic viscosity	m^3/kg (dl/g)

$$[\eta] = \lim_{c \to 0} \frac{\eta_{sp}}{c} = \lim_{c \to 0} \ln \frac{\eta_{rel}}{c}$$

Symbol		Definition	Units
Θ		Flory theta temperature (7.2)	K (°C)
Λ		Molecular weight correction factor for symmetrical band broadening (10.4)	—
λ		Solute diameter/pore diameter (3.4)	—
λ		Branching index; number of branching points per unit MW (12.8)	—
μ_1		First moment (peak retention) (3.3)	cm^3 (ml)
μ_2		Second moment (peak variance) (3.3)	cm^6 (ml^2)
μ_3		Third moment (peak skew) (3.3)	cm^9 (ml^3)
ρ		Density	kg/m^3 (g/cm^3)
σ	•	Standard deviation of a Gaussian instrument spreading function	cm^3 (ml)
σ_x	•	Standard deviation of SEC elution peaks of any shape	cm^3 (ml)
σ_{disp}		σ value due to column dispersion (11.3)	cm^3 (ml)
σ_{inj}		σ value due to sample injection (11.3) (7.4)	cm^3 (ml)
σ_{MWD}		σ value due to sample MWD (11.3)	cm^3 (ml)
σ_t^2		Peak variance of a column set (8.6)	cm^6 (ml^2)
τ	•	Peak skew parameter (decay constant of the exponential modifier for a skewed σ-τ peak model)	cm^3 (ml)
τ		Tortuosity factor (3.4)	—
Φ_0		Universal constant of Flory viscosity theory, $\Phi_0 = 2.86 \times 10^{23}$ (2.4)	cm^3 (dl/mole)
Φ		Volume fraction of extraparticle solvent volume of the total liquid volume in the column (3.3)	—
Φ		Molecular weight correction parameter (10.4)	—
$\Phi(r)$		Volume of pores with radius r (2.5)	cm^2 (ml/cm)

ABBREVIATIONS

ACS	American Chemical Society
AD	Analog-to-digital converter
ASTM	American Society for Testing and Materials
A-term	Eddy-diffusion contribution to plate height
AU, AUFS	Absorbance units, absorbance unit full scale
B-term	Longitudinal diffusion contribution to plate height
C-term	Mass transfer (lateral diffusion) contribution to plate height
CRT	Cathode ray tube
DMF	*N*,*N*′-dimethylformamide
EVA	Poly(ethylene-co-vinyl acetate)
FFF	Field flow fractionation
GC	Gas chromatography
GFC	Gel filtration chromatography
GPC	Gel permeation chromatography
GPCN3	GPC raw-data processing computer program
GPCV2	GPC linear calibration method—version 2
GPCV3	GPC linear calibration method—version 3
HETP	Height equivalent to theoretical plate
HFIP	Hexafluoroisopropanol
HPGFC	High-performance GFC
HPGPC	High-performance GPC
HPSEC	High-performance SEC
i.d.	Inside diameter
IR	Infrared
LALLS	Low-angle laser light scattering
LC	Liquid chromatography
LEC	Liquid exclusion chromatography (SEC)
LLC	Liquid-liquid chromatography
LS	Light scattering
M_H	Hamielec linear calibration plot
MI	Melt index

M_p	Peak position calibration plot
M_t	True linear calibration plot
MW	Molecular weight
MWD	Molecular weight distribution
NATFAT	Sodium trifluoroacetate
ODCB	*o*-Dichlorobenzene
PAN	Polyacrylonitrile
PAN-S	Polyacrylonitrile with sulfonate groups
PE	Polyethylene
PET	Poly(ethylene terephthalate)
PP	Polypropylene
PS	Polystyrene
PS	Pore size
PSD	Pore-size distribution
PSM	Porous silica microsphere
PVA	Poly(vinyl acetate)
RI	Refractive index
RIU	Refractive index unit
SBF	Separation by flow
SEC	Size-exclusion chromatography
S/N	Signal to noise
TCB	1,2,4-Trichlorobenzene
TFFF	Thermal field flow fractionation
THF	Tetrahydrofuran
TLC	Thin-layer chromatography
UC	Universal calibration
UV	Ultraviolet

INDEX

Abstracting services, 16
Adsorption, 245, 282
  effect of ion type, 424
  effects, 237
  elimination, 237, 426, 431
  ionic strength effects, 424, 431
  minimization, 423
  pH effects, 424, 431
  protein, 425, 429
  protein isotherms, 430
  reduction with Carbowax, 425
  temperature effects, 233
Air baths, 143
Air bubbles, 282
Alkanes, 349
Alumina packing, 179
Amino acid buffers, 424
Amino-modified supports, 429
Antioxidant, 351
  high temperature, 387
  solvent, 261
Apparatus:
  column slurry-packing, 192
  criteria, 124
  preparative, 360
  schematic, 124
Area, peak or curve, 318
Association effects, sample/solvent, 222
ASTM, 15, 168
Automated data handling, 326
Automation justification, 249
Axial dispersion, *see* Band broadening

Balanced-density, slurry technique, 190, 195, 199
Band, *see* Peak
Band broadening:
  column parameters, 82, 85
  concentration effects, 239, 243, 245
  concentration overloading, 89
  definition, 53, 322, 323
  distribution coefficient effects, 239, 243, 245
  effect on broad-standard calibration, 294
  effect of K_{SEC}, 78, 82, 85
  effect on SEC-MW error, 107
  elution curve shapes, 105
  excessive extra-column volume, 54
  experimental factors, 82, 85, 89
  extent of permeation, 78, 85, 86
  extra column, 126
  by fractionation, 224
  GPCV2 and GPCV3 calibration, 301
  injection volume effects, 239, 243, 245
  interrupted flow, 77
  longitudinal diffusion, 72, 77
  manual correction, 323
  mass transfer, 78
  minimization, 165
  packing homogeneity, 54
  particle size effect, 82, 85
  rate theory, 78
  recycle, 372
  stationary phase effect, 72
  synonyms, 53
  temperature effects, 89
  viscosity effects, 239, 243, 245
  see also Column dispersion
Band broadening parameters:
  Gaussian peak shape model, 57
  peak standard deviation, 57
  peak variance, 57
  peak width, at base, 57, 58
    at half-height, 57, 59
Band broadening terminology, 57, 71, 72
Baseline:
  definition, 277, 278
  stability, 274, 277
Baths, constant temperature, 143
Beer:
  fingerprinting, 445

separation, 439
Bibliographies, 14
Bimodal calibration, 118
Bimodal columns:
concept, 310
different polymer/solvent systems, 312
wide linear calibration, 311
Bimodal pore size, 118, 267
SEC calibration, 119
Bio-Gel-P, 172
Biopolymer calibration, 443
Biopolymer dissolution:
effect of pH, 221
effect of salt concentration, 221
Blends, GPC analysis, 413
Block copolymer analysis, 407
Blood plasma, 432
μ-Bondagle, 177-179
Bonded phases, 423
Branched polyethylene chromatograms, 389
Branched polymers, intrinsic viscosity, 400
Branching, polymers, 399
by GPC, 400
Branch points, 399
Broadening, definition, 322, 323
Bubbles, 282
Buffers, amino acid, 424

Calculation of MW:
computer, 326
manual, 318, 320
Calibration:
copolymers, 410
drugs, 348
peak dependence, 280
polymer type dependence, 340
standards, 262
see also Molecular weight calibration
Calibration constant:
C_2, additivity, 114
D_2,
bimodal pore size, 109
determination, 109
MW accuracy criterion, 108
relation to calibration constant C_2, 114
resolution, 103, 104
slope of the calibration, 109
Calibration curves:
μ-Bondagel, 178, 179
Du Pont SE, 178
effect of solute shape, 116
EM Gel-OR, 172
EM Gel Type SI, 177
LiChrospher, 176
Porasil, 184
Styragel, 183
μ-Styragel, 168
TSK gel, 169
Calibration fit, 118
Calibration linearity:
bimodal pore size packings, 119
effect of pore size distribution, 118
Calibration method:
accuracy, 302, 307
actual MW, 307
band broadening influence, 294
broad-MWD standards, 294
column property, 286
constant D_2, 108
definition, 286
effect of dispersion, 93
effective linear, 298
elution curves, 307
errors, 299, 302, 304
experimental effects, 286
experimental test, 302, 304
Hamielec, 297
integral, 294, 297
integral-MWD method, 294
linear calibration, 298
MW errors, 299, 302, 304
MW and MWD dependence, 302, 304
needs for recalibration, 286
small molecules, 287
validity for specified polymer/solvent systems, 286
Calibration standards, 262
broad MWD, 294, 299
dextrans, 396
obtaining, 264
sources, 262
Cam, pump, 131
Carbon number values, 346
Carboxylic acid dimerization, 222
Carrier solvent, *see* Mobile phase
C_2 and D_2 values, 114
Cell:
back pressure, 147
detector, 147
photometric detector, 152
tapered, 152
Z-type, 152

Chain-folding, polymers, 401
Chevron computer program, 328
Chromatogram:
  baseline, 277
  comparisons, 317
  cutting, 278
  dextrans, 396
  digitization, 318, 319
  ends, 278-280
  labeled parts, 316
  limits, 278-280
  polyvinyl alcohol, 395
  strip chart, 382
Chromatography, 125
  adsorption, 19
  different forms, 19
  ion-exchange, 19
  gas (GC), 19, 57, 60, 65
  liquid, (LC), 19
  liquid-liquid, 19, 23
  liquid-partition, 19, 23
  liquid-solid, 19
Chromophore, 151
Cleanup:
  fractionation, 357
  procedures, 415
  of samples, 445
Coffee separation, 438
Colligative properties, 13
Column:
  blank cleaning, 194
  blanks, 200
  configuration, 200
  consolidation, 191
  coupling, 272
  degradation, 205
  dimension, 200
  exclusion limit, 26
  extra-column effects, 126
  "good," 202
  handling, 268, 269
  infinite diameter, 201
  internal diameter, 200
  large diameter, 358
  leaks, 282
  length in recycle, 367
  life, 434
  matched, 272
  materials, 199
  order, 273
  overload, 362
  packings, 166
  performance, 268
  permeability, 204
  preparation of, 191, 197, 358
  pressure drop, 204
  purging, 205
  selection guidelines, 117, 268
  stability, 434
  storage, 205
  techniques with small particles, 205
  testing, 205
  total permeation limit, 26
  tubing, 199
  tubing cleaning, 194
Column dispersion:
  diffusion pattern, 56
  effect on elution curve, 309
  effect on MW, 307
  illustration of flow, 56
  terminology, 55
  see also Band broadening
Column dispersion factor, 109
  additivity of connecting columns, 114
  column sets, 113
  determination, 109
  flow rate dependency, 109
  GPCV2, GPCV3 calibration methods, 301
  MW accuracy criterion, 108, 110
  recycle and reverse-flow experiments, 109
  resolution equation, 103, 104
  variations with V_R, 109
  see also Peak standard deviation
Column dispersion and peak capacity, 101
Column dispersion skew factor, 107, 302
Column efficiency:
  particle size effects, 187, 433
  preparations, 359
  temperature effects, 233
Column packing:
  apparatus, 192
  machine, 198
  methods, 186
  particle size, 235, 236
  pore effects, 234, 235
  solvent effect, 255
  see also Packings
Column packing technique:
  balanced-density method, 190, 199
  bed consolidation, 191
  dry-packing, 189, 197
  particle suspension, 190

particle wetting, 191
reservoir for slurry method, 197
semirigid gels, 199
slurry-packing, 190, 191, 199
slurry-packing solvents, 192
soft gels, 199
viscosity effects, 191
viscous slurry, 191
wet-packing, 190
Column parameters, effect on band broadening, 82, 85
Column performance:
parameters, 108, 111
restoration, 205
specification, 203
Column performance criteria:
application, 108
at constant, σD_2, 108
plate height, 97
plate number, 97
temperature effects, 433
Column plate count, *see* Plate number
Column resolution:
definition, 97
determination, 273
effect of internal diameter, 359
see also Resolution
Complexation:
with sodium dodecylsulfonate, 426
solute, 426
Composition:
copolymer, 408
distribution, 405, 408
Compressibility correction, 133
Computer:
calculations of MW, 326
corrections to MWD, 332
curve corrections, 332
Computer programs:
Chevron, 328
Du Pont CR&D, 327
Concentration:
column overloading, 242
effects, 241
sample, 275, 277
Connected columns, 272
Connectors, 142
Continuous preparative SEC, 367
Copolymers:
block/analysis, 405, 407
composition, 410
GPC, 405
MW calibration, 410
Corrected chromatograms, 333
Corrections to molecular weight:
manual example, 323
manual method, 322
Coupling columns, 272
m-Cresol, 268, 390
Crude oil, 355
Crystallinity, 401
polyethylene, 402, 403
C-term plate-height contribution, *see* Mobile phase mass transfer; Stationary phase mass transfer
Curve:
area, 318
cutting, 278
shape, 322
width, 317
see also Peak
Cut-points, 363

Data:
evaluation, 315
real-time calculations, 331
reduction, 315
software conversion, off-line, 328
Data handling, 315
automated, 326
hardware, 144
Data reduction systems, 145
Deconvolution of curves, 333
Degassing solvents, 127, 261
Degradation of polymers, 225
Denaturation, 424
Derivatives, fluorescent, 162
Detection limit, 146
Detector:
bulk-property, 146
cell, 147
drift, 146
ease of operation, 147
flow-sensitive, 148
gas-segmented reaction, 160
general, 146
high-temperature, 159
infrared (IR), 157
linearity, 147
low-angle light-scattering, 156
noise, 146
performance criteria, 146

post-column reaction, 160
preparative, 360
reaction, requirements for, 160
reactor, 159
response, 146
selective, 146, 152
sensitivity, absolute 146
relative, 146
serviceability, 147
simultaneous use, 406
solute-property, 146
stability, 148
temperature control, 148
UV, characteristics, 156
spectrophotometric, 152
variable wavelength, 152
viscometer, 162
Detector response:
repeatability, 147
tuning of, 147
Dextrans:
calibration, 172
chromatograms, 444
molecular weights, 397
standards, 396
o-Dichlorobenzene, 387
Differential chromatography, 376
Diffusion coefficient:
band broadening effects, 82, 85, 86, 88
restricted, 87, 89
solute MW dependence of, 81, 88
solute size, effects, 88
Diffusion model, SEC theory, 46
Digitization:
A-D convertors, 328
chromatograms, 318, 319
Dimerization of carboxylic acids, 222
Diol-modified supports, 428
Dispersion, *see* Column dispersion
Dispersion correction factors, 106
Dispersions, particle, 416
Dispersity, 317
Dissolution:
effect of crystallinity, 210
polymer, 209
sample, 274
ultrasonic, 222
Distribution, MW, *see* Molecular weight distribution (MWD)
Distribution, solute:
free energy difference, 27
enthalpy difference, 27
entropy difference, 27
thermodynamic equilibrium of, 27
Distribution coefficient, 25, 28, 32, 49, 287
definition, 22
effect of sample concentration, 245
D_2-limiting value:
high resolution, 117
pore shape effects, 115
pore size distribution, 117
pore-size effects, 114, 115
solute shape, 115
DNA, 438
polymerase, 445
Double-layer, electrical, 430
Draw-off method in recycle, 374, 375
Drug calibration curve, 348
Dry-packing technique, 189
Du Pont SE packings, 176

Eddy diffusion, 165
definition, 55
Effective linear calibration constants, 298, 300
Electrical double-layer, 430
EM Gel-OR calibration, 172
Enthalpy of mixing, 210, 212
Enzyme:
calibration for, 428, 429
recovery, 429
separation, 436
Epoxy resin, 352
chromatograms, 414
Equipment, 123
criteria, 124
integrated, 125
preparative, 360
Error sources, MW and MWD, 227-330, 326
Etched polyethylene, 402, 404
Expansion factor, 211
Experimental versus true elution curves, 105
Experiment optimization, 249
Extra column effects, 126
skewed peak dispersion, 89, 90
Extraparticle mass transfer, 55. *See also* Mobile phase mass transfer
Extraparticle mobile phase dispersion, 71, 73
Extraparticle plate height:
effect of particle size, 73
effect of polymer MW, 73
extraparticle coupling effect, 71, 73, 74
see also Flow-diffusion coupling

Fast chromatography, 378
Field flow fractionation, 50
Filters, 142
Filtration:
  sample, 275
  solution, 275
  solvent, 261
Fittings:
  compression, 147
  low volume, 142
  zero-dead-volume, 142
Flory MWD:
  condensation polymers, 295
  for integral-MWD calibration, 295
  peak model, 117
Flory theta temperature, 210
Flory universal constant, 35
Flow, Poiseuille, 162
Flow-diffusion coupling:
  Giddings coupling theory, 65, 67
  illustration, 66
Flow feedback, 133
  positive control, 134
Flow rate:
  constancy, 133
  control, 133
  corrections, 231
  dependence of σ, 109
  determination of, 274
  drift, 230
  fluctuations, 227
  integrated total volume method, 143
  maximum, 270
  measurement, 143
  preparative, 361
  specifications, 230
  study, 27, 29, 47
  syphon-counter, 143
  volume measurements of, 143
Flow rate effects:
  drift, 230
  efficiency, 224, 225
  molecular weight errors, 227-230
  pump fluctuation, 227-229
  repeatability, 229
  resolution, 224
  retention, 231
  shear degradation of polymer, 226
Flow tube, 143
Fluoram reagent, 160
Fluorescamine reagent, 160
Fluorescence, 161
Fluorescent derivatives, 162
Fluorimeter, 160
  filter, 161
Formulations analysis, 413
Fraction collector, 144, 357
Fraction purity, 363
Fraction yield, 363
Free energy of mixing, 210
Frits, 142

Gauges, 142
Gaussian peak shape model:
  definition of, 57, 58
  exponentially modified, 107, 302
  GPCV2 calibration method, 301
  peak variance, 59
  standard deviation of, 57, 61
Gel filtration chromatography (GFC):
  cleanup, 445
  isolation, 444
  modern, utility of, 420
  molecular weight determinations, 439
  prefractionation, 445
  preparation of surface-modified substrates, 420
  sample cleanup, 445
  techniques, 419
  utility, 3
  see also Gel permeation chromatography (GPC)
Gel packings:
  hydrophilic, 172
  small pore, 346
  table of, 170
Gel permeation chromatograms:
  additives, 382
  dextrans, 396
  natural rubber, 384
  reaction rate study, 383
Gel permeation chromatography (GPC):
  applications, 381
  blends, 413
  branching of polymers, 400
  cleanup procedures, 475
  copolymer analysis, 406
  differential method, 376
  dispersions, 416
  formulations analysis, 413
  high temperature, 386, 387, 390
  inorganic polymers, 394

literature of, 14
polyelectrolytes, 397
quality control analysis, 414
solvents, 250, 252
utility, 4
vacancy chromatograms, 376
see also Gel filtration chromatography (GFC)
Gels:
Fractogel, 168
MicroPak BKG, 169
OR-PVA, 168
polystyrene, 167
polyvinylacetate, 169
semirigid organic, 167
small-particle, 167
small-pore, 168
temperature, limit, 168
TSK, 169
type SI, 173
General retention equation, 25
Glyceryl:
modified supports, 428
surface modification, 423
Glycidoxypropyltrimethoxysilane, 422
Glycophase-*G*, 432
GPCV2 and GPCV3 calibration methods, 298
Grafted polymer analysis, 409
Guanidine hydrochloride, 440

Hamielec calibration method, 298
Hard sphere model, 31
Hardware:
data handling, 144
miscellaneous, 142
Heart-cut technique, 364
Height equivalent to a theoretical plate, *see* Plate height
Hexafluoroisopropanol, 390, 391, 393
High-speed SEC, 378
High temperature GPC, 386, 387, 390
H versus *v* plot:
classical versus coupling theory, 67
van Deemter plot, 64
of various dispersion processes, 68
Hydraulic capacitor, 134
Hydrocarbon separation, 349
Hydrodynamic chromatography, 50
hydrodynamic volume, 355
universal calibration, 291
volume of molecules, 291
Hydrogel, 173
Hydrogen-bonding solvents, 214
Hydrolysis of silica, 184
Hydrophobic interactions, 431

Infinite-diameter column, 201
Infrared detector, 157
use with organic solvents, 256
Infrared techniques, 411, 412
Inhibitors:
column packing, 387
solvent, 261
Injection:
preparative, 361
reproducibility, 139
sample, 276
septum, 138
stop-flow, 139
syringe, 138
syringe-septumless, 139
valve, 139
volume, 126, 239, 240
Injectors, 138
Inorganic polymer:
GPC, 394
separation, 352
Instrumental dispersion, *see* Band broadening
Instruments:
integrated, 144
microprocessor-controlled, 144
Integral-MWD calibration method, 294-297
Interferometer, 150
Intermolecular association, 220
Intraparticle flow-diffusion interaction, 83, 84
Intraparticle mass transfer, 56. *See also* Stationary phase mass transfer
Intrinsic viscosity:
branched polymers, 400
calculation from SEC, 338, 400
Mark-Houwink equation, 292
Intrinsic-viscosity equation:
Mark-Houwink constants, 35
MW dependence of, 292
Ion-exclusion effect, 430
Ionic polymers, 398
Ionic strength:
effect on proteins, 428
optimization of, 424
IR detection, 157
with organic solvents, 256
stop-flow, 411

Journals, 14-15

K and a values, 252
  table, 336
Kinetic factors, effect on band broadening, 82, 85, 88

Laboratory operations, 249
Lamellae, 401
Lamps, low pressure mercury, 151
LC distribution coefficient, 20, 22, 25, 28, 30
Leaks, 282
LiChrospher, 173
  calibration, 176
Light-scattering:
  detector, 157
  nomograph, 158
Linear calibration:
  approximations, 102
  best fit, 267
  errors, 118
  methods, 298
  range for bimodal columns, 309
Linear columns, 267
Literature, 14, 16
Longitudinal diffusion, 57
Longitudinal dispersion, *see* Band broadening
Loop, sample, 141

Mark-Houwink constants:
  determination, 294
  intrinsic-viscosity equation, 35
  table, 252, 336
  universal calibration, 292, 293
Mark-Houwink equation, 292
Mass balance, 282
Mass transfer dispersion processes, *see* Mobile phase mass transfer: Stationary phase mass transfer
Mass transfer minimization, 166
Material balance, 345
Matrix effects, 235
Mobile phase:
  average linear velocity, 23
  compressibility effects, 133
  computer-compensated flow, 129
  definition of, 19, 22
  degassing, 127
  delivery, 230
  moving mobile phase volume, 22, 24
  pumping errors, 230
  reservoirs, 126
  selection, 222, 238
  stagnant volume, 22, 24
  total volume, 22, 24
  true linear velocity, 23, 25
  velocity effects, 224
  see also Solvents
Mobile phase lateral diffusion, *see* Mobile phase mass transfer
Mobile phase mass transfer:
  definition of, 55
  synonyms for, 55
Models for SEC theory, 31
Molar volume, 345
Molecular association, 222
Molecular size:
  composition dependence, 400, 410
  distribution by GPC, 383, 384
  parameters, 286
Molecular volume, 346
Molecular weight:
  accuracy, 288, 294, 297
  accuracy criterion, 109, 112
  actual, 308
  averages, 105
  calculation, off-line, 327
    real-time, 331
  calculation example, 320
  calculation manual, 318, 320
  calibration, 347
  colligative properties, 13
  corrections, computer, 332
    manual, 322
  dextran standards, 397
  effect on R_g, 288
  effect on RI, 246
  errors, 227-230
  light scattering, 13
  manual corrections, 323
  measurements, 12
  molecular size relationship, 288
  number average, 5
  properties, 8
  property dependencies, 6
  reproducibility in GFC, 442
  retention volume relationship, 347
  sources of error, 326
  true, 308
  universal calibration, 335, 337
  weight average, 8
  Z-average, 11

Molecular weight accuracy:
column sets, 112
concept, 112
criterion, 110
improvement by recycle, 368, 372
Molecular weight accuracy criterion:
equation, 108
independence of sample MWD, 108
in polymer analysis, 104
Molecular weight calibration:
biopolymer, 443
mobile phase effects, 238
sample concentration, 242, 243
solute conformation, 43, 116
solvent effects, 238
standards, 262
see also Calibration; Universal calibration
Molecular weight determination:
computer calculations, 326
manual calculations, 318, 320
peak position method in GFC, 442
small molecules, 346
see also Calibration
Molecular weight distribution (MWD), 285
accuracy of calibration, 297
chromatograms, 382-385
comparative technique, 339
composition, 405, 408
cumulative weight fraction, 319, 321
differential number fraction, 322
differential weight fraction, 321
from elution curve, 285
expressions, 8-10
fractionation method, 13
by GPC, 383, 384
polymer property, 6
polyvinyl alcohol, 395
Molecular weight errors:
column dispersion, 107, 112
flow rate fluctuations, 227
Molecule chain-length, 346
Monomer separations, *see* Small molecule separations
Mononucleotides, 438
Most probable MWD, 295. *See also* Flory MWD

NATFAT, 393
Nickel complexes, 352
Nitric acid etching, 402, 403
Nucleic acids and bases, 435, 438
Nucleosides, 435
Nucleotides, 435
Nylon separations, 390, 393

Objectives, problem, 249
Oil additive, 367
Oil characterization, 355
Oligmer separation, 352
Operating parameters, effect on SEC performance, 122
Operating variables, 432
Optimizing experiments, 249
Optimum separations, 265
Optimum velocity, van Deemter equation, 65
Organic packings:
cationic exchange properties, 424
effect of solvent, 255
Osometry, 13
Overlapped peaks, 363, 369
Overloading effect, 89, 362

Packing consolidation, 191
Packing resolution factor, 104, 110, 113, 203
Packings:
amide-modified, 431
amine and amides, 422
amine-modified, 430
Bio-Gel P, 172
μ-Bondagel, 177
chemically modified, 421
"diol," 422
Du Pont SE, 176
effect of silica silanization, 186
ether-modified, 431
Gel-type SI, 173
Glycophase-G/GPG, 179
hydrocarbon-modified, 184
Hydrogel, 173
hydrophilic, 422
LiChrospher, 173
mechanical stability, 189
MicroPak BKG gel, 168
OR-PVA gel, 168
particle size, 235, 236
photomicrographs, 167
Poramina, 179
Porasil, 181
pore effects, 234, 235, 238
porous silica, table of, 174
preparative, table of, 182
procedure for surface modification, 422

rigid inorganic, 173
Sephadex, 173
Spherosil, 181
μ-Stytragel, 167, 181
surface-effects, 237, 420
surface-modified, 427, 420
wetting of, 427
Synchropak GPC, 179
TSK gel, 169
SW-type, 172
Vit-X, 183
weight in columns, 195
Zorbax, PSM, 176
see also Column packing
Packing stability, 239, 434
Packing techniques, 189
dry packing, 197
slurry preparation, 195
wet packing, 191
see also Column packing technique
Particles:
agglomeration, 189
charge on, 430
GPC analysis, 416
mechanical stability, 192, 270
modification, 183
rigidity of, 189
rigid photomicrographs, 167
settling rates, 190
shear force on, 189
silanization, 183
size segregation, 190
spherical versus irregular, 188
see also Packings
Particle size:
distribution, 235
size range, 188
Particle size effects, 235, 236
advantages of small particles, 186
on band broadening, 82, 85, 86
on separation time, 187
Peak:
area, 318
asymmetry, 92
asymmetry factor, 203
capacity, 101
cutting, 278, 279, 280
Gaussian model, 107, 302
labeled parts, 316
limits, 278, 279, 280
skew, 85, 91, 323
Peak broadening, *see* Band broadening
Peak capacity, 27, 101, 102, 344
factor, 20
in resolution equation, 101
Peak height, definition, 316
Peak position calibration:
biopolymers and proteins, 289, 291, 442
different pore size columns, 290
effect of pore size, 290
lack of narrow MWD standards, 289
small molecules, 346
Peak resolution:
effect of unequal peak sizes, 99
peak separation, 100
standard resolution curves, 98, 99
Peak retention:
effect of solvent goodness, 28
flow rate independence, 29, 47, 83
influencing factors, 42, 49
temperature independence, 28
Peak retention parameters:
SEC solute distribution coefficient, 25
special SEC terminology, 24
stationary-phase loading effect, 21
Peak separation factor, 101
Peak shapes, 283, 322, 323
flow rate effect, 81
MW and diffusion effects, 80
particle diameter, effects, 80
peak skew, 85
Peak skew, 203. *See also* Skewing
Peak standard deviation:
additivity rule, 59
definition of, 57, 58
in resolution equation, 97
Peak tailing, 203
Peak variance, 272
additivity rule, 58
mathematical definition of, 57, 58
Peak volume, 126
Peak width, 57, 59, 91, 97, 317
Pellicular particles, 87
Permeability, 204
Photometers, 151
Plant extract, 447
Plate count, *see* Plate number
Plate height:
additivity rule, 61, 68
column efficiency indicator, 63
definition, 60, 61
diffusion coefficient, 76

effect on column performance, 97
flow rate dependence, 84
independence of retention, 61, 62
K_{SEC} dependence, 86
mass transfer, 72
MW dependence, 88
nonpermeating solute, 76, 77
packing porosity effects, 72, 74
plate theory results, 61
stationary phase contribution, 75
Plate height equations, 72, 82
flow-diffusion coupling, 67
reduced plate height and velocity, 70, 71
van Deemter, 64
Plate height versus velocity plot, 433
Plate number:
column efficiency indicator, 63
column performance, 97, 109, 111
definition of, 60, 61
errors, 91, 93
experimental determination, 91
independence of retention, 61, 62
peak area method, 91
plate theory results, 61
relation to column length, 61, 62
resolution equation, 101
skewed peaks, 91
see also Plate height
Plate theory:
binomial solute distribution, 61
Gaussian peak profile, 61
hypothetical column, 60
plate height, 60
plate number, 60
predicted peak shapes, 61, 62
predictions, 62
random-walk model, 63
van Deemter equation, 63, 64
Plugging, column, 281
Poiseuille, flow, 163
Poisson MWD, 295
Polyacrylonitrile, 399
Polyamides, 390
Polychloroprene, branched, 400
Polydispersity, 318, 339
Polyelectrolytes:
definition, 397
NATFAT suppression, 393
Polyesters, 390
Polyethylene, 386
branched, 389
crystallinity, 402, 403
etched, 402, 404
Polymer branching determination, 400
Polymer nonsolvents, 220
Polymer properties, molecular weight effect, 6
Polymer repeat units, 296
Polymers:
biopolymers, 425
branched, 399
chain-folding, 401
copolymers, 405
grafted, 409
inorganic, 394
ionic, 398
Polymer shear degradation, 225, 227
with pellicular packings, 87
Polymer solubility, 209
parameters, 216
Polymer solutions, refractive index increments, 245
Polymer solvents, 250, 252
selection, 254
Polymer standard:
narrow MWD, 289
preparative isolation, 365
Polyolefins, 386
chromatograms, 388
Polypeptides, 440
calibration, 427
Poly(propylene glycol), 356
Polystyrene:
equivalent MW, 290
Flory temperature, 217
peak position curve, 290
solubility, 215
Poly(vinyl alcohol), 424
Poramina, 179
Pore effects, 234, 235, 238, 239
Pore geometry, effect on column resolution, 114
Pore model, 31
Pore radius (or diameter):
alteration by surface modification, 423
effective, 33, 42
hydraulic radius, 33
ink-bottle structure effect, 40, 41
mercury porosimetry curves, 41, 42, 49
Pore size:
bimodal, 265
optimum, 265
selection, 265

Pore volume effects, 235
  molecule accessibility, 25, 49
  optimization, 266, 267
Porous silica:
  Carbowax-treated, 425
  microspheres, 188
  small-pore, 346
  unmodified, 424
  see also Silica
Prefractionation, 445
Preparative SEC, 180, 357
  column efficiency, 359
  continuous, 367
  process-scale, 366
  by recycle, 374, 375
Pressure:
  excessive column, 270, 281
  feedback, 133
Problem objectives, 249
Process control, 378
Process-scale SEC, 366
Proteins, 440
  adsorption of, 429
  calibration with SDS, 427
  charge, 430
  conformation, 427
  high molecular weight, 437
  molecular weight, calibration, 428
    estimation, 427
  SDS complexes, 426
  separation, 435, 437
Pulsations, 131
Pulse damper, 131, 134, 142
Pumps, 128
  accuracy, 129
  comparison, 137
  constant pressure, 136
  drift, 129
  flow-feedback, 129
  gas-driven, 136
  hydraulic amplifier, 136
  noise, 129
  pneumatic amplifier, 136
  positive-displacement, 135, 136
  positive-feed, 129
  preparative, 360
  pulsations, 131
  reciprocating, advantages, 135
    diaphragm, 135
    simple single-head, 130
    three-headed, 133
  reciprocation, dual head, 131
  repeatability, 129
  resettability, 129
  screw-driven, 135
  specifications, 129
  syringe, 135
Purification:
  by recycle, 368, 374
  solvents, 261
Purity, sample isolation, 100

Q-factor, 340
Quality control, 414
Quantitative analysis, 349
Quenching, 162

Radius of gyration:
  biopolymers, 45, 46
  different shaped solutes, 44, 45
  equivalent hard sphere, 35, 36
  macromolecules, 288
  poly-γ-benzyl-L-glutamate, 45, 46
  polymer MW dependence, 35, 44, 46
  polystyrene, 36, 40, 42, 45, 46
  rigid rod solutes, 44, 45
  universal calibration, 43
Random coil model, 39
Random copolymer solubility, 221
Random walk model, 63
Range of calibration, 118, 119
Rate theory:
  band broadening, 78
  plate height equation, 82
  statistical moments of peaks, 82
Reactions, alkoxyl, 422
Real-time data handling, 331
Reciprocating pumps:
  advantages, 135
  types, 130
Recorder characteristics, 144
Recycle chromatography, 367
  accurate MW, 368, 372
  accurate polydispersity, narrow MWD, 368, 372
  advantages, 368
  band spreading, 368
  chromatograms, 369, 375
  column dispersion calibration, 84
  equipment, 370
  methodology, 370
  optimum resolution, 369

peak separation, 368
theory, 368, 369
Reduced mobile phase velocity, 69
Reduced plate height, 69
Reference materials, 262
Refractive index:
effect of solute molecular weight, 245
increment values, 252
Refractometer:
deflection-type, 148
differential, 148
Fresnel, 149
interferometer, 150
properties, 151
selection of mobile phase, 150
Reservoirs, 126
Resins, 355
Resolution, 112
dependence on column σD_2, 104, 108
dependence on MW separation, 103
dependence on separation parameters, 101
determination of, 273
flow rate effects, 120, 224
maximum, 266
MW accuracy, 102, 104
in MWD analyses, 102
operating parameters, 119, 122
pore geometry, 114
recovery of, 282
sample loading effects, 121
specific, 104, 111
temperature effects, 233
very high, 367
Restricted diffusion, 82, 85, 87, 89
Retention theory:
biological polymers, 39
cylinder-shaped pore, 31, 38, 40, 42
equilibrium theories, 31, 33, 38, 40, 42
exclusion in cylindrical pore, 31, 33, 38, 40, 42
flexible polymers, 31, 39, 40, 42
general statistical theory, 37
hard-sphere solutes model, 31
mean solute external length, 38
nonspherical rigid molecules, 34, 37, 116
once-broken rod solute, 39
pore shapes, 31, 33, 38, 40, 42
pore size distribution, 35, 49, 114
random-coil solutes, 31, 39, 40, 42
random-plane shaped pores, 33, 38
random-rod pore model, 34
random-sphere pore model, 34
rectangular-shaped pores, 32, 38
rigid rod solutes, 37
slab-shaped pores, 32, 38, 40, 42
solute configurational freedom, 37
solute conformations, 31, 33, 38, 42, 116
solute spatial freedom, 37
spherical-shaped pores, 32, 38, 40, 42
variation in pore cross section, 35
Retention time, 316
Retention volume:
definition, 316
flowrate effects, 231
sample concentration effect, 243
temperature effects, 233
Reverse-flow experiment, 94
Reynolds number, 72
RI detector, 148
characteristics, 150
selection of mobile phase, 150
Rigid molecule model, 37
t-RNA separation, 438
Rubber extractables, 353

Safety, 145
solvent use, 258, 260
Sample:
capacity, 358
cleanup, 445
collection, 144
concentration, 241, 277
dissolution, 274
injectors, 138
injection, 276
loop, 141
solubility, 209
volume, 126, 139
weight, 241
Sample injection, 276
analytical, 138
preparative, 362
reproducibility, 139
Sample size:
band broadening, 239
column overloading, 242
effect, 244
effect on retention, 240
monomer, 244
polymer, 244
viscous fingering, 243

Sample solubility, 209
  temperature, 232
Sample/solvent association, 223
Sample volume, 126, 239
  preparative, 360, 362
Sample weight:
  analytical, 241
  preparative, 362
Sampling:
  automatic, 141
  preparative, 361
  reproducibility, 139
  valve, 139
Sedimentation of Particles, 190
Separation:
  capacity, 266
  development, 21
  high-speed, 353
  linearity, 267
  linear molecular weight, 268
  optimizing, 265
  polymer type dependence, 340
  range, 26, 265
Separation capacity, 114
Separation factor, 22
Separation-by-flow, 47, 50
Sephadex, 173
Shear, 225, 227
Shultz-Zimm MWD, 295
Silanization effects on MW, 186
Silanol conversion, 184
Silica:
  adsorption by, 183
  hydrolysis, 184
  in situ silanization, 184
  microspheres, 188
  modification, 183
  reaction with chlorotrimethylsilane, 185
  silanization effect, 186
  small-pore, 346
  solubility in H_2O, 255
  surface-modified, 185
  trimethylsily-modified, 185
  unmodified, 424
Size classification, particles, 434
Size distribution:
  examples, 382, 383, 384, 385
  polymers, 405, 408
  see also Molecular weight distribution
Size of macromolecules, 288
Skewed σ/τ peak model, 91
Skewing:
  corrections, 323
  definition, 323
Skew of peaks, 85
Slurry dispersion, 194
Slurry liquids, 195
Slurry packing:
  apparatus, 192
  down-flow method, 194
  procedure, 194
  reservoir, 197
  safety considerations, 196
  solvents, 192
  up-flow method, 196
Small molecule separations, 343
  advantages, 344
  applications, 349
  calibration curve, 287
  H versus v plot, 345
  preparative, 366
  techniques, 205
Sodium dodecyl sulfonate, 441
Sodium heparin, 442
Sodium lauryl sulfate, 441
Software, off-line, 328
Solubility:
  biopolymers, 221
  consolute temperature, 210
  copolymer, 221
  effect of pH, 221
  enthalpy of mixing, 212
  limits, 216
  mixed solvents, 216, 220
  polymer sample, 209, 215
  polymer/solvent structure, 210
  relation to theta temperature, 216
  salting-in (-out) effect, 221
  in tetrahydrolfuran, 386
  ultrasonics, 222, 274
Solubility parameter, 212
Solute diffusion coefficient, *see* Diffusion coefficient
Solute/mobile phase combinations, 237
Solute recovery, 282
Solute-size parameter, 347
  rod, coil, sphere, 116
Solution filtration, 275
Solvents, 258
  aqueous, 221
  column effects, 223
  convenience, 254

criteria, 254
degassing, 261
delivery, 230
effect on columns, 255
effect of crystallinity, 210
effect on packings, 255
filtration, 261
"good," 211, 274
hydrogen-bonding tendency, 214
inhibitors, 261
for IR detectors, 256
maximum flow rate, 270
mixed, 216, 220
nonsolvents, 220
physical properties, 258
polymer, 250, 252
"poor," 211
preparation, 254
properties, basic, 210
pumping errors, 230
purification, 261
safety, 258, 260
selection, 222
slurry packing, 192
solubility parameters of, 218
μ-Styragel use, 259, 260
theta, 211
UV-transmitting, 151
viscosity, 270
Solvent degassing, 127
Solvent effects on packing, 223
Solvent metering system, 128. *See also* Pumps
Solvent removal, 363
Solvent reservoirs, 126
Solvent selection, 254, 274
Specific column resolution, 104, 111
Spectofluorimeter, 151, 160
schematic, 153
selectivity, 152
Spherosil, 181
Spreading, *see* Band broadening
Stagnant mobile phase (SEC stationary phase), 22, 24
Stagnant mobile phase mass transfer, 57, 71, 72. *See also* Stationary phase mass transfer
Standards:
calibration, 262
dextrans, 396
obtaining, 264
reference materials, 262
sources, 262
see also Calibration
Stationary equivalent liquid volume, 22, 25
Stationary phase mass transfer, 56
Statistical moments:
$\sigma\tau$ model, 92
of peaks, 82
Stochastic theory, 47
Stop flow GPC/IR, 411, 412
Styragel, 181
μ-Styragel:
calibration curves, 168
solvents, 259
Styrene/butadiene copolymer, 405-408
Styrene determination, 354
Substrate effects:
adsorption, 235
matrix effect, 235
pore size, 235
surface modification, 237
surface modified, 420, 421
Supports, 166
amino-modified, 429
glyceryl-modified, 428
hydrophilic-modified, 422, 431
surface-modified, 420, 421
see also Packings
Synchropak GPC, 179
Syphon counter, 143

Tap-fill method, 197
Techniques, laboratory, 249
Temperature:
band broadening, effects, 89
column performance effect, 433
consolute, 210
control, 143
critical miscibility, 210
Flory theta, 210
high temperature GPC, 386, 387, 390
Temperature effects:
on column efficiency, 232
on MW calibration curve, 233
resolution, 233
sample solubility, 232
Tetrahydrofuran, 386
Theoretical plates, *see* Plate number
Theories of SEC:
less successful, 46
preferred, 31
Thermodynamics of retention, 27, 28, 30

Thermostats, 142
Theta temperature, 210, 217
Total permeation peak, 101
Trace component isolation, 366
1,2,4-Trichlorobenzene, 387
TSK gels, 172
Tubing:
  blanks, 199
  cleaning, 194
Tung correction method, 333, 334
Tung's integral equation, 105

Ultrasonic devices, 274
Ultrasonic dissolution, 222, 274
Universal calibration:
  block copolymers, 410
  copolymers, 410
  curve, 292
  errors, 294
  experimental validation, 292, 293
  hydrodynamic volume, 291, 292
  MW calculations, 335, 337
  polymer branching, 399-401
  R_g-separation concept, 293
Urea, 440
UV detector characteristics, 156
UV photometers, 151
UV-transmitting solvents, 151

Vacancy chromatography, 376
Valves:
  automatic sampling, 141
  injection, 276
  microsampling, 139
  recycle, 370
  sample, 276
  schematic, 140
  switching, 141
van Deemter equation:
  A, *B*, and *C* terms, 64, 65
  effect of flow rate, 64
  optimum velocity, 65
  plate height minimum, 65
Viscometer, 162
Viscosity:
  effect of temperature, 222
  intrinsic, 292
  kinematic, 72
  relative, 292
  specific, 292
  see also Intrinsic viscosity
Viscosity average MW, 335, 338
Viscosity effects, 243
Viscous fingering, 243
Vitamin A palmitate, 35
Vit-X, 183
Void volume, 24

Water, 394, 397, 398
Water-soluble polymers, 394
Wetting of packings, 427

Yield, preparative, 363

Zone spreading, *see* Band broadening
Zorbax PSM, 176

Northern Michigan University
3 1854 003 107 763
EZNO
QD272 C444 Y38
Modern size-exclusion liquid chromatogra